What Others Are Saying

"I think this is the best—the most definitive and well documented—book on climate change."

—**David Collum**, *Cornell University Professor of Chemistry*

"Thomas Kurz's book, *Why There Is No Climate Crisis*, could hardly be better named. Indeed, there is no climate crisis. The book guides readers with little or no prior familiarity with climate science through the basic issues. It explains the mechanism of warming by greenhouse gases and shows that these gases make a minor contribution to the natural warming that began with the end of the Little Ice Age, and coincidentally, with the beginning of the Industrial Revolution."

—**William Happer,** *Professor Emeritus of Physics, Princeton University*

"*Why There Is No Climate Crisis* is well-researched pushback against the most massive scientific fraud in human history."

—**Tom Nelson**, *Host of the* Tom Nelson Podcast *and Producer of* Climate the Movie

Why There Is No Climate Crisis

The Surprising Evidence You Should Know

Thomas Kurz

Dedicated to my grandchildren for whom we must be good stewards of the Earth that they will inherit.

Published by Clovercroft Publishing, Franklin, Tennessee

Published in association with Shane Crabtree of Clovercroft Publishing
www.clovercroftpublishing.com

Edited by Bob Irvin

Interior Design by Suzanne Lawing

ISBN: 978-1-968127-29-9 (print)

Printed in the United States of America

Contents

Introduction

Evaluating Priorities

When I learned that a friend, without grandchildren, had encouraged his children to not become parents because of a perceived impending climate crisis, I realized just how profoundly the climate change issue has impacted people's lives. My friend believes in science, and we have all been told that the "scientific consensus" tells us there is a climate crisis due to global warming brought on by the burning of fossil fuels. I respect my friend, and I wanted to know why he and others say climate change is the "existential crisis" of our times. I knew I had to delve into this topic myself to better understand the severity of the climate crisis. As I dug into the science and history of climate change, I was shocked by the surprising and suppressed information that I uncovered.

The so-called "scientific consensus" tells us that *nearly all* warming today is caused by the greenhouse effect of anthropogenic (human-made) greenhouse gas emissions and that temperatures today are unprecedented.[1] A simple test of this hypothesis is to look at the Earth's temperatures before the burning of fossil fuels. Fortunately, such information is provided from historical records, glacial advance and decline, past tree line altitudes, animal migrations, and paleoclimate science which analyzes chemical isotopes in the layers of ice cores, tree rings, ocean sediments, and speleothems (stalactites and stalagmites). You never hear about the scientific consensus from this evidence which tells of several periods over the last 10,000 years when temperatures were as warm or warmer than today. Climate has always been cyclical, and we are now at the peak of several convergent natural warming cycles.

Because such information goes against the climate crisis narrative, much of it has been suppressed. Respected scientists who have spoken out against the climate crisis have been villainized as "climate deniers" and have had their careers destroyed. In this book, I provide the surprising and suppressed historical and scientific evidence that you should know.

Recent global warming is a reality. To determine the appropriate response, we need to have a rational understanding of the risks of climate change and the tradeoffs and impacts of potential solutions. I enjoy the outdoors and firmly believe in environmental protection for our Earth. There are several crucially important environmental and social goals which must be met to protect our environment, safeguard biodiversity, and ensure the health and the very existence of humanity.

It is our sacred duty and utmost priority to protect this Earth for future generations and ensure continued prosperity for life. To achieve this, it is imperative that we eliminate dangerous pollution of our land, sea, and air; control the use and disposal of harmful materials; and protect ecosystems and endangered species.

Additionally, bringing the populations of poor countries out of poverty is another crucially important objective. Not only is this humane, it is a critical step in protecting the environment. A person consumed with finding enough food to feed their children and avoid starvation is not someone who will be concerned with the environment. The loss of endangered species is greatest in developing countries.[2] To have any hope of creating global engagement in the cause of conservation and to fight pollution and environmental destruction, we must lift all people, from around the globe, out of dire poverty.

I believe it is our responsibility to find solutions to these issues. For many years I elected to incur higher utility prices by receiving much of my power from solar and wind sources. I assumed climate change was an element of environmentalism. However, in recent years, I have seen the climate crisis hijack the environmental movement and the war on poverty. The climate crisis has diverted attention, media coverage, investment, research, and funding away from potentially more pressing and more dangerous environmental and poverty problems.

The Congressional Budget Office (CBO) released a summary of the Inflation Reduction Act signed into law by then President Biden. In this legislation, the CBO reported that $391 billion has been allocated for climate and related energy initiatives. A report about this legislation from Goldman Sachs "projects green subsidies will cost $1.2 trillion—more than three times what the law's supporters claim."[3] President Biden also reentered the United States in the Paris Agreement that sets a target of limiting temperature increase, since 1850, to just 1.5°C. According to a paper by Bjorn Lomborg, "The Paris Agreement, if fully implemented, will cost $819-$1,890 billion per year in 2030, yet will reduce emissions by just 1% of what is needed to limit average global temperature rise to 1.5°C."[4] More recently, Janice Yellen, the Secretary of the Treasury under then President Biden, said that the global transition to a low-carbon economy requires $3 trillion in new capital each year through 2030.[5]

More than ever, we need to determine if mitigating climate change is worth this massive investment and, crucially, if it is more pressing than spending this money to solve other urgent environmental and social issues.

An example of the climate crisis diverting conservation efforts is seen in the Audubon Society. The Audubon Society is an organization established to protect birds. It now states, "Audubon strongly supports wind power and recognizes that it will not be without some impact." The Society also admits on one of its web pages that an impact of wind energy is an estimated 140,000 to 679,000 bird deaths per year from turbine collisions.[6] **Birds killed by wind turbines include endangered species.**

I suppose Audubon's support of wind power is due to a claim that climate change will cause the extinction of species. Yet data on extinctions does not support this claim. Birds can migrate, and observations show species have survived in habitats that are outside their well-established geographic range but meet their requirements.[7] The rate that extinctions occur has been declining rapidly as temperatures have risen. The number of documented animal species that went extinct was more than 50 in the decade 1900-1909. During the decade that started in 2000, a period of global warming, the documented number of animal extinctions had fallen ten-fold—to about 5![8] A study by the Department of Ecology and Evolutionary Biology of the University of Arizona analyzed rates and patterns of extinctions over the past 500 years. They found extinctions peaked about 100 years ago had have since declined.[9] "The researchers found that in the last 200 years, there was no evidence for increasing extinction from climate change."[10]

Scientific studies have confirmed that colder climates, not warmer climates, were responsible for most past extinction events.[11] Furthermore, it is well established in the fields of biology and ecology that regions with warmer climates tend to have greater biodiversity. A recent paper in *Nature* that explores the fatalities and wildlife behavior of birds and bats near windmills states, "Harvesting wind energy can have negative consequences for biodiversity."[12] The Audubon Society is choosing to accept the certain slaughter of hundreds of thousands of birds, including endangered species, to support a hypothesis of mass extinction from warming that, to the present, has not been supported by observational data or paleontological studies.

Another example of the climate crisis hijacking the conservation effort is reflected in the Sierra Club. The Sierra Club has significantly shifted its focus toward climate change in recent decades. Its primary mission was once preserving wilderness; today, climate change is a centerpiece of its work. Climate change now permeates nearly every aspect of the Sierra Club's activities. Their major campaigns of "Beyond Coal" and "Beyond Gas" are all framed around reducing greenhouse gas emissions through a 100% commitment to renewable energy. The Sierra Club 2022 Annual Report highlights the organization's "push for bold climate legislation" as a core priority across lobbying, legal action, and grassroots campaigns, suggesting significant spending diverted from conservation efforts to renewable energy and reducing greenhouse gas emissions.[13]

In 2009 the U.S. Environmental Protection Agency (EPA) classified carbon dioxide (CO_2) as a form of pollution under the Clean Air Act, thereby diverting attention, staffing, and funding away from mitigating toxins and real pollutants. CO_2 is the invisible, odorless, non-toxic part of our breath which we exhale, and it is an essential gas in photosynthesis and for life on Earth. If CO_2 levels dropped below 150 parts per million (ppm), plants would stop growing and there would be no food for all of humanity. **CO_2 is an essential gas for life, and it is not a toxin.** Instead, we need solutions to air pollution including the use of clean burning natural gas and nuclear power and investments in catalytic converters, scrubbers, filtration systems, and chemical processes to remove dangerous particulate toxins.

In cities like Beijing and Los Angeles, where smog is trapped by surrounding mountains or temperature inversions, the air pollution is clearly visible and has serious adverse effects on health. This isn't CO_2. As scientist William Happer said, "If you can see it, or smell it, it isn't CO_2."[14] We should be asking companies about their "pollution footprint," but instead we ask about their "carbon footprint." **Even the term "carbon footprint" is misleading as it sounds black and dirty**

like soot, but CO_2 is *not* dirty; it is the clean, harmless, and transparent bubbles in the carbonated beverages that we drink.

On September 20, 2023, the EPA announced $4.6 billion in grants to cut "climate pollution." The announcement states: "As part of its evaluation of applications, EPA will prioritize measures that achieve the greatest amount of GHG [greenhouse gas] emissions reductions." On August 16, 2024, the EPA announced another $27 billion in grants to reduce greenhouse gases.[15] Together, these two initiatives represent $31.6 billion *not* spent on the reduction of toxic pollutants, which is the true mission of the EPA. By February 2025, $20 billion of these funds had been distributed to non-governmental climate change organizations.[16]

The government of Sri Lanka banned synthetic fertilizers to reduce greenhouse gases and improve the country's Environmental Societal Governance (ESG) score. Corruption and years of irresponsible borrowing by President Gotabaya Rajapaksa's Sri Lanka government, combined with the impact of COVID lockdowns to Sri Lanka's tourism industry, drained the country's foreign exchange reserves so that the country was unable to make payments on its debt.[17] The production of synthetic fertilizers has a high "carbon footprint," and its application releases nitrous oxide, a powerful greenhouse gas. Sri Lanka's government banned synthetic fertilizers in a move to organic farming which would improve the country's ESG score, thereby enhancing its position to participate in the debt-for-nature swaps initiative of the World Bank. The debt-for-nature swaps involve the cancelling of debt for environmental-related action by the debtor nation.[18] Sri Lanka succeeded in obtaining a high ESG score of 98 (above 70 is considered good; Sweden is 96, and the United States is 51).[19] However, in Sri Lanka, the result was a decrease in agricultural yields of 40% to 50%, food inflation of 57.4%, hardship, and devastation to the economy.[20] Much suffering was caused by this "climate change mitigation" action.

Figure 1 – Aerial View of the Deforestation in Haiti. *The aerial view of the border of Haiti on the left and the Dominican Republic on the right demonstrates the contrast between Haiti, which relies heavily on wood as fuel for heating and cooking, and the Dominican Republic, which primarily uses fossil fuels. Poor countries with limited energy pollute more and cause more environmental damage than energy-rich countries.* Image courtesy of NASA: Kekesi, Alex, "Haitian Deforestation," NASA, 25 October 2002. https://svs.gsfc.nasa.gov/2640?utm_source=chatgpt.com.

Much of the world is energy poor, and residents rely on burning wood, charcoal, or dung-fired stoves for cooking and heating. It is estimated by the World Health Organization that 3.2 million people die each year in developing countries due to the health effects of indoor air pollution, primarily from respiratory diseases.[21] Expanded use of natural gas or propane in these countries could save many lives, but climate change initiatives are discouraging the deployment of these clean burning fuels. Using natural gas or propane would also help the environment in these areas as the use of wood for heating and cooking leads to deforestation. Such deforestation has adverse impacts on the ecosystem and endangered species.

Deforestation is one reason poor countries have the highest rate of endangered and threatened wildlife. Brian Gritt reported: "Haitians burn wood charcoal for 60% of their domestic energy production. A hundred years ago, over 60% of Haiti's land was forested. Today, less than 1% is. By contrast, forests in the neighboring Dominican Republic remain largely intact, even though it consumes eight times more energy per capita than Haiti. The big difference is that the Dominican Republic gets almost all of its energy from fossil fuels."[22] Such deforestation is clearly seen in aerial views of Haiti (see Figure 1) where wood continues to be a major source of heating and cooking for the populace. According to NASA, "The border between Haiti and the Dominican Republic (D.R.) is more than just a political boundary. It also reflects the large amount of deforestation that has occurred on the Haitian side of the border. One can easily see from satellite imagery the lush forests still thriving on the D.R. side of the border, which is in sharp contrast to the Haitian side of the border."[23]

According to the United Nations, 1.18 billion people live in energy poverty.[24] The International Energy Association (IEA) estimates that 750 million people in developing countries have no electricity at all.[25] Coal and natural gas power plants could raise the standard of living of the poor significantly, but Western banks refuse to issue loans to third world countries for coal or natural gas power plants over concerns about climate change. Consequently, these people continue to live in energy poverty. Such power plants would move the fuel burning out of peoples' homes, saving millions of lives. A central power plant is also more energy efficient, meaning it produces less overall pollution and lowers costs.

Journalist Kathryn Hall tells the sad story of how premature babies died needlessly in a hospital in the West African country of Gambia because medical professionals did not have enough electricity to run an incubator or provide sufficient ultrasound diagnostics.[26] The reality of energy poverty is often tragic. Natural gas power plants have a low pollution footprint, and even coal power plants with proper pollution-control technology produce low levels of pollution. It was reported that developed countries may have financed $100 billion to $116 billion in developing countries for climate action in 2022.[27] One-fifth of aid to developing countries is for climate mitigation.[28] We need to ask the question of whether this massive investment could have been better used to help the poor in developing nations with financial aid and solving energy poverty.

Countries that include China and India have been moving out of poverty as they industrialize. The economic miracle has been amazing, but the pollution in these countries is a problem. The smog in Beijing, Shanghai, Delhi, and other Chinese and Indian cities is unacceptable. According to the World Health Organization, about 2 million people die each year from air pollution in China.[29] It is estimated that there are 1.1 million deaths each year in India from air pollution.[30] But instead of focusing political pressure and research in technologies such as catalytic converters, scrubbers, filtration systems, and chemical processes to reduce pollution, the West is encouraging these countries to reduce non-toxic CO_2 emissions. In 2024, seventy-four cities and towns in India averaged air pollution levels *ten times* the World Health Organization safe guidelines.[31] In the Chinese city of Datong, air pollution is so bad that vehicle operators at times need to drive with their headlights on during the day.[32]

Reducing coal and oil emissions will lower particulate matter (PM). Such PMs are significant pollutants and the major hazard to human health from air pollution. PMs penetrate deep into the lungs and bloodstream and cause respiratory and cardiovascular diseases. We should focus investment on electric generation from natural gas and nuclear power and support natural gas-powered vehicles and EV use in urban areas. **Despite these obvious solutions, climate change**

organizations such as Greenpeace and 350.org oppose both natural gas and nuclear power.

Natural gas is a clean and viable alternative solution to pollution as its combustion has virtually no PMs or sulfur dioxide (SO_2). Further, it releases less nitrogen oxide gases (NO_x) than burning hydrogen. NO_x formation is primarily via the Zeldovich mechanism as nitrogen in the air is oxidized from heat. The Zeldovich mechanism increases exponentially with temperatures above 1,600°C, to 1,700°C. Hydrogen has a flame temperature of 2,210°C, gasoline is 1,980°C, and natural gas is 1,947°C. With its lower flame temperature, natural gas as a fuel produces less NO_x than gasoline or hydrogen.

Natural gas can fuel the generation of electricity and be used to power internal combustion engines (ICE). As an ICE fuel, it is clean, practical, abundant, and affordable. According to the U.S. Department of Energy, the Gasoline Gallon Equivalent (GGE) of gasoline is about 122,000 British Thermal Units (BTU); hydrogen GGE is about 135,000 BTU; and natural gas GGE is about 125,000.[33] A GGE of gasoline costs about $3.25, hydrogen is about $15.00, and natural gas GGE costs about $2.75. Adjusting for energy content and assuming 27 miles per gallon, natural gas costs $0.10 per mile versus $0.12 per mile for gasoline and $0.50 per mile for hydrogen. And because natural gas burns cleaner than gasoline, it reduces engine wear and oil changes.[34] Nevertheless, because its use releases CO_2 and its production and transport increases methane in the atmosphere, it has been vehemently opposed by many environmental groups due to fears of global warming. Such climate change-based opposition to natural gas use comes from the Sierra Club, the Friends of Earth International, the Natural Resources Defense Council, and the World Wide Fund for Nature.

We should also increase investment in air pollution controls. **The trillions of dollars spent to reduce and sequester CO_2 emissions would be better spent directly on natural gas infrastructure,**

nuclear power, and technologies to reduce real and dangerous air pollution. China has been implementing pollution technologies in its coal plants, but India has not yet focused on pollution controls, except for power plants near densely populated areas.

The United States has demonstrated the effectiveness of implementing pollution control measures. Since 1970, the amount of fossil fuel emissions has increased significantly. However, in this same period, **air pollution from fossil fuel burning has been reduced from 300 million tons (U.S. tons; i.e., 2,000 pounds) per year to less than 150 million tons per year, according to the U.S. EPA.**[35] These pollutants include ammonia (NH_3), PM, SO_2, NO_x, volatile organic compounds (VOC), and carbon monoxide (CO). But we can do still more if we focus investment on the development and deployment of innovative technologies in this area. This investment needs to be greater in the developing world, yet we focus our efforts on setting up treaties to reduce the carbon footprint of these nations rather than their pollution footprint. We need to determine whether this is the appropriate priority.

The climate crisis urgency has also led to the investment and deployment of countless solar panels and windmills. Wind turbines and solar panels use massive amounts of mined materials, including copper and rare earth metals, the mining of which is detrimental to the environment. Copper and rare earth metals are in limited supply and are also needed for the expansion of the electrical grid, electronics, and EVs. Ore grades are in decline for these metals, so as demand grows, an ever-larger amount of rock needs to be excavated, crushed, and processed to produce each ton of refined metal.[36] The mining intensity of these technologies is a major source of pollution.

A windmill requires 542 U.S. tons of steel and concrete for each megawatt of electricity. This compares with 5.2 U.S. tons of steel and concrete to produce one megawatt of electricity from a natural gas power plant.[37] It takes 10,000 to 20,000 watt-hours to produce one kilogram of steel from iron ore, and it takes 2.1 million to 2.2 million watt-hours of energy to produce 1 kilogram of electronic grade silicon (Si).[38] "Electronic grade Si is generally 99.99% pure. The Si used in the manufacturing of solar cells and solar components must be even more pure. A purity of 99.9999999% is required for the most advanced solar cells. This is often referred to as 9N, for 9 nines, a process which requires repeated refining."[39] Most windmills and solar panels are made in China using electricity made predominately from coal-generated power, which often lacks the anti-pollution technologies used in the United States.

Although nearly all coal plants in China have anti-pollution scrubbers, the sheer volume of coal burned in China still poses a pollution challenge. China is the largest producer of coal in the world, and most of the power plants in China use local coal. It is also known that coal from China is of low quality and contains elevated levels of impurities such as sulfur and ash. These impurities contribute to increased pollution when coal is burned, and they have less combustion efficiency. Because of the lower combustion efficiency of Chinese coal compared with coal from other countries, more needs to be burned to produce equivalent energy. This additional burning of poor-quality coal contributes even further to air pollution. Therefore, the production of solar panels and windmills indirectly leads to increased air pollution in China.

Both solar panels and windmills also require rare earth metals. Most of these resources are from China. Near the Chinese city of Baotau, what was once fields of wheat and corn are now the largest rare earth mines in the world. Beyond the city is a five-mile-long tailings lake. Seven million tons of rare earth tailings are dumped into this lake each year. The rare earth metals extraction process uses hazardous substances such as hydrofluoric acid. In visiting the tailing lake near Baotau, Simon Parry reported: "The lake instantly assaults your senses. Stand on the black crust for just seconds and your eyes water and a powerful acrid stench fills your lungs. … villages around breathe in the same poison every day." Nearby villagers have suffered severe skin and respiratory diseases, and cancer rates have exploded.[40]

The life of windmills and solar panels is about 25 years. Windmills are made of durable aerospace materials that are difficult if not impossible to recycle. Solar panels typically contain small amounts of toxic-heavy metals including cadmium, lead, and gallium arsenide. The deployment of solar panels and windmills to fight climate change today may result in an environmental recycling challenge within a generation as these installations reach the end of their useful life.

As a lover of the outdoors, I find no greater beauty than what can be seen around us in nature. Artist Andy Warhol said, "I think having

Figure 2 – The Blight of Wind and Solar Energy on the Natural Landscape. *Windmills and solar farms are like "sour notes" which destroy the beauty of our landscapes and seascapes. Extreme examples of destroying natural beauty include the wind farm on the San Gorgonio Pass in California (left) and Taihang Mountain in China (right). Images:* San Gorgonio Pass, photo by Erik Wilde, "Wind turbines in Southern California 2016.jpg," Wikimedia Commons, provided under the Creative Commons Attribution-Share Alike 2.0 Generic License. https://commons.wikimedia.org/wiki/File:Wind_turbines_in_southern_California_2016.jpg; Taihang Mountain solar farm, photo by Yanga, iStock, provided under the iStock Content License, https://www.istockphoto.com/photo/photovoltaic-power-generation-equipment-scene-on-the-mountain-gm1927214805-555747541.

land and not ruining it is the most beautiful art that anyone could ever want." I have always enjoyed the beautiful countryside of England where the stone walls and stone cottages with slate roofs are in beautiful harmony with the green rolling hills. However, in my recent trips to England, I have been annoyed by the visual "sour notes" in the landscape of windmills and solar farms. Offshore windmills are also destroying beautiful seascapes. As stewards of the Earth, we should also protect its beauty. **Climate alarmism has appropriated our conservation efforts so that the Sierra Club, whose original mission was to protect the beauty of our natural environment, now promotes blight on our landscapes from wind and solar energy.**

Extreme examples of the blight of wind and solar energy can be seen in the wind farm of San Gorgonio Pass, California, near Palm Springs, and the solar farm on Taihang Mountain in China (see Figure 2). The naturalist and founder of the Sierra Club, John Muir, said, "Going to the mountains is like going home." This can no longer be said about Taihang Mountain. I can only image what Emerson, Thoreau, and Muir would have to say about such destruction of the beauty of nature. Is climate change so urgent that we should destroy the natural beauty of our landscapes and seascapes?

The wind and solar energy transition has also disproportionately hurt the poor. The promise of green energy has been to lower costs since there is no fuel required to generate electricity from a wind or solar farm. It turns out, however, that fuel cost is not the main expense-driver. The modern world is dependent on reliable power 24 hours, every day, without interruption. However, the Sun only shines in the day, and wind generally falls sharply during the night and at random intervals. Power is needed throughout the night, and the charging of electric vehicles, primarily overnight, will significantly increase night electricity demand. Consequently, wind and solar power need to be backed up with either expensive and inadequate energy storage solutions or redundant nuclear or fossil fuel power stations. **Investment**

in such redundant power sources drives up costs. Transmission line costs to connect the electrical grid to distant solar and wind farms have also driven up the price significantly.

Germany has been one of the most aggressive nations in transitioning to wind and solar energy, approaching 50% of its total electric generation. The result has been a massive increase in the cost of electricity. Around the year 2000, the cost per kilowatt hour of electricity in Germany was about .05 to .06 Euros; by 2020, the cost had skyrocketed to over 0.40 Euros. This compares to just above 0.10 Euros in the United States and less than 0.20 Euros in the nuclear power dominant power grid of France.[41] Some of this energy cost increase is no doubt due to the cost of natural gas after the Russian invasion of Ukraine. However, much of the cost increase is due to the deployment of green energy.

The result of high energy prices has been the closure of energy-intensive industries in Germany. For example, the chemical industry, once the crown jewel of German industry, has seen a decline in production of over 20% since 2021.[42] The emerging technology of the future is artificial intelligence (AI). AI data centers require massive amounts of steady and affordable electric power, 24 hours every day. Solar panels provide power only in the day and the wind generally doesn't blow much at night. Because backup battery technology is expensive, renewable energy will not be able to provide power that is stable, uninterrupted, and affordable (see Chapter 14). According to energy industry expert Steve Goreham, "Unless Europe abandons Net Zero and efforts to convert their power grid to wind and solar, AI will be a failure."[43]

The United Kingdom has also been aggressive at implementing green energy. As a result, UK electricity prices have nearly tripled since 2003 from 11.6 pence to 33.4 pence per kilowatt hour. Electricity now costs £59 billion more than in 2003, which is 2.1% of GDP and £3,543 more ($4,401) per family per year.[44] California is the most aggressive green energy state in the U.S. with 17.5 GW of utility-scale solar, 6 GW of wind, and 14 GW of residential rooftop solar.[45] The cost of California electricity has increased by 98.2% over the last 15 years, the highest rise in the U.S. Average energy prices increased by 30.6% in other states. According to the EIA, as of May 2024 the average cost of 1 kilowatt hour of electricity in the U.S. is $0.1643, but the cost in California is more than double at $0.3441.

These green energy price increases hurt the poor and middle class. Families in the lowest income households spend 35% of their income on energy while those in high-income households spend only 10% of their income on energy.[46] The deployment of wind and solar energy is thus disproportionately hurting the poor and middle class. Government subsidies are often provided to help the poor, but this places a further burden on the middle class with both higher energy costs and higher taxes.

Another example of climate change policy proving detrimental to the poor and vulnerable is the eviction of the Ogiek people from their native homes in Kenya. Since November 2, 2023, over 700 Ogiek hunter-gatherer people have been evicted from their ancestral lands and seen their homes destroyed. Human rights lawyer Dr. Justin Kendrick has stated that the global carbon credit market allows rich nations to purchase carbon credits from poor nations with uninhabited forested areas. Dr. Kendrick says the government of Kenya is evicting people from the Mau forests of Kenya to maximize these lucrative carbon credits.[47]

Let me be clear: we do live in an era of a warming climate. Crucial questions include: will this warming be significant, is warming dangerous, and is it more important than these other pressing environmental, social, and natural preservation issues? Many politicians and news outlets state "climate change" is *the* existential threat of our time. Former President Joe Biden said, "The only existential threat humanity faces, even things more frightening than a nuclear war, is global warming going above 1.5 degrees in the next 10 to 20 years."[48] This all sounds so serious.

I desired to know if climate change is such a critical and urgent issue that these other immensely important environmental and social challenges should become secondary issues. Climate alarmist politicians and news outlets repeatedly have said the science is "settled" and "97% of scientists agree." If true, such a large consensus sounds convincing, but as Mark Twain said, "The best way to get a sure thing on a fact is to go and examine it for yourself, and not take anybody's say-so."[49]

I decided to study climate change to determine the truth of the matter using an evidence-based scientific method. I believe an investigation of climate change is important to ensure we direct all attention, spending, and research to address the most devastating and urgent environmental and related social issues facing our planet. I was shocked with what I found and felt compelled to write this book.

Up until now, the whole story has not really been told.

Use and Abuse of the Scientific Method

The foundation of accepted science is to postulate a hypothesis and then perform experiments using acceptable methods to prove or disprove the hypothesis. This is the scientific method. Richard Phillips Feynman provided a brilliant description of the scientific method. Professor Feynman was an American theoretical physicist known for his work in quantum mechanics, quantum electrodynamics, the physics of the superfluidity of supercooled liquid helium, and particle physics, for which he proposed the Parton model. For his contributions to the development of quantum electrodynamics, Feynman received the Nobel Prize in Physics in 1965 jointly with Julian Schwinger and Shin'ichirō Tomonaga.

In a lecture at Cornell University in 1964, the great physicist Feynman said, "In general we look for a new law by the following process. First, we guess it—no, don't laugh, that is the truth. Then we compute the consequences of the guess, to see if this is right, if this law we guess is right, to see what it would imply, and then we compare the computation results to nature or, we say, compare to experiment or experience, compare it directly with observations to see if it works. If it disagrees with experiment, it's wrong. In that simple statement is the key to science. **It doesn't make any difference how beautiful your guess is, it doesn't matter how smart you are, who made the guess, or what his name is. … If it disagrees with experiment, it is wrong.**"[50]

In the following pages, I present reputable observational data, historical records, paleoclimate reconstructions, and archeological findings of previous climate cycles. I have used accepted physical principles and proven formulas to calculate greenhouse gas warming. I think you will find this data as surprising as I did. Nearly everything I was told about climate change and the climate crisis was wrong. Mark Twain wisely said, "What gets us into trouble is not what we don't know. It's what we know for sure that just ain't so."[51]

I think you will be shocked as I was to learn from this book that: 1) there is no confirmed scientific consensus that climate change is dangerous and that nearly all warming is from greenhouse gas emissions (see Chapter 1); 2) observational data demonstrates there is no trend, or declining trend, in severe weather in recent years as the climate has warmed (see Chapter 3); 3) meteorological theory predicts rising temperatures will moderate the severity of storms, which is confirmed in modern measurements and historical records going back 1,000 years (see Chapters 3 and 6); 4) sea level rise is less than a foot (31 centimeters) per century, and today's trend is in line with historical records of the past 150 years (see Chapter 3); 5) as the temperature has warmed in recent years, the extinction of species has plummeted, polar bear populations have grown, and the Great Coral Reef has witnessed record growth (see Chapter 3); 6) the warming and increased CO_2 in the atmosphere has been greening, not browning, the Earth, and this has led to growing harvests to feed an expanding population (see Chapter 4); 7) cold is far more harmful to humankind than warmth, and we should welcome, not fear, a warming climate (see Chapters 4 and 6); 8) current warming is not unprecedented, as climate has always been cyclical (see Chapters 5, 6 and 13); 9) quantum physics of greenhouse gas radiative forcing (warming) calculations have the temperature by end of the 21st century of less than 1°C warmer than today, which will be more beneficial than harmful (see Chapters 9 and 10); 10) the power of CO_2 to warm declines logarithmically as concentrations increase (see Chapter 9); 11) the net warming of the oceans is not from heat in the atmosphere since the oceans are, on average, nearly 4°C warmer than the air at sea level (see Chapter 13); and 12) several natural climate cycles move into cold phases after about 2030, suggesting temperatures will decline after that date (see Chapters 7 and 13). The perspective that anthropogenic (human-caused) increases of greenhouse gases leads to a "climate crisis" is counter to what science, history, and measured observational data tell us.

Archaeology, historical records, and paleoclimate reconstructions reveal that the 1,000-year Eddy Solar Cycle propels the Earth into a warm period every millennia. This was experienced during the Roman Warm Period 2,000 years ago, the Medieval Warm Period 1,000 years ago, and with current Modern Warming (see Chapters 5 and 6). The media and the United Nations International Panel on Climate Change (IPCC) ignore the overwhelming evidence supporting these climate cycles (see Chapter 5). All of these warm cycles coincide with solar maximums of the Eddy Solar Cycle (see Chapter 7). We are currently in the largest solar maximum in 10,000 years.[52] Since 1850/1900, the Earth's average temperature has increased by 1.1°C.[53] Despite experiencing a solar maximum in the 20th century and early 21st century, the IPCC and climate alarmists attribute nearly all of this warming to increased emissions of CO_2, methane, and other anthropogenic greenhouse gases.[54] However, the accepted radiative forcing equations for greenhouse gases can only account for less than one half of this warming due to increasing CO_2 concentrations—from 280 parts per million (ppm) in 1850 to 427 ppm today (see Chapters 9 and 10).

The United Nations International Panel on Climate Change (IPCC) RPCP6.0 scenario (moderate-to-high emissions) forecasts a CO_2 level of 700 parts ppm by the end of this century.[55] In this scenario, CO_2 levels will increase from 427 ppm today to 700 ppm by the year 2100. The United Nations IPCC also forecasts a temperature increase of 3.2°C between 1850 to 2100.[56] The IPCC also claims that temperatures have increased by 1.1°C from 1850/1900 to 2021.[57] This means they are forecasting a temperature increase of 2.1°C from now to the end of the century. However, using the accepted ERF adjusted Myhre radiative forcing equations[58] produces a warming of only 0.80°C for increases in CO_2, methane, and all other anthropogenic greenhouse gases from today through the year 2100 (see Chapter 9). A similar increase in temperature is also calculated using MODTRAN software and lapse rate methods of estimating radiative forcing of greenhouse gases (see Chapter 9). **IPCC climate-model forecasts are 2.2 times larger than supported by physics (2.1°C vs. 0.94°C).** Similar to this 2.2-fold overestimation, IPCC climate models forecast temperatures between 1979 and 2019 that are 2.4 times larger than actual measurements from highly calibrated weather balloons and satellites (see Figure 60 in Chapter 10).[59]

Since accepted physics equations for radiative forcing calculate only 0.94°C in greenhouse gas warming by the year 2100, why do these models forecast 2.1°C, which is 2.2 times more than 0.94°C? The answer: they postulate that warming of 0.94°C from greenhouse gases will increase water vapor, clouds, and melting ice caps, which will have a 2.2 times warming multiplier effect on temperature. This assumption is not derived or observed in measurements, yet it is the basis of the "climate crisis." Studies and observational data demonstrate these factors have less than a 0.5 times (×) warming multiplier effect (see Chapter 10). Clouds have both a warming and cooling effect, with the cooling effect of low clouds being much larger than the warming effect (see Chapter 11). Low clouds cool the Earth by reflecting solar energy off their white surface and back out to space. The net impact of clouds is cooling. The net impact of clouds produced from increased humidity and increased sulphate aerosols from phytoplankton growth caused by higher temperatures has a 0.7× cooling multiplier effect, so it more than offsets any of the warming impact of water vapor, melting ice caps, and clouds themselves (see Chapters 10, 11, and 12). **The 2.2× warming multiplier hypothesis is not supported by physics or observational data and is, therefore, wrong. And without this imaginary 2.2× warming multiplier, there is no climate crisis.**

The temperature of the Earth is determined by the net amount of shortwave radiation from the Sun received by the Earth and the longwave radiation sent from the Earth out to space; this is known as the *Earth's energy budget.* When the amount of energy in and out are equal, the temperature is stable, but when more radiation energy comes in than goes out, the Earth warms. Greenhouse gases can impact this balance, as they absorb outgoing longwave radiation. Satellite observational measurements in recent years have given us a tremendous amount of new observational data about the Earth's energy budget. If greenhouse gases were the major source of warming, then the amount of outgoing longwave radiation at the top of the atmosphere would be decreasing as it is trapped by the greenhouse effect.[60] However, that is not what is being observed. Clouds and the Earth's Radiant Energy System (CERES) satellite measurements since 2000 have recorded an increase in outgoing longwave radiation.[61] These observations suggest recent warming is primarily driven by an increase in incoming shortwave radiation from the Sun that is caused by decreased reflection of sunlight out to space from lower albedo (low cloud cover).[62] Confirmation comes from the Moderate Resolution Imaging Spectroradiometer (MODIS) and CERES satellite measurements over the past 24 years which show both reduced albedo and increased incoming shortwave solar radiation.[63] Applying Feynman's definition of the scientific method, it doesn't matter who tells you that recent warming is primarily from greenhouse gases, this hypothesis disagrees with observations and is therefore wrong.

Climate is complex and influenced by a multitude of factors including solar cycles, cosmic rays, cloud cover, ocean temperature oscillations, atmospheric and oceanic heat transport to polar regions, volcanoes, aerosols, ice caps, and greenhouse gases. All of these need to be considered in the investigation of climate change. Recent warming is due to a combination of these and other factors. **Overwhelming scientific evidence establishes that about half of the warming in recent years is from natural causes, leaving about half of the warming from increased levels of CO_2**. Scientists who cite this evidence and challenge the climate alarmist narrative (i.e., that an increase in CO_2 levels is the dominant driver of climate) are villainized as "climate deniers." This should be appalling to anyone who relies on scientists and the scientific method to discover truth. *All* evidence needs to be examined, and hypotheses need to be challenged and evaluated using observational data and appropriate scientific methods.

Respected scientists who report evidence that is not in alignment with the climate alarmist narrative have had their research funding reduced or eliminated and consequently had their careers destroyed. Some of these scientists include Richard Lindzen, Judith Curry, Willie Soon, Sallie Baliunas, Henrik Svensmark, Peter Ridd, Susan Crockford, and William Gray. On the other hand, other researchers who are supporters of the climate alarmist narrative have received generous funding and appointments to prestigious positions which facilitate their public exposure as well as their narratives. Carl Sagen said, "Science is generated by and devoted to free inquiry: the idea that any hypothesis, no matter how strange, deserves to be considered on its merits. The suppression of uncomfortable ideas may be common in religion and politics, but it is not the path to knowledge; it has no place in the endeavor of science. We do not know in advance who will discover fundamental new insights."[64] **Promoting climate alarmists while "cancelling" climate alarm skeptics is perhaps the most egregious episode in the history of modern science.**

Publishing scientific papers in peer reviewed journals is the lifeblood of an academic career. Scientist Patrick T. Brown from Johns Hopkins has admitted he had to distort the findings of his studies to appeal to the editors of prestigious science journals including *Nature* and *Science*. Referring to the climate alarmist narrative, he said, "... editors of these journals have made it abundantly clear, both by what they publish and what they reject, that they want climate papers to support a certain preapproved narrative..."[65] Such actions feed a dysfunctional culture where scientists are afraid to speak out, thus manufacturing an illusionary "consensus."

Climatologist Judith Curry was a professor and chair of the School of Earth and Atmospheric Sciences at Georgia Institute of Technology. Like me, Professor Curry supported the climate change narrative as she thought it was the responsible thing to do. In 2009, she began to question the narrative after leaked "Climategate" emails[66] revealed the manipulation that went into building a climate crisis consensus. She labeled belief in an existential threat from global warming caused primarily from anthropogenic greenhouse gases "a manufactured consensus."[67] She criticized the narrative that recent climate change has been dominated by human causes versus natural variability. She said, "... we have been misled in our quest to understand climate change, by not paying sufficient attention to natural causes of climate change, in particular from the Sun and from the long-term oscillations in ocean circulations."[68] For taking this fact-based position, she was labeled a "climate denier," ostracized by her university, and attacked by climate alarmists. The resulting negative environment in her academic career led to her eventual resignation.

Physicist Henrik Svensmark is a professor in the Division of Solar Physics at the Danish National Space Institute in Copenhagen. His experimental research on cloud formation and the impact of low clouds on climate has uncovered what is probably one of the dominant drivers of climate change today and in past ages of the Earth (see Chapter 7). Because Professor Svensmark's research does not support greenhouse gases as the control knob of climate change, he found it nearly impossible to find funding for his research despite the great insights on climate change revealed in his work.[69] William Gray was a professor at Colorado State University and one of the world's leaders in hurricane research. Al Gore invited him to attend a climate conference. After Professor Gray responded that he did not support the climate crisis narrative, he never again received funding from government grants.[70] Professor Gray had to donate $500,000 of his own money to keep the hurricane forecast program, known as the Tropical Meteorology Project, under his direction, from shutting down.[71]

I had naively assumed that the mission of universities was to pursue truth. I was puzzled why respected scientists like Judith Curry would be maligned by their own institutions for presenting credible data that did not support the climate crisis narrative. Universities depend heavily upon research grants and on those professors who secure grant funding. An estimated $44.6 billion of grants for climate research was awarded between 1990 and 2018.[72] That figure has likely gone up significantly since. Universities and professors benefit from research that supports the climate crisis narrative, while respected professors like Henrik Svensmark and William Gray, whose research goes against the crisis narrative, have difficulty finding funding for their research. As Upton Sinclair wrote, "It is difficult to get a man to understand something if his salary depends on him not understanding."[73]

Examining Climate Change

I began my examination of climate change through watching an online physics course on *YouTube* by Michel van Biezen titled "Astronomy Chapter 9.1—Earth's Atmosphere."[74] This is an excellent course of 61 five-minute lectures; it covers the physics of the greenhouse effect in detail. After this thorough look at the physics behind climate change, van Biezen concludes that the data and science reveal that global warming is being exaggerated in climate models. I also watched instructive videos from a number of prominent scientists including: renowned physicist Freeman Dyson from Princeton and formerly Cornell University;[75] one of the most respected atmospheric physicists, Richard Lindzen from MIT;[76] Nobel laureate of physics, John Clauser;[77] respected physicist William Happer from Princeton;[78] prominent climatologist Judith Curry from Georgia Tech;[79] and physicist and former Provost for Cal Tech, Steven Koonin.[80]

These lectures helped me understand that climate change involves far more than just CO_2. Princeton physicist Dyson said that warming from CO_2 is modest and, overall, more positive for humankind than harmful. MIT atmospheric scientist Lindzen claims it is ridiculous to assume CO_2 controls climate when one considers the many known natural climate forcings. He also stated that claims of increases in severe weather lack supporting data and are contrary to accepted meteorological theory which predicts declining severe weather as the Arctic warms. **Nobel laureate Clauser said that climate models miss the key driver of climate: cloud cover.** He said cloud cover is far more powerful than CO_2 in forcing the climate and is a powerful thermostat that keeps the Earth from overheating.

Princeton physicist Happer covers the physics that explains why CO_2-induced warming is minimal, and that there will be far less CO_2

warming in the future as the warming power of CO_2 is increasingly saturated (see Chapter 9). Georgia Tech climatologist Curry describes how fundamental natural drivers of climate such as ocean oscillations (see Chapter 13) and solar cycles (see Chapter 7) are not accounted for in climate alarm forecasts. Caltech physicist and former Under Secretary for Science in the Obama Energy Department, Koonin, describes how CO_2 exerts a small warming effect of about 1%, which is low compared to the entire climate system. Koonin says predictions of climate and weather are based upon climate models that are totally unfit for this purpose and too sensitive to CO_2 impacts. He also clarifies that climate is driven by the oceans (see Chapter 13).[81]

I appreciated the critical role of the Sun in climate change after watching videos from several solar/astrophysicists including Willie Soon, formerly from the Center for Astrophysics Harvard & Smithsonian;[82] Nir Shaviv from Hebrew University;[83] Henrik Svensmark from the Danish National Space Institute;[84] and Valentina Zharkova from Northumbria University.[85] I read the well-researched and meticulously footnoted book *The Neglected Sun: Why the Sun Precludes Climate Catastrophe*, by respected German scientist and environmentalist, Fritz Vahrenholt and Sebastian Lüning.[86]

Former Harvard & Smithsonian astrophysicist Soon makes a connection between solar cycles and temperature. He points out that the coldest period of the Little Ice Age in the seventeenth century was during the deep Maunder Solar Minimum (see Chapters 5 and 6). Solar physicist Shaviv shows the excellent correlation between solar cycles and ocean temperature (see Chapter 7). Danish astrophysicist Svensmark describes his work in identifying cloud formation and how clouds impact climate. He provides experimental and observational data to show how solar cycles, and their modulation of cosmic rays, play a central role in the formation of cloud nucleating aerosols (see Chapters 7, 11, and 12). Solar physicist Zharkova uses principal component analysis of the magnetic fields of the Sun to predict past and future solar cycles and climate. Her calculations foresee a cold climate cycle beginning in the 2030s (see Chapter 7). German scientists Fritz Vahrenholt and Sebastian Lüning demonstrate the critical role of the Sun in past climate cycles and in current warming. **Vahrenholt and Lüning also show that the current cooling phase of the Sun renders impossible the climate catastrophes put forward by climate alarmists.**

The scientists listed above present scientific data which confirms modest warming from increases in CO_2 and the role of other natural climate variability. They all agree the climate is changing and warming in recent years, but they dispute the narrative that there is a climate crisis. These sources furthered my curiosity on the climate alarmist narrative, and I dug deeper.

I have outlined in this book the truth I uncovered from this research. To assume CO_2 is the primary control knob of the climate is counter to climate-related measurements, principles of physics, paleoclimate reconstructions, archaeology findings, and the historical record. The real climate and science deniers are those who discount evidence for global warming from natural causes and deny climate change cycles of the past such as the Little Ice Age, 500 years ago, and the Medieval Warm Period, 1,000 years ago. Despite efforts by climate alarmists to erase these climate cycles, the evidence is overwhelming that these past climate cycles did indeed occur (see Chapters 5 and 6).

Increased emissions of CO_2 and other anthropogenic greenhouse gases raise atmospheric temperature, but only modestly. This is convincingly demonstrated in both observations and radiative forcing equations (see Chapters 9 and 10). A fundamental driver of climate is the transport of heat in the atmosphere and oceans. The oceans are a massive solar collector, and they collectively absorb about 90% of all solar energy striking the Earth (see Chapter 13). Climate change is driven in large part by the variation of the exchange of heat from the oceans to the atmosphere. Increased CO_2 emissions have little or no impact on ocean temperatures (see Chapter 13). Ocean temperatures are primarily driven by the distance and tilt of the Earth to the Sun (Milankovitch cycles), solar variation (solar cycles), and low cloud cover (albedo).

Low clouds reflect 70% to 95% of solar energy back out to space,[87] so low cloud cover is a substantial driver of ocean temperatures and climate change (see Chapters 11 and 12). Ocean currents move heat mostly to the Northern Hemisphere and the Arctic, and this is where most of global warming has occurred today and in past climate cycles. The North Pole warmed 25 times (×) more than the South Pole between 1978 and 2022 (see Chapter 13).[88] Warm climate cycles of the past include the Medieval Warm Period (900–1250), the Roman Warm Period (250 BC–400 CE), the Minoan Warm Period (1500 BC–1100 BC), and the Holocene Climatic Optimum (4000 BC–7000 BC). In all of these climate cycles, warming was more pronounced in the Northern Hemisphere, just as we are experiencing today (see Chapters 5 and 6).

Observations and historical reconstructions show that CO_2 emissions are poorly correlated with temperature cycles (see Chapter 8). Over the past one million years, there has been no correlation—or a negative correlation, 87% of the time—between CO_2 and temperature.[89] More recently, in the period between 1945 and 1976, CO_2 levels increased five-fold while temperatures declined (see Figure 49 in Chapter 8). This was also a period of cooling ocean temperatures that were clearly much more powerful than the radiative forcing of anthropogenic greenhouse gases. On the other hand, factors driving cloud cover and ocean temperatures are reported in countless peer reviewed papers that correlate extremely well with past warm and cool climate cycles over days, years, decades, centuries, thousands of years, and even hundreds of millions of years of the Earth's history (see Chapter 7).

The following pages cover previous climate cycles and examine each of the climate drivers to facilitate a science-based rational discussion on climate change.

Chapter 1:

The Scientific Consensus

Opinions of Respected Scientists

Climate alarmists would have you believe that no credible scientist would deny the reality of a climate crisis from global warming due to anthropogenic greenhouse gases (GHGs). In June 2013 President Obama gave a speech on the dangers of climate change. Referring to those who challenge the climate crisis narrative, Obama said, "We don't have time for a meeting of the flat-earth society."[1] In a speech on February 29, 2024, President Biden referred to those who are skeptical of a climate crisis as "Neanderthals."[2] Despite such propaganda, the fact remains that there are many extremely respected scientists, including Nobel Laureates, who are skeptical of the climate crisis. Many of these scientists are physicists (Table 1). The mechanism of GHG warming is from quantum physics. It is therefore telling that many prominent physicists have publicly stated there is no climate crisis. Several climate alarm skeptics are astrophysicists. This is also insightful since the climate alarmist narrative often underestimates the contribution of the Sun in climate change.

Table 1 Scientists Who Are Climate Crisis Skeptics			
Scientist	**Focus**	**University**	**Awards, Appointments/ Additional information**
Freeman Dyson	Physics	Cornell, Princeton	Erico Ferme Award, Max Plank Medal
Syun-Ichi Akasofu	Space Physics and Climatology	Arctic Research Center at the University of Alaska, Fairbanks	Chapman Medal, John Adam Fleming Medal, Japan Academy of Sciences Award, Order of the Sacred Treasure (awarded by the Emperor of Japan), Hannes Alfvèn Medal, named one of the "World's Most Cited Authors in Space Physics," Edith R. Bullock Prize for Excellence
Ivar Giaever	Physics	Rensselaer Polytechnic Institute	Nobel Prize, 1973
John Clauser	Physics	UC Berkley, Lawrence Livermore Lab	Nobel Prize, 2022
William Happer	Physics	Princeton	Davisson-Germer Prize
Richard Lindzen	Atmospheric Sciences	Chicago, Harvard, MIT	Alfred P. Sloan Fellowship, AMS Charney Award
Steven Koonin	Physics	Caltech, NYU	Caltech Provost; Under Secretary for Science, Dept. of Energy, Obama Administration

Judith Curry	Atmospheric Sciences	University of Colorado-Boulder, Penn State University, University of Wisconsin, Georgia Tech	Henry G. Houghton Research Award; NASA Advisory Council
John Christy	Atmospheric Sciences	University of Alabama in Huntsville	NASA Medal for Exceptional Scientific Achievement, AMS Special Award
Henrik Svensmark	Astrophysics	Danish National Space Institute, University of California, Nordic Institute for Theoretical Physics, Niels Bohr Institute	Energy-E2 Research Prize, Knud Hojgaard Research Prize
Roy Spencer	Meteorology	University of Alabama in Huntsville, NASA, University of Wisconsin	AMS Special Award; Director, George C. Marshall Institute
Nir Shaviv	Astrophysics	Hebrew University, Canadian Institute for Theoretical Astrophysics	Caltech Lee DuBridge Prize; Chairman, The Racah Institute of Physics
Ian Plimer	Geologist	University of Melbourne, University of Adelaide, University of Newcastle	Daley Prize, Eureka Prize, Clarke Medal; 'Plimerite' (a new mineral named in honor of Dr. Plimer)
Willie Soon	Astrophysics	Harvard & Smithsonian, Mount Wilson Observatory	Petr Beckman Award

Sallie Baliunas	Astrophysics	Harvard & Smithsonian, Mount Wilson Observatory	Petr Beckman Award, Wesson Fellow, Donald E. Billings Award, Bok Prize, Newton Lacy Pierce Prize; Deputy Director of the Mount Wilson Observatory
Valentina Zharkova	Astrophysics	Northumbria University, Glasgow University	SOHO/MDO Award, NSF U.S. Award
Fritz Vahrenholt	Chemistry, Environment	Repower Systems AG (a wind power company), RWE (an energy company)	Honorary Doctorate, University of Hamburg; Director, Deutsche Shell AG
Nicola Scafetta	Astronomy, Climatology, Physics	University of Napoli Federico II, Duke University, ACRIM Lab (solar irradiance measurement satellite)	Developed "Diffusion Entropy Analysis"
S. Fred Singer	Astrophysicist	University of Maryland, Johns Hopkins, University of Miami, University of Virginia	Director for Atmospheric and Space Physics; Deputy Administrator, EPA; Deputy Secretary, Dept. of Interior; Dean, School of Environmental and Planetary Sciences, University of Miami; Pioneered satellite weather measurements

To assume that all scientists believe there is a climate emergency is just not true. The organization Climate Intelligence (CLINTEL)

has prepared a statement that there is no climate emergency, and it has collected the signatures of over 2,000 brave scientists and professionals who support this statement.[3] This list includes physicist and Nobel laureate Professor Ivar Giaever, physicist and Nobel laureate Dr. John Clauser, and Princeton physics professor and expert in GHG radiation, Dr. William Happer. Professor Giaever said, "I would say that basically global warming is a non-problem."[4] Dr. John F. Clauser, the recipient of the 2022 Nobel Prize in Physics, recently made this statement: "There is no climate crisis."[5] Princeton physicist Happer has said, "Current alarm over carbon dioxide is mistaken … Fears about man-made global warming are unwarranted and are not based on good science."[6] Atmospheric science professor, Dr. John Christy, has said, "The evidence is clear that we are not in a climate crisis."[7]

Even renowned physicist and the father of the hydrogen bomb, Edward Teller, who was once a climate alarmist, signed a statement before his death that stated, "There is no convincing scientific evidence that human release of carbon dioxide, methane, or other GHGs is causing[,] or will, in the foreseeable future, cause catastrophic heating of the Earth's atmosphere … " More than thirty-thousand American scientists (31,497), including Teller and Freeman Dyson, signed this petition.[8] Some have speculated that Teller's signature on this petition is not valid. However, the sponsors of the petition maintain its legitimacy and their verification process, the signature matches the writing style of existing Teller autographs, and no forensic evidence has ever been presented to cast doubt on it authenticity.

In 2021, Professor Steven Koonin authored and published a book titled *Unsettled*. Its aim is to demonstrate with clear data that the climate crisis narrative is not "settled science," as is often claimed by climate alarmists. Dr. Koonin is an MIT educated physicist who was a professor and the provost at Caltech before working in the Obama Administration as Under Secretary for Science in the Department of

Energy. If you look at videos of his talks from his time in the Obama Administration, Koonin was an ardent supporter of global warming alarmism. In 2013 he was asked by the American Physical Society to lead the drafting of an updated statement on climate. As part of that process, he convened a workshop with six leading climate experts and six leading physicists to stress test the state of climate science. Koonin went on to write, "I came away from the APS workshop not only surprised but shaken by the realization that climate science was far less mature than I had supposed." He concluded that humans exert a growing, but physically small, warming influence on climate.[9]

Based on these findings, Koonin prepared a statement and submitted it to a committee of the American Physical Society. This statement was not approved by the committee. Koonin commented that one of the members of the committee said, "We can't say that even if it is true because it give too much ammunition to deniers."[10] True to his belief that scientists have a special responsibility to bring objective science to a discussion, Koonin resigned from this committee and authored an article in the *Wall Street Journal* arguing that the science on climate change is not settled.[11] In his 2021 book he uses data from accepted reports and sources to show dozens of climate alarmist claims are not supported by data.

Since publishing his book, Koonin has been a frequent guest on podcasts and other forums.[12]

What 97% of Scientists Agree On

President Obama said, "97% of scientists agree: climate change is real, man-made, and dangerous."[13] The 97% argument has persisted to this day. **I was aghast to learn that the quoted "97% of scientists agree" is nothing more than propaganda and a misrepresentation of the data.** This number comes from a 2013 paper by John Cook, and colleagues, which reviewed abstracts of 11,944 papers on climate.[14] Of those thousands of abstracts, 7,930—66% of the total—

were excluded as they gave no opinion on man-made global warming, leaving only 4,014 papers reviewed, or 34%, which expressed an opinion on the issue. Of these remaining 4,014 papers, 97% stated humans have caused recent warming, but only 105 of these more than four thousand papers said humans caused *most* of the warming, and the remaining abstracts stated humans are a cause of the warming.[15] **So only 105 of the included 4,014 papers reviewed—or less than 3%—stated climate change was mostly man-made.**

This paper focuses primarily on the cause of global warming. Whether recent warming is dangerous or catastrophic was not addressed by the Cook study.

Other surveys and studies have been cited to claim a scientific consensus that global warming is mostly caused by humans. Bart Verheggen et al. published a survey in 2014 asking if greater than (>) 50% of recent warming could be attributed to anthropogenic GHGs.[16] The survey was sent to scientists who published papers that included "global warming" and "global climate change" in the titles. About 6,000 surveys were sent out and 1,869 returned.

About 66% of respondents from returned surveys agreed that GHGs have caused more than 50% of recent warming. The 66% figure is a majority, but the term "consensus" is generally not applied to a majority, but rather to an overwhelming agreement of opinion on a matter. While 66% of those surveyed agreed that recent warming is mostly human caused, this figure is not close to a 97% consensus.

Most significant, we do not know if these scientists believe 51%, or 100% —or somewhere in between—of warming is from human causes. Because of this wide range of 51% to 100%, the findings of this survey are not very informative. Many of the scientists who are climate crisis skeptics listed in Table 1 agree that about half of the 1.1°C of warming since 1850/1900 is from anthropogenic greenhouse gases and the other half is from natural causes. Judith Curry said, "… it's roughly half natural, half human-caused at this point is sort of the way I think about it."[17] Richard Lindzen said, "The claim that 51% of the small warming over the past 60 years is due to man's activities is completely consistent with there being no problem worth bothering about."[18] William Happer cites 1°C in warming during the past two centuries and calculates CO_2 warming to be 0.51°C of that warming.[19] Nir Shaviv said, "...at least half, and possibly two-thirds, of the 20th century's warming is related to increased solar activity."[20]

So, claiming that 50% or more of climate change is caused by humans is not very meaningful. The so-called "consensus" pushed by the United Nations Intergovernmental Panel on Climate Change (IPCC) and climate alarmists is, "Nearly all of this can be attributed to human influence. The contribution to global warming of natural factors, such as the Sun and volcanoes, is estimated to be close to zero."[21] The Verheggen survey does not support this position as the "consensus" of scientists.

The paper states that of the scientists responding to the survey who had published ten or more peer-reviewed papers on climate, a greater majority agreed that more than 50% of recent warming is human caused than did the remainder of the scientists. This is not surprising given the bias of the journals and funding sources of climate research from which the papers are based. Both publishing and research funding are heavily skewed in conformity with the climate crisis narrative. The survey did not ask if recent warming is dangerous.

A 2014 study by Neil Stenhouse et al. for the American Meteorology Society (AMS) states: "Research conducted to date with meteorologists and other atmospheric scientists has shown that they are not unanimous in their views of climate change." The authors cite studies by Farnsworth and Lichter, and Doran and Zimmerman. In the 2009 Farnsworth and Lichter survey, it was found that 64% of meteorologists surveyed "are convinced humans have contributed to global warming."[22] In the 2009 survey by Doran and Zimmerman, it was found that 90% of meteorologists surveyed were convinced climate change is occurring, and 82% agreed that human activity is a significant contributing factor in changing the global mean temperature. However, the study did not define the term "significant."[23] **Please note that neither the Doran and Zimmerman or Farnsworth and Lichter studies conclude there is a consensus that *nearly all* recent warming is caused by humans, as is stated by the IPCC.**

The Stenhouse study summarizes results of 7,062 surveys sent to AMS members, of which 1,854 surveys were returned. This survey specifically asked if recent warming over the last 150 years was primarily caused by humans. Of those surveyed, 52% agreed that the warming was mostly human induced.[24] Like the Verheggen study, this is a majority, but 52% is certainly not a "97% consensus," and we do not know if they believe 51% or 100% (or somewhere in between) of warming is caused by humankind. **Therefore, the Stenhouse study provides no basis to confirm the IPCC "consensus" view that *nearly all* recent warming is human caused.**

Like the Verheggen survey, the authors of the Stenhouse study point out that a higher percent of those meteorologists who publish papers on climate change attribute most of the warming to human causes. This is not surprising, since climate journals, and climate research funding, on which the papers rely, are significantly aligned with the climate alarmist narrative. The AMS survey asked if warming would be harmful or beneficial over the next one hundred years, but the results to this question were not reported.

A 2016 study by Ed Maibach from George Mason University, of 7,682 professional members of the American Meteorological Society, had a 53.3% participation rate (or a return of the survey) from 4,092 respondents. Of those returned surveys, 96% agreed that climate change is happening and 46% agreed that *most* of the warming is anthropogenic. More significantly, the study found that only 33% agreed that warming would be primarily harmful. Another 60% agreed that the impact would be equally beneficial and harmful.[25]

A 2021 study by Lynas et al. reviewed 3,000 peer-reviewed papers on climate and declared that 99% of scientists agree with the "consensus" of anthropogenic global warming.[26] A paper published in the journal *Climate* on October 30, 2023, reexamined this study and its methodology.[27] Israeli physicist David Dentelski and his five coauthors point out that the hypothesis for which these 3,000 papers are classified is "the existence of human-caused climate change." In other words, 99% of scientists agree that humans contribute to recent global warming. Interestingly, the Dentelski paper reviewed several of these 3,000 papers from well-known skeptics of the climate crisis, including Dr. Willie Soon, whose papers were classified in the 99% consensus in this study. This is not surprising since most climate crisis skeptics do not deny an influence of anthropogenic warming, and rather only question the amount of the warming and its consequences.

This "99% consensus" study by Lynas also did not address the question of whether climate change is dangerous or a crisis.

There are additional surveys and studies that claim that 97% to 100% of scientists agree on this issue. But what do they agree on? All these studies and surveys show scientists agree that humans contribute to warming, and yet they reveal that there is no consensus that warming is dangerous (see Table 2). **Based on these studies and surveys, there is no basis to say there is a "consensus" among scientists that recent global warming is dangerous, or that *nearly all* of it is caused by humans**. A nice summary of the climate change consensus studies has been compiled by John Robson (from the Climate Nexus Discussion) in his video, "The 97 Percent Consensus Myth Revisited."[28]

Table 2 Scientific Consensus: Human Contribution to Warming				
Study	**Humans Contribute to Warming**	**Humans Cause 50% or More of the Warming**	**Humans Cause Nearly All of the Warming**	**Global Warming Is Dangerous**
Cook Study of published climate papers	97%	3%	NA	NA
Verheggen Survey of climate paper authors	NA	66%	NA	NA
Farnsworth and Lichter meteorologists	64%	NA	NA	NA
Doran and Zimmerman meteorologists and geophysical scientists	82%	NA	NA	NA
Stenhouse Survey of meteorologists	83%	52%	NA	NA
Maibach Survey of meteorologists	68%	46%	NA	33%
Lynus Study of published climate papers	99%	NA	NA	NA

Since many prominent scientists challenge the climate crisis narrative and surveys show there is no consensus that warming is dangerous or that *nearly all* recent warming is from anthropogenic greenhouse gases, where is the climate crisis narrative coming from? The United Nations is a leading source of this story line. In Chapter 2, we will discuss the United Nations and its International Panel on Climate Change (IPCC) and how they have attempted to manufacture a climate crisis through public announcements of a boiling planet, falsely claiming an increase in severe weather, the telling of half-truths, and hiding past climate cycles.

Chapter 2: IPCC

The United Nations has formed a climate science group known as the International Panel on Climate Change (IPCC). The "climate crisis" narrative echoed by politicians and the media is primarily driven by policymakers of the United Nations and its IPCC. The bias of the IPCC is evident in its original mission statement. Its role, as defined in the "Principles Governing IPCC Work," is "to assess on a comprehensive, objective, open, and transparent basis the scientific, technical, and socio-economic information relevant to understanding the scientific basis of risk of *human-induced climate change*, its potential impacts, and options for adaptation and mitigation . . . "[1] If human-induced climate change were determined to be a minor risk, then the very existence of the IPCC would be moot. "The objective of the IPCC is to provide governments at all levels with scientific information that they can use to develop climate policies."[2]

Every few years the IPCC publishes a detailed report on the scientific, technical, and socio-economic impacts of climate change, with the most recent report being over 10,700 pages. These reports are known as Assessment Reports (AR) and to date, the IPCC has issued 6 reports, (i.e., *AR1, AR2, AR3, AR4, AR5,* and *AR6*). Many respected scientists and economists have served in various subgroups of the

IPCC, and the detailed scientific "assessment reports" are generally of high quality. However, policymakers of the IPCC have an agenda, to promote the anthropogenic global warming narrative and to paint a climate crisis in the "Summary for Policymakers" document, even though *there is no mention of "climate catastrophes" at all in the scientific sections of the IPCC assessment reports.*[3]

On July 27, 2023, U.N. Secretary-General António Guterres said, "Climate change is here, it is terrifying. And it is just the beginning. The era of global warming has ended; the era of global boiling has arrived."[4] This theme echoed rhetoric from the 2023 World Economic Forum meeting in Davos, where Al Gore said climate change is "boiling the oceans."[5] In August 2021 U.N. Secretary-General Guterres also said, "Extreme weather and climate disasters are increasing in frequency and intensity."[6]

As we shall see in Chapter 3, *these statements have no basis in fact or science.* So what is driving the climate crisis fearmongering of the U.N. and IPCC? Stephen H. Schneider, co-author of the 2001 IPCC report, provides an explanation. In a *Discover* magazine interview he said, "And like most people, we'd like to see the world a better place, which in this context translates into our working to reduce the risk of potentially disastrous climate change. To do that we need to get some broad-based support, to capture the public's imagination. That of course means getting loads of media coverage. So, we have to offer up scary scenarios, make dramatic, simplified statements, dramatic statements, and make little mention of any doubts we might have."[7] Perhaps this explains former Vice President Gore's statement that climate change is "boiling the oceans." **Such a claim is pure scaremongering as the Earth has had CO_2 levels above 7,000 ppm, versus the 427 ppm today, and the oceans did not boil.**

A United Nation panel stated, "It is unequivocal that human influence has warmed the atmosphere, ocean, and land … Human-induced climate change is already affecting many weather and climate extremes in every region across the globe."[8] Climate and environmental policies enacted by the United Nations are based on two suppositions: 1) there is a climate crisis, and 2) the climate crisis is human-induced, almost entirely from an increasing level of CO_2 from burning fossil fuels. Promoting human-induced damage to the Earth as a narrative is central to building public support for United Nations policies. In a November 14, 2010 interview with the Swiss newspaper *Neue Zürcher Zeitung*, Dr. Ottmar Edenhofer, co-chair of the U.N. IPCC Working Group 3, said, "First of all, developed countries have basically expropriated the atmosphere of the world community. But one must say clearly that we redistribute de facto the world's wealth by climate policy. Obviously, the owners of coal and oil will not be enthusiastic about this. One has to free oneself from the illusion that international climate policy is environmental policy. This has almost nothing to do with environmental policy anymore, with problems such as deforestation or the ozone hole."[9]

The scientific sections of the IPCC assessment reports include, for the most part, a relevant summary of scientific knowledge relating to climate change. However, most of these scientific reports are highly technical and cover thousands of pages, so they are almost never read by the media or the public. The "Summary for Policymakers" provides a more concise layman's overview for the media and public. Unfortunately, the "Summary for Policymakers" promotes a sensational, human-induced climate crisis and is guilty of stating half-truths and cherry-picking data to support this narrative. Because human-induced climate change is central to the United Nations narrative and policies, any evidence that suggests a major contribution to climate change coming from natural climate variability is ignored. Since increases in human-induced greenhouse gas emissions occurred only after the start of the Industrial Revolution in the nineteenth century, well-established historical climate cycles before the nineteenth century are omitted.

In the "Summary for Policymakers" section of the most current IPCC assessment report, *AR6*, there is a graph, from the European-based PAGES 2k network, of historical temperatures that erased the historical Roman Warm Period, the cold Dark Ages, the Medieval Warm Period, and the Little Ice Age and portrays modern warming zooming up like the blade of a hockey stick (see Figure 32 in Chapter 5).[10] As you will see in Chapters 5 and 6, there are dozens of credible scientific and historical studies that document the existence of these past climate cycles including papers from scientists associated with the PAGES 2k network (e.g., Fredrik Ljungqvist and Ulf Büntgen). The PAGES 2k hockey stick graph is an anomaly, but it is the only depiction of past climate cycles presented in the *AR6* "Summary for Policymakers." This anomalous PAGES 2k hockey stick graph supports the human-induced climate crisis narrative, so the IPCC includes it in its "Summary for Policymakers." Shockingly, no study documenting evidence of these historical climate cycles is mentioned in the *AR6* "Summary for Policymakers," not even those published by PAGES 2k network members Ljungqvist or Büntgen.

An example of telling a half-truth to promote a climate crisis, the IPCC "Climate Change 2023, Summary for Policymakers" states, "Hazards and associated risks expected in the near term include an increase in heat-related human mortality and morbidity (high confidence)."[11] This is a true statement and is taken from the Zhao 2021 *Lancet* study.[12] However, it is misleading because it tells only half of the story. **What is not said in the "Summary for Policymakers" is the fact that nine times more people die from cold than heat, and many more lives have been saved due to warming temperatures.** The Zhao 2021 *Lancet* paper reports that between 2000 and 2019, 4,594,098 people died globally from cold, whereas only 489,075 deaths are attributable to warmer temperatures (see Figure 21 in Chapter 3). Furthermore, the study concludes, "Most excess deaths were linked to cold temperature, whereas fewer were linked to hot temperatures in 2016-19, although cold-related deaths have decreased over time." Furthermore, the *Lancet* paper states, "Globally, from 2000-03 to 2016-19, the cold-related excess death ratio changed by negative 0.51 percentage points, and the heat-related excess death ratio increased by 0.21 percentage points, leading to a net decline of 0.30 percentage points." **In other words, global warming has *led to saving more lives overall*.** Not only have nine times more people died from cold than heat between 2000 and 2019, deaths from cold are declining more than twice as fast as deaths from heat are increasing. By hiding the decline in deaths from cold due to climate change, IPCC policymakers have implied the false message that climate change is causing death rates to increase. The opposite is true.

In public announcements, the United Nations policymakers take every opportunity to sound the alarm of dangerous weather events caused by increasing levels of CO_2, methane, and other anthropogenic greenhouse gases. Surprisingly, many of their alarmist claims contradict what scientists have written in the detailed scientific sections of the IPCC assessment reports. For example, a December 5, 2023, press release from the World Meteorological Organization (WMO) of the United Nations, titled, "Rate and impact of climate change surges dramatically in 2011–2020" states: "Our weather is becoming more extreme, with a clear and demonstrable impact on socio-economic development."[13] The press release goes on to list extreme weather events including floods and tropical cyclones (hurricanes) even though these claims are not supported in the scientific sections of the most recent *2021 IPCC Assessment Report, AR6*. Regarding floods, IPCC *AR6* states, "There is low confidence in the human influence on the changes in high river flows on the global scale."[14] Regarding hurricanes, the scientific section of IPCC *AR6* concludes, "There is low confidence in most reported long-term (multi-decadal to centennial) trends in Tropical Cyclone frequency- or intensity-based metrics due to changes in the technology used to collect the best-track data."[15]

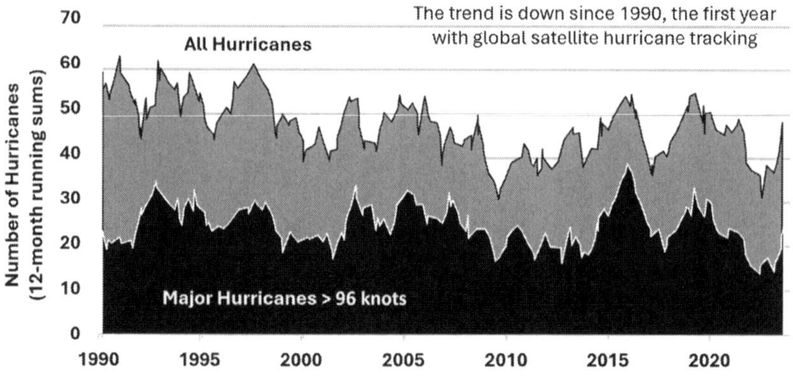

Figure 3 – Number of Global Hurricanes, 1990 to 2023. *Representatives of the United Nations announced that hurricanes have become worse in recent years due to climate change. This claim is not supported by data or in the conclusions of the scientific sections of the IPCC Assessment Reports. Globally, there have been fewer hurricanes, and there is no trend in severe hurricane frequency as measured by satellites since 1990, the first year with complete global satellite coverage.* Source: Meteorologist Ryan Maue extracting data from the Tropical Cyclones, Radar, Atmospheric Modeling, and Software Team (TC-RAMS) at Colorado State University, August 2023.

In other words, observational data shows no trends for increased floods or hurricanes.

Yet another example of cherry-picking data to fit a climate crisis narrative in the *AR6* "Summary for Policymakers" is its description of hurricane trends. It states: "The proportion of intense tropical cyclones (Category 4–5) and peak wind speeds of the most intense tropical cyclones are projected to increase at the global scale with increasing global warming (high confidence)."[16] This statement is counter to historical trends since the Accumulated Cyclone Energy index measuring hurricane strength is declining (Figure 5 in Chapter 3). This prediction is partially true because, as oceans warm, the associated moisture content of the air feeds the strength of hurricanes.

However, wind is a more powerful mechanism of evaporation than temperature. Everyone knows this principle. If you hang laundry out on a clothesline, a few degrees of warmer temperatures hardly changes the time required to dry the clothes. We all know, however, that if a brisk wind comes up, the laundry dries very rapidly. The *AR6* "Summary for Policymakers" section of *AR6* conveniently omits that the accelerated warming of the Arctic due to climate change is reducing wind speeds thereby decreasing ocean evaporation associated with winds (see Chapter 3). It also fails to mention that climate change warms the upper troposphere more than the lower troposphere, and this lowers the energy available for hurricanes (see Chapter 3). These forces counteract increased evaporation from warm oceans in driving hurricane frequency and intensity.

Chinese Academy of Sciences atmospheric scientist Ning Ma and coauthors recount in their paper, "Recent Decline in Global Ocean Evaporation Due to Wind Stilling": "More pronounced warming at high latitudes compared to mid- or low-latitudes in recent decades has weakened the meridional temperature gradient, thereby attenuating the thermal wind in the free atmosphere. This, in turn, somewhat limits evaporation because of decreased aerodynamic transfer. As a result, the net effect of rising temperatures on evaporation is also influenced by changes in wind speed indicating that global evaporation may not necessarily increase despite significant warming in recent decades."[17] In this paper the authors show how global measurements of evaporation from four satellite data sets show global evaporation declined from over 40 mm per year in 2008 to about 20 mm per year in 2017—despite rising temperatures.[18]

According to a paper by Old Dominion University atmospheric scientist Robert Tuleya and coauthors in the *Journal of Atmospheric Sciences*: "... results confirm that a scenario (e.g., global warming) in which the upper troposphere warms relative to the surface will have less tropical cyclone intensification than one with a uniform warming with height."[19] Tuleya et al. conclude: " … a climate model warming scenario in which the upper troposphere warms relative to the surface warming will offset a significant amount of the effect of surface warming alone on intensity."[20] Despite abundant evidence, the *AR6* "Summary for Policymakers" conveniently omits these other tropical cyclone drivers. Data shows hurricanes have not become more severe or frequent as temperatures have increased over the past 30 years (see Figures 3, 4, and 5). These counteracting forces are important meteorological principles for understanding hurricane trends, yet the *AR6* "Summary for Policymakers" section is silent on these additional drivers of tropical cyclones because they destroy the crisis narrative of climate change increasing hurricane frequency and severity.

Much of the fear of climate change, egged on by the United Nations and the IPCC, has been due to false claims of climate catastrophes from warming. In Chapter 3 we examine observational data which shows extreme weather events, wildfires, extinctions, and sea level rise have declined or have been stable during recent warming. We also discuss why warmer weather produces less severe storms, which is true today and in the past.

Chapter 3: Climate Catastrophes

Extreme Weather Events

The media has brainwashed the population into believing every extreme weather event of our time is caused by climate change. *Data proves this supposition is wrong.* In his book *Unsettled*, Professor Steven Koonin uses data from the IPCC Assessment Reports, which clearly show that despite the recent increases in CO_2 concentrations in the atmosphere, catastrophic weather events have not increased.[1] Data from Koonin's book and other credible sources show that hurricanes have not become more severe (see Figures 4, 5); incidences of strong tornadoes are actually significantly down (Figure 6); heat waves have moderated since the 1930s (Figure 7); the percentage of the Earth in drought is declining (Figure 8); fires and acreage burned have dramatically decreased since the 1920s and 1930s (Figure 9); and there is no identifiable trend in floods. Of all the claims by climate alarmists of increasing extreme weather events, none can be seen in the data.

Table 3 Extreme Weather Events Are Not Increasing			
Event	Increasing/More Severe	No Trend	Fewer/Less Severe
Hurricanes	–	–	Slight Decline
Strong Tornadoes	–	–	Significant Decline
Heat Waves	–	–	Significant Decline
Droughts	–	–	Less Severe
Wildfires	–	–	Significant Decline
Floods	–	No Trend	

A paper published in the *European Physical Journal Plus* on 13 January 2022, by University of Milan physicist Gianluca Alimonti and coauthors reviewed data on extreme weather events globally. The paper shows a lack of trend, or decline, in hurricanes, tornadoes, extreme precipitation events, floods, and droughts. It also presents an increase in global greening and agricultural production. The authors concluded that observational data does not support a climate emergency.[2] This publication goes against the climate alarmist narrative, and a notice was issued on September 30, 2022, that "Conclusions in this manuscript are currently under dispute." On 23 August, 2023, the article was retracted. The reason for retraction cited: "Concerns were raised regarding the selection of the data, the analysis, and the resulting conclusions of the article." The authors disagree with the retraction, and it is worth noting that all data cited in the article are from reputable public sources and other peer-reviewed papers. This type of censorship is detrimental to true scientific discovery. Nevertheless, to avoid criticism of using data from a retracted publication, I have used sources in this chapter from my own research that are not provided in the retracted Alimonti et al. article.

Hurricanes: Tropical cyclones are the most destructive storm systems on Earth. They can inflict significant damage to ecosystems and infrastructure and threaten human life. Tropical cyclones are known by several names depending on where they occur. Tropical cyclones occurring in the Atlantic Ocean are known as hurricanes. Typhoons are tropical cyclones in the western North Pacific Ocean. Cyclones is the term used in the South Pacific and Indian oceans. The distinction is purely regional and linguistic, and not scientific. IPCC policy makers and the media have echoed a message that hurricanes are becoming more frequent and damaging due to climate change. The data on tropical cyclones, however, does not support these claims.

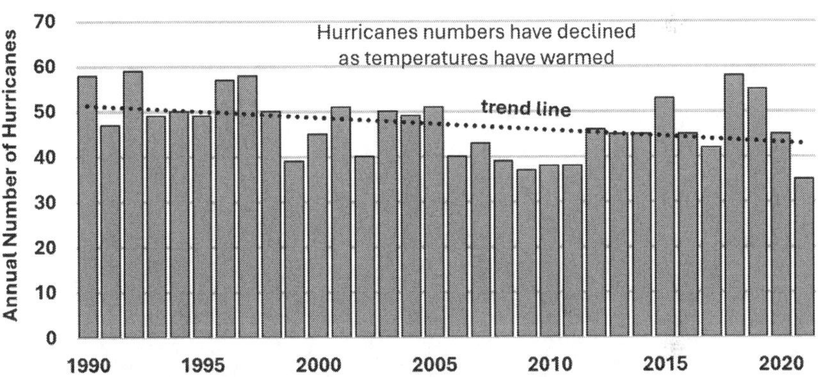

Figure 4 - Number of Hurricanes Globally, 1990 to 2021. *Globally, there have been fewer hurricanes as measured by satellites since 1990. The dotted line represents the linear trend.* Source: Klotzbach, Philip J. et al., "Trends in Tropical Cyclone Activity 1990–2021," *Geophysical Research Letters*, Volume 49, Issue 6, e2021GL095774, 28 March 2022. https://doi.org/10.1029/2021GL095774.

Using satellite data, it has been shown that the frequency and strength of tropical cyclones has declined globally since 1990. Colorado State University atmospheric scientist Philip Klotzbach and coauthors published a paper in 2022, "Trends in Global Tropical Cyclone Activity: 1990-2021," that uses satellite data to track hurricane frequency and strength. Although satellite data has been available since the late 1970s, no data on tropical cyclones was available for the North and South Indian oceans until 1990, so only satellite data after 1990 provides a complete global record. In this study the authors track hurricane numbers and Accumulated Cyclone Energy (ACE). Developed by Colorado State University professor and atmospheric scientist, William Gray in the late 1980s, ACE is a widely accepted metric used to measure the total kinetic energy released by tropical cyclones over their lifetimes. It takes into account the strength and duration of the storm. In their paper, Klotzbach et al. reported, "Global hurricane counts, and Accumulated Cyclone Energy, have significantly decreased since 1990. . ." (see Figures 4 and 5).[3] A similar decline in hurricanes has been reported by the National Hurricane Center for the United States.[4]

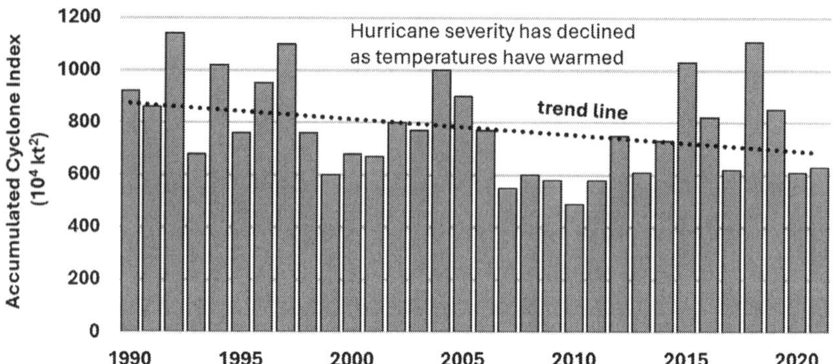

Figure 5 – Accumulated Cyclone Energy, 1990 to 2021. *Satellite data since 1990 reveal fewer hurricanes are occurring globally (see Figure 4), and tropical cyclones are producing less Accumulated Cyclone Energy—a metric accounting for hurricane frequency, intensity, and duration. The dotted line represents the linear trend.* Source: Klotzbach, Philip J. et al., "Trends in Tropical Cyclone Activity 1990–2021," *Geophysical Research Letters*, Volume 49, Issue 6, e2021GL095774, 28 March 2022. https://doi.org/10.1029/2021GL095774.

Perhaps a more important indicator of the destructive power of tropical cyclones is the Power Dissipation Index (PDI). This index goes a step further than ACE in that it combines hurricane intensity, duration, and total power dissipated by a cyclone's winds. As PDI includes a measure of how wind energy interacts with the environment, it provides a good measure of the destructive power of tropical cyclones. This index was proposed by MIT atmospheric scientist Kerry Emanuel in 2000 and is now a widely accepted measure. A study of tropical cyclones since the mid-1990s by Guangdong Ocean University atmospheric scientist Shifei Tu and coauthors in *Nature Communications, Earth & Environment*, shows the decline in global tropical cyclone PDI from 1994 to 2021. The paper states, "The annual cumulative PDI in the Southern Hemisphere has decreased significantly since the mid-1990s, while the PDI in the Northern Hemisphere decreases to a lesser extent and shows strong interdecadal variability throughout the study period."[5] The study shows a dramatic decline in PDI in the Southern Indian Ocean. In this region, cyclone intensity has remained stable, showing a 0.38% increase in intensity, duration that has declined by 12.5%, frequency declining by 16.1%, and PDI declining significantly, by 28.5%.[6] In the most recent 2021 IPCC *Climate Assessment Report,* Table 12.12 states "evidence is lacking or the signal is not present, leading to low confidence of an emerging signal" with tropical cyclones.[7]

Tornadoes: Tornadoes are one of the more violent and destructive weather phenomenon. Tornadoes are formed from the energy of cold air colliding with warm moist air. They only occur in regions with the distinctive geographic and climatic factors required for tornado formation. The United States experiences the highest frequency and intensity of tornadoes worldwide. **About 75% of all global tornadoes are observed in the United States. The number increases to 80% to 85% if you include the U.S. and Canada.** This is because warm, moist air moves northward from the Gulf of America, and the Rocky Mountains channel cold, dry air from Canada into the U.S. They are especially prevalent in the Great Plains states, known as "Tornado Alley," and the Southeastern states, known as "Dixie Alley."

The United States has kept the best global records on tornadoes, so the history of U.S. tornadoes is a relevant focus of many studies. Tornadoes are rated for strength on the Fujita Scale (F-scale) introduced in 1971 by Tetsuya Fujita. The scale rates tornadoes from weak (F0) to the strongest at F5. In 2007, the U.S. switched to the Enhanced Fujita Scale (EF-scale) that continues to use a 0-5 scaling.

A study of tornado activity in the U.S. from 1954 to 2018 was authored by Canadian meteorologists Zuohao Cao and Huaqing Cai, titled, "Trend Analysis of U.S. Tornado Activity Frequency." In this paper they conclude: "Here, we show that over the last six decades (1954–2018), U.S. (E)F1 tornadoes have a statistically significant upward trend but (E)F2–(E)F4 tornadoes have statistically significant downward trends based on both solid trend analyses of three independent methods and robust verifications of reported tornado data using a recently developed approach called sample generation by replacement (SGR)."[8] **The apparent increase in F0 and F1 tornadoes in the United States since 1954 is primarily due to improved reporting rather than an actual increase in occurrence.** WSR-88D Doppler Radar was implemented in the early 1990s. Before Doppler Radar, many of the weaker tornadoes went unreported. Population density in the United States has grown by 150 million people since 1954, with expansion into previously uninhabited areas. So there are more people in more locations to observe weak tornadoes. According to a study in *Nature Scientific Reports*, "Population density explains the long-term trend in Dixie Alley. The step-increase induced due to the installation of the Doppler Radar systems explains the long-term trend in Tornado Alley."[9] Furthermore, before the 1990s, weak tornadoes were unreported by the National Weather Service because they often did not cause damage or attract attention.

The history of strong tornadoes F3 and above is the important metric as these are the destructive storms, and they have been more consistently observed and reported in the United States. **Since 1954, there has been a significant decline in F3, F4, and F5 tornadoes.** Summarizing data from NOAA, meteorologist Roy Spencer has shown the dramatic decrease in F3+ tornadoes between 1954 and 2018 (see Table 6). This data agrees with the findings of the Cao and Cai study. This declining trend is also seen in the declining severity of losses from tornadoes in the United States between 1954 to 2018. The

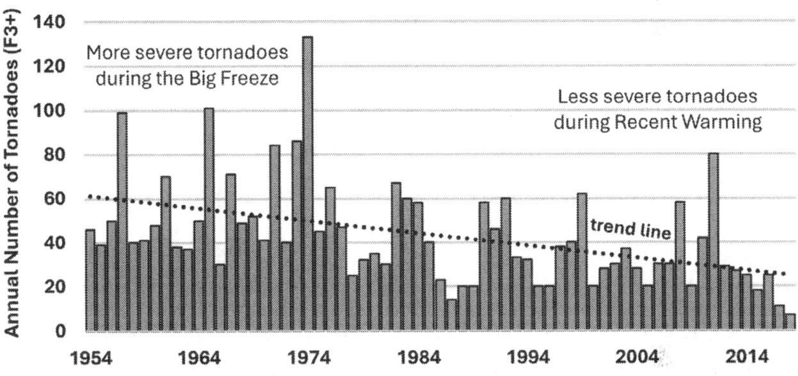

Figure 6 – Strong-to-Violent Tornadoes (F3+) in the U.S., 1954 to 2018. *Data shows strong-to-violent tornadoes are less severe than in previous years. This lower trend continued through 2025. There is no trend if you include moderate F-1 through violent F-5 tornadoes. The trend is up if you include mild F-0 tornadoes, but that is believed to be because of better detection technology, including WSR-88D Doppler Radar introduced in the early 1990s and increased rural population densities which result in more local reports.* Data Source: The U.S. National Oceanic and Atmospheric Administration (NOAA). Data extracted by Dr. Roy Spencer, University of Alabama, Huntsville. Citation: Perry, Mark J., "Inconvenient Weather Fact for Earth Day: The Frequency of Violent Tornadoes Fell to a Record Low in 2017," *AEI*, 19 April 2018. https://www.aei.org/carpe-diem/inconvenient-weather-fact-for-earth-day-the-frequency-of-violent-tornadoes-fell-to-a-record-low-in-2017/.

authors of a study, "Time trends in losses from major tornadoes in the United States" conclude: "Our findings suggest an overall national significant decline in normalized losses from tornado events."[10] The notable absence of EF5 tornadoes in recent years is also telling. The last confirmed EF5 tornado in the U.S. struck Moore, Oklahoma on May 20, 2013. As of January 2026, more than 12 years have passed without an EF5-rated tornado in the U.S. This is the longest such period since reliable records were started in 1950. In the most recent 2021 IPCC *Climate Assessment Report,* Table 12.12 states "evidence is lacking or the signal is not present, leading to low confidence of an emerging signal" with severe windstorms.[11]

Interestingly, known as the "Big Freeze" (see Figure 49 in Chapter 8 and 76 in Chapter 13), temperatures declined from the mid-1940s to the mid-1970s and severe tornadoes began to decline after recent warming ensued in the 1980s.

Heat Waves: According to the IPCC, the Earth has warmed by 1.1°C since 1850/1900.[12] With this additional warmth it is expected that heat waves will become more frequent, hotter, and last longer than in the past. However, the study of heat waves provides contradictory information. Whether heat waves have become more or less severe is determined by the metrics used to measure heat waves and the date range. The U.S. Environmental Protection Agency (EPA) uses a Heat Wave Index (HWI) to assess the severity and geographic extent of extreme heat events across the contiguous United States from 1895 onward. Since 1960, heat waves, as measured by HWI, are growing in severity. However, if you look at the entire record going back to 1895, you find that heat waves during the 1930s were three to four times more severe than in recent years (see Figure 7).[13]

Every forty years or so, the Atlantic Ocean experiences a natural temperature oscillation, known as the Atlantic Multidecadal Oscillation (AMO). Likewise, the Pacific Ocean experiences a twenty-to-thirty-year natural temperature oscillation, known as the Pacific Decadal Oscillation (PDO). Warming since the mid-1990s has been during a time when the warm periods of the AMO and PDO converged. The last time these two warm oscillation cycles converged was during the 1930s and early 1940s, a time of significant heat waves across the globe. The AMO and PDO cold cycles converged during the 1960s and 1970s, and accordingly, heat waves were less severe at this time (see Chapter 13). An analysis of heat waves beginning in 1960 would logically show an increasing trend, since the 1960s and 1970s saw temperatures fall during a cold period. **Heat wave analysis without a record going back to the 1930s and 1940s does not tell the entire story**.

The HWI measures heat waves in the United States, which has the most complete historical temperature record of any global region. However, hot temperatures were not limited to the United States in the 1930s and 1940s. During the 1930s, 23 of the 50 states of the U.S., six of 13 providences of Canada, two of the eight states and territories of Australia, Iceland, and Sweden all set record temperatures that have not yet been exceeded (see Chapter 13). Although temperature records from Asia in the 1930s and 1940s are sparse, a study on temperatures from China revealed that, "In Northwest China, the highest temperature appeared over the period 1930s–1940s. Along the Yangtze River Valley in Central Eastern China and Southwest China, interdecadal high temperature occurred from 1920s to 1940s and in 1990s, but the drought climate mainly appeared from 1920s to early 1940s. In South China, temperature remained at a high value over the period 1910s–1940s."[14]

There are two other heat wave indices in the literature, the Excess Heat Factor (EHF) and the Heat Wave Magnitude Index (HWMI). Both the EHF and the HWMI report an increase in the frequency and duration of heat waves in recent decades. However, neither capture the heat waves of the 1930s and 1940s as the data set for the HWMI starts in 1979 and, due to sparse temperature data, the EHF focuses on data after 1950 in Europe and after 1957 in Australia. Climate change is the long-term study of meteorological events. Without data from the 1930s and 1940s, the EHF and HWMI provide a limited view of heat waves as compared to HWI data. This is especially true in light of the ocean oscillations of the AMO and PDO that likely had an influence on high temperatures in the 1930s-1940s, low temperatures in the 1960s-1970s, and higher temperatures since 1980 (see Chapter 13).

Whereas the EPA Heat Index captures the severity and geographic reach of heat waves, the EHF and HWMI both focus on the growing trend in the number and duration of heat waves in smaller regional areas, especially urban centers. As these measures often concentrate on urban areas, factors like the *urban heat island effect* contribute to more frequent and prolonged heat waves. As cities continue to expand, they remove trees and other natural landscapes and replace them with materials that absorb and emit heat. Known as the urban heat island effect, anyone who has driven into a city from the countryside can observe the temperature increase on the thermometer in their car as they enter the city. I have noticed a 10°F (5.6°C) increase in temperature as I drove from the countryside from New Jersey into New York City. Urban areas have asphalt, concrete, and roofs that absorb solar energy and raise the temperature of the surrounding area. They also have more heat from air conditioning, heating, and transportation exhaust.

The primary purpose of the EHF is to assess heat wave impacts on human health, especially in urban areas. It is best for assessing urban heat stress but is not a good measure of the impacts of climate change. It is highly influenced by the urban heat island effect. The HWMI is primarily used to quantify the short-term intensity and severity of heat waves across different locations and time periods. Because the HWMI is calculated at a local level, it captures localized temperature anomalies, including those caused by urbanization. It is therefore highly sensitive to temperature spikes in urban areas. Because of its focus on short-term spikes in heat and its sensitivity to the urban heat island effects, it is not a good metric to determine the long-term impact of climate change on heat waves. This is especially true as its data only goes back to 1979, which was approximately the start of the recent warming cycle. Despite the deficiencies of the EHF and HWMI indices in evaluating heat waves due to long-term climate change, they are used as the sources of observational data in the IPCC *AR6* report in Chapters 11 and 12 to assess heat waves due to climate change and, they conclude, it is virtually certain that the frequency and intensity of hot extremes, including heat waves, have increased globally since 1950.[15] Media reports echo these findings.[16] By using the EHF and HWMI heatwave indices, they amplify the heat island effect and ignore crucial data from the 1930s and early 1940s. Consequently, they fail to give a full account of the long-term impact of climate change on heat waves.

The principal purpose of the EPA's HWI is to track long-term trends in heat waves across large areas of the United States. Because it measures historical temperature norms over larger regions, it is less influenced by urban warming trends. Its long-term focus and lower impact from the urban heat island effect makes the HWI a more stable metric for tracking heat wave trends relating to the impacts of climate change. Figure 7, *U.S. Annual Heat Wave Indices, 1895 to 2021* provides a better understanding of the impact of climate change on heat waves than the EHF or HWMI indices as it minimizes the urban heat effects and covers sufficient years to include the impact of multidecadal ocean temperature oscillations that impacted heat waves in the 1930s and today. The EPA's HWI shows heat waves have declined significantly in recent decades versus the 1930s and early 1940s, the period that was, other than the period 1995 to 2024, the last time the warm ocean oscillations of the AMO and PDO converged. The

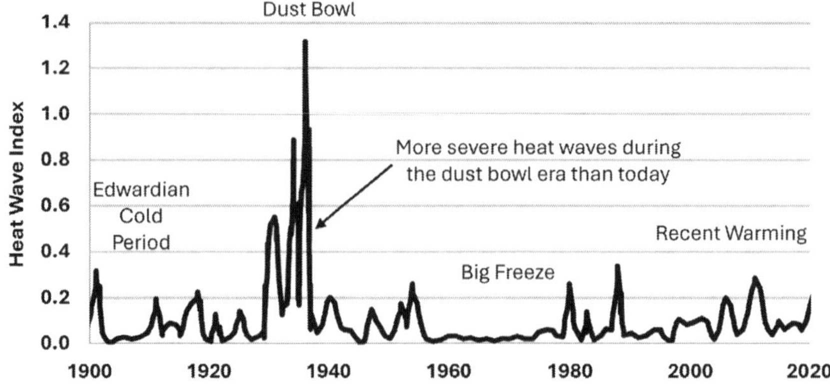

***Figure 7 – U.S. Annual Heat Wave Index, 1895 to 2021.** Despite increases in CO_2 emissions, heat waves are dramatically lower than during the "Dust Bowl" 1930s. Please note the ghost of the natural Atlantic Multidecadal Oscillation, which was cool in the 1910s, warm in the 1930s, colder in the 1970s, and warm again after 1980.* Source: *The United States Environmental Protection Agency (EPA) Climate Change Indicators Heat Waves.* Source: Courtesy of the EPA, "Climate Change Indicators: High and Low Temperatures," 19 January 2021. https://19january2021snapshot.epa.gov/climate-indicators/climate-change-indicators-high-and-low-temperatures_.html.

EPA HWI tracks with general temperature cycles that mirror natural ocean oscillations. Known in the United Kingdom as the "Edwardian Cold Period," it was cold in 1910, warm in the 1930's "Dust Bowl" (see Figure 76 in Chapter 13), cold during the "Big Freeze" of the 1970s, and warm in recent years. You can see these temperature trends in the EPA's HWI.

Droughts: An August 2022 paper from *Royal Society Publishing* titled, "Global drought trends and future projections" concludes: "Droughts do not show any substantial changes at the global scale in at least the last 120 years."[17] The paper closely looks at various measures of drought, including the Standard Precipitation Index (SPI), the Atmospheric Evaporation Demand (AED), vegetation conditions, and soil moisture. SPI is a measure of precipitation; AED is a measure of evaporation from the surface. Not surprisingly, evaporation has increased as temperatures have risen, and this has resulted in more precipitation. With more CO_2 in the atmosphere, plants close their stomata resulting in less evapotranspiration, higher soil moisture, and improved vegetation conditions. Accordingly, global vegetation has grown by 20% to 30% over the past 40 years (see Figure 18 in Chapter 4).[18]

Of these various metrics, the World Meteorological Organization of the United Nations recommends SPI as the standard to gauge global droughts. The Vicente-Serrano paper cites SPI from two sources, including the UK-based Global Precipitation Climatology Center (GPCC) and the Climatic Research Center (CRU). Operated by the

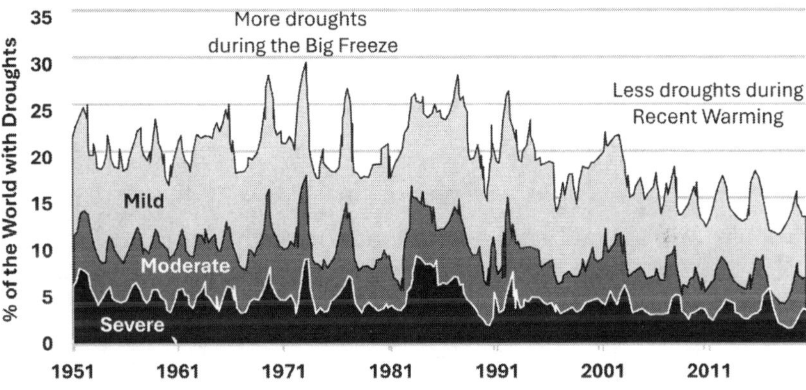

Figure 8 – Global Droughts, 1950 to 2020. *Data shows global droughts are declining. The United Nations World Meteorological Organization recommends the Standard Precipitation Index (SPI) as the standard drought metric. Measurements of SPI from the Climate Research Unit (CRU) in the UK show a decline in global droughts. The chart above shows the total percentage of the world in drought. According to CRU, the percentage of the world in severe, moderate, and mild droughts has declined over the past 70 years.* Source: Vicente-Serrano, Sergio M. et al., "Global drought trends and future projections," *Philosophical Transactions*, Royal Society A, 380:20201085, 18 August 2022, Figure 2. https://doi.org/10.1098/rsta.2021.0285.

German meteorological service, the GPCC is an international data center which provides global precipitation data. Part of the School of Environmental Sciences at the University of East Anglia, the CRU is a research center focused on climate change research, and some of its scientists are known for having a bias toward the climate crisis narrative, as evidenced in the "Climategate" emails. As shown in Figure 8, the CRU show a statistically significant decline in droughts, as measured by SPI over the past 70 years. GPCC also shows a decline in mild, moderate, and severe global droughts over this period.

Interestingly, droughts increased from 1970 through the mid-1990s. A natural temperature cycle of the Atlantic Ocean, known as the Atlantic Multidecadal Oscillation (AMO), was in a cold phase from the early 1960s through the mid-1990s (see Figure 74 in Chapter 13). Increased droughts from 1970 to the early 1990s makes sense as warmer temperatures hold more atmospheric moisture, and colder air holds less water vapor leading to less precipitation in cold times.

In the most recent 2021 IPCC *Climate Assessment Report,* Table 12.12 states "evidence is lacking or the signal is not present, leading to low confidence of an emerging signal" with aridity, hydrological drought, and agricultural and ecological drought.[19] The media would have you believe the Earth is drying up and turning brown due to global warming. The opposite is true. As the temperature warms, the atmosphere holds more, not less, humidity. Anyone who takes care of a swimming pool well knows that more water evaporates on hotter days; the pool owner needs to replenish the water. When the atmospheric temperature increases, more evaporation from the oceans occurs resulting in more moisture in the air. Thus, humidity and rain has increased with global warming, and the Earth has become greener. In addition, warming allows plants to grow better at higher latitudes due to the moderated temperatures.

In the past 35 years, the world has become 20% greener, with some measurements listing it as more than 30%. Scientific papers

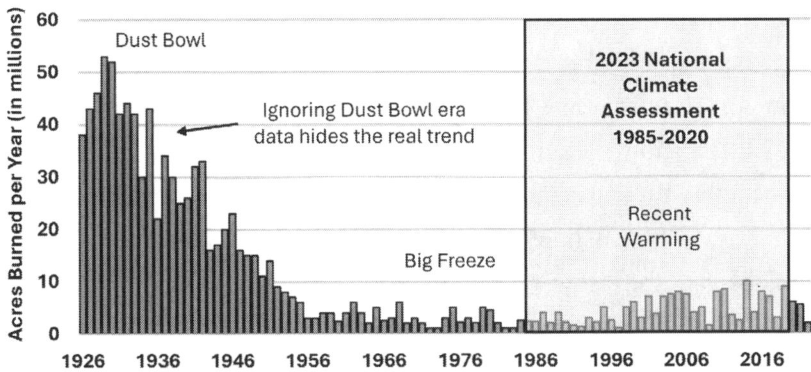

Figure 9 – U.S. Forest Area Burned, 1926-2023. *Wildfires are dramatically down from the 1920s and the "Dust Bowl" days of the 1930s. However, the recent 2023 National Climate Assessment report hid data that did not foster a climate crisis narrative of increased fires. The authors began their chart in 1985, which was during the lowest five-year period on record. The area in the shaded box is what was covered in the report. The chart above shows the data missing from the report before 1985 and after 2020, when there was a declining trend. When the entire dataset is included, the truth is revealed: fires have declined substantially. Wildfires in the 1920s saw five times more acreage burned than currently experienced in the United States.* Data Source: National Interagency Fire Center. Data extracted by Heller, Tony, "Burn Acreage Fraud in the 2023 National Climate Assessment," *Real Climate Science*, 30 November 2023. https://realclimate-science.com/2023/11/burn-acreage-fraud-in-the-2023-national-climate-assessment/#gsc.tab=0.

have attributed 8% of this greening to global warming.[20] A recent paper confirms that the greening of the Earth has continued. This paper looked at greening from 2001 to 2020. Highlights of the paper include: "The global greening is an indisputable fact"; "The rate of global greening increased slightly"; and "The drought has only slowed the global greening but not caused global browning."[21] **The world is getting wetter and greener as it warms, not dryer and browner as climate alarmists would falsely have you think**. If you want a green Earth, you should welcome the warming.

Wildfires: The media would also have you believe the Earth is burning up, and that wildfires have increased in recent years. This also is a false narrative. To provide evidence of the increased burning, they often show data after the 1970s, which does show an increase. But the 1970s were a period of cooling, known as "The Big Freeze" (see Figure 49 in Chapter 8 and 76 in Chapter 13), and you need to look back to the last warm climate cycle of the 1930s during the "Dust Bowl" period. When you look at the past 100 years, you can see that wildfires have declined significantly (Figure 9). The decline in fires during the cold period of the 1960s through the 1980s was during the cold period of the Atlantic Multidecadal Oscillation natural ocean temperature cycle.

The 2023 *National Climate Assessment* report was issued in 2024 under the Biden Administration. To feed the narrative that fires are increasing, the report only showed area burned from 1985 to 2020. The starting year, 1985, was during the lowest five-year period on record, and their end year, 2020, was just before a declining trend began. By cherry-picking the start and end date, they have painted a crisis picture of ever-increasing fires. The authors hid data that shows fires have declined significantly over the past 100 years.[22] Figure 9 shows the hidden data outside the shaded area of 1985 to 2020. This longer-term chart shows the five-fold decline in wildfires in the United States since the 1920s. The decrease in wildfires is not just a trend in the United States. The website for CO2 Science has posted 23 scientific papers which show declining fire trends in Canada, Australia, Turkey, Europe, Siberia, and the United States.[23] NASA satellite fire detection data shows global areas burned dropped by 25% between 2003 to 2019.[24] In the most recent 2021 IPCC *Climate Assessment Report,* Table 12.12 states "evidence is lacking or the signal is not present, leading to low confidence of an emerging signal" with fire weather.[25]

Floods: One surprise in the data is the fact that floods have not increased in modern times. In the IPCC *AR6* report, it states, "Seneviratne et al., 2012 assessed low confidence for observed changes in the magnitude or frequency of floods at the global scale. This assessment was confirmed by *AR5*."[26] The elevated moisture content of the atmosphere in the current recent warming period has resulted in more rain, but the measured data available has not detected any trend in floods. The IPCC *AR6* report indicates an increase in precipitation, and the NOAA "Annual 2022 Global Climate Report" cites precipitation in 2020 at 2.70 mm per day versus the past 40-year average of 2.69 mm per day. Perhaps adaptation to climate is the reason that floods have not increased. Preventative actions such as dikes and other flood control measures may well be the reason there is no trend in floods, despite more rain. **This may be another example of the prudence of investing in measures to adapt to climate, which is**

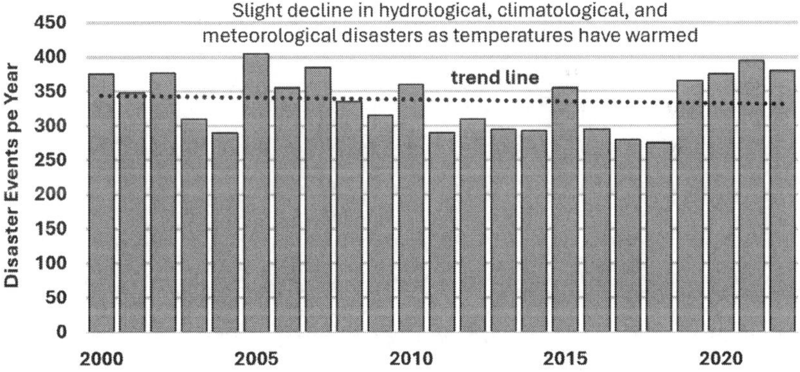

Figure 10 – Global Hydrological, Climatological, and Meteorological Disasters, 2000 to 2022. *Roger Pielke, using data from the EM-DAT of the International Disaster Database, shows no trend in hydrological, climatological, and meteorological disasters over the past 22 years.* Source: Pielke, Roger, Jr., "21st Century Global Disasters: A fresh update on global disasters since 2000," *The Honest Broker.* https://rogerpielkejr.substack.com/p/21st-century-global-disasters?utm_campaign=post&utm_medium=web.

far more affordable and prudent than vain attempts to *control* the climate.

Using data from the International Disaster Database, Roger Pielke has shown that there is no trend in global hydrological, climatological, and meteorological disasters from 2000 to 2022 (see Figure 10).[27] In the most recent 2021 IPCC *Climate Assessment Report*, Table 12.12 states "evidence is lacking or the signal is not present, leading to low confidence of an emerging signal" with river floods and coastal floods.[28] The planet is warming, but such warmth is not resulting in more extreme weather.

Reasons for the Decline in Extreme Weather Events

There are in fact scientific reasons why many of these climate disasters are declining. Meteorological theory provides several reasons why severe storms should decline as temperatures rise. Below is a description of several of these mechanisms.

Meridional Temperature Gradient: Known as Arctic Amplification, recent global warming has been most significant in high northern latitudes. The Arctic is warming faster than the global average.[29] Between 1978 and 2022, temperatures at the North Pole increased by 0.25°C per decade, while the temperature increases in the Tropics measured only 0.12°C in warming per decade, less than half the warming experienced in high latitudes (see Figure 68 in Chapter 13).[30] This means less drastic temperature contrasts between high northern latitudes and the Tropics. As MIT Professor Emeritus of Atmospheric Sciences, Richard Lindzen, points out, severe storms are caused by warm moist air colliding with cold air, so moderating the frigid air should lessen severe storms.[31] Known as the *meridional temperature gradient*, this trend of lowering the gradient because of a warming Arctic is certainly a factor contributing to the decline of severe weather over the past 30 years (see Figures 4, 5, and 6).

Several studies support the principle that a moderating meridional temperature gradient leads to fewer and less severe cyclones. According to MIT professors Charles Gertler and Paul O'Gorman, "The weakening of the meridional temperature gradient is thought to contribute to the observed weakening of eddy kinetic energy (EKE) and cyclone activity levels..."[32] A paper in *Geophysical Research Letters* by Stony Brook University professor and atmospheric scientist Edmund Chang and coauthors show that between 1979 and 2014, the number of strong extratropical cyclones in the Northern Hemisphere in summer decreased at a rate of 4% per decade. The paper goes on to describe how "Extratropical cyclones ... develop due to baroclinic instability associated with the equator-to-pole temperature difference." Chang et al. explain that near-surface temperature is amplified in the Northern Hemisphere, reducing equator-to-pole temperature gradient, which "gives rise to weaker baroclinic waves and fewer or weaker extratropical cyclones."[33] A paper in *Nature Geoscience* by University of Chicago professor and atmospheric scientist Tiffany Shaw and coauthors states, "Fundamentally, cyclones exist due to baroclinic instability, a fluid dynamical instability characteristic of rotating, stratified fluids. Baroclinic instability requires a horizontal temperature gradient, created by differential solar heating that makes the Equator warmer than the poles, and planetary rotation ... A commonly used measure of baroclinicity (when the gradient of pressure and the gradient of density are not aligned) is the maximum Eady growth rate. The Eady growth rate is proportional to the Equator-to-pole temperature gradient..."[34]

Wind Speed: The primary reason cited in the 2021 IPCC *AR6* "Summary for Policymakers" for global warming as a driver of severe hurricanes is the increase in ocean temperatures, which enhances evaporation leading to more moist air. Increased warm moisture in the air is the fuel that feeds and strengthens hurricanes. Warm oceans do not directly strengthen hurricanes; rather, it is energy of latent heat from evaporation off the surface of warm oceans. Warmth does enhance evaporation, but there are other factors, especially wind. Wind is much more powerful than warmth in enhancing evaporation. MIT professor and atmospheric scientist, Kerry Emanuel in his paper, "The Theory of Hurricanes": "The rate of evaporation from the sea surface increases linearly with wind speed … doubling the wind speed approximately doubles the latent heat flux, whereas the dependance on sea surface temperature is weaker and nonlinear."[35] According to the Clausius-Clapeyron relationship, a 1°C increase in temperature can increase evaporation by 7%. Increasing wind speed from 5 meters per second (m/s) to 10 m/s can double evaporation (a 100% increase). Increasing wind speed from 5 m/s to 15 m/s can triple evaporation (a 200% increase). We all know that if we have a wet towel and we increase the temperature in the room by 5°C the towel will dry a little faster. However, if we place the towel in front of a fan, it will dry much more rapidly. This is because, during evaporation, a layer of humid air forms just above the surface, and this layer inhibits evaporation. Wind blows away this moisture-saturated layer of air and replaces it with dry air that significantly increases the rate of evaporation.

In addition to reducing storms' frequency and intensity by lowering baroclinicity, moderating the Arctic-to-Tropics temperature gradient lowers wind speed. Atmospheric scientists Aiguo Dai and Jiechun Deng describe why reduced wind speeds are a result of Arctic Amplification.[36] The IPCC is reported to have estimated that winds will be stilled by 10% by the year 2100.[37] Chinese Academy of Sciences atmospheric scientist Ning Ma and coauthors recount in their paper, "Recent Decline in Global Ocean Evaporation Due to Wind Stilling": "More pronounced warming at high latitudes compared to mid- or low-latitudes in recent decades has weakened the meridional temperature gradient, thereby attenuating the thermal wind in the free atmosphere. This, in turn, somewhat limits evaporation because of decreased aerodynamic transfer. As a result, the net effect of rising temperatures on evaporation is also influenced by changes in wind speed indicating that global evaporation may not necessarily increase despite significant warming in recent decades."[38] In this paper the authors show how global measurements of evaporation from four satellite data sets show global evaporation declined from greater than 40 mm per year in 2008 to about 20 mm per year in 2017, *despite rising temperatures.*[39] High winds play a crucial role in the formation and severity of storms because they are a major contributor to the moist air that feeds the storm. Winds can transport more warm moist air into storms to sustain them and increase their severity. The dampening of winds due to climate change leads to a decline in storm strength, which counteracts the positive impact of warm oceans.

In the *Bulletin of the American Meteorological Society*, NOAA hurricane researcher Peter Black and coauthors state, "Evaporation rates in hurricanes are primarily controlled by wind speed, with fluxes increasing sharply above 15 m/s, far exceeding variations due to sea surface temperature alone."[40] Field data from hurricanes presented in this paper confirm that wind speed dictates evaporation surges, outstripping the contribution of sea surface temperature in storm conditions. A paper in the *Journal of Climate* by NCAR atmospheric scientist Kevin Trenberth states, "The moisture flux ($F = \rho * q * V$) is dominated by wind speed (**V**) in regions of strong flow, whereas temperature effects on humidity are secondary over short time scales."[41] Trenberth's equation and analysis highlight wind as the primary driver of evaporation, with ocean heat in **q** being less powerful in the short time scales of storms. Both the Black and Trenberth papers affirm that wind speeds are more powerful than ocean heat in driving evaporation to feed a storm, especially in dynamic short-term scenarios like hurricanes. The stilling of wind speeds from Arctic Amplification and the decline in the meridional temperature gradient has a powerful impact on lessening the power of storms and counteracting sea surface temperature's role in contributing to hurricane and storm strength.

Lapse Rate (Vertical Temperature Gradient): As climate change causes the oceans to warm, an increased amount of latent heat is removed from the ocean surface through evaporation. This latent heat is released into the atmosphere high in the troposphere when the water vapor condenses into water. The result is more extreme warming in the upper troposphere than in the lower troposphere.[42] Radiosonde data from weather balloons shows temperatures in the upper troposphere are warming 1.4 times faster than the lower troposphere.[43] This means the temperature gradient between the upper troposphere and lower troposphere is moderating. Such moderation of the vertical temperature gradient changes the lapse rate. The lapse rate refers to the rate at which atmospheric temperature decreases with an increase in altitude. A steeper lapse (colder upper troposphere) indicates a rapid temperature drop with height, and this leads to greater atmospheric instability, which can enhance storm development. Conversely, a moderated lapse rate (warming upper troposphere) signifies a slower temperature decrease with altitude, contributing to atmospheric stability and reduced likelihood of storm formation. Climate change is thus moderating the lapse rate gradient, and this may be yet another reason why there is no trend, or a declining trend, in the observation of severe storms. This trend is clearly seen in the reduction of severe tornadoes over the past 40 years (see Figure 6).

Storms are created by convective energy. A paper by Sorbonne University (France) atmospheric scientist Sandrine Bony and coauthors states, "A decrease in the lapse rate due to upper-tropospheric warming reduces the depth and intensity of convective systems."[44] Convective Available Potential Energy (CAPE) is a measure of energy for thunderstorms. CAPE measures the convective energy available for updrafts in thunderstorms, so if the number is lower, thunderstorms are less severe. CAPE is impacted by a combination of several factors, including high moisture content; wind shear; and the lapse rate gradient as the primary drivers. If isolated, higher moisture content due to warming oceans has potentially strengthened CAPE by about 10% to 20% since 1980. Climate alarmists focus on this one driver to predict that climate change will lead to more severe weather. However, we have already seen that climate change is moderating the lapse rate gradient, which also lowers CAPE. According to the *Learning Weather* by the Penn State Meteorology Department, "... it should come as no surprise that the environmental lapse rate plays a key role in the calculation of CAPE. So, when lapse rates are steep, CAPE tends to be high."[45] Conversely, when the lapse rate moderates, as is occurring with global warming, the contribution to CAPE from higher moisture levels is counteracted. As CAPE has declined from a lower lapse rate, atmospheric stability has increased. Guangdong Ocean University (China) atmospheric scientist Shifei Tu and coauthors document in their *Nature Communications, Earth and Environment* paper that between 1994–2021 there was a 28.5% decline in tropical cyclone PDI in the Southern Indian Ocean. PDI is an index which measures the destructive power of tropical cyclones. The authors conclude: "The decrease in tropical cyclone frequency is influenced by increased atmospheric stability."[46]

According to a paper in the *Journal of Atmospheric Sciences* by Old Dominion University professor and atmospheric scientist Robert Tuleya and coauthors, " ... results confirm that a scenario (e.g., global warming) in which the upper troposphere warms relative to the surface will have less tropical cyclone intensification than one with a uniform warming with height."[47] Tuleya et al. conclude, "... a climate model warming scenario in which the upper troposphere warms relative to the surface warming will offset a significant amount of the effect of surface warming alone on intensity."[48] John Shewchuk estimates that moderation of the vertical troposphere temperature gradient has decreased the strength of CAPE by 17% in tropical regions since 1980.[49] A moderating lapse rate gradient counteracts moister content of CAPE and could explain why severe weather is not increasing from global warming. According to Stony Brook University atmospheric scientist Edmund Chang and coauthors, "In the subtropics, maximum warming is projected to occur in the upper troposphere, increasing the static stability…" The authors go on to say that this effect gives "… rise to weaker baroclinic waves in the lower troposphere and fewer or weaker extratropical cyclones." Chang et al. describe the moderation of the meridional temperature gradient and the moderation of the lapse rate as reasons why "Climate models have projected a decrease in extratropical cyclone activity in the Northern Hemisphere in all four seasons."[50]

The most common cause cited for the predicted increase in severe storms from climate change is due to the increase in warm, moist air that increases the strength of CAPE. For example, an article by the Royal Meteorological Society says, "Rising global temperatures due to climate change means warmer air, which allows it to hold more moisture, roughly 7% more moisture per 1°C of warming. This boosts the chance of thunderstorms, leading to more violent storms and more lightning strikes."[51] Although it is true that moister air increases the severity of storms, such predictions fail to include the counteracting forces of moderating the Arctic to Tropics meridional temperature gradient, lower wind speeds, and moderating the upper troposphere to lower troposphere temperature gradient or lapse rate. There are sound scientific reasons why severe weather predictions due to climate change have failed to materialize in observational data.

History Confirms Cold Drives Extreme Weather. Historical records of past climate cycles confirm that severe weather is less extreme during warm periods and more extreme during cold periods as expected from meteorological theory of the impact of the Meridional Temperature Gradient and the Lapse Rate. According to University of Helsinki (Finland) professor and climate historian Sam White, "...the mid-1500s to early 1600s actually witnessed the highest average levels of hurricane activity for at least the past 500 years. In particular, counts of all storms, storms identified as hurricanes, total shipwrecks, and the rates of shipwrecks found in the Spanish colonial records all peaked during the late 1500s."[52] This was during the cold period known as the Little Ice Age (see Chapter 5 and Chapter 6). The British Royal Navy kept meticulous records of weather on its ships, recording hourly wind speeds and weather. An examination of these ship logbooks by University of Sunderland (UK) geographer and climatologist Dennis Wheeler and coauthors reveals a markedly enhanced gale frequency during the period from 1685 to 1700, which was during the Maunder Solar Minimum (1645-1715), one of the coldest periods of the Little Ice Age. In the 1730s, it became warmer just after the end of the Maunder Solar Minimum, which coincides with a significant decline in gale activity.[53]

University of Arizona professor and paleoclimatologist Valerie Trouet and coauthors cite 15 scientific papers that reconstruct historical severe weather from proxies and show less severe weather during the Medieval Warm Period (1000 to 1250), more severe during the cold Little Ice Age (1300-1850), and less severe in "Modern Warming" (1850 to today).[54] Greenland ice cores show storminess increasing during the cold Little Ice Age. Salt spray in ice cores is a proxy for storminess. Breaking waves create microscopic salt aerosols in the air. These droplets are typically removed quickly from the atmosphere. To reach the Greenland Ice Sheet, they must be driven by storm systems that produce strong winds and turbulent atmospheric disturbances. The Greenland GISP2 ice cores show a marked transition from reduced levels of storminess during the Medieval Warm Period to severe storminess during the Little Ice Age cold period after about 1400.[55]

Normal coastal breezes don't transport large volumes of sand very far. Storm-force winds can lift and push large quantities of sand inland. More frequent dune-building is produced from more extreme episodes of storm-force winds. By analyzing coastal dune layers, scientists can read the past record of storminess. Radiocarbon dating of organic material trapped between dune layers allows researchers to place these storm events in time. Several coastal dune studies have confirmed increased storminess during the Little Ice Age (1300 to 1850) including the coast of Denmark,[56] the British Isles,[57] Sweden,[58] the Netherlands,[59] Scotland,[60] northern France,[61] southwestern France,[62] western Iberia Peninsula (Portugal and western Spain),[63] the French Mediterranean coast,[64] the western North Pacific,[65] the Guangxi Province in China,[66] and the Bahamian Archipelago.[67]

A paper by geologist Saskia Jelgersma of the Geological Survey of the Netherlands and coauthors states, "An increasing level of storm surge elevations over the recent Holocene is observed with a particular maximum during the Little Ice Age."[68] A paper on storm intensity over the past 1500 years by Woods Hole Oceanography Institute climate scientist Tyler Winkler and coauthors finds more hurricanes passing through the Bahamian Archipelago "during the Little Ice Age (LIA: 1300-1850 CE) relative to the prior millennium and the past 170 years."[69] In a study of storms in the western North Pacific by MIT and Woods Hole Oceanographic Institute marine geologist James Bramante and coauthors, they conclude, "Combined with existing records, our reconstruction demonstrates that low-baseline typhoon activity prior to 1350 CE was followed by an interval of frequent storms during the Little Ice Age."[70] A paper by Université de Caen Normandie (France) professor and paleoclimatologist Laurent Dezileau and coauthors states, "The apparent increase of the superstorm activity during the latter half of the Little Ice Age was probably due to the thermal gradient increase."[71] Georgetown University associate professor and climate historian Dagomar Degroot discusses how the coldest periods of the Little Ice Age increased the frequency and severity of storms.[72] Cold periods increase the Meridional Temperature Gradient. In modern times, the Meridional Temperature Gradient has weakened from Arctic warming, which as expected, has led to less extreme weather in recent years.

University of Bern (Switzerland) associate professor and climate and environmental physics scientist Christoph Raible and coauthors found that extreme cyclone intensity was enhanced during the very cold Maunder Solar Minimum compared to today.[73] Trouet et al. made the following comments about this study, "In areas of northern Europe, where the number of cyclones showed a decrease, during the Maunder Minimum, the intensity of cyclones showed a significant increase. The reason for such a behavior was further investigated by the authors (Raible et al.) and they were able to identify that an enhanced meridional temperature gradient during the Maunder Minimum (due to polar amplification) was the major driver of this cyclone intensification, via increased lower tropospheric baroclinicity and a decrease in the static stability."[74] A colder Arctic strengthens the meridional temperature gradient resulting in more severe storms. A warmer Arctic, as experienced recently, weakens the meridional temperature gradient and leads to less extreme weather. **Fear mongering propaganda that claims global warming is causing more severe weather goes against science, the historical record, and modern weather measurements.**

The Impact of CO_2 on Soil Moisture: The decline in fires can be directly attributed to higher levels of CO_2. Plants use pores, called stomata, in their leaves to breathe in CO_2, but these stomata dry out the plant as water evaporates out of the pores. Known as evapotranspiration, "the pathway for transpiration in plants is the same one that allows for plant intake of carbon dioxide."[75] With more CO_2 in the atmosphere, plants partially close their stomata and lose less water. Over time, plants also evolve with fewer stomata to adjust to higher concentrations of CO_2 in the atmosphere. A study on this topic was conducted by Indiana University-Bloomington and Utrecht University in the Netherlands, titled, "Rising carbon dioxide is causing plants to have fewer pores, releasing less water to the atmosphere." The study finds, "As carbon dioxide levels have risen during the last 150 years, the density of pores that allow plants to breathe has dwindled by 34 percent, restricting the amount of water vapor the plants release to the atmosphere." The paper also reports: "... doubling of today's carbon dioxide level—from 390 parts per million to 800 ppm—will halve the amount of water lost to the air."[76]

In a separate paper, published by *PNAS*, many of the same scientists describe a model they devised to look at the decline in stomatal density with increasing amounts of $CO_{2.}$ In this paper, the authors state, "Here we reconstruct a 34% (±12%) reduction in maximum stomatal conductance (gsmax) per 100 ppm CO_2 increase as a result of the adaptation in stomatal density (D) and pore size at maximal stomatal opening (amax) of nine common species from Florida over the past 150 y."[77]

According to a Colorado State University paper, "More than 99.9% of the water used by an irrigated crop or turf is drawn through the roots and transpires through the leaves."[78] Therefore, because of reduced evapotranspiration, increased levels of CO_2 result in plants taking less water out of the soil resulting in the ground retaining moisture, which deters fires. Researchers used the NASA Global Inventory Modeling and Mapping Studies (GIMMS) Global Agricultural Monitoring (GLAM) system to obtain soil moisture data to predict fire risk in Australia and California. These researchers stated, "We quantified the relationship between observed fire activity and soil moisture conditions and analyzed the soil moisture conditions for two extreme fire events. Our findings show that fire activity is strongly associated with soil moisture anomalies. Lagged correlation analysis demonstrated that a remote-sensing based soil moisture product could predict fire activity with a 1–2-month lead time. Soil moisture anomalies consistently decreased in the months preceding fire occurrence, often from normal to drier conditions, according to a spatiotemporal analysis of soil moisture in two extreme fire events. Overall, our findings indicate that soil moisture conditions prior to large wildfires can aid in their prediction and operational satellite-based soil moisture products such as the one used here have real value for supporting wildfire susceptibility and impacts."[79] Soil moisture impacts wildfires, and the USGS monitors soil moisture in its TOPOFIRE system to predict fire risk since low soil moisture is an indicator of greater fire vulnerability.[80]

Increased Moisture in the Air: The fact that droughts are declining is also not surprising. Historically, periods of cold are when the Earth has been more arid. Wolfgang Behringer, in his book *A Cultural History of Climate*, conducted a comprehensive study of historical climate cycles and concludes, "Increased aridity may be regarded as the typical feature of global cooling."[81] A study of the climate in North Central China over the past 1800 years by Chinese Academy of Sciences professor and geoscientist Liangcheng Tan and coauthors reveals the close correlation between temperature and precipitation. The authors found that around 300 CE to 400 CE the temperature anomaly was about -0.3°C and the precipitation index was about -1; around 900 CE to 1100 CE the temperature anomaly was about +0.4°C and the precipitation index grew to about +1.5; around 1600 CE to 1700 CE the temperature anomaly was about -0.5°C, and the precipitation index declined to about -2; around 2000 CE, the temperature anomaly was about +0.5°C and the precipitation index grew to about +0.4.[82]

When global temperatures increase, the atmosphere will hold more moisture. According to the Clausius-Clapeyron relationship, a 1°C increase in temperature can increase moisture content in the air by 7%. When temperature declines, humidity also declines, as the water condenses out of the air. Historical records, archaeological findings, and paleoclimate reconstructions all confirm the Earth was more arid during the cold periods of the Greek Dark Ages, the Dark Ages, and the Little Ice Age, and it was moister and lusher during the warm periods of the Holocene Climatic Optimum, and the Minoan, Roman, Medieval, and Modern Warm Periods (see Chapter 6). As the Earth has warmed in recent years, specific humidity and precipitation have increased, and droughts have declined.

Other Climate Disasters

Other climate crisis disasters reported in the media are not getting worse. These reports include: extinctions of endangered species; decline of polar bear populations; decline of coral in the Great Barrier Reef; sea level rise; and climate-related deaths. **Not only are these disasters not happening, a warming world is lessening the severity of many of these events.**

Table 4
Other Climate Crisis Events Are Not Getting Worse

Event	Getting Worse	No Trend	Less Severe
Endangered Species	-	-	Fewer extinctions
Polar Bears	-	-	Growing populations
Great Barrier Reef	-	-	Record coral growth
Deaths from Heat/Cold	-	-	Cold kills far more
Climate-Related Deaths	-	-	Far fewer deaths
Sea Level Rise	-	Same for 150 years	

Endangered Species: Protecting endangered species is an important obligation of humanity. Climate crisis activist Greta Thunberg spoke at the United Nations Climate Action Summit on September 23, 2019, where she said, "Entire ecosystems are collapsing. We are in the beginning of a mass extinction..."[83] This was a follow-on to her December 2018 *Ted Talk* speech titled, "School Strike for Climate." In that talk she said, "Furthermore, does hardly anyone speak about the fact that we are in the midst of the sixth mass extinction with up to 200 species going extinct every single day."[84] Such rhetoric has motivated countless numbers of climate crisis activists, but these claims are not supported by documented evidence.

The International Union for Conservation of Nature and Natural Resources documents the extinction of 529 animal species over the past 500 years, between the years 1500 to 2000, in its Red List of extinct species.[85] Similar data is provided by Endangered Species International.[86] This data shows that extinctions have been declining even though temperatures have warmed up. The Earth has warmed since the mid-19th century, and the number of extinctions has been in decline. Between 1870 and 2009, the trend line of the number of extinctions shows a decline, and this rate of decline has been steep since we entered a warm period after 1980 (see Figure 11).

A study by the Department of Ecology and Evolutionary Biology of the University of Arizona analyzed rates and patterns of extinctions over the past 500 years including almost 2 million species. They found extinctions in plants, arthropods, and land vertebrates peaked about 100 years ago and have since declined.[87] **"The researchers found that in the last 200 years, there was no evidence for increasing extinction from climate change."[88]**

It is likely that most of the decline in species extinctions in recent years has been due to restrictions on hunting and other conservation efforts rather than climate change. It is difficult to tease out the impact of climate and the impact of intervention through conserva-

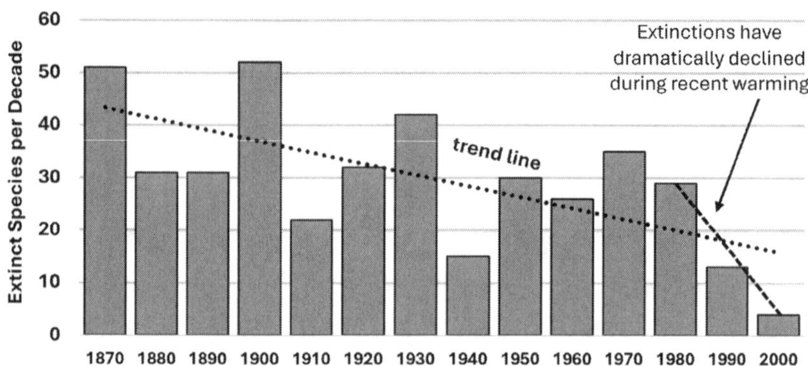

Figure 11 – Red List of All Extinct Species by Decade, 1870 to 2009. *As temperatures have warmed since 1870, the trend line of documented species extinctions each decade has declined. This decline has been especially rapid since the warming that began in 1980. The decline is likely primarily driven by restrictions on hunting and other conservation efforts, not climate change.* Source: Data extracted by Gregory Wrightstone from the International Union for Conservation Nature's Red List of Threatened Species. Wrightstone, Gregory, "Mass extinction lie exposed: life is thriving," *Inconvenient Blog*, 13 May 2019. https://inconvenient-facts.xyz/blog/f/mass-extinction-lie-exposed-life-is-thriving.

tion efforts as the cause of this decline. However, one fact is clear: the data shows warm temperatures are not hastening the extinction of species, particularly since 1980. It is not a surprise since scientific studies published in *GeoScience World*, *Earth and Planetary Science Letter*, *Journal of Paleogeography*, *Nature Geoscience*, and *Global and Planetary Change* have all shown past extinctions were caused by a *cooling* Earth, not a warming Earth.[89] Furthermore, biodiversity, in general, tends to be higher in warmer climates.

The alarm about species extinctions due to climate change was legitimized in a statement in IPCC *AR4* that states, "Approximately 20 to 30% of plant and animal species assessed so far (in an unbiased sample) are likely to be at increasingly high risk of extinction as global mean temperatures exceed a warming of 2 to 3°C above pre-industrial levels (medium confidence)."[90] This conclusion of the IPCC comes from a single paper by University of Leeds biologist Chris Thomas and coauthors.[91] Yet, as University of Virginia economist and legal scholar Scott Johnston has pointed out, the Thomas et al. paper has been heavily criticized by other scientists.[92] For example, Oxford professor and ecologist Owen Lewis has noted that the species-area relationship method used by Thomas et al., "conceals a number of assumptions and complications and problematic uncertainties concerning the usefulness of climate envelope models for predicting ranges under different climate change scenarios."[93] According to Lewis, the Thomas et al. study is from specific geographic areas and does not consider the migration of species and the "widespread ability of species to persist if transplanted or introduced outside their current range."[94] Due to migration, Lewis concludes, "The net effect on diversity at any one locality might well be positive, as species spread towards the poles from the most species-rich habitats near the equator."[95] Furthermore, Lewis states, the Thomas et al. conclusions "were specific to the regions and species included in the study" and "interpret their results as though they are global estimates ... even if predictions for the specific taxa and regions included in the study are accurate, the extrapolation to a global scale may be misleading."[96]

A summary article by Yale University/UC Santa Barbara professor and biologist Daniel Botkin and a team of 18 biologists note limitations of the model used in the Thomas et al. paper.[97] Regarding species migration in a changing climate, the Botkin et al. paper explains how the model assumes species are "in equilibrium with their current environment, and that therefore species become extinct outside the region where the environment, including the climate, meets their present or assumed requirements," but this assumption contradicts existing data and observations that "show species have survived in small areas of unusual habitat, or in habitats that are outside their well-established geographic range but actually meet their requirements."[98] The Botkin et al. article concludes: "The Thomas et al. study in particular may have greatly overestimated the probability of extinction."[99]

The model used in the Thomas et al. paper is further criticized by professor and statistical ecologist Carsten Dormann from the University of Freiburg (Germany), who states, "The problems associated with the present distribution of species are so numerous and fundamental that common ecological sense should caution us against putting much faith in relying on their findings for further extrapolations."[100] Biologists Miguel Araújo and Carsten Rahbek also note that the bioclimatic models used by Thomas et al. for future predictions are "based on some problematic ecological assumptions."[101] Scott Johnston concludes, "Given the extensive and foundational criticism by biologists of the methodology underlying the species loss probability prediction generated by Thomas et al. the IPCC's publication of that probability without qualification seems dangerously misleading,

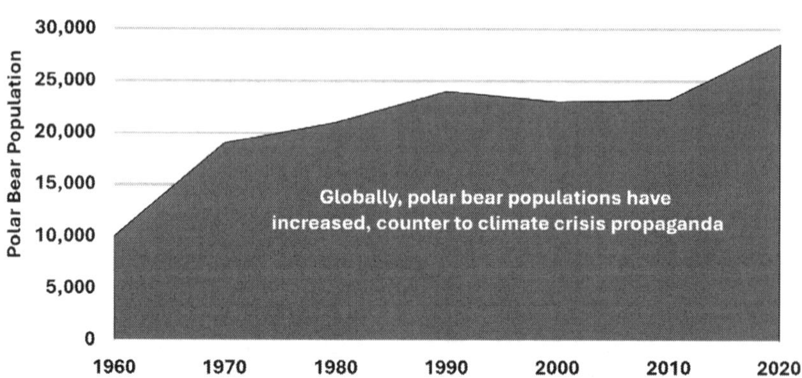

Figure 12 – Polar Bear Populations, 1960 to 2020. *Polar bears, the poster mascot for climate change alarmism, have prospered in recent years.* Source: Crockford, Susan J., "Testing the hypothesis that routine sea ice coverage of 305 mkm2 results in a greater than 30% decline in population size of polar bears (*Ursus maritimus*)" *PeerJ Preprints*, 19 January 2017. https://doi.org/10.7287/peerj.preprints.2737v1.

and in any event clearly exemplifies the rhetoric of adversarial persuasion, rather than 'unbiased' assessment."[102]

The IPCC *AR4* statement that plant species assessed thus far are likely to be at increased risk of extinction from increases in global temperature is counter to observational data. A study published in *Nature,* titled "Accelerated increase in plant species richness on mountain summits is linked to warming" by University of Bayreuth (Germany) professor and ecologist Manuel Steinbauer and coauthors studied studied plant diversity from 302 mountain summits in Europe spanning 145 years of observations to assess biodiversity changes. They found "a continent-wide acceleration in the rate of increase in plant species richness, with five times as much species enrichment between 2007 and 2016 as fifty years ago, between 1957 and 1966. This acceleration is strikingly synchronized with accelerated global warming and is not linked to alternative global change drivers. The accelerating increases in species richness on mountain summits across this broad spatial extend demonstrate that acceleration in climate-induced biotic change is occurring even in remote places on Earth, with potentially far-ranging consequences not only for biodiversity, but also for ecosystem and functioning and services."[103]

Polar Bears: The polar bear has been the poster child of climate change advocates, but facts reveal that polar bear populations have increased in recent years, not decreased as dramatically depicted in climate alarmist propaganda. Zoologist Susan Crockford's research and preprint paper on polar bears demonstrates polar bear populations are growing and even thriving despite warming temperatures (see Figure 12).[104] **Her findings gave a death blow to the climate change narrative promoting polar bears as victims of anthropogenic global warming, but it may have cost her job**. In May 2017, Crockford's lectures were shut down and she was subsequently fired from her position as an adjunct professor at the University of Victoria, a position she had held for 15 years. She claimed she was fired because of telling students politically incorrect facts about polar bears.

The climate crisis narrative claims polar bear populations are in decline due to climate change with only a few thousand remaining. Although regional populations have experienced declines, such as in Western Hudson Bay, overall, the global polar bear population has seen growth since the 1960s. Between 1972 to 1977, polar bear populations were estimated to be between 10,000 to 20,000 according to a study by renowned polar bear experts Douglas DeMaster from the University of Minnesota and Ian Stirling from the University of Alberta (Canada).[105] In 2015, the International Union for Conservation of Nature's 2015 Red List of Threatened Species puts polar bear numbers between 22,000 and 31,000,[106] a number which agrees with Crockford's research.

Great Barrier Reef: Another poster child of climate alarmism has been the decline of coral in the Great Barrier Reef because of climate change. Bleaching of the reef was experienced in 1998 and 2002, and a decline in coral cover was seen between 2000 to 2012. This decline fits nicely into the climate alarmist narrative. Hurricanes are known to be especially harsh on coral; and Cyclone Hamish was no exception with the Great Barrier Reef experiencing more decline. There are many species of coral in the Great Barrier Reef, but Porites and Acropora corals are the two most prominent genera. In the Great Barrier Reef, water temperatures at depths of 0-30 meters, where Porites and Acropora corals grow, range from 23°C to 30°C, with an annual average around 26°C to 27°C. Both Porites and Acropora corals' ideal temperature ranges are between 25°C to 29°C. It is believed that these corals may see declines in growth when temperatures exceed 32°C.

A study in the *Journal of Experimental Marine Biology and Ecology* shows how colonies of Porites corals in the Great Barrier grow faster in warmer temperatures.[107] According to this paper, in the Great

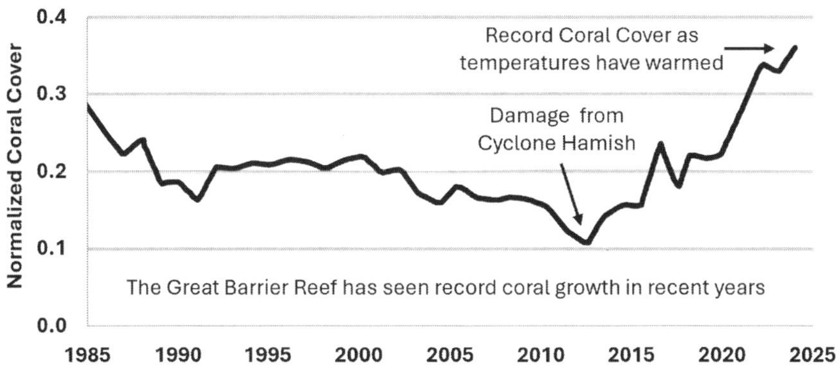

Figure 13 – Coral Cover of the Great Barrier Reef, 1985 to 2024. *In 2021, 2022, and 2024 the Australian Institute of Marine Science (AIMS) reported the highest overall coral cover on record of the Great Barrier Reef.* Source: The Australian Institute of Marine Science as published in joannevova.com.au. and Ridd, Peter, "Peter Ridd: The Vibe Shift," *Tom Nelson Podcast,* #273, 17 January 2025. https://www.youtube.com/watch?v=a5YXrNUpGfE.

Barrier Reef, "For each 1°C rise in Sea Surface Temperature, average annual calcification increased by 0.39g cm^{-2} $year^{-1}$ and average annual extension increased by 3.1 mm $year^{-1}$." Most of the corals in the Great Barrier Reef also live in Indonesia and Papua, New Guinea where the water is 29°C. These temperatures average 2°C warmer than in the Great Barrier Reef in general, and the corals grow approximately twice as fast in Indonesia and Papua, New Guinea as corals of the southern Great Barrier Reef.[108]

The Australian Institute for Marine Science (AIMS) has been tracking coral growth in the Great Barrier Reef since 1985. Data from AIMS shows that, overall, coral cover has been growing in the Great Barrier Reef for more than 10 years (see Figure 13). Since records began in 1985, coral cover in 2021 was a record, and coral cover in 2022 exceeded 2021 to set a new record. **Coral cover in 2023 was similar to 2021, and was again near record levels, and a new record was set, yet again in 2024.**[109] And although some declines from record 2024 levels were reported in 2025, coral levels remain well above lows reported between 2010 and 2015 in the Northern, Central and Southern Great Barrier Reef.[110] Since corals in the Great Barrier Reef are known to grow faster in Indonesian and Papua waters that are warmer than found in the Great Barrier Reef in totality, the warming of the Pacific may have contributed to the fast recovery of the reef from the damage of Cyclone Hamlish. Nevertheless, the climate crisis narrative is stubbornly entrenched. Although record coral cover levels were reported in the AIMS 2023/2024 report, the report goes on to say, "Despite current high levels of hard coral cover, the GBR remains exposed to the consequences of climate change, particularly more frequent and intense marine heatwaves."[111]

Geophysicist Peter Ridd from James Cook University (Australia) wrote about the recovery of the Great Barrier Reef and stated that coral is the "least endangered of any ecosystem to future climate change." He said, "Corals are particularly well adapted to temperature changes ... the warmer the better. It seems odd that coral scientists are worrying about global warming because this is one group of organisms that

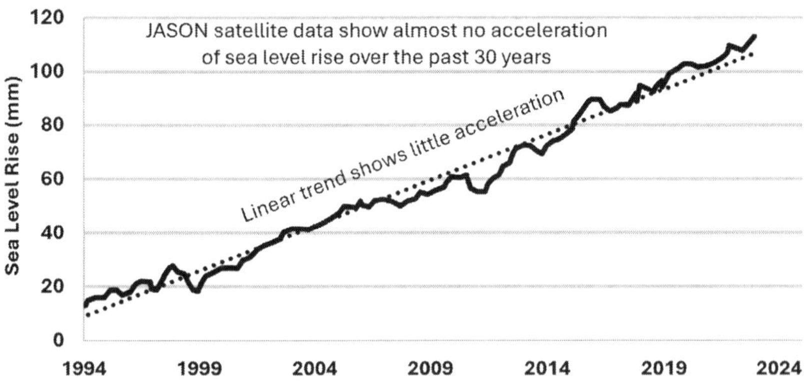

Figure 14 – JASON Satellite Measurements of Sea Level Rise, 1994 to 2023. *JASON satellite measurements of mean sea level (MSL) rise since 1994, as analyzed at the University of Colorado, averaged 3.4 mm/year with an acceleration of 0.084 mm per year^2.* Source: University of Colorado, "Most Recent GMSL Release," Sea Level Research Group, 2023_rel2. https://sealevel.colorado.edu.

like it hot. Corals are most abundant in the tropics, and you certainly do not find fewer corals close to the equator. Quite the opposite, the farther you get away from the heat, the worse the corals. A cooling climate is a far greater threat."[112] **For challenging the climate narrative, he was fired from his position at James Cook University**.

Sea Level Rise: Sea level rise is another climate alarmist claim, with some predicting a greenhouse gas-induced sea level rise of two meters by the end of the century.[113] This claim is not supported by trends seen in observational data. The oceans have been rising since the Earth emerged from the cold Little Ice Age at the end of the 18th century (see Figure 15). There has been almost no acceleration in sea level rise in recent years other than cyclical variations, which have followed temperature swings from tidal cycles and natural ocean temperature oscillations (see Chapter 13). Because of tectonic shifts, land masses on which tide gauges are placed rise and sink (Sweden

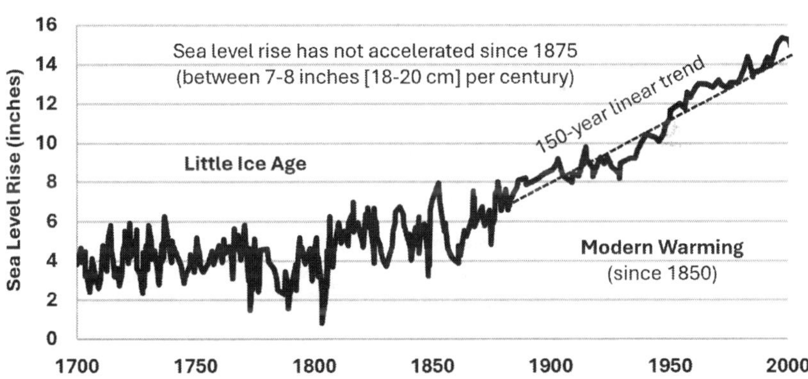

Figure 15 – Global Sea Level Rise, 1700 to 1998. *Sea rise has seen no appreciable acceleration since 1850. Most of the acceleration occurred before this date as the Earth emerged from the Little Ice Age. The oceans have risen by about 7 to 8 inches (~170 mm to 203 mm) per century over the past 150 years.* Source: Jevrejeva S. et al,, "Recent global sea level acceleration started over 200 years ago?" *Geophysical Research Letters*, Volume 35, Issue 8, 30 April 2008. https://doi.org/10.1029/2008GL033611.

is rising and Venice, Italy is sinking). Therefore, measurements of sea level in one location cannot be extrapolated globally. An aggregate of tide gauges shows an average sea level rise of 1.7 mm to 1.8 mm per year (6.7 to 7.1 inches per century).[114] Tulane University assistant professor and oceanographer Sönke Dangendorf and coauthors used GPS corrections for land motion to estimate sea level rise from 1993 to 2012 by tide gauges at 3.1 mm per year (12.2 inches per century).[115]

There are two satellite sea level measures, the GRACE (Gravity Recovery and Climate Experiment) and JASON (Joint Altimetry Satellite Oceanography Network) satellite systems. Between 2003 and 2014, GRACE satellites measured global sea level rise of 2.14 mm per year.[116] The GRACE satellite measures the Earth's gravity field, so it measures mass changes from melting ice sheets, glaciers, and changes to terrestrial water storage. It does not measure thermal expansion of the ocean, which represents about one third of the total sea level rise, or another 1 mm, increasing 2.14 mm per year to 3.14 mm per year. JASON satellites measurements from 1994 to 2023 show an average sea level rise of 0.084 mm/yr_2. (see Figure 14).[117] JASON satellite measures of sea level rise acceleration in this period of only 0.084 mm/yr^2.[118]

The JASON satellite directly measures sea surface height, so it includes thermal expansion. According to NASA, sea level rise was 9.1 cm over the past 30 years, or an average of 3.0 mm per year. However, in March 2023, NASA reported sea level rise in 2022 was 2.7 millimeters, lower than the past 30-year average, so in 2022, sea level rise actually decelerated.[119] Using the JASON satellite data of the rate of sea level rise of 3.4 mm per year and acceleration of 0.084 mm/yr^2, oceans would rise by 491 mm (19.3 inches) between 2025 and the end of this century. This assumes sea level rise continues to accelerate at the current rate of 0.084 mm/yr^2. However, sea level rise is not linear, it is cyclical.[120] We are currently at the height of a cyclical warm period of the natural Atlantic Multidecadal Oscillation (AMO)

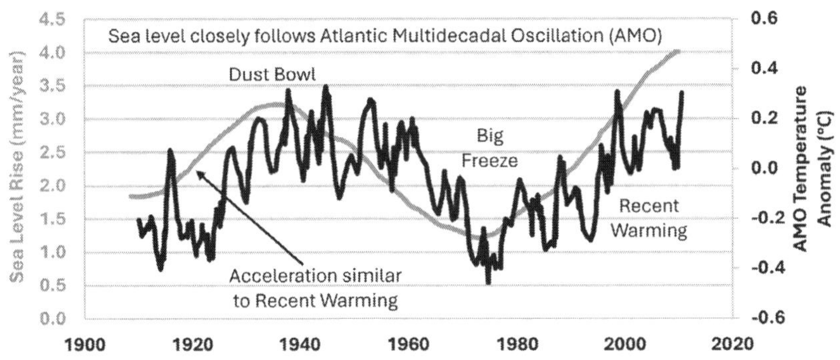

Figure 16 – Cyclicality of Sea Level Rise and the Atlantic Multidecadal Oscillation, 1910 to 2010. *The rate of sea level rise has varied with ocean temperatures. Atlantic Ocean temperatures naturally oscillate between warm and cool every 30 to 40 years in an event known as the Atlantic Multidecadal Oscillation (AMO). The figures above show various estimates of sea level rise from 1910 to 2010 and monthly Atlantic Ocean sea surface temperature anomalies. The line in gray of sea level change depicts the sum of elements of sea level rise from models (e.g., glaciers, Greenland ice sheet, etc.). The correlation of sea level rise with the rise and decline in ocean temperatures of the AMO is insightful. The sea level rise of 3 mm per year in 2000 is similar to the rise in 1940, and is to be expected in the current warm oscillation of the AMO.* Sources: *(Sea Level Rise)* Frederikse, Thomas et al., "The cause of sea-level rise since 1900," Nature, Volume 584, pp. 393-397, 19 August 2020. https://doi.org/10.1038/s41586-020-2591-3. *(AMO Temperature)* 1910 to 2010, May, Andy, "Musings on the AMO," Andy May Petrophysicist, 26 May 2025, HadSST4.1 Detrended AMO Anomaly. https://andymaypetrophysicist.com/2025/05/26/musings-on-the-amo/.

and sea level rise today is similar to 1940. From 1940 to 1980 sea level rise declined from about 3.3 mm per year to 1.3 mm per year. If you consider cyclicality since 1940, sea level rise will likely average about 2.3 mm per year or less than 7 inches (17 cm) by 2100. This figure is in line with historical sea level rise over the past 150 years (see Figure 15). About 7 inches of sea level rise (~170 mm) with 75 years to adapt is hardly a crisis.

Sea level rise has followed ocean temperatures. It has been documented that temperatures of the Atlantic Ocean oscillate naturally between warm and cool every 30 to 40 years in an event known as the Atlantic Multidecadal Oscillation (AMO). The correlation between the AMO and sea level rise is insightful. The Atlantic Ocean was cool in the 1910s, warm in the 1940s, cool in the 1970s, and warm in the 2000s. **Climate alarmists often cite the 3 mm of sea level rise in recent years as high compared with the sea level rise in the 1970s, thus falsely inferring an acceleration of sea level rise.** However, the 1970s was a cold period. No less than three *Time* magazine cover articles in the 1970s warned of "The Big Freeze" (see Figure 76 in Chapter 13). The 1970s was a time of declining temperatures, growing Arctic ice, and slower sea level rise.

A paper by Delft University of Technology (the Netherlands) oceanographer Thomas Frederikse and coauthors provides a detailed assessment of sea level rise from 1900 to 2000. This paper clearly shows the cycles of sea level rise and confirms that the 3 mm per year sea level rise over the last 30 years is not unusual during a warm period of the AMO and is consistent with sea level rise before 1940.[121] A comparison of the Atlantic Multidecadal Oscillation with the reconstruction of sea level by Frederikse et al. reveals a striking similarity in cyclicality (see Figure 16). Both sea level rise and the AMO accelerated from 1910 to 1940, decelerated from 1940 to 1970, and have accelerated since 1980. Today, we are in a warm cycle of the AMO and accelerating sea level rise. As the AMO returns to a cold phase, estimated to come after 2030, we should expect sea level rise will decelerate as was experienced from the mid-1940s to 1970. To assume sea level rise will continue at the current rate ignores historical cyclicality. **Any competent scientist knows that *one should not extrapolate off the top of a curve in a known cyclical system, yet this is exactly how catastrophic sea level rise forecasts are produced.***

Sea level rise can only properly be analyzed over long periods due to several natural cycles. Natural cycles that impact short-term sea level rise include ocean temperature oscillations such as the Atlantic

Multidecadal Oscillation and the Pacific Decadal Oscillation, solar cycles, and multi-year tidal forces including the 8.9-year perigean tidal cycle and the 18.6-year nodal tidal cycle. Due to these multi-year cycles, sea level measurements using satellite altimetry, which began in 1993, is too short of a time frame to reliably detect sea level acceleration.[122] Dutch hydraulic engineer Hessel Voortman and Dutch data analyst Rob De Vos conducted a comprehensive study on global sea level acceleration by looking at tide gauge records from a large number of locations all over the globe. They limited the sites to only those that had a minimum history of 60 years or more. They found that "approximately 95% of the suitable locations show no statistically significant acceleration of the rate of sea level rise. The investigation suggests that local, non-climatic phenomena are a plausible cause of the accelerated sea level rise observed at the remining 5% of the suitable locations. On average, the rate of rise projected by the IPCC is biased upward with approximately 2 mm per year in comparison with the observed rate." [123]

Because there is no detectible acceleration over longer-term periods, if sea level rise continued at the current rate of about 3 mm per year, the rise in sea level over the next century would be about one foot (~300 mm). However, since we are currently in the warm period of the Atlantic Multidecadal Oscillation, we would expect the average rate will be lower than 3 mm per year. Using tide gauge sea level measurements with a history of 60 years or more, Voortman and De Vos have found average sea level rise is 1.5 mm per year, far lower than the current rate or the rate from the Frederikse et al. paper.[124] Average sea level rise of 1.5 mm per year is about 6 inches per century, which is similar to the sea level rise over the past 150 years.

Sea level rise is caused by several factors. According to the IPCC, sea level rise is influenced by melting of glaciers and ice sheets (63%), and the thermal expansion of seawater due to increased temperature (33%).[125] Other smaller impacts (<5%) include pumping and use of water from underground aquifers, storage of water in man-made reservoirs, and changes in humidity which move water from oceans to the atmosphere. Figure 16 shows the close correlation between sea level rise and 60-to-80-year ocean temperature oscillations. This is not a surprise as the melting of ice sheets and glaciers is impacted by ocean temperatures.

A study by one of the world's leading glacier experts, Professor Hanspeter Holzhauser, from the University of Bern, Switzerland, reveals that glacier melt has been cyclical and in line with the warm periods of 1,000-year cycles (see Figure 27), which includes the Roman, Medieval, and Modern warm periods.[126] Paleoclimate ocean sediment samples have confirmed warmer seas during these warm periods (see Chapter 5). There is a clear connection between sea level rise and ocean temperature. Historical tide gauge records also confirm these 1000-year sea level cycles (see Figure 30).[127] It just so happens that these 1,000-year cycles are in sync with millennial Eddy Solar Cycles. These solar cycles are times of higher solar activity and low cloud cover which result in more solar radiation reaching and heating the oceans. Observational data suggests sea level rise is significantly impacted by ocean warming due to natural causes. The warmest period of the Holocene (from ~11,870 years ago to today) was the Holocene Climatic Optimum (HCO), 7000 BC to 3000 BC.[128] Studies show sea level was higher during the HCO than today.[129] **Attributing sea level rise entirely to increased concentrations of CO_2 is not supported by evidence.**

Surprisingly, island atolls, which climate alarmists falsely claim are disappearing, are generally seeing shores stabilize or expand as the growth of coral accelerates with warmer temperatures. According to Chinese Academy of Sciences marine scientist Yuxin Zhang and coauthors, despite rising sea levels, many island shorelines have maintained relative stability or equilibrium without significant alterations. A comprehensive reexamination of data on 30 Pacific and Indian

Ocean atolls, encompassing a total of 709 islands, has demonstrated that none of the atolls experienced an alarming loss in land area. Furthermore, 88.6% of the islands were either stable or increased in size, while only 11.4% showed a decrease in area.[130] Furthermore, the study on atoll shorelines found: "Human activities, particularly reclamation and land filling, were identified as primary drivers of local shoreline transformations, while natural factors have a comparatively minor impact. Moreover, the ongoing rise in sea levels is identified as an exacerbating factor for coastal erosion rather than the primary cause."[131]

The Guardian (online newspaper) proclaimed on June 28, 2023, "Rising sea levels already pose an existential threat to the populations of Tuvalu…"[132] Yet the facts are that the 101 islands of Tuvalu have *gained* 2.9% in land area over the last 50 years. The Maldives has long been a poster child of sea level rise, and the government cabinet held an underwater cabinet meeting in 2009 to publicize their "plight." Ironically, the Maldives has seen a net increase in land area over the past 30 years.[133]

Unfulfilled Climate Prophecies

Doomsday sayers predicting the end of the world have always attracted adherents. It is a seductive message that the Earth is being destroyed by carbon dioxide, and it's attractive to posit that we can become heroes by driving electric vehicles to prevent this calamity. In the past, a person calling for the end of the world was called a religious fanatic; today they are known as climate alarmists. In 1831, religious leader William Miller began to preach the end of the world would occur with the Second Coming of Christ in 1843. He attracted as many as 100,000 followers. When the date failed to materialize, he reset the date to 1844. After his second prophecy was missed, most of his followers left the movement.

Repeated failed predictions is a similar pattern seen in the climate movement. Countless prophecies of pending climate doom are stated over and over by climate alarmists, yet they fail to materialize (see Table 5). It seems only a matter of time before the climate alarmist apocalypse movement loses credibility and followers.

In 1988, Hussein Shihab, the Maldives Environmental Affairs Director, said, "A gradual rise in average sea level is threatening to completely cover this Indian Ocean nation of 1196 small islands within the next 30 years."[134] In hindsight, this prediction is not in line with reality as the islands of the Maldives have increased in land area since 1988.[135] At the current rate of global sea level rise of 3 mm per year, and with the growing land mass of these islands, there appears to be no viable threat to the Maldives from sea level rise. Consequently, developers are now planning to build 120 luxurious beachfront villas in the Maldives.[136]

In the year 2000, Avid Viner, senior research scientist said, "Within a few years, winter snowfall in the UK will become a very rare and exciting event. Children just aren't going to know what snow is."[137] This is yet another false prediction. Snowstorms in winter have continued in the UK through the present. Furthermore, data from the reputable Rutgers University Snow Lab reveals a slight increasing trend in snowfall in the Northern Hemisphere from 1967 to 2023 (see Figure 17).

In 2020, rangers at Glacier National Park in Montana stealthily began to remove signs that said, "Glaciers … are now rapidly shrinking due to human-caused climate change. Computer models indicate the glaciers will all be gone by the year 2020." The signs have been removed because the glaciers remain intact. In fact, some glaciers, including the Grinnell Glacier, have increased in size in recent years.[138]

A December 2001 article reported that climate scientist George Hurt said changes in climate could potentially extirpate the sugar maple industry in New England within 20 years.[139] However, maple

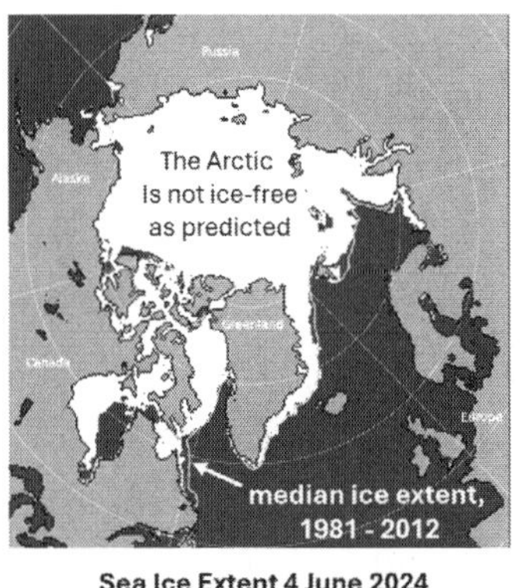

Sea Ice Extent 4 June 2024

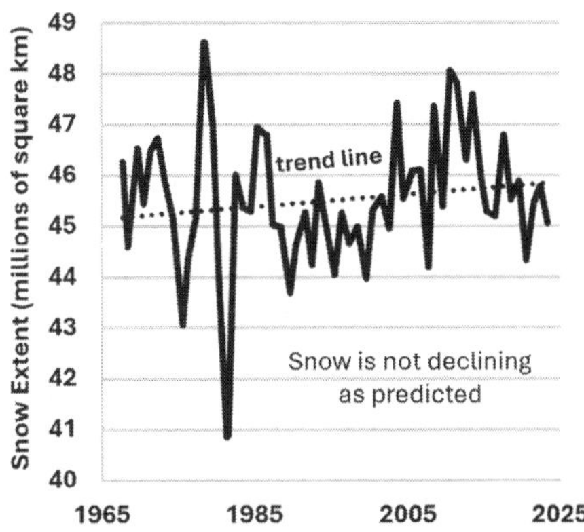

Figure 17 – The Arctic Is Not Ice-Free (left) and Snow has Not Disappeared (right)*. Arctic Ice Extent (white) on June 4, 2024, was similar to the ice extent of the 1981 to 2010 median (outline). Total snowfall in the Northern Hemisphere has increased slightly since 1967.* Sources: (*Sea Ice Extent*) Courtesy of the National Snow and Ice Data Center. https://nsidc.org/sea-ice-today; (*Snow Extent*) Rutgers University Global Snow Lab. https://climate.rutgers.edu/snow-cover/chart_seasonal.php?ui_set=nhland&ui_season=1.

syrup industry statistics show production of maple syrup in all New England states was up significantly over the past 20 years. Volume grew between 1988 to 2008 from 9,000 gallons to 19,000 gallons in Connecticut; from 44,000 gallons to 65,000 gallons in Massachusetts; from 74,000 gallons to 95,000 gallons in New Hampshire; from 310,000 gallons to 328,000 gallons in New York; and from 370,000 gallons to 640,000 gallons in Vermont.[140] The volumes have only grown further since 2008.

Al Gore has also made several alarming climate predictions that have not been fulfilled. During his 2007 Nobel Peace Prize acceptance speech, Gore said the following about Arctic Sea ice: "One study estimated that it could be completely gone during summer in less than 22 years. Another new study, to be presented by U.S. Navy researchers later this week, warns it could happen in as little as seven years—seven years from now."[141] However, Arctic Sea ice remains throughout the year to this day. In 2008 Gore said the melting Arctic would jeopardize polar bear habitats and have severe impacts on polar bear populations. The reality is, polar bear populations have increased (see Figure 12).

On January 15, 2018, *Forbes* published an article in which Harvard Professor James Anderson is reported to say, "The chance that there will be any permanent ice left in the Arctic after 2022 is essentially zero."[142] Between 2007 and 2012, other climate alarmists including scientists James Hansen,[143] Jay Zwally,[144] Peter Wadhams,[145] and Wieslaw Maslowski[146] made predictions of an ice-free Arctic by or before 2020. Since these predictions, the Arctic has not become ice-free. Ice extent for June 4, 2024, as reported by the National Snow and Data Center, shows ice coverage similar to the median average between 1981 to 2010 (see Figure 17). In addition, the minimum sea ice extent was 3.39 million square kilometers on September 17, 2012, while the minimum sea ice extent in 2024 on September 11, 2024 was 4.28 million square kilometers, a 26% increase.[147] Harry Stern from the Polar Ice Center at the University of Washington has reported that the decline in Arctic Ice was significant until 2007, but has been stable since, despite a drop in 2012. According to Stern, "We say that a regime shift occurred in 2007 because September sea-ice area changed from declining to stable."[148] Ice extent declined between 1979 to 2007 in the Arctic as expected during the warm cycle of the Atlantic Multidecadal Oscillation (AMO), but is now stable as the AMO is past its peak warm phase and Arctic ice is far from "completely gone" as predicted.

In June of 2018, climate crisis activist Greta Thunberg shared a story on Twitter titled, "Top Climate Scientist Says Humans Will Go Extinct If We Don't Fix Climate Change By 2023." She added a top climate scientist is warning that climate change will wipe out all of humanity unless we stop using fossil fuels over the next five years.[149] Having not learned anything from Greta, U.S Congresswoman Alexandra Ocasio-Cortez (commonly known as AOC) issued a warning in 2019 that "the world is going to end in 12 years if we don't address climate change."[150] As a result of this quote, a doomsday clock is now placed in Union Square in New York City, which counts down the time left until the end of the world, according to AOC. We will see how that works out. For more information on failed climate prophecies, see the video by Roman Balmakov.[151]

Table 5
Failed (and Likely to Fail)
Apocalyptic Predictions of Climate Prophets

Date	Climate Prophet	Role	Prediction
1988	Hussein Shihab	Environmental Affairs Director	Maldives under water by 2028
2000	David Viner	Climate Scientist	Winter snowfall in the UK will soon be very rare
2001	George Hunt	Climate Scientist	End of New England maple syrup by 2008
2007	Al Gore	Former U.S. Vice President	Ice-free Arctic by 2014
2007	Jay Zwally	NASA Climate Scientist	Ice-free Arctic by 2012
2008	Al Gore	Former U.S. Vice President	Severe polar bear decline by 2014
2008	James Hanson	NASA Climate Scientist	Ice-free Arctic by 2018
2008	James Lovelock	Scientist	Ice-free Arctic by 2012
2009	Peter Wadhams	Ocean Physics Professor	Ice-free Arctic by 2015
2012	Wieslaw Maslowski	Navy Scientist	Ice-free Arctic by 2016
2018	James Anderson	Harvard Professor	Ice-free Arctic by 2022
2018	Greta Thunberg	Climate Activist	Extinction of Humans by 2023

We have seen in this chapter that climate catastrophes, claimed by climate alarmists, have not occurred. In Chapter 4, we will examine the significant benefits of warming and carbon dioxide.

Chapter 4:

Benefits of CO_2 and Warming

CO_2 Is Beneficial to Plant Growth and Agriculture

Climate alarmists and the media falsely claim the Earth is getting browner due to climate change. Interestingly, climate alarmists warn of browning and weather disasters without comprehensive data to support their narrative, yet scientific data demonstrates many benefits of increased CO_2 and warming. Measuring total leaf area of the Earth, NASA satellites have confirmed that the world greened by more than 20% in 35 years (Figure 18).[1] The observation that the world has greened since 1982, as seen through NASA satellites, indicates an overall increase in vegetation cover on Earth. This is a land area equal to more than twice the size of the United States.

Known as CO_2 fertilization (see, as example, Figure 19), peer-reviewed papers have attributed 70% of this greening to increased CO_2, which is an essential plant food.[2] Most of the greening has occurred in arid lands that were formerly too dry to grow vegetation. Plants breathe CO_2 into tiny pores in their leaves, known as stomata. Plants become more drought resistant with more CO_2 as they close their stomata and evolve over time with fewer stomata.[3] Plants lose water from their stomata, so the closed stomata reduces plant water loss.

Other studies on greening have been conducted, and all of these have confirmed global greening between 1982 and 2000. However, some have questioned whether global greening was a short-lived occurrence or a long-term trend that is continuing. To address this issue, a new study was conducted to look at greening and browning of the world between 2001 and 2020. The study used the latest versions of the LAI satellite datasets, which have been updated. The updates reduce uncertainty in the analysis of global vegetation change trends after 2000. All four LAI datasets showed significant global greening between 2001 and 2020. The data showed accelerated greening in 55.1% of the areas, compared with 7.28% of browning. The paper highlights that "global greening is an indisputable fact."[4]

The great benefit of more CO_2 is the tremendous increase in agricultural production to feed a growing population. That is why some commercial greenhouses triple CO_2 levels to 1,200 parts per million (ppm) to stimulate crop growth. There is an almost exact correlation between the increase in CO_2 to the upsurge in agricultural production (see Figure 20). **It is ironic that we call "green energy" the very energy which will limit CO_2 and result in less greening.**

Climate alarmists just cannot seem to accept good news. Instead of welcoming the greening, they point out that plants growing in higher levels of CO_2 is problematic since such plants lack nutrients needed for faster growth, which leads to eroding nutritional value of staple crops.[5] However, this fact is well known and is addressed by fertilizers. According to Oklahoma State University horticulture professor Bruce Dunn, it is the practice of CO_2-supplemented greenhouses to increase the amount and mixture of fertilizers to adjust for the increased levels of CO_2. Increasing CO_2 levels require fertilization with both macronutrients, mainly nitrogen, phosphorus, potassium, calcium, magnesium, and sulfur; and micronutrients such as iron, zinc, boron, copper, manganese, and molybdenum. Micronutrient fertilizing is especially important.[6]

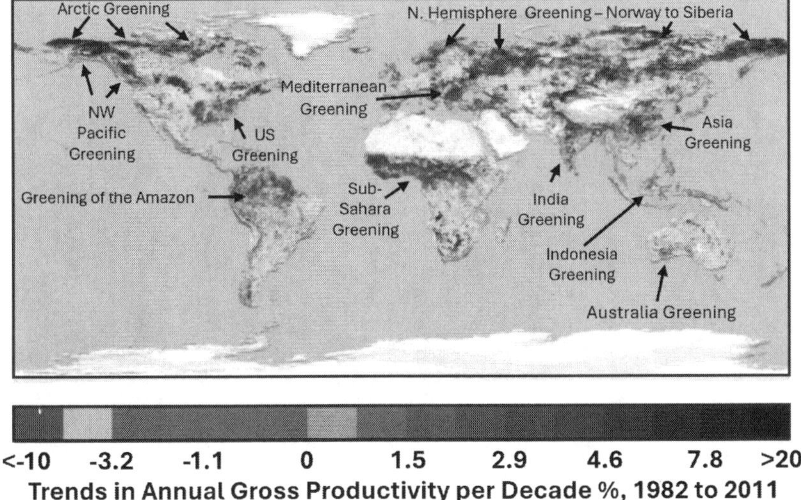

Figure 18 – Greening of the Earth, 1982 to 2011. *CO_2 has greened the Earth by more than 20% between 1982 to 2011 as recorded by NASA satellites. This is an area equal to more than twice the size of the United States.* Source: "Global Trends in Gross Vegetative Productivity 1982-2011 from Satellite Measurement [35]. The increase of plant growth is especially noticeable in arid regions such as south of the Sahara Desert. At higher CO_2 levels, the stomata in plant leaves are smaller, reducing the loss of water [32]." Figure license via Creative Commons Attribution 4.0 International. https://www.researchgate.net/figure/Global-Trends-in-Gross-Vegetative-Productivity-1982-2011-from-Satellite-Measurements_fig6_370633740.

A January 2023 study by Harvard University economists Charles A. Taylor and Wolfram Schlenker concludes: "We find a significant and robust CO_2 fertilization effect by linking Orbiting Carbon Observatory-2 (OCO-2) satellite-measured CO_2 fluctuations to yield fluctuations of corn, soybeans, and winter wheat from 2015 to 2021. Our results suggest that a significant proportion of observed yield gains for corn, soybeans, and winter wheat since 1940 may be attributable to increases in CO_2, an important driver of agricultural productivity growth."[7] It is well established that plants use sunlight

Figure 19 – Plants Love CO_2. *Experiments by Arizona State University geologist, Craig Idso demonstrate the increased growth of plants that receive higher levels of CO_2. This is really no surprise as CO_2 is food for plants and is essential for life on Earth. AMB stands for ambient, which is the CO_2 concentration in the atmosphere at the time this experiment was conducted. Source:* "Growth of Eldarica Pine Trees at CO_2 Concentrations of 385, 535, 685, and 835 ppm [31]." Figure license via *Creative Commons Attribution 4.0 International.* https://www.researchgate.net/figure/Growth-of-Eldarica-Pine-Trees-at-CO-2-Concentrations-of-385-535-685-and-835-ppm-31_fig5_370633740.

to convert CO_2 into glucose through photosynthesis. This is the reason greenhouses enhance the level of CO_2 to increase crop yields. In their study, Taylor and Schlenker used 2015 to 2021 data from NASA's OCO-2 satellite to estimate the contribution of CO_2 fertilization to crop yields.

From these results they consistently found a large CO_2 fertilization effect, which they used to estimate the CO_2 fertilization effect going back to 1940. The study estimates that for each 1 ppm increase in CO_2 in the atmosphere, there is an increase in corn yields of 0.4%, an increase in soybean yields of 0.6%, and an increase in wheat yields of 1.0%.[8] Since CO_2 levels have increased from about 310 ppm in 1940 to about 427 ppm by 2025, this 117 ppm increase has contributed to an enhancement of crop yields since 1940 of about 47% for corn, 70% for soybeans, and 117% for wheat.

Hardly a crisis, increased levels of CO_2 are helping us feed a growing population. Stanford University professor Paul R. Ehrlich published a book titled *The Population Bomb* in 1968; in it, he predicted food supplies would become increasingly dire as the world's population grew. Ehrlich's predictions of doom have not materialized; in fact, the world's food security is far better today than in 1968. During the 1960s, India was one of the areas most threatened with food insecurity and was a net importer of food, all of this as its population continued to grow. Today, India is a net exporter of food, a

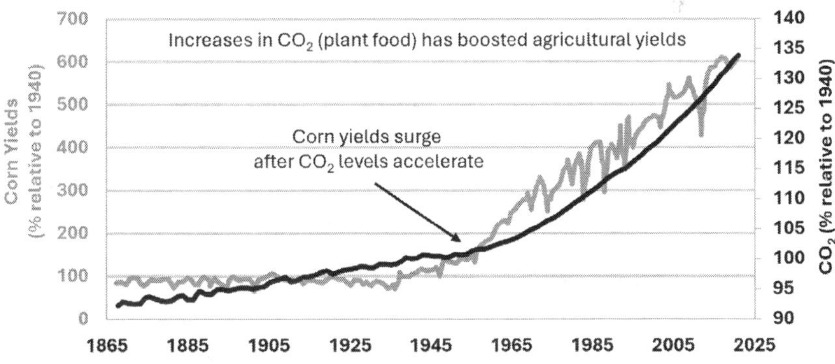

Figure 20 – Record-Setting Corn Harvests Track CO_2 Increases: U.S. Corn Crop Yields and CO_2 Levels, 1866 to 2021. *Global crop yields have grown tremendously in recent years. A considerable proportion of the yield gains observed for corn since 1940 may be attributable to increases in CO_2. The figure above displays U.S. corn crop yields and average annual CO_2 levels. Concentrations of CO_2 are normalized relative to 1940 (1940 value = 100). Aggregate corn yields are normalized relative to 1940 (1940 value = 100).* Sources: (*Corn Yield*) National Agricultural Statistics Service (NASS) under the USDA. (*CO_2*) Taylor, Charles A. and Schlenker, Wolfram, "Environmental Drivers of Agricultural Productivity Growth: CO_2 Fertilization of US Field Crops," National Bureau of Economic Research, Cambridge, MA, January 2023, p. 34. https://www.nber.org/system/files/working_papers/w29320/w29320.pdf.

transformation that would have been astonishing to imagine in the 1960s.

Increasing levels of CO_2 fertilization have improved crop yields. This has contributed to improved food security and reduced famines in the world (see Figures 20 and 37 in Chapter 6). *With populations continuing to grow, reducing CO_2 emissions is a catastrophic policy as it would adversely impact crop yields and curtail global food supplies.*

Plants also like warm climates. A *Nature* study titled "Accelerated increase in plant species richness on mountain summits is linked to warming" looked at plant biodiversity on mountaintops. The authors found global warming caused five times more plant species enrichment, and they concluded that global warming has accelerated plant diversity.[9]

Global Warming Has Saved Many Lives

A recent paper studying climate-related deaths by Shandong University (China) epidemiological researcher Qi Zhao and coauthors in *The Lancet* provides data to show that cold kills nine times more people each year than heat (see Figure 21).[10] This is especially true of countries in Africa and Asia. The study looked at temperature-related deaths throughout the world from January 1, 2000 to December 31, 2019. The study found that there were 4,594,098 cold-related deaths and 489,075 heat-related deaths globally in this twenty-year period. The study further found that these deaths translated into 67 deaths per 100,000 inhabitants for cold-related morbidity and seven deaths per 100,000 inhabitants for heat-related deaths.

The study also found the following: "From 2000–03 to 2016–19, the global excess death ratio changed by −0.51 percentage points (95% eCI −0.61 to −0.42) for cold temperatures and increased by 0.21 percentage points (0.13–0.31) for hot temperatures, resulting in a net decline of −0.30 percentage points." This means the rate of deaths from cold is declining more than twice as fast as the rate of deaths from heat is increasing. Therefore, global warming is accelerating the decline of climate-related deaths. A warming planet has saved many lives. Furthermore, data from this study showed global warming saved more lives in Asia and Africa than in Europe or North America. **Counter to the social justice climate change narrative, nations with high populations of "people of color" are benefiting far more than so-called "white" Western nations from reduced morbidity that is a direct result of global warming.**

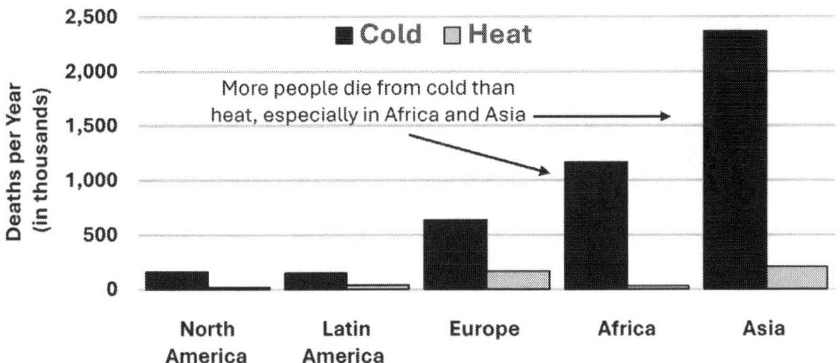

Figure 21 – Cold Kills Far More People Than Heat. *Data from around the world confirms that nine times more people died from cold than heat between 2000 to 2019. Global warming saves many lives and should be celebrated. This is particularly true in countries of Africa and Asia.* Source: *The Lancet*: "More Cold Death Than Heat Death Everywhere," *Electroverse.* https://i0.wp.com/electro-verse.co/wp-content/uploads/2022/12/image-25.png?ssl=1 For the full paper see Zhao, Qi et al., "Global regional and national burden of mortality associated with non-optimal ambient temperatures from 2000 to 2019: a three-stage modeling study." *The Lancet*, Volume 5, Issue 7, E415-E425, July 2021. https://doi.org/10.1016/S2542-5196(21)00081-4.

Climate- and weather-related deaths today have plummeted to one-fiftieth the number experienced in 1920.[11] **Although this may suggest warmer temperatures have been good for humankind, the larger implication is the power of adapting to climate**. Of great help are advancements in forecasting severe weather events and other adaptations, such as flood control measures, improved heating systems in cold weather, and air conditioning in heat waves. Ironically, many of these lifesaving adaptations are provided by the burning of fossil fuels. This 100-year history is convincing evidence of the power and benefits of adaptation to climate change. ***Money would be far better spent adapting to climate rather than vainly trying to control it.***

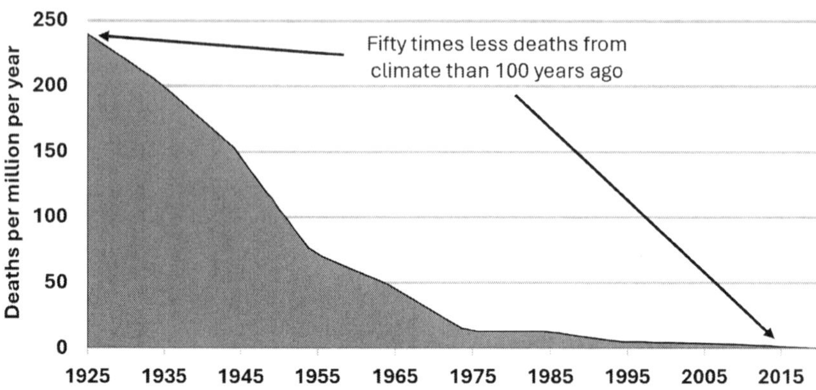

Figure 22 – Climate-Related Death Risk Has Plummeted, 1920 to 2020. *Climate-related deaths are significantly down from historical levels.* Source: OFDA/CRED International Disaster Database (2021). Lomborg, Bjorn, "Climate Change Is Not an Apocalyptic Threat—Let's Address It Smartly," Hoover Institution, Stanford University, September 2024. https://www.hoover.org/sites/default/files/research/docs/Lomborg_ClimateChange_web_240903.pdf.

Global warming can sound frightening. Climate alarmists cite rising *average global temperatures* to paint a picture of gloom. In reality, recent climate change has been one of moderating temperatures. Temperature records reveal tthat we currently have far less heat waves than the 1930s and more moderate winters due to the recent warming.[12] Average temperatures have gone up, but in a positive way, mostly from a reduction in cold rather than an increase of heat. Record high temperatures for 23 of 50 states of the U.S., six of 13 Canadian providences, and two of eight states and territories of Australia were set in the 1930s. Iceland and Sweden also set high temperature records in the 1930s that have not been surpassed (see Chapter 13). Meanwhile, cold spells have been declining. **Since cold winters lead to more deaths and damaged crops due to late frosts, we should welcome the recent warming, which is moderating the climate** (see Figure 23).

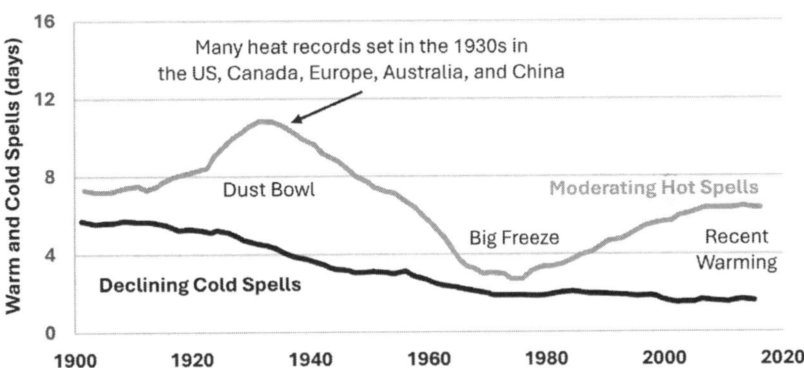

Figure 23 – Moderating Temperatures from Climate Change: Cold Spells and Hot Spells, 1900 to 2020. *Climate alarmists use average temperature increases in recent years to suggest a trend of ever-increasing dangerous heat. The data shows average warming has been mostly from milder winters not hotter summers. Note the dip in warm spells in the 1970s, which was during the cold cycle of the Atlantic Multidecadal Oscillation.* Source: Courtesy of U.S. National Oceanic and Atmospheric Administration (NOAA) and the U.S. Global Change Research Program (USGCRP): "Climate Science Special Report: Fourth National Climate Assessment, Volume 1, Chapter 6" ("Temperature Changes in the United States"), USGCRP, Washington, D.C. https://science2017.globalchange.gov/chapter/6/.

Economic Benefits of Warming

The historical benefits of a warming Earth are expected to continue. University of Sussex (UK) economist and IPCC contributor Richard Tol analyzed 22 studies on the economic impact of climate change.[13] The mean results of these 22 papers conclude that warming from climate change would be net positive for the global economy until the warming exceeded 1.7°C from the temperature at the beginning of the industrial era (see Figure 24). The temperature has warmed by 1.1°C since the beginning of the industrial era (1850/1900).[14] Therefore, the world must warm by over 0.6°C from today's temperature before experiencing negative economic impacts. This is about the same increase in temperature as is expected from the radiative forcing of CO_2 between today and the end of the 21st century (see Chapter 9).

Tol states, "If we take the confidence interval at face value, the impact of climate change does not significantly deviate from zero until 3.5°C warming."[15] To achieve 3.5°C warming from industrial times means 2.4°C in warming from today's temperature. Exceeding a 2.4°C temperature increase solely from fossil fuel emissions is very unlikely (see Chapter 9), so CO_2 emissions are not likely to have any significant negative impact on the economy throughout the twenty-first

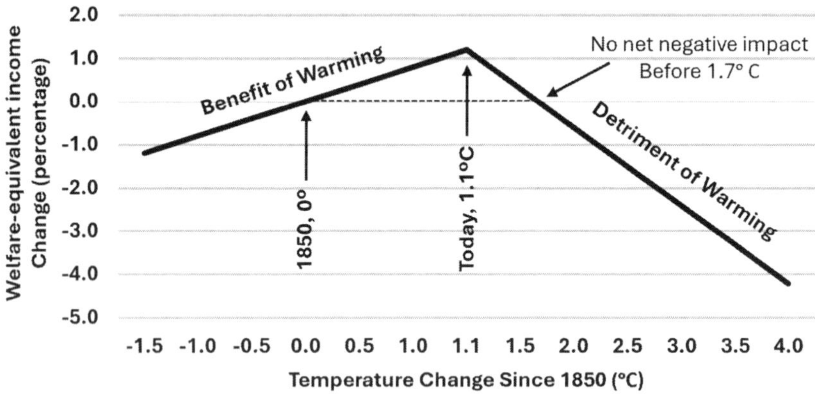

Figure 24 – Global Economic Cost of Climate Change. *Economist Richard Tol compiled the results of 22 economic studies on the impact of climate change. The mean of these studies shows increased temperatures are expected to provide net economic benefits with up to a 1.7°C increase in warming from preindustrial times, which is 0.6°C from today's temperature. This is about the same warming expected from CO_2 warming through the end of the twenty-first century.* Source: Tol, Richard S., "The Economic Impacts of Climate Change," *Review of Environmental Economics and Policy*, Volume 12, Number 1, Winter 2018, pp. 4-25. https://doi.org/10.1093/reep/rex027.

century, and only minimum impact thereafter. And this assumes the unlikely event that there are no natural cooling events in this period. Tol concludes: "Thus climate change would appear to be an important issue primarily for those who are concerned about the distant future, faraway lands, and remote probabilities."[16]

Yale University economist and Nobel laureate William Nordhaus has studied the economics of climate change in detail. Nordhaus' DICE model, including his 2023 update, estimates the economic damage of climate change from carbon dioxide for various temperature increases since 1850. Using his updated DICE model, Nordhaus estimates the damage from carbon-induced climate change to GDP to be -3.1% for a 3°C in warming experienced since 1850 and -7.0% of GDP for a 4.5°C increase in temperature since 1850.[17] Extrapolating from his model suggests damages to GDP of between -1.5% to -2.0% for a 2°C increase in temperature. Since 1850, temperatures have increased by 1.1°C. To reach 2°C will require heating of 0.9°C from today. In a worst-case scenario where no action is taken to limit CO_2 emissions, Nordhaus predicts the CO_2 concentrations will reach 800 ppm by about 2105.[18] As will be covered in Chapter 9 and 10, an increase to 800 ppm of CO_2 will raise temperatures by 0.9°C from the present, or 2.0°C from 1850. Future average global GDP growth is expected to be about 2% per year.[19] In the 80-year period between today and 2105 and 2% per year average GDP growth, the compounded total GDP growth will be nearly 400%. Under the Nordhaus estimates, the cost of carbon emissions results in losing only 9 months to one year of GDP growth in 80 years! Several economists have criticized Nordhaus' work as being too conservative. But then, when did economists ever agree? As the popular joke goes, "If you put 10 economists in a room, you'll get 11 opinions." However, even if we were to double the impact to -3% to -4% of GDP growth, this results in losing only about one-and-one-half to two years of average GDP growth in 80 years, or growth of 396% instead of 400% which is hardly a crisis.

Tipping Points

Some economists including Nicholas Stern[20] from the London School of Economics and Steve Keen[21] from Kingston University, (UK) forecast significantly more damage from climate change as they include "tipping points" in their models. Tipping points assume temperatures continue to rise with no abatement or they accelerate causing catastrophic change. They assume there will be continuous melting of ice caps and glaciers leading to constantly rising sea levels and changing ocean circulation patterns impacting regional climates significantly. Or they assume warming accelerates, triggered by the release of methane from melting permafrost and ocean sediments. The weakness of these apocalyptic tipping point models is that they assume nearly all temperature increase is from anthropogenic greenhouse gases, primarily CO_2, and that temperature rise, ice melt, sea level rise, CO_2 warming, and methane release continue to increase in a linear or accelerated fashion.

The fact is, less than one-half of the 1.1°C in temperature rise between 1850/1900 to 2021 can be attributed to CO_2 emissions (see Chapter 9). Furthermore, the power of CO_2 to warm the atmosphere decreases logarithmically as concentrations increase. Steve Keen clearly does not understand the physics of greenhouse gas warming for he said, "If you increase the amount of CO_2 in the atmosphere linearly, you get a logarithmic *increase* in temperature."[22] This is completely opposite of reality. As set forth in the accepted Myhre radiative forcing equations,[23] if you increase the amount of CO_2 linearly you get a logarithmic *decrease* in marginal temperature (see Chapter 9)!

Natural drivers of climate are cyclical and much of the warming since 1850 and going forward was and will be from natural drivers of climate (see Chapters 10 and 13). Sea level rise and glacier melt are cyclical events (Figures 16, 29, and 30). We are currently at the peak of the warm cycle of natural ocean oscillations (Chapter 13) and solar cycles (Chapter 7). Any competent economist should know you never extrapolate off the top of a curve in a cyclical system, yet that is exactly what they have done to produce catastrophic tipping point induced forecasts. Tipping points are not found in the historical record. During the Cambrian Era (~452 million to ~485 million years ago), CO_2 levels were at 7,000 ppm[24] as compared to 427 ppm today. Yet, although temperatures were warmer in the Cambrian Era, the climate did not enter a runaway greenhouse state or any catastrophic instability. The Earth's natural thermostat of shade from clouds keeps the Earth from overheating (see Chapters 11 and 12).

Characterizing the climate feedback of the release of methane from melting permafrost and ocean sediments as a tipping point is a gross exaggeration. The North Pole is warming 25 times faster than the South Pole (see Chapter 13).[25] This has resulted in methane concentrations of about 1825 ppb at the South Pole and about 1950 ppb at the North Pole.[26] The ERF adjusted Myhre et al. radiative forcing equation[27] tells us that the increase in heating from 1825 ppb to 1950 ppb of methane has produced a temperature increase of 0.024°C (see Chapter 13). And this small change in temperature has not been instantaneous but has taken many decades. According to University of Chicago professor and geophysicist David Archer and coauthors, to reach any meaningful levels of methane in the atmosphere from melting ocean sediments would take millennia.[28] "No one has thought of a mechanism that would release a significant fraction of the methane in the ocean on a human time scale of the coming century. A more plausible scenario would be a slow chronic release of methane from ocean hydrates comparable to the release from high latitude peats."[29]

Methane decomposes easily and has a relatively short lifetime of about 12 years. Therefore, the removal of methane will scale with the slow increase of methane. This short lifespan limits runaway risks of methane. Furthermore, it is unrealistic to assume temperatures will continue to increase unabated with continual melting of permafrost and ocean sediments, since natural cyclical drivers of climate will drive future cold periods over the next few decades, centuries, and millennia (see Chapters 7 and 13). In fact, the decline in Arctic Ice, and associated methane release from permafrost, was significant until 2007, but has been stable since, despite a drop in 2012.[30]

One of the tipping points discussed by Steve Keen is the change of ocean circulation from decreased salinity by the dilution of ocean salt content from melting polar ice caps. This hypothesis suggests saltier water is heavier and sinks and this drives currents. According to this hypothesis, less salinity will slow the Gulf Stream Ocean Current (aka Atlantic Overturning Meridional Circulation) and result in a colder Europe.[31] Evidence for this tipping point is not convincing. Measurements over 50 years show an increase rather than decrease in salinity since "the North Atlantic as a whole has exhibited a decrease in freshwater content from 1955 to 2006."[32] In addition, the thermohaline force (sinking of heavier saline water) was shown by MIT professor and oceanographer, Carl Wunsch to be far too weak to drive ocean currents.[33] Observations support Wunsch's conclusions. A study of the Gulf Stream by NOAA oceanographer Dan Seidov and coauthors between 1965 to 2017 found long-term stability and resilience of the current.[34] This period includes the significant melting of the polar ice caps between September 1980 to September 2007 (from ~7 ½ million to ~4 ¼ million square kilometers sea ice extent) and the extreme ice melt in September 2012 (~ 3 ½ million square kilometers sea ice extent).[35] If polar ice melt in recent years has not had an impact on the stability of the Gulf Stream Ocean Current, as Seidov et al. have observed, it seems illogical to expect this to be a major tipping point. This is especially true since despite the low in 2012, Arctic polar ice melt has been stable since 2007[36] and the warm phase of the Atlantic Multidecadal Oscillation (AMO) peaked around 2010 will likely be moving into a cold phase after 2030,[37] which will likely reverse the trend of melting of the Arctic polar ice sheet.

Keen also mentions the loss of the Amazon rainforest as another tipping point of the climate. If he is truly concerned about this tipping point he should push for more CO_2 in the atmosphere. The entire Amazon is one of the areas that has benefited from CO_2 fertilization (see Figure 18). Since 1982, NASA satellites have shown the Amazon has greened.[38] This is not to say we should not restrict human deforestation of the Amazon rainforest from logging and land use. The point is, climate change from CO_2 is not hastening the destruction of the Amazon rainforest.

Another economic benefit of global warming has been the reduction in global weather disaster losses over the past 43 years. As discussed in Chapter 3, global warming has moderated the Meridional Temperature Gradient and slowed winds, both of which have led to less severe storms. Global warming has moderated extreme temperatures between the Arctic and the Tropics since the Arctic has warmed twice as fast as the Tropics. This moderation of the Meridional Temperature Gradient weakens eddy kinetic energy (EKE).[39] Severe storms are produced when warm, moist air collides with cold air, and are driven by EKE; the cold air has moderated with climate change. The weaker Meridional Temperature Gradient in a warm climate also stills winds,[40] which feed moist air into storm tracks, thereby weakening the storms. The IPCC is reported to have estimated winds will be stilled by 10% by the year 2100.[41] Global warming also produces more warming at the top of the troposphere from the condensation of water vapor. Known as the Lapse Rate or Vertical Temperature Gradient, this decrease in the Lapse Rate during warmer times produces less Convective Available Potential Energy (CAPE), which is a major driver of storms.[42] Accordingly, both severe weather and the cost of losses from global weather disasters has decreased since 1990.[43] As discussed in Chapter 3, data shows hurricanes and severe tornadoes have declined in numbers and intensity since 1990.

Climate alarmists often post data that shows such costs increasing, but they rely on unadjusted dollars or cherry-pick nations that exhibit a trend opposite of the global trend. Over time, more building has occurred, resulting in larger property losses. As an example, building on the coast of Florida has exploded since 1990. U.S. Census records show the population of Florida grew from 12.9 million in 1990 to 23.4 million by 2024. Most of these Floridians reside in coastal regions, so a hurricane today will destroy much more property than in 1990. Adjusting the disaster losses by global Gross Domestic Product (GDP) is a reasonable way to account for this growth. When adjusted for GDP, the cost of global weather disaster losses has declined (see Figure 25).

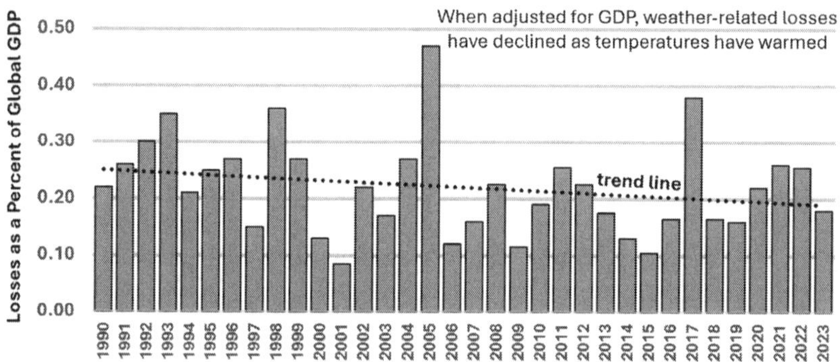

Figure 25 – Global Weather Disaster Losses as a Percent of Global GDP, 1990 to 2023. *Global weather disaster losses have declined since 1990. Total losses need to account for additional infrastructure built each year that is subject to severe weather damage. Global Gross Domestic Product (GDP) is a reasonable adjustment to account for such growth. When adjusted for global GDP, the losses from global weather shows a noticeable decline.* Source: Pielke, Roger Jr., "Global Weather Disaster Losses as a Percent of Global GDP: 1990-2023," Cited by Dr. Matthew M. Wielicki @MatthewWielicki X post. https://x.com/matthewwielicki/status/1889678185136038250?s=51&t =qWx0UbKJuQfOx7mMEjSxsw.

Climate change has always been cyclical. In Chapter 5 we will examine the history of past climate cycles to better understand Modern Warming.

Chapter 5: Past Climate Change Cycles

Historical Climate Change Cycles

Climate alarmists claim current warming is unprecedented over the past 125,000 years.[1] This statement goes against the findings of paleoclimate analyses. Studies have shown that temperatures during the Holocene Climatic Optimum (7000 BC to 3000 BC) were 1°C to 2°C warmer than today.[2] Scientific studies have also established that the Roman Warm Period, 2,000 years ago, and the Medieval Warm Period, 1,000 years ago, were comparable to temperatures today, and warmer than today in many regions, particularly in Greenland, Iceland, Europe, China, and Indonesia. Studies also show temperatures plunged during the Little Ice Age about 500 years ago. In his book *The Little Ice Age*, UC Santa Barbara professor and anthropologist Brian Fagan state that during the Medieval Warm Period, "Average summer temperatures were between 0.7°C and 1.0°C above twentieth century averages."[3]

These historical climate cycles have been long established with dozens of scientific papers providing evidence of their occurrence. Woods Hole Oceanographic Institution paleoceanographer, Lloyd

Keigwin, published a paper in *Science* that reconstructed Sea Surface Temperatures (SST) in the North Atlantic. The results show "SST was ~1°C cooler than today [approximately] 400 years ago (the Little Ice Age) and 1700 years ago, and ~1°C warmer than today 1000 years ago (the Medieval Warm Period). Thus, at least some of the warming since the Little Ice Age appears to be part of a natural oscillation."[4]

Historical Temperature Proxies

Isotopes can be used as proxies to determine past temperatures before thermometer measurements were available. Isotopes are variations of atomic elements that have a varied number of neutrons. Isotopes of an atomic element possess the same chemical properties because they have the same number of protons and electrons. Accordingly, isotopes look and behave the same in nature, but they have different masses. Mass spectrometry can be used to measure and differentiate isotopes. For example, a sample of carbon can be analyzed by mass spectrometry to determine its composition of Carbon-12 (^{12}C), Carbon-13 (^{13}C), and Carbon-14 (^{14}C). Likewise mass spectrometry can analyze water to determine the mixture of water ($^{1}H_20$), deuterium heavy water ($^{2}H_2O$), and tritium heavy water ($^{3}H_2O$) in the sample.

Past temperatures have been constructed based on the ratio of isotopes Oxygen-16 to Oxygen-18 ($\delta_{18}O$) in the Greenland and Antarctica ice cores, ocean sediment cores, lake sediment cores, stalactites, stalagmites, and tree rings. The ratio of isotopes Oxygen-16 to Oxygen-18 in these samples, as measured by mass spectrometry, can be used as a proxy to estimate past temperatures. Water (H_2O) molecules containing heavier Oxygen-18 (^{18}O) require more heat to evaporate than water molecules containing Oxygen-16 (^{16}O). As a result, during cold periods, a higher amount of the lighter Oxygen-16 water ($H_2{}^{16}O$) evaporates out of the oceans into the atmosphere, leaving the oceans more enriched with heavier Oxygen-18 water ($H_2{}^{18}O$).

During warm times, more Oxygen-18 water evaporates increasing Oxygen-18 water vapor in the atmosphere and reducing Oxygen-18 water in the ocean.

Since there is a higher ratio of Oxygen-16 to Oxygen-18 water vapor in the atmosphere during cold times, snowfall contains less Oxygen-18. Therefore, during cold times, polar ice cores contain lower values of Oxygen-18. Cave stalactites are formed from rainwater percolating through soil and limestone into caves. Like ice cores, the calcite (a form of $CaCO_3$) will contain less Oxygen-18 in cold periods. The same is true of tree rings as trees absorb rainwater absorbed through roots. On the other hand, samples such as marine sediments record higher values of Oxygen-18 in their shells during cold times. As more Oxygen-16-based water evaporates in cold periods, it leaves the ocean water enriched with Oxygen-18, which is incorporated into shells contained in marine sediments as shells are primarily made up of calcium carbonate ($CaCO_3$).

Temperature proxy reconstructions have uncovered semi-millennial climate cycles including the Minoan Warm Period (~ [approximately] 1500 BC to ~1100 BC); followed by the cold Greek Dark Ages (~1100 BC to ~800 BC); then the Roman Warm Period (~250 BC to ~400 CE); followed by the cold Dark Ages (~400 CE to ~900 CE); then the Medieval Warm Period (~950 CE to ~1250 CE); followed by the cold Little Ice Age (~1300 CE to 1850 CE); and, finally, Modern Warming (1850 to today).

Medieval temperature reconstructions from GISP2 ice cores in Greenland are about 1°C warmer than modern temperatures.[5] Spikes in temperature are clearly found during the Minoan, Roman, Medieval, and Modern Warm periods in Greenland ice cores (see Figure 26).[6] Ocean sediments were used by Sorbonne University paleoclimatologist Marie-Alexandrine Sicre and coauthors in 2008 to reconstruct sea surface temperatures near Iceland. This Iceland sea temperature study confirms the same spikes in temperature for the Roman and

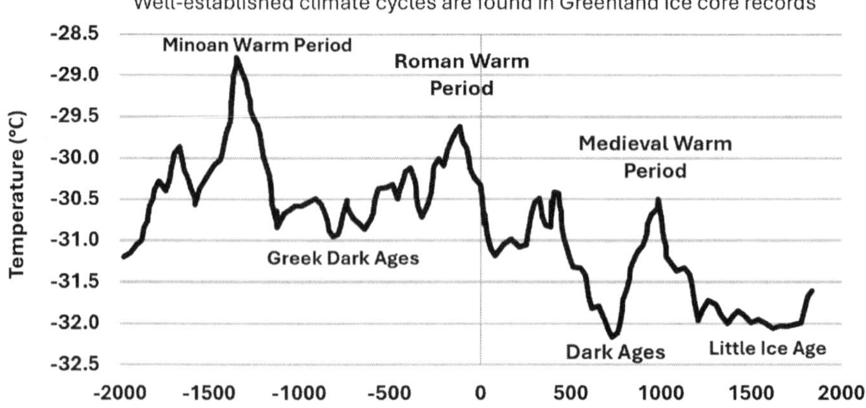

Figure 26 – Central Greenland Surface Temperatures, 2000 BC to 1900 CE. *Ice cores oxygen isotope proxies reveal that temperatures in Greenland during the Medieval Warm Period, Roman Warm Period, and Minoan Warm Period were warmer than the present. The Little Ice Age, the Dark Ages, and the Greek Dark Ages, where temperatures dropped, is also seen in the ice cores. Note the downward trend in temperature since the Minoan Warm Period. Historical temperatures come from GISP2 Greenland ice cores.* Source: Humlum, Ole et al., "Identifying natural contributions to late Holocene climate change," *Global and Planetary Change*, Volume 79, Issue 1, pp. 145-156, 9 September 2011, https://doi.org/10.1016/j.gloplacha.2011.09.005.

Medieval warm periods and a decline in temperature in the Little Ice Age. The paper by Sicre et al. observed 1.5°C in warming during the Medieval Warm Period.[7] A study by University of Saskatchewan (Canada) professor and isotope geochemist William Patterson and coauthors states, "On the basis of $\delta^{18}O$ data, reconstructed water temperatures for the Roman Warm Period in Iceland are higher than any temperatures recorded in modern times."[8] A study by Stockholm University (Sweden) professor, paleoclimatologist, and climate historian, Fredrik Ljungqvist in 2010 of the Northern Hemisphere "show[s] a distinct Roman Warm Period c. AD 1–300, reaching up to the 1961–1990 mean temperature level, followed by the Dark Age

Cold Period c. AD 300–800. The Medieval Warm Period is seen c. AD 800–1300 and the Little Ice Age is clearly visible c. AD 1300–1900, followed by a rapid temperature increase in the twentieth century. The highest average temperatures in the reconstruction are encountered in the mid to late tenth century and the lowest in the late seventeenth century. Decadal mean temperatures seem to have reached or exceeded the 1961–1990 mean temperature level during substantial parts of the Roman Warm Period and the Medieval Warm Period"[9] (see Figure 58 in Chapter 10).

Stockholm University (Sweden) paleoclimatologist Anders Moberg and coauthors used sediment samples and tree rings to reconstruct Northern Hemisphere temperatures. Moberg data shows a spike in temperature during the Medieval Warm Period and drop in temperature during the Little Ice Age (see Figure 27).[10] The Moberg et al. paper states, "According to our reconstruction, high temperatures—similar to those observed in the twentieth century before 1990—occurred around 1000 CE to 1100 CE, and minimum temperatures that are about 0.7 K below the average of 1961–90 occurred around 1600 CE."[11] Moberg et al. concludes, "This large natural variability in the past suggests an important role of natural multi-centennial variability that is likely to continue."[12] **In light of the Moberg et al. *Nature* paper, it is not rational to assume that *nearly all* of the 1.1°C of modern warming since 1850/1900 is from increases in CO_2 emissions.** It is more likely that about one half of modern warming is from increases in greenhouse gases and about half is from natural variation.

Interestingly, the Ljungqvist and Moberg temperature reconstructions both capture the timing of significant and known solar minima, most of which occurred during the Little Ice Age. These solar minima include the Oort, Wolf, Spörer, Maunder, and Dalton solar minima (see Figures 27 and 58 in Chapter 10). These solar minima have been identified from other studies of sunspot activity and cosmic ray flux proxies including $\Delta^{14}C$ and ^{10}Be isotopes found in tree rings, ice cores, and sediment samples. Ljungqvist and Moberg et al. used $\delta^{18}O$ and other temperature proxies, but not cosmic ray proxies, so the close correlation between the Ljungqvist and Moberg temperature reconstructions and these solar minima is an excellent external

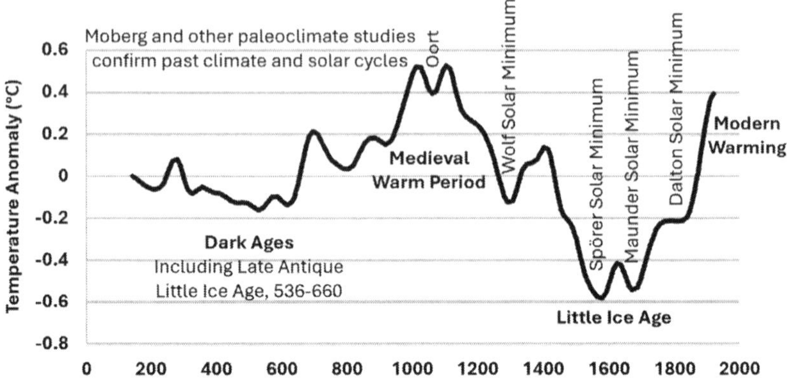

Figure 27 – Moberg Northern Hemisphere Temperature Reconstruction. *Anders Moberg et al. used ocean sediment cores and tree ring data to reconstruct temperatures in the Northern Hemisphere over the past 2,000 years. These temperature reconstructions show with clarity the cold Dark Ages, Medieval Climate Optimum, the Little Ice Age, and Modern Warming just as these same climate cycles are seen clearly in Greenland ice cores. The impact on temperature from the Oort, Wolf, Spörer, Maunder, and Dalton solar minima are also found in the Moberg temperature reconstruction. Moberg also captures the temperature drop during the Late Antique Little Ice Age (536 BC to 660 BC).* Source: Moberg A. et al., "Highly variable Northern Hemisphere temperature reconstructions from low- and high-resolution proxy data," *Nature*, Volume 433, pp. 613-617, 10 February 2005. https://doi.org/10.1038/nature03265.

confirmation of the reliability of these temperature reconstructions. Both the Moberg and Ljungqvist reconstructions also capture the very cold period known as the Antique Little Ice Age that occurred between 536 CE to 660 CE (see Chapter 6).

NCASI ecologist Craig Loehle published a paper where he used non-tree proxies to reconstruct the climate over the past 5,000 years from 18 separate locations.[13] Loehle concludes, "The mean series shows the Medieval Warm Period (MWP) and Little Ice Age (LIA) quite clearly, with the MWP being approximately 0.3°C warmer than 20th century values at these eighteen sites." In a 2020 *Nature* paper by National Research Council of Italy paleoclimatologist Giulia Margaritelli and coauthors reconstruct temperatures in the Mediterranean. They conclude, "The most solid image that emerges of this trans-Mediterranean comparison is the persistent regional occurrence of a distinct warm phase during the Roman Period. This record comparison consistently shows the Roman as the warmest period of the last 2,000 years, about 2°C warmer than average values for the late centuries for the Sicily and Western Mediterranean regions."[14] Woods Hole scientist Lloyd Keigwin used isotope ratios in marine organisms to reconstruct past sea temperatures in the North Atlantic. The cold Dark Ages, the warm Medieval Warm Period, and the Little Ice Age are clearly seen in his data.[15]

University of Hong Kong (China) paleoclimatologist YuXin He and coauthors conducted a paleoclimate study using sediments from two lakes on the northern Tibetan Plateau. They found that "temperatures during the MWP [Medieval Warm Period)] were slightly higher than the modern period in this region. Further, our temperature reconstructions, within age uncertainty can be well correlated with solar irradiance changes, suggesting a possible link between solar forcing and natural climate variability, at least on the northern Tibetan Plateau."[16]

Oxygen isotopes and growth patterns in tree rings is another proxy that confirms the existence of the Medieval Warm Period, the Little Ice Age, and Modern Warming. In the 2004 *Encyclopedia of Forest Sciences*, University of Göttingen (Germany) professor and dendrochronologist Martin Worbes published a study of climate reconstructions using tree rings.[17] Using 1,205 individual tree ring series from different tree species across the entire northern hemisphere,

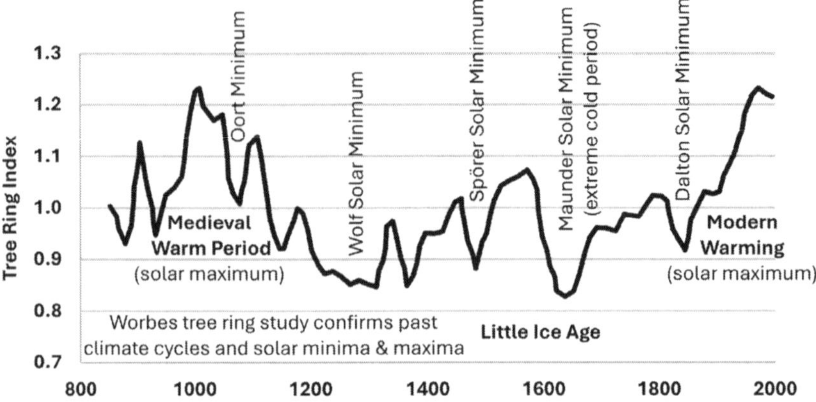

Figure 28 – Historical Climate Cycles from Tree Ring Reconstructions. *A chronology of temperatures that conform to known solar minima and maxima dating back to 800 CE are constructed by Martin Worbes using 1205 individual tree ring series across the entire northern hemisphere. The Medieval Warm Period was as warm and, in some areas, warmer than today, and the Little Ice Age is shown as being colder than both Modern Warming and the Medieval Warm Period. The coldest period is found during the Maunder Solar Minimum.* Source: Worbes, Martin, "Mensuration—Tree-Ring Analysis," *Encyclopedia of Forest Sciences,* pp. 586-599, 2004. https://doi.org/10.1016/B0-12-145160-7%2F00059-4.

Worbes reconstructs temperatures back to 800 CE Worbes concludes, "The Medieval Warm Period and the Little Ice Age as well as recent global warming are clearly shown." **In Worbes's reconstruction, temperatures in the Medieval Warm Period were as warm, and in some locations warmer, than Modern Warming of today**. Worbes also reconstructed cold drops in temperature that align perfectly with solar minima that have been found from ^{10}Be and Δ^{14}C isotope analysis and sunspot observations. These solar minima include the Oort Solar Minimum, Wolf Solar Minimum, Spörer Solar Minimum, Maunder Solar Minimum, and Dalton Solar Minimum (see Figure 28). The Worbes reconstruction shows the time of the Maunder Solar Minimum as the coldest times during the Little Ice Age, which is confirmed by historical records (see Chapter 6).

In yet another study, prominent Yunnan University (China) professor and paleontologist Yu Liu and coauthors used tree rings from China to reconstruct temperatures over the past 2485 years. Once again, the Roman Warm Period, Medieval Warm Period, and Little Ice Age climate eras are confirmed.[18] The paper states, "The results showed that extreme climatic events on the [Tibetan] Plateau, such as the Medieval Warm Period, Little Ice Age, and 20th Century Warming appeared synchronously with those in other places worldwide." The study goes on to say, "The long-term trends (>1000 a) of temperature were controlled by the millennium-scale cycle and amplitudes were dominated by multi-century cycles. Moreover, cold intervals corresponded to sunspot minimums. The prediction indicated that temperature will decrease in the future until 2068 AD and then increase again." [19]

Historical Evidence

We have additional evidence beyond oxygen isotopes and patterns in tree rings to confirm the warm temperature cycles of the Minoan, Roman, and Medieval Warm Periods, and the cold periods of the Greek Dark Ages, Dark Ages, and Little Ice Age. Additional proof of these climate cycles include: 1) glacial growth and retreat in Norway, Greenland, Peru, New Zealand, China, and the Alps; 2) historical sea level rise; 3) the settlement of Greenland made possible by the warm climate in the Medieval Warm Period, and the abandonment of Greenland forced by the cold climate in the Little Ice Age; 4) the tree line in the Medieval Warm Period in Canada, Europe, and Russia were farther north or at higher altitudes than today; 5) the freezing over of the Thames River in London, the Venetian Lagoon, Boston Harbor, Chesapeake Bay, and the Rio Grande River during the Little Ice Age; 6) entomology and animal migrations; and 7) agricultural records in Europe, Greenland, and China confirm these climate cycles.

Historical records confirm these semi-millennial climate cycles over the past 3,000 years. Saarland University (Germany) professor and climate historian Wolfgang Behringer has written an excellent and meticulously footnoted book which reviews the historical record on these climate cycles.[20] Behringer writes, "In a number of research projects, Hubert Horace Lamb in England, Christian Pfister in Switzerland, Rüdolf Brazdil in the Czech Republic, [and] Rudiger Glasier in Germany have so clearly demonstrated climatic fluctuations in European History that their evidence is now beyond all doubt."[21] Behringer also provides historical evidence of these climate cycles from Greenland, China, the Americas, and other geographical areas.

Glacial Record: Glacial records, including moraines (dirt, rocks, and debris transported by glaciers) and ice cores, indicate glacier advances or retreats, and glacial sediments provide evidence of climate cycles over the past 9,000 years, including the Holocene Climatic Optimum (~7000 BC to ~3000 BC), the Roman Warm Period (~250 BC to ~400 CE), the Medieval Warm Period (~950 CE to ~1250 CE), the Little Ice Age (~1300 CE to 1850 CE), and Modern Warming (1850 CE to today). We have already seen the clear climate cycles of the Minoan Warm Period, cold Greek Dark Ages, Roman Warm Period, cold Dark Ages, Medieval Warm Period, Little Ice Age, and Modern Warming in the ice cores from Greenland (see Figure 24).[22] From approximately 6000 BC to 2000 BC, during and just after the Holocene Climatic Optimum, most glaciers in Norway were completely melted.[23]

The glacial record from the Tibetan Plateau confirm the warmth of the Medieval Warm Period and the cooling of the Little Ice Age in China.[24] Glaciers in the Peru Andes show warmer and wetter conditions in the Medieval Warm Period and confirm colder, drier conditions during the Little Ice Age.[25] Moraines from the Andean glaciers in Peru show glacial retreat during the Medieval Warm Period and

Modern Warming, and glacial advance during the Little Ice Age.[26] Moraines and sediment records from the Southern Alps in New Zealand show advances during the Little Ice Age.[27] Ice cores and moraines from Rocky Mountain glaciers in Canada show cooling in the Little Ice Age with glacier advances.[28]

As glaciers have retreated during Modern Warming, artifacts such as tools and clothing, human remains, and buried forests have been uncovered. Such evidence of prior warm periods has come from the Alps, Andes, Himalayas, and Arctic. In Switzerland, the melting of the Schnidejoch and Tschierva glaciers uncovered artifacts from the Bronze Age around 2000 BC. The discovery of the remains of the Ötzi Iceman suggests ice-free conditions during the Holocene Climatic Optimum (3,200 BC). Artifacts found from the Holocene Optimum, Roman Warm Period, and Medieval Warm Period in the Schnidejoch Pass in Switzerland indicate passable routes during these warm periods.[29] Buried forests under the Pasterze Glacier in Austria date to the time of the Holocene Climatic Optimum and "indicate that between cal. BC 8100 and cal. BC 6900 the Pasterze Glacier was smaller than present."[30] Hunting arrow dart points from the Medieval Warm Period were found under a Yukon glacier in Alaska, suggesting melted ice at that time.[31]

One of the world's most renowned experts on glaciers, Professor Hanspeter Holzhauser of the University of Bern, Switzerland, measured the advances and retreats of the Great Aletsch glacier in the Alps over the past 3,500 years (see Figure 29). **The growth and melt of this glacier clearly show the warmth of the Minoan, Roman, Medieval, and Modern warm periods and the cold of the Greek Dark Ages, Dark Ages, and Little Ice Age.**[32]

During the Minoan Warm Period, the Great Aletsch Glacier in the Alps melted to its lowest point in 3,500 years, around 1350 BC.[33] It rebounded to a maximum size in 600 BC during the cold Greek Dark Ages and retreated again during the Roman Warm Period, reaching a minimum size around 50 BC.[34] During the Roman Warm Period other glaciers retreated, and passage through the Alps, such as the Schnidejoch Pass, were open all year for the Romans. The Dark Ages saw the regrowth of glaciers. In the Alps, glaciers advanced from the early fifth century to the mid eighth century. The Val de Bagnes Glacier in Switzerland grew, making an old Roman Road impassible.[35] The Great Aletsch Glacier peaked in size around 600 CE and then began to shrink rapidly into the Medieval Warm Period.[36] During the Medieval Warm Period, between 900 and 1,300 glaciers retreated in Europe, North America, and throughout the world, a sure sign of warming.[37] During the Little Ice Age glaciers began to grow again.

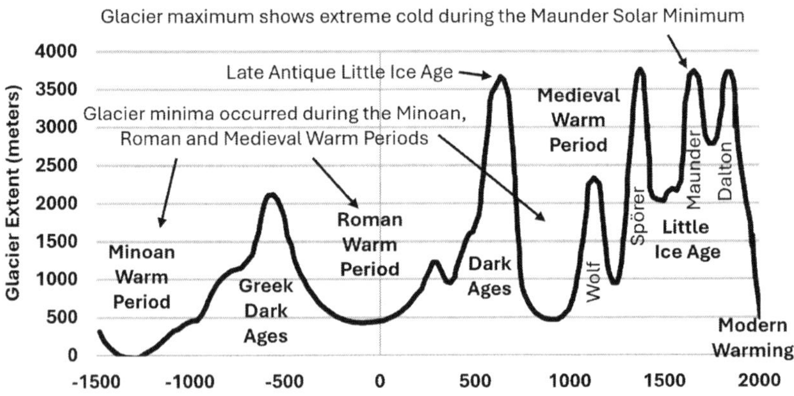

Figure 29 – Growth and Decline of the Great Aletsch Glacier During the Last 3,500 Years. *Growth and decline of the Great Aletsch Glacier in the Alps clearly show the climate cycles of the Minoan Warm Period (Bronze Age Optimum), the cool Greek Dark Ages, the Roman Warm Period (Roman Age Optimum), the cool Dark Ages, the Medieval Warm Period (Medieval Climate Optimum), and the Little Ice Age. Maximum glacial size is seen during the cold Maunder Solar Minimum (1645 to 1715). Glacier advance is also documented during the Wolf, Spörer, and Dalton solar minima. The cold period known as the Late Antique Little Ice Age* (536 CE to 660 CE) *is also evident.* Source: Holzhauser, Hanspeter et al., "Glacier and lake-level variations in west-central Europe over the last 3,500 years," *The Holocene*, Volume 15, Number 6, pp. 789-801, 30 November 2004. https://doi.org/10.1191/0959683605hl853ra.

The Great Aletsch Glacier grew to its largest size in 3,500 years in about 1650,[38] which was during the Maunder Solar Minimum (1645 to 1715). In 1644, during the Little Ice Age, the Bishop of Geneva led a procession near Chamonix to plead with God to halt the advancing Des Bois Glacier. The Des Bois Glacier had already swallowed up two villages, and a third was in imminent danger.[39] The chapel of Saint Pétronelle in the Alps also succumbed to ice during the Little Ice Age. Hans Rebmann wrote, "On the side of the mountain, at Sainte-Pétronelle, there was once a chapel, a place of pilgrimage, but a great glacier now hangs there and has entirely covered the site."[40] The Great Aletsch Glacier and other glaciers once again began to melt as Europe entered the Modern Warming Period after 1850.[41] Not only is the impact of the Maunder solar minimum evident in the glacier record, glacier advance during the Wolf, Spörer, and Dalton solar minima are also documented (see Figure 29). The cold period known as the Late Antique Little Ice Age from 536 CE to 660 CE is also reflected in the glacier record.

Sea Level: According to Wolfgang Behringer, "A warm period in the high Middle Ages can scarcely be disputed."[42] As a result of melting glaciers during the Medieval Warm Period, sea levels rose. Niels Bohr Institute (Denmark) associate professor and paleoclimatologist Aslak Grinsted and coauthors published a paper to reconstruct sea level rise over the past 2,000 years using tide gauge, satellite, and proxy data. The authors concluded, "Over the last 2,000 years, minimum sea levels occurred around 1730 and maximum sea level around 1150."[43] The paper states that the sea level maximum during the Medieval Warm Period is 12 to 21 centimeters higher than the 1980-1990 average.[44] These sea level reconstructions also echo the Medieval Warm Period and the Little Ice Age with a lag between extreme temperatures of each cycle and the actual sea level increase (see Figure 30). This lag is due to the delay in melting glaciers as temperatures increase.

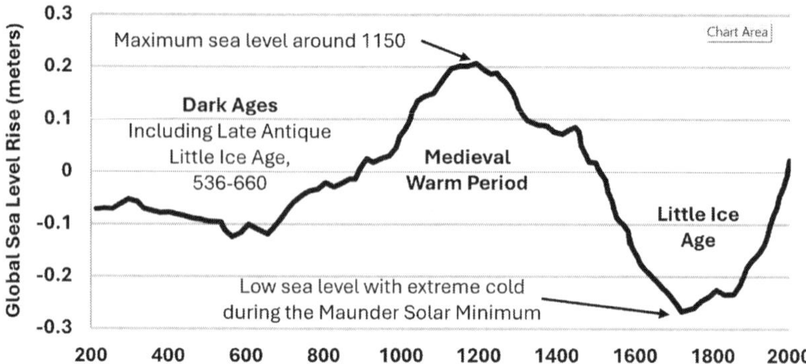

Figure 30 – Reconstructed Sea Level Rise, 200 CE to 2000. *Reconstructed sea level rise using observed sea level and proxy reconstructions align with temperature cycles of the Medieval Warm Period, the Little Ice Age, and Modern Warming. Lowest sea level was in 1730, just after the extreme cold of the Maunder Solar Minimum (1645 to 1715).* Source: Grinsted A. et al., "Reconstructing sea level from paleo and projected temperatures 200 to 2100 AD," *Climate Dynamics*, Volume 34, pp. 461-472, 6 January 2009. https://doi.org/10.1007/s00382-008-0507-2.

A paper by Yale/Tufts University professor and coastal geologist Andrew Kemp and coauthors studied sea level using salt-marsh sediments and foraminifera (single-cell marine organisms) deposits in North Carolina to infer sea level rise. They found a 400-year period of rising sea levels starting in 950 CE and another trend of sea level rise beginning in 1865 CE.[45] They show through modeling that their sea level reconstructions are consistent with global temperatures fluctuations of the Medieval Warm Period, Little Ice Age, and Modern Warming climate cycles.

The warmest period of the Holocene (from ~11,800 years ago to today) was the Holocene Climatic Optimum (HCO) from 7000 BC to 3000 BC.[46] Studies show sea level was higher during the HCO than today. University of Pisa (Italy) coastal geomorphologist Matteo Vacchi and coauthors show sea levels during the HCO to be 0.8 meters to 4.0 meters higher than present.[47] Texas A&M assistant professor and geophysicist Roger Creel and coauthors have sea level 0.24 meters higher during the HCO than they are today.[48]

Agricultural Records: Agricultural records confirm warming during the Medieval Warm Period. "Grapes are presently grown in Germany up to elevations of about 560 meters, but from about 1100 AD to 1300 AD, vineyards extended up to 780 meters, implying temperatures warmer by about 1.0-1.4°C."[49] In January 1187, during the Medieval Warm Period, trees blossomed in Strassburg, Germany.[50] In the Baltic, grapes were grown 500 kilometers farther north during the Medieval Warm period than they are today.[51] In England during the Medieval Warm Period, "Vineyards flourished 300 to 500 kilometers north of their twentieth-century limits."[52]

Millet was cultivated in Scandinavia in the Minoan Warm Period.[53] Pollen analysis has shown that Norway had crops in the Medieval Warm Period that disappeared with the onset of a cold climate.[54] During the Medieval Warm Period, wheat and oats were grown as high north as Trondheim, Norway, and barley in high latitudes of Norway.[55] Trondheim is at a latitude of 63.4 degrees north and has a short growing season. The Arctic Circle starts at 66.5 degrees north where there is at least one day a year in which there is continuous darkness. Barley was also grown in Greenland in the Medieval Warm Period.

Archaeological evidence, pollen, grape seeds, and vine remnants have been discovered in Roman times in Britain in Northamptonshire and even farther north. Following the Roman period, the climate cooled into the Dark Ages, and although grapes continued to be grown in Southern England monasteries, it is unlikely grapes were grown in Northern England as there is little evidence to support such a claim. During the Medieval Warm Period, wine grapes were once again grown in Northern England, farther north than is common today. According to Hurbert Lamb of the British Meteorological

Office, "In the *Domesday Book* (1085), 38 vineyards were recorded in England besides those of the king. The northernmost vineyards were near York, but the most favoured country was from Northants and the Fenland southwards. This implies summer temperatures perhaps 1 and 2°C higher than today."[56] Wine grapes grown in France and Germany during the Medieval Warm Period were 500 kilometers north of present-day vineyards.[57]

"During the Little Ice Age (approximately 1300-1850), vineyards in England significantly declined, and grape cultivation became much more challenging due to the colder and more unpredictable climate."[58] Wine cultivation was abandoned in England in about 1450 due to the cold climate in the Little Ice Age.[59] Vineyards have returned to England today during our current warming period, but unlike vineyards during Roman times, English vineyards today are primarily in southern regions of Kent and Sussex, due to their more favorable climate.

The warm climate of the Medieval Warm Period also allowed Vikings to settle Iceland and later Greenland under Eric the Red (c. 950–1005). A study of $\delta^{18}O$ isotopes in marine mollusks by University of Saskatchewan (Canada) professor and isotope geochemist William Patterson and coauthors show temperatures near Greenland to be cold before 240 BC, warm from 230 BC to 140 CE, cold from 410 CE to 640 CE, warm from 640 CE to 1280 CE and cold from 1380 CE to 1420 CE.[60] Not known as a farming area today, recent excavations have uncovered about 450 Viking farms in Greenland.[61] Viking graves on Greenland are in areas of permafrost in the twentieth century, although not evidently at the time of burials.[62] Three churches, one large estate, and 95 farms have been excavated on the west cost of Greenland, mostly under permafrost.[63] By the fourteenth century, Greenland was abandoned as the world plunged into the Little Ice Age. As the climate cooled, the vegetation period grew dramatically shorter and growing cereal became impossible.[64] Excavations of bone and teeth showed famine, and disease ravaged the island during the Little Ice Age.[65]

Subtropical plants spread to Northern China in the Medieval Warm Period. Citrus fruits and Chinese grass (*Boehmeria nivea*) have never grown as far north in China as they were during the thirteenth century. Records show their cultivation in 1264 was several hundred kilometers farther north than in the twentieth century.[66] Zhang De'Er of the Chinese Academy of Meteorological Sciences concludes, "On the basis of knowledge of the climatic conditions required for planting these species, it can be estimated that the annual mean temperature in South Henan Providence in the thirteenth century was 0.9-1.0 higher than at present."[67] Japanese monks kept meticulous records of cherry blossoms for more than seven centuries. According to these records, cherry blossoms arrived early during the Medieval Warm Period. Osaka Prefecture University (Japan) environmental scientists Yasuyuki Aono and Keiko Kazui found the cherry blossom dates also document the existence of four cold periods during the Little Ice Age. They conclude, "These cold periods coincided with the less extreme periods, known as the Wolf, Spoerer, Maunder, and Dalton minima, in the long-term solar variation cycle, which has a periodicity of 150-200 years."[68]

Entomology and Animal Migrations: Animal and insect remains have been studied to reconstruct climate conditions including the Roman Warm Period, the Medieval Warm Period, and the Little Ice Age. Such studies indicate migrations northward in warm periods and southward during cold periods in the Northern Hemisphere. Many of these studies estimated warming in Europe from insect and animal remains during the Roman Warm Period. Entomologist Stephen Brooks of the National History Museum in London studied insect fossil *chironomids* (non-biting midges) from lakes in northern Europe. The study estimates temperatures were warmer in the

Roman Warm Period than during the Dark Ages Cold Period.[69] University of Lausanne (Switzerland) paleoclimatologist Basil Davis and coauthors integrated mammal immigration data (red deer and wild boar) with pollen data to estimate temperatures in Europe were warmer in the Roman Warm Period than in the subsequent Dark Ages Cold Period.[70] The insect *A. quanriguttatus* is no longer found in Britain, but fossil deposits have been found in wells, which "may imply a warmer or a least more continental climate c. 2000 Before Present."[71] Archeological evidence also establishes that the warm-loving nettle ground bug *Heterofaster urticae* was found as far north as York, England during the Roman Warm Period and Medieval Warm Period, but absence during the cold Dark Ages and Little Ice Age. Despite recent warming, this nettle ground bug is only found today in areas of Southern England.[72]

Insect remains provide evidence of warmer conditions during the Medieval Warm Period through species distributions. University of Bern (Switzerland) associate professor and paleoclimatologist Oliver Heiri and coauthors studied *chironomid* (midge) larval from lake sediments in the Swiss Alps to reconstruct summer temperatures. By counting the total *chironomids*, they found the Medieval Warm Period temperatures were warmer than the Little Ice Age.[73] The small rove beetle *Euaesthetus laeviusculus* was in Greenland from 1000 CE to 1350 CE but is no longer present in Greenland, which reflects sufficient cooling to remove the species.[74] The whirligig *bettel Gyrinus colymbus* has been found in deposits from Medieval times in Leicester, England, but is no longer found in the British Isles, suggesting a 1°C decline in July temperatures.[75] Traces of the heat-seeking *Aglenus brunneus* beetle are found in York during the Medieval Warm Period, which also confirms the warm temperatures at that time.[76] A study by The National History Museum in London entomologist Stephen Brooks and University College London paleoclimatologist Hillary Birks analyzes *chironomid* remains from lakes in Norway and Svalbard and reconstructs temperatures in the Medieval Warm Period that are warmer than the Little Ice Age.[77]

Insect and animal remains also document the drop in temperature during the Little Ice Age. A paper by University of Sheffield (UK) paleoentomologists Paul Buckland and Pat Wagner states in the abstract, "There is presently little convincing evidence of climatic rather than human impact upon the faunas during the Little Ice Age."[78] Such denials of natural warming are common in paper abstracts to conform with the climate crisis narrative condoned by editors of prestigious journals. Johns Hopkins climate scientist Patrick T. Brown said, "... editors of these journals have made it abundantly clear, both by what they publish and what they reject, that they want climate papers to support a certain preapproved narrative..."[79] The Buckland and Wagner paper goes on to outline several insects' fauna that might confirm the Little Ice Age, and they conclude: "Is there a fossil insect signal for the Little Ice Age? The answer has to be a guarded, Yes, probably."[80] They discuss how the species *Helophorus gacialis* in the Alps migrated to lower altitudes during the Little Ice Age.[81] *Hydraena britteni* is a widespread fossil in samples in southern Iceland until the Late Medieval Period, when it became extinct, suggesting a 1°C decline in July temperatures.[82] The Heiri et al. paper show that temperatures were below modern levels in the Swiss Alps during the Little Ice Age based upon shifts in *chironomid* species during colder conditions.[83]

Tree Lines: Tree lines are another indication of climate. Trees cannot grow in extremely cold weather, and the variation in both altitude and latitude of tree lines indicates climate changes. As temperatures warm, tree-line elevations increase, and tree lines are at higher latitudes. The reverse is true in cold times. Because temperature is the primary cause of tree lines, both altitude and latitude play a role. The atmosphere declines in temperature by about 6.5°C for each kilome-

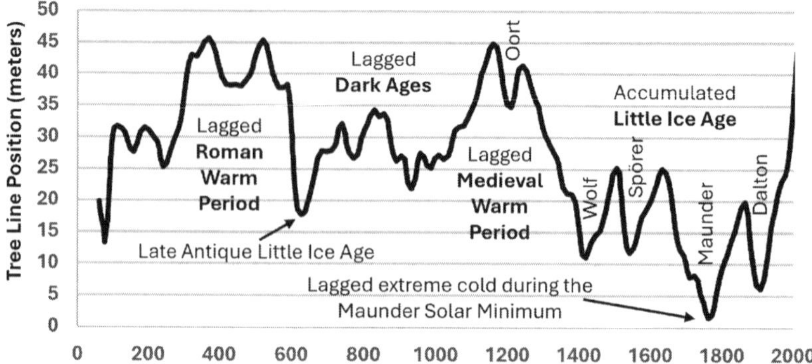

Figure 31 – Büntgen et al., Elevational Position of Tree Lines Reveals Past Climate Cycles. *Global changes in alpine tree line elevations is another verification of the climate cycles of the Roman Warm Period, cold Dark Ages, Medieval Warm Period, Little Ice Age, and Modern Warming. The impact of the Oort, Wolf, Spörer, Maunder, and Dalton solar minima are clearly seen in the tree line data. The temperature drop during the Late Antique Little Ice Age (536 CE to 660 CE) is also present. This study shows reconstructions of alpine tree line positions across the Northern Hemisphere.* Source: Ulf Büntgen et al., "Common Era tree line fluctuations and their implications for climate reconstructions," *Global & Planetary Change*, Volume 219, December 2022, https://doi.org/10.1016/j.gloplacha.2022.103979.

ter of altitude, although variations in humidity and other factors can impact the lapse rate. At some altitude it is too cold for trees to grow. This altitude defines the tree line. Latitude can also impact temperature, as it is hot in the Tropics and very cold in polar regions. For this reason, the tree line in tropical mountains is about 4,000 to 5,000 meters. In mid-latitude mountains such as the Rocky Mountains, the tree line is about 3,000 to 4000 meters. In high latitudes in or near the Arctic, the tree line is much lower, sometimes only a few hundred meters above sea level. Local climate must also be taken into account. For example, the tree line in the Alps is about 2,000 meters, much less than in the Rocky Mountains.

Studies of tree lines indicate Medieval and Roman times were "characterized by climates that were relatively favorable for high-altitude tree growth in the wide geographical perspective of the Northern Hemisphere."[84] During the cold Dark Ages tree lines fell in altitude by 200 meters in Central Europe.[85] During the Medieval Warm Period, the tree line in the Alps rose to an elevation of over 2,000 meters, much higher than today.[86] In northwestern Russia, the pine forests grew at least 100-140 meters above the modern pine tree line.[87] The presence of pine trees from 1000 CE to 1200 CE is found 100 meters above the current pine tree limit in Sweden.[88] The northern tree line in the Medieval Warm Period in northern Québec, Canada was 130 kilometers farther north than it is today.[89] Cambridge University pro-

Figure 32 – The London Winter Fair Market on the Thames River, 1683. *Winter fair markets were held on the ice of the Thames River in London during the cold Little Ice Age. The image above depicts London and the winter fair market on the Thames River in 1683. London Bridge over the Thames River is seen in the distance.* Source: Public Domain by Wikimedia Commons, File:Frost Fair of 1683.JPG. https://commons.wikimedia.org/wiki/File:Frost_Fair_of_1683.JPG#Licensing.

fessor and dendroclimatologist Ulf Büntgen and coauthors published a paper on the fluctuation of tree line elevations over the past 2,000 years (see Figure 31). The authors highlighted, "Tree lines were highest after the Roman and Medieval warm periods." Their paper also shows the lowering of altitude of tree lines during both the cold Dark Ages and the Little Ice Age.[90] They also show significant increases in the altitude of tree lines in recent years. The impact of the Oort, Wolf, Spörer, Maunder, and Dalton solar minima during the Little Ice Age are also clearly seen in the tree-line data. The cold of the Late Antique Little Ice Age from 536 CE to 660 CE is also seen in the tree line record.

Frozen Rivers, Lakes, and Ocean Bays: Historical records of rivers, lakes, and ocean bays freezing over are another indication of historical climate. During the second French expedition of Cartier up the St. Lawrence Seaway in 1535, they found that the whole St. Lawrence had frozen near modern day Montréal, Canada.[91] The St. Lawrence River is very wide with strong currents, making it difficult to freeze over. The fact that it froze over in1535, reveals how cold the weather was during the sixteenth century. Sam White, Professor of Political History at the University of Helsinki (Finland), recounts the annals of the Spanish Franciso Vázquez de Coronado expedition which documented that the Rio Grande River in New Mexico was frozen over for several weeks in the 1540s and afforded loaded horse crossings on the ice. [92] There is no record of the Rio Grande in this area freezing over in modern times. In1591, the Spanish expedition of Nuevo León Gaspar Castano de Sosa reported the Sante Fe River near Santa Fe, New Mexico was "frozen so hard that the horses crossed the ice without breaking through.[93] White also recounts the narratives of the Vizcaíno expedition to California in 1602. In December of that year at Monterrey Bay, Vizcaíno recorded, "the well… was frozen to the thickness of a palm's width."[94] This is far colder than any observations in the past century in Monterrey Bay.

In London during the Little Ice Age, winter Frost Fair markets were held on the ice of the frozen Thames River (see Figure 32).[95] "Between January and mid-February 1684 thousands of people spanning the entire social spectrum, from King Charles II and members of the royal family to the lowliest pauper, venture upon this 'Freezeland.' They did so in carriages, on horseback, or by foot. At its height the fair extended from London Bridge to Vauxhall."[96] The last winter market on the Thames River was 1814; the ice has not frozen solid enough since then to hold a fair with stalls, entertainment, and various activities on the ice.[97] During the 1814 Frost Fair on the Thames River, an elephant was marched across the river alongside Blackfriars Bridge.[98]

The Thames River was wider and shallower in the Middle Ages, and the Medieval Stone London Bridge, completed in 1209, slowed the flow of the river; both of which enhanced ice formation.[99] Nevertheless, the freezing of the Thames River clearly shows a contrast between the climates of the Dark Ages, Medieval Warm Period, and the Little Ice Age. The first Frost Fair was in 695 CE during the cold Dark Ages.[100] Although there is mention of a frozen Thames River in 1536 and 1564,[101] there is no record of a frozen Thames River during the Medieval Warm Period despite the fact the London Bridge was in place for nearly a century before the start of the Little Ice Age. The Frost Fairs did not resume until 1608 during the cold seventeenth century of the Little Ice Age.[102] During the decade of 1660, Robert Boyle provided observations as to the thickness of the ice: about eight inches near the middle of the river and more than two yards on the banks.[103]

During the cold Dark Ages, the river Danube froze so solidly in winters that it afforded cart crossings.[104] During the winter of 1607/1608, the Rhine River in Germany froze all the way up to Cologne, and the Main River iced up all the way past Frankfurt.[105]

The freezing of the Rhine River to Cologne is extremely rare today and freezing of the Main River near Frankfurt is virtually unheard of in recent years. In France, the Seine River iced over for nearly two months so that carriages could pass over the ice.[106] Diarist Pierre de l'Estolie wrote, "The cold was so extreme and the freeze so great and bitter, that nothing seemed like it in the memory of man."[107] Although occasional ice flows are reported in the Seine River, freezing over has not been reported in recent years. Italy experienced unusually cold weather between 1594 to 1598. The Arno River in Florence and the Lagoon of Venice froze over.[108] The author Toaldo mentions that the Venetian Lagoon was frozen in the years: 859, 860, 864, 1492, 1594, and 1755.[109] These are all years in the cold Dark Ages and Little Ice Age climate cycles. The Arno River is fast flowing, so accounts of it freezing over seem unbelievable today. The Dutch artist Hendrick Avercamp painted several winter scenes in the early 1600s depicting massive crowds skating on the canals in Amsterdam, Antwerp, and surrounding areas.[110] Today, the canals in Amsterdam and Antwerp only gets a thin layer of ice that isn't safe for skating on.

In China during the Dark Ages, the Yangzi froze over more than once.[111] According to Chia-chang Chang, "It has been concluded from the total freezing of major lakes in China that the average temperature there between 1470 to 1850 must have been one degree colder than in the late twentieth century."[112]

In the winter of 1642, Boston Harbor froze over.[113] The last time Boston Harbor completely froze over was in 1844,[114] which was near the end of the Little Ice Age and the Dalton Solar Minimum. During that winter, a merchant of Boston paid for a canal to be cut in the ice 7 mile long and 100 feet wide in which the steam ship Britannia left the dock at East Boston on the 3rd of February 1844 on her voyage to Liverpool, England.[115] During the Little Ice Age, European merchants would leave Boston Harbor before winter to avoid the ice.[116] John Hull, the first mint master of Boston wrote the following in the winter of 1654, "The frost was extreme and suddenly froze the Bay over, that in a very few days it was firm to pass betwixt the town and Long Island, and a constant passage to Charlestowne and Noodles Island and this continued for about a month."[117] Merchant, legislator, and colonial magistrate Samuel Sewall recorded in his diary in Boston on January 24, 1685, "Friday night and Satterday were extreme cold, so the Harbour frozen up, and to Castle [Island]."[118] Boston's severe cold in the winter of 1740/41 was so newsworthy that the *Pennsylvania Gazette* reported the words of a Bostonian as follows: "Our Harbour and Rivers are entirely froze up, and people go down to the Castle and other Islands upon the Ice. On Charles River is erected a Tent, for the Entertainment of Travellers ... the Ice is continued for a Space of above twenty Miles, and that the Sea cannot be discerned from the Shoar, so that it looks as if the whole Bay was a solid Body of Ice." [119] The diary of another Boston resident Deacon John Tudor wrote the following on January 18, 1752, "Capt. Atkias and I rode in a sleigh to the Castle on our return we rode oute to the channel whare a Number of Men was cutting the Ice to open the Channel way for Vessels to go oute. The ice was 9 inches thick." [120]

It was also very cold further south in Virginia. During the winter of 1607/1608, one of the Jamestown settlers wrote in a letter that the winter grew "so intense that one night the river at our fort froze almost all the way across."[121] The freezing of the James River is almost unimaginable in today's climate.[122] In the winter of 1642, Chesapeake Bay, about 40 miles downstream from Jamestown, also froze over.[123] Chesapeake Bay froze over again during the winter of 1779-1780.[124] This was the infamous cold winter of Valley Forge where Washington encamped with his Continental Army. French colonel, Duc de Lauzun, who was assigned to assist the Americans the following year, compared Lebanon, Connecticut, where he encamped, as being sent to Siberia.[125] During the winter of 1779/1780, "it was possible to walk across the firmly frozen Chesapeake Bay in Maryland and the Albemarle Sound in North Carolina. Harbors froze in Virginia. Sleights traveled the ice from Staten Island to Manhattan Island, and for a time people traversed the sound from Connecticut to Long Island."[126] George Washington had an icehouse at his home in Mount Vernon in Virginia; he regularly harvested ice from the Potomac River.[127] Freezing of the Potomac River was routine during the Little Ice Age, but it has been a very rare event since.

Eddy Solar Cycles

So what caused the temperature decline of the Little Ice Age? CO_2 levels did not change much during this period, so CO_2 radiative forcing can be ruled out as a cause.[128] CO_2 levels were 185 ppm during the warmest period of the Medieval Warm Period and 175 ppm during the coldest period of the Little Ice Age.[129] Using the ERF adjusted Myhre radiative forcing equation, this 10 ppm decrease would only produce a temperature change of 0.24 W/m^2 or 0.07°C.

Papers by University of Wisconsin scientist Yafang Zhong and coauthors[130] and University of Colorado professor and paleoclimatologist Gifford Miller and coauthors[131] propose the Little Ice Age was primarily caused by volcanic activity. Although volcanic eruptions may have contributed to short-term cooling spikes in temperature, they cannot explain the 500-year Little Ice Age between 1350 to 1850. The dust and sulfate aerosols from volcanic eruptions are known to dissipate within one year or less.[132] Even large volcanic eruptions are known to have no substantial impact on climate after three years.[133] Except for volcanic eruptions around 1450, volcanic activity was low from 1300 to almost 1600.[134] Yet these years, during the Wolf and Spörer solar minima, experienced cold temperatures. Climate historian Sam White wrote that the winter of 1540-1541 "bought more consistent cold than are recorded in modern instrumental records" and the Rio Grande River near Albuquerque froze over for a month and afforded crossings on the ice by loaded horses.[135] Yet volcanic activity was low in 1541 and 1542 (see Figure 65 in Chapter 12). Volcanic activity was higher from 950 to 1050 during the Medieval Warm Period (see Figure 65).[136] And although volcanic activity increased during the Maunder Solar Minimum (1645 to 1715), it was lower than volcanic activity between 1800 to 1900 (see Figure 65),[137] yet temperatures were warmer in the nineteenth century than during the Maunder Solar Minimum. Although volcanoes contributed to cold times during the Little Ice Age, evidence suggests they are not the primary driver of cooling during the Little Ice Age (see Chapter 12, Volcanoes for a more in-depth discussion).

The Little Ice Age saw the lowest solar activity in thousands of years.[138] The cooling was most pronounced in the Northern Hemisphere, which also suggest the decline was due to cooler oceans transporting less heat to the Arctic via the Meridional Overturning Circulation, the Gulf Stream current in the Atlantic Ocean, and the Kuroshio current in the Pacific Ocean (see Chapter 13). Grand Solar Minimums, including the Wolf, Spörer, Maunder, and Dalton solar minima; and heavy cloud coverage over the oceans were likely causes of the temperature drop. A study from NASA suggests the lower ultraviolet radiation during the Maunder Solar Minimum caused a drop in temperature and ozone in the stratosphere, and this weakened the polar vortex and resulted in chilly temperatures.[139] Evidence thus suggests that the Little Ice Age was caused by lower cloud cover, less Total Solar Irradiance, a cooling stratosphere, and a weaker polar vortex that allowed more heat to escape out to space from polar regions, all of which are driven by solar minima (see Chapter 7).

Scientists Nicola Scafetta and Fritz Vahrenholt cite several scientific papers from *Nature*, *Science*, and *Solar Physics* that present compelling evidence of a quasi-millennial solar cycle, known as the Eddy Cycle, and how it impacts climate.[140] The Eddy Solar Cycle switches back and forth from a grand solar maximum to a grand solar minimum about every 500 years (see Figures 40 and 41). History and paleoclimate proxies all confirm millennial climate cycles which switch from warmth to cooling and back again about every 500 years, in sync with millennial Eddy solar cycles, including the Little Ice Age. Columbia University climatologist Gerald Bond and colleagues first described these climate cycles in the North Atlantic and explicitly stated that they were synchronous with solar activity.[141] Scafetta and Vahrenholt cite 15 scientific papers that establish a connection with solar activity and past climate cycles in the USA, Brazil, Patagonia, Peru, Antarctica, South Africa, Morocco, Oman, India, China, Australia, Spain, Austria, and Finland.[142] Carbon isotopes are a proxy of solar activity (see Chapter 7). University of Bern Professor Hanspeter Holzhauser's study of glaciers in the Alps connects glacier advance and decline to solar cycles. He concludes, "A comparison between the Great Aletsch Glacier and the residual ^{14}C records

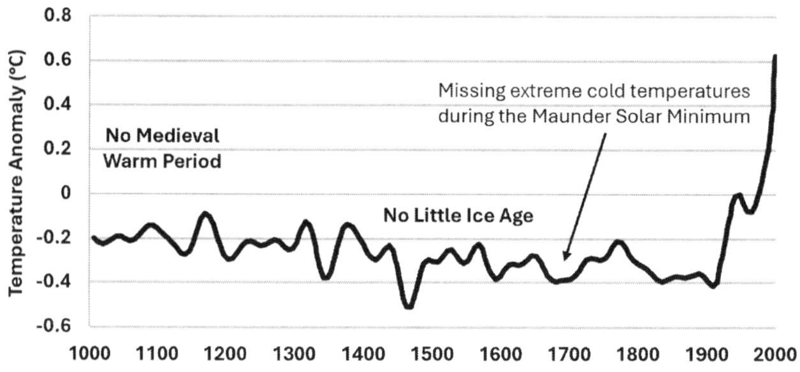

Figure 33 – The Mann Hockey Stick Graph of Climate. *The Mann hockey stick graph above was featured in the IPCC Assessment Report 3. This one graph erased the Medieval Warm Period and Little Ice Age, despite dozens of scientific papers that confirm these climate periods from ice core, sediment cores, stalactites, and other proxy and historical records.* Source: IPCC, *Climate Change: 2001: The Scientific Basis, Contribution of Working Group I to the Third Assessment Report of the Intergovernmental Panel on Climate Change,* "Summary for Policymakers," Cambridge University Press, Cambridge, UK, January 2001, p. 3, Figure 1(b). https://www.ipcc.ch/site/assets/uploads/2018/02/WG1_TAR-FRONT.pdf.

supports the hypothesis that variations in solar activity were a major forcing factor of climatic oscillations in west-central Europe during the late Holocene."[143]

Cloud cover which shades the Earth and reflects sunlight back out to space is a significant factor in climate change (see Chapters 7, 11, and 12). Astrophysicists Henrik Svensmark and Nir Shaviv have provided convincing evidence of how solar cycles modulate cloud cover (see Chapter 7).[144] Variations of cloud cover, in sync with solar cycles, have a major impact on the climate. It is interesting to note that the cold Little Ice Age was a period of the Wolf, Spörer, Maunder, and Dalton minimum solar cycles and a time of cloudiness. The Anglican bishop Robert Burton (1577-1640) described the times in the British Isles as one of "endless days when dark clouds obscure the sunlight."[145] During the Little Ice Age, the Reverend Daniel Schaller, of Stendal in the Prussian Alps, wrote, "There is no real constant sunshine, neither a steady winter nor summer; the earth's crops and produce do not ripen, are no longer as healthy as they were in bygone years."[146]

IPCC Erased Historical Climate Cycles

The Medieval Warm Period and Little Ice Age do not align with the anthropogenic global warming narrative. Since fossil fuels were not in use in those times, these warm and cool cycles must have occurred from natural variation. Despite overwhelming evidence of the Medieval Warm Period, an email was sent by a major IPCC climate science researcher, who wrote: "We have to get rid of the Medieval Warm Period."[147] In another email, climate scientist Michael Mann wrote: "...it would be nice to try to 'contain' the putative 'MWP'[Medieval Warm Period]..."[148] **The *Third IPCC Assessment Report*, "Summary for Policy Makers," featured as a centerpiece the infamous "hockey stick" graph of Michael Mann which completely removed the Medieval Warm Period and Little Ice Age.**[149]

This hockey stick graph was produced from data in a paper by Michael Mann, Raymond Bradley, and Malcom Hughes.[150] The curve resembles a hockey stick, where the first 850 years show stable temperatures, void of the Medieval Warm Period and Little Ice Age. This part of the graph was the flat handle of the hockey stick, while in the last hundred years the temperature increased rapidly, forming the blade.

The Mann hockey stick graph has been discredited as Mann et al. used a flawed statistical methodology. Using the R^2 statistical test to validate the chart, the Mann temperature graph fails.[151] The Mann Hockey Stick was heavily criticized for major deficiencies in its paleoclimatic proxies and statistical methods. Crok references six such critiques in scientific journals including Soon and Baliunas in *Climate Research*, 2003; McIntyre and McKitrick in *Energy & Environment*, 2003 and 2005, and *Geophysical Research Letter*, 2005; McShane and Wyner in *Annals of Applied Statistics*, 2011; and the book *The Hockey Stick Illusion* by Montford.[152] Mann used primarily tree ring proxy data to predict temperature by year based on the growth rings of the trees. Where the tree ring data fits the anthropogenic warming narrative, Mann et al. included it in the graph. However, the tree ring data showed cooling in the second half of the 20th century, which did not fit the narrative. So he discarded tree ring proxy data and switched to thermometer data in the 20th century to create the blade of the hockey stick.[153] IPCC colleague, Phil Jones referred to hiding the decline in tree-ring derived data as "Mike's *Nature [Journal*] trick."[154] In one of the Climategate emails, Phil Jones admitted he had also used Mike's trick.

Although not the original Mann Hockey Stick, in 2004, Phil Jones and Michael Mann published a paper to summarize climate reconstructions over the past millennia. Not surprisingly, their conclusions mirror the Mann Hockey Stick with less than ±0.2°C variation in any century before modern warming.[155] A reconstruction of sea level rise by Aslak Grinsted from the Niels Bohr Institute (Denmark), John Moore of the University of Lapland Arctic Centre (Finland), and Svetlana Jevrejeva of the National Oceanography Centre (UK) compared historical sea level rise to both the Jones and Mann paper and the historical climate reconstruction by Moberg et al. The Moberg data includes a noticeable Medieval Warm Period and Little Ice Age (see Figure 27). Grinsted et al. found the Moberg temperature reconstruction gives fits that, on average, have a factor 23 greater likelihood than the Jones and Mann paper.[156] They concluded, "We further find that the Moberg et al. temperature reconstruction is more consistent with observed sea level rise than the Jones and Mann reconstructions which conclude does not have a cold enough Little Ice Age."[157]

Despite the scientific flaws in the hockey stick, Mann was rewarded since his work supports the anthropogenic climate change narrative. He was appointed by the IPCC as lead author on climate variability for the IPCC's *Third Assessment Report*. Professor of Meteorology of the University of Hamburg, Hans Von Storch, was also a lead author in the IPCC *Assessment Report 3*. He strongly opposed any use of Michael Mann's hockey stick graph in the report. He was overruled and not invited to participate in the subsequent IPCC *Assessment Report 4*.[158] The inclusion of this faulty graph in IPCC *AR3* shows the bias of the IPCC.

Michael Mann took aggressive action against the Soon and Baliunas paper that was published in *Climate Research*. Leaked Climategate email reveal that "Mann and his team speak of encouraging colleagues to stop treating *Climate Research* as a valid publication for scientific findings, and of 'getting rid of' both the paper's editor, Chris de Freitas, and the journal's editor in chief, Hans von Storch. Another plan involved trying to precipitate a mass resignation from the editorial board."[159] The pressure of boycotting the journal appears to have worked. *Climate Research* was criticized publicly for its peer review process that allowed the Soon and Baliunas paper to be published, and the editor in chief, von Storch, resigned. In addition, the doi reference for the Soon and Baliunas paper has been disabled. In an interview in *Climate the Movie,* Harvard & Smithsonian astrophysicist Sallie Baliunas recounted how she retired early due to the pressure she felt for not supporting the climate crisis narrative.[160]

Despite being discredited, the Mann Hockey Stick was used significantly in the media and internet posts to substantiate anthropogenic greenhouse warming. The Mann Hockey Stick was not included in subsequent IPCC reports *AR4* and *AR5*. **Surprisingly, in the**

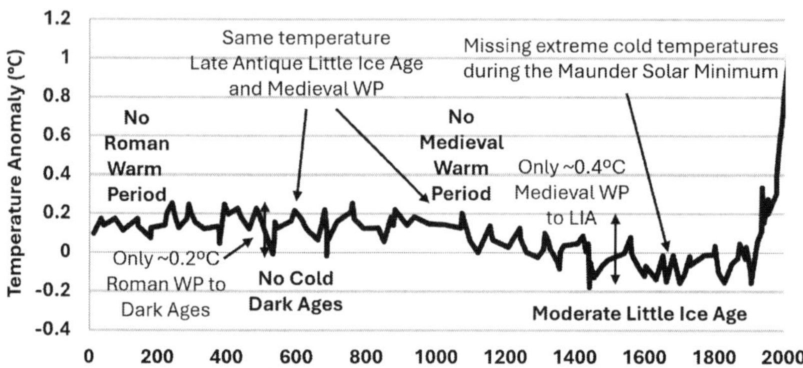

Figure 34 - The PAGES 2k Hockey Stick Graph of Climate. *The PAGES 2k hockey stick graph above was featured in the Summary for Policymakers in the IPCC Assessment Report 6 in 2021. Going beyond the Mann Hockey Stick, this chart erases the Roman and Medieval Warm Periods, the cold Dark Ages, and the severity of the Little Ice Age. Contrary to the historical record, this graph shows the same temperature during the Late Antique Little Ice Age and the Medieval Warm Period. It also moderates the frigid Little Ice Age and erases the extreme cold during the Maunder Solar Minimum. This graph ignores evidence to the contrary from dozens of scientific papers and numerous historical records.* Source: IPCC, "2021: Summary for Policymakers," *Climate Change 2021: The Physical Science Basis. Contribution of Working Group I to the Sixth Assessment Report of the Intergovernmental Panel on Climate Change*, Cambridge University Press, Cambridge, United Kingdom and New York, NY, USA, Section A.1.8, Figure SPM.1, p. 6, 9 August 2021. https://www.ipcc.ch/report/ar6/wg1/downloads/report/IPCC_AR6_WGI_SPM.pdf.

"Summary for Policymakers" of the *2021 IPCC Assessment Report 6*, a similar hockey stick graph was presented that goes beyond the Mann Hockey Stick—as it erases the Roman Warm Period, the cold Dark Ages, the Medieval Warm Period and the severity of the Little Ice Age.[161] Furthermore, this new hockey stick graph eliminates the documented severe cold during the Antique Little Ice Age and the Maunder Solar Minimum. It has the same temperature during the Medieval Warm Period and the Antique Little Ice Age, which goes against all historical knowledge (see Figure 34). In recent years, the Arctic has warmed 25 times faster than Antarctica, with Antarctica hardly warming at all (see Tables 68 and 69 in Chapter 13). Yet, "Antarctica borehole temperature data point to the Little Ice Age as the largest climate anomaly of the last 1000 years with temperatures on average 0.52 ±0.18°C colder than those of the last century."[162] The *AR6* IPCC hockey stick graph has the global temperature anomaly for the Little Ice Age less than what was experienced in Antarctica (see Figure 34)!

The new hockey stick graph is from the PAGES 2k group headquartered at the University of Bern, where Thomas Stocker chairs the Climate and Environment Physics department. Stocker co-authored the "Summary for Policy Makers" of the IPCC *AR3* report, which featured the Mann Hockey Stick. With strong links to the IPCC, Stocker also ran for the IPCC chairmanship in 2015. **It is interesting to note that in 2013, scientists from this same PAGES 2k group published a reconstruction of temperatures over the past 2,000 years that shows parts of the first millennium were as warm as present day.**[163] Referring to the *AR6* hockey stick graph, CLINTEL investigators write, "Evidence suggests that a significant part of the original PAGES 2k researchers could not technically support the new hockey stick and seem to have left the group in dispute. Meanwhile, in 2020, the dropout scientists published a competing temperature curve with significant pre-industrial temperature variability. **On the**

basis of the thoroughly verified tree rings, the specialists were able to prove that summer temperatures had already reached today's level several times in the pre-industrial past."[164] This new paper by Ulf Büntgen et al. includes 22 authors.[165] The work of Cambridge University professor and dendroclimatologist Ulf Büntgen and coauthors was not included in the IPCC report, which shows the bias of the IPCC.

The *AR6* PAGES 2k Hockey Stick uses poorly documented tree ring data that tree ring specialist Büntgen had previously cautioned are too complex to be used as overall temperature records.[166] In contrast, the Büntgen 2020 paper validates every tree ring data set individually.[167] In some cases, the *AR6* PAGES 2k Hockey Stick erroneously used proxies that turned out to reflect hydroclimate and not temperature.[168] Canadian statistician Steve McIntyre studied the *AR6* PAGES 2k Hockey Stick proxy data and summarized his critique, which includes results from questionable data processing.[169] McIntyre points out the blade of the *AR6* hockey stick disappears when some data sets are analyzed with a dplR statistical package or a single Hugerhoff curve fit.[170] The *AR6* PAGES 2k Hockey Stick makes the same error as the Mann Hockey Stick in combining paleoclimate and thermometer records. The proxy data should be validated against thermometer data, not replaced by thermometer data.

Despite its flaws and the fact that its temperature reconstruction is counter to a large body of peer-reviewed papers and historical documents that confirm past climate cycles, the *AR6* PAGES 2k Hockey Stick is prominent in media, internet posts, and AI queries. It is ironic that climate alarmists claim to be on the side of the *scientific consensus*, yet they put their trust in the anomalous *AR6* PAGES 2k Hockey Stick and thereby reject the vast body of peer-reviewed literature that confirms the historical reality of the Roman Warm Period, Medieval Warm Period, the Late Antique Little Ice Age, and the Little Ice Age. University of Helsinki professor and climate historian Sam White has written, "The Little Ice Age is not a dogma. It is an increasingly firm consensus backed by considerable evidence across a variety of sources."[171] **Those that push a narrative that eliminates these past climate cycles are the true "climate deniers."**

The chart in IPCC *AR6* also suggests that temperatures today are warmer than during the Holocene Climatic Optimum that occurred 9,000 to 5,000 years ago. They correctly identify that the Holocene Climatic Optimum was the "warmest multi-century period in more than 100,000 years." We are currently living in the Holocene, and multiple studies show that the temperatures experienced during the Holocene Climatic Optimum were the warmest during the entire Holocene and much warmer than today, by as much as 2°C.[172] This makes sense as the Milankovitch cycle of obliquity (the axial tilt of the Earth) has been declining from 24.2 degrees 9,000 years ago to 23.5 degrees today, which, as predicted by Milankovitch, has resulted in cooling (see Figure 38).[173] Evidence from many studies also show that temperatures during the Medieval Warm Period and the Roman Warm Period were as warm or warmer than today, and the Minoan Warm Period was warmer than today.[174]

Paleoclimate reconstructions can be debated, but glacier advance and decline, sea level measurements, tree line altitudes and latitudes changes, entomology and animal migrations, frozen bodies of water, and agricultural records do not lie (see Figure 29, Figure 30, and Figure 31). "The Holocene Climatic Optimum was the period in the last 100,000 years when glaciers were at their smallest, while the Little Ice Age was the period in the last 7,000 years when glaciers were at their largest."[175] University of Kentucky glaciologist Caleb Walcott-George and coauthors analyzed soil samples 509 meters (1,670 feet or nearly a third of a mile) under a Greenland ice sheet and concluded: "the ground below the summit was exposed to sunlight 7.1 ± 1/1 thousand years ago."[176] This was during the Holocene Climatic Optimum.

Such evidence presented in this Chapter 5 and Chapter 6 confirms the Holocene Climatic Optimum was warmer than today and that the Roman Warm Period, cold Dark Ages, Medieval Warm Period, and Little Ice Age climate cycles are beyond doubt. The inclusion of the *AR6* PAGES 2k Hockey Stick graph and the exclusion of the extensive number of paleoclimate studies and other evidence of these climate cycles discredits the IPCC and shows its bias in perpetuating the "Climate Crisis" narrative.

Since the overwhelming evidence of natural climate cycles discredits the hypothesis that nearly all recent warming is from anthropogenic greenhouse gases, the IPCC and Michael Mann have attempted to erase history and falsely eliminate these past temperature swings. Because the evidence of past global warming cycles in Europe and Greenland is so overwhelming, Mann and the IPCC argue that the Medieval Warm Period and Little Ice Age were not global, but regional, primarily in Europe. **Claiming past climate cycles were regional seems a strange argument against the reality of these past warming cycles, since climate change today is mostly regional**. Warming between 1978 and 2022 saw a temperature increase of 0.25°C per decade in the Arctic and 0.01°C in Antarctica (see Figures 68 and 69 in Chapter 13).[177]

Current global warming is also localized, predominantly in the Northern Hemisphere, especially Greenland and Europe. This localized warming is due to the heating of oceans by the Sun and the transport of this heat to northern latitudes by the Meridional Overturning Circulation, Gulf Stream, and Kuroshio Ocean currents. Heat from these currents is released into the atmosphere in the North Atlantic and North Pacific (see Chapter 13). The transport of heat by these ocean currents is most pronounced in the oceans off Greenland and Europe (see Figures 71 and 72 in Chapter 13), but also in China and Japan. Mann and the IPCC try to discredit the reality of past climate cycles by insisting past cycles were limited to Europe.

Claiming global warming is localized is actually an argument against anthropogenic greenhouse gas warming. Greenhouse warming would be more uniform globally since CO_2 emissions disperse rapidly throughout the atmosphere, and the concentration of CO_2 in the Arctic and Antarctica are virtually the same. However, ocean currents move water warmed by the Sun to northern latitudes, especially Greenland and Europe. The warming of the Minoan, Roman, Medieval, and Modern warm periods have coincided with the peak of the warm periods of the Eddy Solar Cycle. Since Greenland ice cores also show significant heating in the Minoan, Roman, and Medieval warm periods (see Figure 26), these same ocean currents likely brought warm seawater, heated by the Sun, to northern latitudes in these periods, just as is happening today.

Climate historian Sam White has written, "Finally, lest we forget, long-term climate reconstructions, both proxy and documentary, have spread far beyond Europe. Suffice it to say that climatologists and historians have gathered data from a range of physical and written sources across Asia and now increasingly in the Americas, Africa, Australia, and Oceania. Chinese and Japanese sources have supported particularly strong reconstructions of Little Ice Age climate changes, as well as solid historical studies of their impacts." [178] Most of the evidence for past global warming has come from the Northern Hemisphere, including Europe, China, Japan, Greenland, and North America where the warming has been most pronounced. About 90% of the world's populations today live in the Northern Hemisphere.[179] Ocean currents that created more warming in the Northern Hemisphere in past climate cycles are relevant today. Due to population densities, Northern Hemisphere warming has by far the greatest impact on humanity.

Nanjing University (China) paleoclimatologist Bao Yang and coauthors used a combination of multiple paleoclimate proxy records including ice cores, tree rings, lake sediments, and historical documents and found China had "a warm stage in AD 0-240, a cold interval between AD 240 and 800, a return to warm conditions from AD 800-1400, including the Medieval Warm Period between AD 800-1000, the cool Little Ice Age period between 1400-1920, and the present warm stage since 1920."[180] A paleoclimate study by Chinese Academy of Sciences paleoclimatologist Ming Tan and coauthors show the same temperature cycles in northern China.[181] They are also found in northeast China in a paleoclimate study by Chinese Academy of Sciences paleoclimatologist Quansheng Ge and coauthors.[182] Lanzhou University (China) paleoclimatologist Pingzhong Zhang and coauthors show a relationship between the Asian Monsoon and the Swiss record of Alpine glaciation. They point out the climate cycles shown in Alpine glaciation of the Dark Ages, Medieval Warm Period, and Little Ice Age are in sync with temperatures and monsoon strength in China as well as in Mesoamerica, Minnesota, and France.[183] A study by Seoul National University (South Korea) paleoclimatologist Jungjae Park used pollen samples to reconstruct temperature changes of the past on the east coast of Korea "in which the 'Medieval Warm Period,' 'Little Ice Age,' and 'Migration Period' were clearly shown."[184]

Historical records, archeology, and paleoclimate reconstructions from Greenland, Europe, the North Atlantic, North America, China, Japan, Africa, South America, Antarctica, New Zealand, the South and Central Pacific, and the Indian Ocean all confirm these climate cycles were experienced globally. Vahrenholt and Lüning in their book *The Neglected Sun* references sixteen papers which document the Medieval Warm Period in Africa, Antarctica, Asia, Oceania, North America, and South America (see Table 6).[185] **Regarding evidence of the Medieval Warm Period outside of the North Atlantic, Vahrenholt and Lüning write, "The volume of data today is overwhelming."**[186]

Table 6
Medieval Warm Period: Regions outside the North Atlantic as cited by Vahrenholt and Lüning in their book *The Neglected Sun: Why the Sun Precludes Climate Catastrophe*

Region	Scientific Publication Reference
Africa	**Africa 1:** CO2 Science (2011) Medieval Warm Period (Regional: Africa) – Summary. http://www.co2science.org/subject/m/summaries/mwpafrica.php **Africa 2:** Powers, L. A. et al., "Organic geochemical records of environmental variability in Lake Malawi during the last 700 years, Part I: The TEX86 temperature record," *Paleogeography, Palaeoclimatology, Paleoecology,* Volume 303, Issues 1-4, pp. 133-139, 1 April 2011. https://doi.org/10.1016/j.palaeo.2010.09.006
Antarctica	**Antarctica 1:** CO2 Science (2011) Medieval Warm Period (Regional: Antarctica) – Summary. http://www.co2science.org/subject/m/summaries/mwpantarctica.php **Antarctica 2:** Li, Y. et al., "Glaciochemical evidence in an East Antarctica ice core of a recent (AD 1450-1850) neoglacial episode," *Journal of Geophysical Research: Atmospheres,* Volume 114, Issue D8, pp. 1-11, 28 April 2009. https://doi.org/10.1029/2008JD011091 **Antarctica 3:** Bertler, N. A. N. et al., "Cold conditions in Antarctica during the Little Ice Age – implications for abrupt climate change mechanisms," *Earth and Planetary Science Letters*, Volume 308, Issues 1-2, pp. 41-51, 1 August 2011. https://doi.org/10.1016/j.epsl.2011.05.021 **Antarctica 4:** Orsi, A. J. et al., "Little Ice Age cold interval in West Antarctica: evidence from borehole temperature at the West Antarctic Ice Sheet (WAIS) divide," *Geophysical Research Letters*, Volume, Issue 9, L09710, 9 May 2012. https://doi.org/10.1029/2012GL051260 **Antarctica 5:** Rhodes, R. H. et al., "Little Ice Age climate and oceanic conditions of the Ross Sea, Antarctica from a coastal ice core record," *Climate of the Past*, Volume 8, Issue 4, pp. 1223-1238, 30 July 2012. https://doi.org/10.5194/cp-8-1223-2012

	Antarctica 6: Lu, Z., et al., "An ikaite record of late Holocene climate at the Antarctic Peninsula," Earth and Planetary Science Letters, Volumes 325-326, pp. 108-115, 1 April 2012. https://doi.org/10.1016/j.epsl.2012.01.036
Asia	**Asia 1:** CO2 Science (2011) Medieval Warm Period (China) – Summary.http://www.co2science.org/subject/m/summaries/mwpchina.php **Asia 2:** Zhu, H.F. et al., "August temperature variability in the southeastern Tibetan Plateau since AD 1385 inferred from tree rings," *Paleogeography, Palaeoclimatology, Paleoecology,* Volume 305, Issues 1-4, pp. 84-92, 15 May 2011. https://doi.org/10.1016/j.palaeo.2011.02.017 **Asia 3:** Mackay, A. W. et al., "1000 years of climate variability in central Asia: assessing the evidence using Lake Baikal (Russia) diatom assemblages and the application of a diatom-inferred model of snow cover on the lake," *Global and Planetary Change,* Volume 46, Issues 1-4, pp. 281-97, April 2005. https://doi.org/10.1016/j.gloplacha.2004.09.021 **Asia 4:** Yamada, K. et al., "Late Holocene monsoonal climate change inferred from Lakes Ni-no-Megata and San-no-Megata, northeastern Japan," *Quaternary International,* Volume 220, Issues 1-2, pp. 122-32, 15 June 2010. https://doi.org/10.1016/j.quaint.2009.09.006 **Asia 5:** Tan, L. et al., "Climate patterns in North Central China during the last 1,800 years and their possible driving force," *Climate of the Past,* Volume 7, Issue 3, pp. 685-692, 4 July 2011. https://doi.org/10.5194/cp-7-685-2011 **Asia 6:** Park, J., "A modern pollen-temperature calibration data set from Korea and quantitative temperature reconstructions for the Holocene," *The Holocene,* Volume 21, Issue 7, pp. 1125-1135, 16 May 2011. https://doi.org/10.1177/0959683611400462 **Asia 7:** Kaniewski, D. et al., "The medieval climate anomaly and the Little Ice Age in coastal Syria inferred from pollen-derived paleoclimatic patterns," *Global and Planetary Change,* Volume 78, Issues 3-4, pp. 178-187, August-September 2011. https://doi.org/10.1016/j.gloplacha.2011.06.010 **Asia 8:** Hao, Z-X. et al., "Winter temperature variations over the middle and lower reaches of the Yangtze River since AD

	1736," *Climate of the Past,* Volume 8, Issue 3, pp. 1023-30, 2 June 2012. https://doi.org/10.5194/cp-8-1023-2012 **Asia 9:** Selvaraj, K. et al., "Late Holocene monsoon climate of northeastern Taiwan inferred from elemental (C, N) and isotopic (°13C, 171. °15N) data in lake sediments," *Quaternary Science Reviews*, Volume 37, pp., 48-60, 22 March 2012. https://doi.org/10.1016/j.quascirev.2012.01.009 **Asia 10:** Kitagawa, H. and Matsumoto E., "Climatic implications of °13C variations in a Japanese cedar (Cryptomeria japonica) during the last two millennia," *Geophysical Research Letters,* Volume 22, Issue 16, pp. 2155-8, 15 August 1995. https://doi.org/10.1029/95GL02066 **Asia 11:** Treydte, K. S. et al., "Impact of climate and CO_2 on a millennium-long tree-ring carbon isotope record," *Geochimica et Cosmochimica Acta,* Volume 73, Issue 16, pp. 4635-4647, 15 August 2009. https://doi.org/10.1016/j.gca.2009.05.057 **Asia 12:** Oppo, D. W., Y. Rosenthal and Linsley, B. K., "2,000 year-long temperature and hydrology reconstructions from the Indo-Pacific warm pool," *Nature*, Volume 460, pp. 1113-1116, 27 August 2009. https://doi.org/10.1038/nature08233
Oceania	**Oceania 1:** Wilson, A. T. et al., "Short-term climate change and New Zealand temperatures during the last millennium," *Nature*, Volume 279, pp. 315-317, 24 May 1979. https://doi.org/10.1038/279315a0 **Oceania 2:** Cohen, T. J. et al., "A pluvial episode identified in arid Australia during the medieval climatic anomaly," *Quaternary Science Reviews,* Volume 56, pp. 167-71, 21 November 2012. https://doi.org/10.1016/j.quascirev.2012.09.021
North America	**North America 1:** Clegg, B. F. et al., "Six millennia of summer temperature variation based on midge analysis of lake sediments from Alaska," *Quaternary Science Reviews,* Volume 29, Issues 23-24, pp. 3308-3316, November 2010. https://doi.org/10.1016/j.quascirev.2010.08.001
South America	**South America 1:** CO2 Science (2011) Medieval Warm Period (South America) – Summary. http://www.co2science.org/subject/m/summaries/mwpsoutham.php

	South America 2: Bird, B. W. et al., "A 2,300-year-long annually resolved record of the South American summer monsoon from the Peruvian Andes," *PNAS*, Volume 108, Issue 21, pp. 8583-8588, 9 May 2011. https://doi.org/10.1073/pnas.1003719108
	South America 3: Ledru, M.-P. et al., "The medieval climate anomaly and the Little Ice Age in the Eastern Ecuadorian Andes," *Climate of the Past Discussion*, Volume 9, Issue 1, pp. 307-321, 5 February 2013. https://doi.org/10.5194/cp-9-307-2013
	South America 4: Fletcher, M. S. and Moreno, P.I., "Vegetation, climate and fire regime changes in the Andean region of southern Chile (38°S) covaried with centennial-scale climate anomalies in the tropical Pacific over the last 1500 years," *Quaternary Science Reviews*, Volume 46, pp. 46-56, 16 July 2012. https://doi.org/10.1016/j.quascirev.2012.04.016
	South America 5: Neukom, R. J. et al., "Multiproxy summer and winter surface air temperature field reconstructions for southern South America covering the past centuries." *Climate Dynamics*, Volume 37, pp. 35-51, 28 March 2010. https://doi.org/10.1007/s00382-010-0793-3

Paleoclimatologist and Distinguished Professor at Rutgers University, Yair Rosenthal, published a reconstruction of temperatures in the Pacific Ocean in Indonesia with Braddock Linsley and Delia Oppo. They analyzed ocean sediment core samples. Based upon the ratio of Oxygen-16 to Oxygen-18 in these samples, Rosenthal et al. reconstructed temperatures of the Northern and Southern Pacific and Indian Oceans going back 10,000 years. The Little Ice Age, the Medieval Warm Period, the Roman Warm Period, and the Holocene Climatic Optimum are seen clearly in this data (see Figure 35). They are consistent with glaciation and tree line records.[187] They found temperatures during the Holocene Climatic Optimum, 6,000 to 7,000 years ago in the Pacific Ocean, to be 2°C warmer than today.[188] In Figure 35, it is instructive to note the decline in temperature over the past 9,000 years as the Milankovitch cycle of obliquity (the axial tilt of

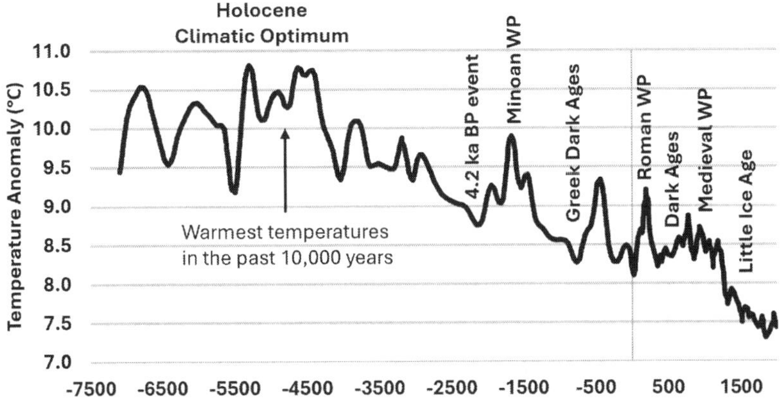

Figure 35 – Rosenthal et al. Holocene Temperature Reconstruction of the Northern and Southern Pacific and Indian Oceans. *A temperature reconstruction using ocean sediment samples from Indonesia show the climate swings of the Little Ice Age (LIA), the Medieval Warm Period (MWP), the Roman Warm Period (RWP), and the Holocene Climatic Optimum. Clearly these climate periods were not limited to Europe and Greenland as claimed by the IPCC, Michael Mann, and other climate alarmists. The temperature reconstructions above include the Northern and Tropical Pacific, and portions of the Indian and Southern Oceans. Note the downward trend in temperature since the Holocene Climatic Optimum.* Source: Rosenthal, Yair et al., "Pacific Ocean Heat Content During the Past 10,000 Years," *Science,* Volume 342, Number 617, 1 November 2013. https://doi.org/10.1126/science.1240837.

the Earth) has been declining from 24.2 degrees in about 7000 BC to 23.5 degrees today (see Chapter 7 for additional details).

For additional evidence on the global nature of the Medieval Warm Period and the Little Ice Age, see "Geologic Evidence of Recurring Climate Cycles and Their Implications for Cause of Global Climate Change—the Past is the Key to the Future," by Don J. Easterbrook from the Department of Geology, Western Washington University.[189] Also see "More evidence that the Medieval Warm Period was global," by Anthony Watts. Watts maps climate changes during the Medieval Warm Period as cited in more than 1,000 temperature and moisture proxy studies.[190] These maps show that temperatures during the Medieval Warm Period were primarily in the Northern Hemisphere, as they are today. However, a number of temperature proxies show warming also in the Southern Hemisphere during the Medieval Warm Period.

In Chapter 6 we will delve deeply into the history of climate change. A study of history shows us the reality and consequences of past climate cycles. We will explore climate catastrophes of the past to better understand the implications of current warming.

Chapter 6:
Civilizations and Climate Change

Warming Is Good, Cold Is Bad

Norman Cousins is reported to have said, "History is a vast early warning system." It is instructive to study the history of climate to glean lessons for the future. Is warmth really a danger, or is cold the real menace? History clearly teaches us that civilizations prosper in warm ages and decline in cold times. This should come as no surprise, as crop yields go up in warmer, humid climates and down in colder, arid ones. In addition, warm periods produce more land for crop cultivation at higher latitudes and higher elevations. Greater food production has repeatedly gone hand-in-hand with prosperity, growing populations, fewer episodes of societal unrest, and less malnutrition and disease.

The benefits of warmth are seen in the prosperity of the Holocene Climatic Optimum, Bronze Age Optimum (Minoan), Roman Age Optimum, Medieval Climate Optimum, and Modern warm periods. Depopulation and collapse of civilizations of the Bronze Age during the cold Greek Dark Ages, the decline of civilization in the

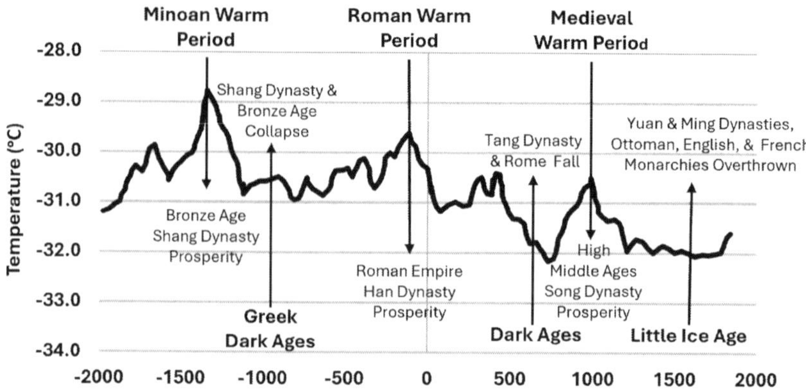

Figure 36 - Temperature Impacts the Rise and Fall of Civilizations. *Civilizations have thrived during warm times and declined during cold times. The chart above correlates estimated Greenland temperatures from ice core data to the rise and fall of major civilizations.* Source: (Temperature) Humlum, Ole et al., "Identifying natural contributions to late Holocene climate change." *Global and Planetary Change*, Volume 79, Issue 1, pp. 145-156, 9 September 2011. https://doi.org/10.1016/j.gloplacha.2011.09.005.

cold Dark Ages, and the massive starvation of the Little Ice Age teach us the detriment of cold climates. **There is a reason historians call these warm climate cycles *Optimums*.**

The Paris Climate Agreement seeks to limit a temperature increase from the end of the Little Ice Age in 1850 to just 1.5°C. Adopted in December 2015, the limit of 1.5°C in warming was arrived at primarily from unfounded concerns of sea level rise threats to island nations. The delegates at the Paris Agreement meeting initially felt a limit of 2°C should be the goal, but a consortium known as the Alliance of Small Island States (AOSIS) pushed to lower the number to 1.5°C due to concerns about sea level rise. These island nations— including the Maldives, Marshall Islands, and Tuvalu—feared they faced an existential threat due to sea level rise. In reality, temperature has increased by 1.1°C between 1850/1900 to 2021, and none of these island nations face existential threats—88.6% of these islands have seen land areas stable or increasing (see Chapter 3).[1] Fear mongering led to the goal of 1.5°C, yet this target persists to this day *despite* growing evidence that the sea level rise concerns were greatly exaggerated. **Key drivers of this goal—Tuvalu, the Marshall Islands, and the Maldives—have seen their net land area increase in recent years.**[2] Temperatures have already risen by 1.1°C since 1850/1900, so it is difficult to see how the mild warming of an additional 0.4°C will result in calamity. Lessons of history teach us that civilizations prosper in warmer times, and it is far better for humankind to warm by 1.5°C or more out of the frigid temperatures of the end of the Little Ice Age in 1850. To assume a return to the cold temperatures of the Little Ice Age is irrational and ignores the lessons of history.

In his book *A Cultural History of Climate*, Saarland University (Germany) professor and climate historian Wolfgang Behringer chronicles civilizations and how the climate impacted their rise and fall. Citing archeological, paleoclimate, and historical records, Behringer documents the rise of advanced civilizations of the Bronze Age, Roman Empire, and Medieval High Period during warm temperatures with lush growing conditions.[3] These are all times of great prosperity and growing populations. He also details the collapse of societies during the cold Greek Dark Ages, the cold Dark Ages, and the Little Ice Age.

A common theme is that cold climates lead to harvest failures, and this results in famines, increased disease due to malnutrition, migrations of peoples to warmer regions, wars, rebellions, and declining populations. A study by University College London professor and archaeologist Andrew Bevan and coauthors looks at the population of Britain and Ireland in relation to historical climate change. In their study, they "demonstrate multiple instances of human population downturn over the Holocene (9700 BC to today) that coincide with periodic episodes of reduced solar activity and climate reorganization as well as societal responses in terms of altered food-procurement strategies."[4] **The real climate crisis times have repeatedly been cold periods. Warm periods, including today, have always been times of prosperity**.

Climate is, of course, only one factor impacting the prosperity of nations. The Renaissance and American Constitution occurred during the Little Ice Age. Unleashing the power of fossil fuels and machines and the spread of capitalism and liberty during the Industrial Revolution has had a far greater impact on prosperity than climate. As mechanization and science have improved agricultural productivity, climate impacts on civilization have been less pronounced. Technological advances help us more easily adapt to a changing climate. Nevertheless, **a survey of history teaches us that warmer climates have positive benefits and colder climates have severe negative impacts**. A paper by UC Santa Barbara professor and anthropologist Douglas Kennett and head of the Potsdam Institute for Climate Impact Research (Germany), Norbert Marwan, states, "Climate change on decadal and centennial time scales is now recognized to play a role in the rise and fall of some preindustrial societies. Persistently warm centuries in Europe in the last millennium fostered geometric population growth and the aggregation of populations, whereas cooling resulted in declining population, the spread of disease, political instability, and war."[5]

History teaches us there is a remarkable correlation between cool and arid climates and the fall of civilizations. Although the fall of many of these civilizations may stem primarily from political and military causes, a colder, more arid climate may have indirectly contributed to their downfall. A cold and dry climate places stress on societies. **Cold and dry conditions lead to crop failure, famine, disease, population decline, social instability, and climate-forced migrations**. The economic impact on agrarian societies is vast, resulting in economic decline and lower tax collections. Such impacts lead to social unrest, rebellion, invasion from climate migrants, and weakened governments. Thus, colder climates have indirectly contributed to the vulnerability of civilizations to decline and overthrow.

Holocene Climatic Optimum

The Holocene Climatic Optimum (HCO) from about 7000 BC to 3000 BC was a period of a relatively warm and stable climate, especially in the Northern Hemisphere. Evidence from $\Delta_{14}C$ and ^{10}Be isotopes infer a notable peak in solar irradiance and reduced cosmic ray flux around 5000 BC to 4000 BC.[6] Known as the Milankovitch obliquity cycle, the tilt of the Earth varies from 22.1° to 24.5° every 41,000 years, and when the tilt is at its highest angle, the Northern Hemisphere receives more sunshine in the summer, melting ice sheets and providing general warming. The Milankovitch obliquity cycle was at its peak of 24° in about 7000 BC (see Figure 38 in Chapter 7).[7] The combination of high obliquity, low cosmic ray flux, and higher solar irradiance resulted in warmer temperatures.

Several temperature reconstructions of the HCO have temperatures 1°C to 2°C warmer than today.[8] This is particularly true in high northern latitudes such as Greenland and Scandinavia. The warming in polar regions and melting ice sheets raised global temperatures. Speleothem (cave stalactites and stalagmites) and lake sediment records also confirm wetter conditions during the HCO.[9] Analysis of ancient pollen shows a northward shift in vegetational zones during the HCO, indicating warmer temperatures. Pollen samples from northern latitudes suggest summer temperatures 1°C to 2°C higher than today.[10] A paleoclimate study by Fengling Yu at Durham University (and now a professor at Xiamen University in China) reconstructs historical precipitation in China from 0 CE to 4500 BC. The wettest time in China during this period was about 4400 BC.[11]

The warmer and wetter conditions of the HCO supported the development of agriculture in early civilizations in the Fertile Crescent and Yellow River Valley. The warm, wet, and stable climate of the HCO allowed for the domestication of plants and animals which supported the emergence of complex societies. This created an environment conducive for transitioning human societies from nomadic hunter-gatherer lifestyles to established agricultural communities. Warmer conditions extended growing seasons and improved crop yields. The surplus food production supported the development of early urban centers like Catalhöyük (modern-day Turkey) and Jericho (West Bank). These settlements emerged around 7000 BC to 6000 BC. Archaeological evidence from Mesopotamia shows that cities like Urik developed as agricultural productivity increased during the HCO's stable climate of 5000 BC. In northern China, the Yangshao culture thrived in the Yellow River Valley around 5000 BC due to favorable climatic conditions.

In the Fertile Crescent, the increased rain during the HCO resulted in abundant growth of wild cereals such as wheat and barley. University of Wisconsin professor and climatologist John Kutzbach and coauthor estimated the warm climate-induced water vapor flux and precipitation increased by 25% in North Africa in this period.[12] Known as the "Cradle of Civilization," this region in the Middle East saw the emergence of some of the earliest human civilizations. The area encompasses parts of modern-day Iraq, Syria, Turkey, Iran, Egypt, Jordan, Lebanon, Israel, and Palestine. Sumer is known as the world's first advanced civilization and dates to about 5000 BC. Key cities in Sumer included Uruk, Ur, Eridu, Lagash, and Nippur. These cities rose as the population mastered the cultivation of cereals, and the abundant food supply allowed populations to grow and agrarian city-states to emerge. Uruk (present-day southern Iraq) is notable for its monumental mud-brick architectural construction. The favorable climate conditions of the HCO were instrumental in the rise of civilizations in the Fertile Crescent. These civilizations and their innovations became the forerunner of city-states of the subsequent Bronze Age.

The increased rainfall and warm temperatures of the HCO in northern China increased the growth of grains, particularly millet and rice. The Yanshao people cultivated millet and began to form extensive villages in about 5000 BC in the Yellow River Valley. This led to growing populations and emergence of the first civilizations in China. The abundance of food due to favorable climate conditions during the HCO enabled the Yangshao culture to flourish, leading to population growth, advancements in pottery, architecture, and social organization. Evidence from archaeological sites indicates the presence of communal buildings, social stratification, and specialized crafts. The Yangshao culture is renowned for its painted pottery, which reflects both artistic expression and technological advancement. The warm, wet, and stable conditions of the HCO enabled the Yangshao culture to emerge and thrive in Neolithic China. This laid the foundation for subsequent Chinese civilizations.

The warm climate of the HCO was the catalyst that allowed the creation of the first human civilizations in the Middle East and China. Since many of these civilizations began and prospered with temperatures which were up to 1°C to 2°C higher than today, it is irrational that the target set in the Paris Climate Agreement is to limit the temperature increase to 1.5°C since the cold period of the Little Ice Age in 1850. We have already seen 1.1°C in warming since 1850/1900, **so the real target of the Paris Agreement is to limit warming to 0.4°C from today's temperature. This is 0.6°C to 1.6° cooler than the prosperous temperatures of the HCO. History teaches us this target is foolish**, yet nations blindly follow this unjustifiable target as though our very existence depends upon it. When President Joe Biden entered office, he reentered the United States in the Paris Agreement. Ignoring lessons of history, his administration then proceeded with its climate plan, committing $391 billion to climate initiatives in the

Inflation Reduction Act, in part to provide commitments to meet the nonsensical Paris Agreement target. President Biden's Secretary of the Treasury, Janet Yellen, said that the global transition to a low-carbon economy requires $3 trillion in new capital each year through 2050.[13] According to a paper by Bjorn Lomborg, "The Paris Agreement, if fully implemented, will cost $819 billion–$1,890 billion per year in 2030, yet will reduce emissions by just 1% of what is needed to limit average global temperature rise to 1.5°C."[14] **It's not a stretch to say this would be perhaps the worst cost-to-benefit expenditure ever proposed in the history of humankind.**

Minoan Warm Period

The period from approximately 3300 BC to 1200 BC is known as the Bronze Age. During this period, many regions of the world experienced warmer and wetter climatic conditions that facilitated the rise and prosperity of civilizations. The period from 3300 BC to 3000 BC came at the end of the HCO where temperatures were 1°C to 2°C warmer than today.[15] From 3000 BC to 1200 BC, temperatures fluctuated but were generally warmer than modern times. The emergence of the early Bronze Age coincided with the relatively stable and humid climate in regions such as the Nile in Egypt, Tigris and Euphrates rivers in Iraq and Iran, the Indus in Pakistan and India, and the Yellow River valleys in Northern China. This climatic stability and abundant precipitation associated with warmth supported the development of agriculture, enabling surplus food production and the growth of complex societies.

This warm and humid climate boosted agricultural production and allowed passage from agrarian societies to advanced civilizations. People were freed from basic food production, and this advanced a division of labor. New administrative, economic, and military positions were created, including royal courts, state officials, priests, craftsmen, merchants, and soldiers. Great ancient urban civilizations began at this time in Mesopotamia, Egypt, Pakistan, India, and China. Advancements were made in metallurgy, and it was discovered that a mixture of copper, tin, arsenic, and other metals formed bronze, a hardened metal capable of holding its edge. The invention of bronze tools revolutionized laboring activities, significantly increasing the production of food.

Palynological (fossil pollen analysis) records indicate that the Early Bronze Age (~3300 BC to ~2100 BC) was generally warm and wet, supporting robust vegetation.[16] During this period, Sumerian civilization progressed in modern Iraq including the city-states of Uruk, Ur, and Lagash. The abundance of food allowed time to innovate in areas such as metallurgy, complex social structures, and legal systems. In 3200 BC, Uruk was the birthplace of cuneiform writing. The Akkadian Empire in modern Iraq emerged later in this period under Sargon of Akkad. The Old Kingdom Egypt, known for its pyramid construction in Giza, also emerged. Early Minoan civilizations in Crete and the Aegean Sea came into existence as evidenced from early palatial centers like Knossos. Pre-Mycenaean settlements rose in Greece. Elamite Civilization emerged in Iran including Susa and Anshan. In Pakistan and India, the Harappan Civilization emerged and prospered in the Indus Valley.

The climate turned colder and drier between 2200 BC and 1900 BC leading to stress on civilizations. Archaeology and paleoclimatic data from soil and sediment cores in the Habur Plains in Syria show a severe drought at 2200 BC, where precipitation in Mesopotamia dropped from 600-800 mm per year before 2500 BC to ~300 mm per year in 2200.[17] The Akkadian Empire declined in 2154 BC, potentially linked to a severe drought around 2200 BC; the drought is known as the "4.2ka BP" event.[18] The Old Kingdom Egypt declined in 2181 BC, possibly due to the 4.2 ka BP drought, although political fragmentation was a key factor. Sumerian Civilization also declined in 2100 BC. The pre-Mycenaean settlements were abandoned in 2200 BC. The

Harappan Civilization in Pakistan and India declined in 1900 BC, potentially linked to weakening monsoon rains.

The Middle Bronze Age (2100 BC to 1600 BC) started out cooler and dry, but warmed, and humidity increased after 1750 BC. With renewed warmth and wetter conditions, new Bronze Age civilizations emerged and prospered. The Babylon Civilization in modern Iraq was founded in 2000 BC. The Middle Kingdom of Egypt came into being in 2050 BC and flourished. The Minoan Civilization in Crete emerged in 2000 BC. Canaanite cities of the Levant in Israel, Lebanon, and Syria appeared in 2000 BC. Hittite cities in Turkey emerged around 1800 BC. The drier conditions early on, from 2200 to 1750, may have stressed agricultural systems, but wetter conditions later supported urban growth. In China, the Elitou culture, associated with the Xia Dynasty, developed around 2000 BC. This Elitou culture was likely driven by favorable climate and agricultural surplus. This civilization saw the rise of urban settlements and bronze castings. The warm, wet climate supported Elitou's growth of millet, and populations grew.

During the Late Bronze Age (1600 BC to 1200 BC), a highly integrated network of world powers emerged in the Middle East and Mediterranean. Economic trade and diplomatic relations developed between the New Kingdom of Egypt, the Hittite Empire in Turkey, Mycenaean Civilization in Greece, Assyrian Empire in Iraq, Minoan Civilization in Crete, and the Civilizations of the Levant in Israel, Lebanon, and Syria. From 1500 BC to 1200 BC, Beryllium-10 isotopes from ice cores and Carbon-14 isotopes from tree rings show relatively high levels of solar activity.[19] Known as the Minoan Warm Period, this was a time of warm temperatures (see Figure 36). The great civilizations of the Bronze Age appeared and prospered during the Minoan Warm Period. Historians point to the warm and stable climate of the Minoan Warm Period as an environment that fostered cultural development and trade. Prosperity peaked in this Minoan period (1500 BC to 1200 BC). During this period civilizations of Mesopotamia and the Mediterranean formed, and alliances and trading relationships also prospered. **The lack of severe weather during this warm period may have been conducive to shipping and increased trading relationships in the Mediterranean Sea.** It is well-documented that cold produces more severe weather and warmth produces less severe weather (see Chapter 3).

In China, the Shang Dynasty emerged in about 1600 BC and lasted until 1046 BC. Like their counterparts in the Middle East, the Chinese Bronze Age civilizations were periods of urban centers, bronze metallurgy, and early writing, primarily in the Yellow River and Yangtze River basins. The temperature in Northen China was warm during this period with a gradual decline in precipitation late in the Shang Dynasty.[20] A weakening monsoon (~1400 BC to 1100 BC) may have reduced crop yields, contributing to social stress and decline of the Shang Dynasty, although textual records emphasize political and military causes for the collapse.

Cold Greek Dark Ages

The sudden fall of the great Bronze Age nations of the Hittites (Turkey), Trojans and Hittites (Turkey), New Kingdom Egypt, Kassites (Babylon), Minoans (Crete), and Mycenaeans (Greece) occurred as the Earth climate cycle became cold and dry.[21] Historical records from the Hittites and New Kingdom Egypt record multi-year droughts and declining agricultural production. Many scholars have attributed the collapse of these Bronze Age civilizations to attacks by foreign invaders known as the "Sea Peoples." Some historians believe climate calamities led to migrations and wars with the Sea Peoples and eventual collapse of these civilizations.[22]

A study by Paul Sabatier University (France) paleoclimatologist David Kaniewski and coauthors reveals the relationships of climate-driven famine, seaborne-invasion by the Sea Peoples, and politico-economic collapse in the late Bronze Age.[23] George Washington

University classics professor Eric Cline has stated: "The Sea Peoples may well have been responsible for some of the destruction that occurred at the end of the Late Bronze Age, but it is much more likely that a concatenation of events, both human and natural—including climate change and drought, seismic disasters known as earthquake storms, internal rebellions, and 'systems collapse'—coalesced to create a 'perfect storm' that brought this age to an end."[24]

During the entire cold period, known as the Greek Dark Ages, hallmarks of civilization such as written records and palace building went dormant. According to Wolfgang Behringer, the "climate plunge" was first discovered in archaeological excavations with confirmation by paleobiologists. He describes conditions during this period, including snow remaining longer in the season and over large areas, glaciers growing, and tree lines moving lower, some 300 to 400 meters lower in the Alps.[25]

A paper by University of New Mexico adjunct assistant professor and anthropologist Brandon L. Drake titled "The Influence of Climatic Change on the Late Bronze Age Collapse and the Greek Dark Ages" studies radiocarbon-dated pollen in the area, and this confirms the decline in agriculture.[26] The paper also covers oxygen and carbon isotopes in Mediterranean Sea sediments and concludes: "Mediterranean Sea surface temperatures cooled rapidly during the Late Bronze Age, limiting freshwater flux into the atmosphere and thus reducing precipitation over land. These climatic changes could have affected Palatial centers that were dependent upon high levels of agricultural productivity. Declines in agricultural production would have made higher-density populations in Palatial centers unsustainable. The Greek Dark Ages that followed occurred during prolonged arid conditions that lasted until the Roman Warm Period."[27] **History repeatedly shows how it is more arid during cold times, and agricultural productivity subsequently declines.**

Around 1200 BC, there is evidence from ice cores and tree rings that indicates a decrease in solar activity.[28] Temperatures also cooled (see Figure 36). It may not be a coincidence that all of the Bronze Age civilizations in the Middle East and Mediterranean, including the Minoans in Crete, the Mycenaeans in Greece, the Hittites in Turkey, the Assyrians and Babylonians in Persia, the Cypriots in Cypress, and the Milannians and Canaanites in Israel, Palestine, Syria, and Jordan all collapsed around 1177 BC after years of cold temperatures and droughts.

The temperature also declined in Northern China during the Greek Dark Ages, approaching modern levels. Speleothem (stalactites and stalagmites) data from caves confirms increased aridity from 1000 BC in Northern China.[29] Pollen and sediment records from the Loess Plateau show reduced vegetation cover in Northern China around 1000 BC.[30] A paleoclimate study by Xiamen University (China) professor and paleoclimatologist Fengling Yu reveals that, between 4500 BC and 0 CE, the driest period in China was about 1100 BC.[31]

The cooler and drier climate after 1100 BC likely caused social unrest. Bronze inscriptions and historical texts, such as the Shijing, reference barren fields and suffering, potentially indicative of drought or crop failure. The fall of the Shang Dynasty in Northern China, traditionally dated to around 1046 BC with the Zhou conquest at the Battle of Muye. The demise of the Shang Dynasty is attributed to multiple factors, including political, military, social, and possibly environmental influences. There is evidence from paleoclimate studies suggesting that there was a change to a cold and dry climate near the end of the Shang Dynasty. A colder and drier climate could have impacted agriculture, particularly millet-based farming, which was crucial to the Shang economy. Reduced crop yields may have led to food shortages, economic strain, lower tax collections, and social instability, indirectly contributing to the dynasty's vulnerability.

Roman Warm Period

After the Greek Dark Ages, the climate once again warmed into the Roman Climatic Optimum. This warm cycle saw a resurgence of civilization and the establishment of the prosperous Roman Empire in Europe and Hun Dynasty in China. Once again, a warm climate provided prosperity to the regions. A 2020 *Nature* study by University of Perugia (Italy) paleoclimatologist Giulia Margaritelli and coauthors describes the link between the warm climate of the Roman period and the expansion of the Roman Empire.[32]

During the warm Roman Optimum, more people lived on Earth than at any time before. This level of population would only be reached again a thousand years later in the Medieval Warm Period.[33] It is estimated there were 60 million inhabitants in the Roman Empire when Augustus died in 14 CE—and 75 million 150 years later.[34] **The early Roman Empire was blessed with a warm and stable climate for several centuries. The warm climate allowed Rome to have bountiful harvests and begin to expand northward.** It has been noted that the prosperity of the Roman Empire coincided with the great prosperity in China under the Hun Dynasty. In China the Hun Dynasty saw populations grow to approximately 60 million by 2 CE.[35]

Beryllium-10 and Carbon-14 records tell us the Roman Warm Period was a time of high and stable solar intensity.[36] Warmth ensued as evidenced through Oxygen-18 temperature proxies and the writings of Roman citizens. Pliny the Elder (23 CE–79 CE) noted that beech trees which used to grow only in the lowlands had become a mountain plant.[37] The cultivation of vines and olives advanced farther north than ever before.[38] The Romans recognized that it was possible to cultivate grapes and olives, susceptible to frost, in regions where cultivation was once futile.[39] The warm and humid weather fueled the agricultural engine of the economy.[40] Pliny the Elder praised the excellence of Italian wheat, including wheat grown in the mountains.[41]

According to University of Oklahoma Professor of Classics, Kyle Harper, "The climate was the enabling background of the Roman miracle. The Roman Climatic Optimum turned the lands ruled by Rome into a giant greenhouse. If we only count the marginal land rendered susceptible to arable farming in Italy by higher temperatures, on the most conservative estimates, it could account for more than all the growth achieved between Augustus (63 BC–43 CE) and Marcus Aurelius (121 CE–180 CE)."[42] One aspect of the Roman Warm Period is that it had been anomalously humid across the Mediterranean.[43] This was similar to the prosperous, warm, and humid Minoan Warm Period of the Bronze Age. This humidity limited droughts and was conducive to agricultural productivity. Since Northern Africa was the breadbasket of the Roman Empire, the humid conditions in Northern Africa contributed to the prosperity of Rome.

The temperature also warmed in China during the Roman Warm Period. Nanjing University (China) professor and paleoclimatologist Bao Yang and coauthors used a combination of multiple paleoclimate proxy records including ice cores, tree rings, lake sediments, and historical documents and found a warm stage in China between 0 CE to 240 CE.[44] Chinese Academy of Sciences paleoclimatologist Ming Tan and coauthors analyzed a stalagmite from the Shihua Cave in northern China in detail. The layers of stalactite TS9501 revealed temperatures going back to 665 BC. They found the temperature in northern China began to climb in about 200 BC and warmed by about 1°C between ~200 BC and ~300 CE.[45] A paleoclimate study by Chinese Academy of Sciences research professor and paleoclimatologist Quansheng Ge and coauthors shows the temperature in northeast China in about 300 CE to be the highest in the past 2,000 years.[46] Lanzhou University (China) professor and paleoclimatologist Pingzhong Zhang and coauthors analyzed a stalagmite in Wanxiang Cave in northern China. They found the conditions around 200 CE to be wetter than normal.[47]

The Han Dynasty (202 BC–220 CE) is often referred to as the "Golden Age" in Chinese history, a period marked by economic prosperity, growing population, cultural achievements, and territorial expansion. The warm and wet climate of the Roman Warm Period likely played an important role in economic prosperity and technical advancements during the Han Dynasty by providing agricultural abundance. Paleoclimate studies such as those cited above indicate that the Han Dynasty prospered during a period of relatively warm and wet conditions. These conditions of extended growing seasons and increased rainfall boosted agriculture productivity for crops like millet, wheat, and rice in the Yellow River and Yangtze River basins.

Cold Dark Ages

Since the publication of Edward Gibbon's *History of the Decline and Fall of the Roman Empire*, many reasons have been given for the collapse of the Roman Empire. However, since the year 2000, there is an increasing number of papers which use paleoclimate reconstructions using carbon, oxygen, and beryllium isotopes to assess the impact of climate on the fall of Rome. These paleoclimate proxies reveal falling solar activity and temperature cooling after 250 CE, leading to droughts and mass migrations. By the fifth century, the climate grew colder and Rome's traditional granary in North Africa dried up.[48] Eugippius (c. 465-533) describes the collapse of the Roman Empire, and he constantly refers to cold, famine, and disease.[49] This is also a period when great cities fell into decline, including Ephesus, Antioch, and Palmyra, and some six hundred settlements in Arabia were abandoned.[50] By the sixth century the population of the Roman Empire declined, and only half its inhabitants were left.[51] Most settlements north of the Alps were abandoned, and pollen analysis testifies to a general decline in agriculture.[52]

This period of colder temperatures from 250 CE to the seventh century is known as the "Migration Era Pessimum." Some scholars, including Kyle Harper, believe colder temperatures and droughts drove Rome's neighbors to the northeast to move southward. The raids by the Huns and Goths were not mere raids, but migrations. The Huns defeated the Roman army, and the Goths conquered the city of Rome in 410 CE.[53] A paper on climate and the fall of the Western Roman Empire by Werner Marx, Robin Hunschild, and Lutz Bornmann of the Max Planck Institute summarizes 85 papers dealing with climate and the fall of the Roman Empire.[54]

Wolfgang Behringer mentions colder winters set in by 250 CE and lasted in the main part of Europe until the ninth century. Beryllium isotopes show a precipitous drop in solar radiation beginning in the 240s CE, and cooling followed.[55] Glaciers grew, including the Great Aletsch and Mer de Glace glaciers.[56] Tree lines fell by as much as 200 meters in Central Europe.[57] The 240s CE stand out as a period of piercing droughts in the southern area of the Mediterranean.[58]

Colder weather induced lower humidity which brought drought to Palestine and the breadbasket of the Roman Empire in Northern Africa. The two decades from 350 CE to 370 CE were the worst multidecadal drought in the last two millennia.[59] Such droughts forced the Huns and Goths to migrate and invade Rome.[60] According to Kyle Harper, "The Huns were armed climate refugees on horseback"; and, the "Goths appeared en masse." In 378 CE the Romans suffered the worst military loss in their history to the Goths, and on August 24, 410, the Goths entered the city of Rome. Although the Roman Empire in the West was sacked, the Roman Empire in the East, out of Constantinople, survived. However, the continued cooling of the climate contributed to the decline of the Eastern Roman Empire as well.

A cooling climate in late antiquity also had a relationship with epidemics. Food shortages was a corollary of disease outbreak.[61] In addition, epidemics gain momentum in the spring, but the high heat of summer can suddenly quelch an outbreak.[62] Kyle Harper points to the bubonic plague in India in the early twentieth century that was knocked down to virtually nothing during the heat of late summer.[63] The cooling in late antiquity was accordingly accompanied with multiple outbreaks of the bubonic plague that devastated the Empire.

The 530s CE through 540s CE were not only frosty, they were the coldest decades of the late Holocene.[64] In 542 CE the great Plague of Justinian struck the Empire with devastating impacts. This epidemic was the feared bubonic plague, or the Black Death. Outbreaks in the Empire occurred in 542 CE, 558 CE, 573 CE, 586 CE, 599 CE, 619 CE, 698 CE, and 747 CE. Accounts by the Bishop of Gregory of Tours (c. 538-594) tell of late frosts, mountain avalanches, harvest failures, famine, and epidemics.[65] After 500 CE, a steep decline in solar activity occurred, reaching a low in the late seventh century, as measured by beryllium isotopes. A great solar minimum centered in the late seventh century was the greatest solar minimum in 2,000 years, lower than the Maunder Minimum that occurred during the Little Ice Age.[66] Temperatures plunged and alpine glaciers reached their millennial maximum.[67] This period is known as the "Late Antique Little Ice Age."

The Late Antique Little Ice Age began with volcanic eruptions beginning in 536 CE followed by reduced solar activity. Volcanoes spew large amounts of sulfur dioxide into the atmosphere. Increased cosmic rays from low solar activity help convert this gas into sulfate aerosols (see Chapters 10 and 11). Sulfate aerosols both scatter sunlight, blocking it from reaching the Earth and nucleate clouds, which reflect more sunlight back to space. Historical statements around 536 BC seem to support volcanic activity. The Roman historian Procopius recorded in 536 CE, "During this year a most dread portent took place, for the sun gave forth its light without brightness... and it seemed exceedingly like the sun in eclipse, for the beams it shed were not clear."[68] Furthermore, ice cores from Greenland and Antarctica show substantial sulfate deposits around 534± 2 CE, confirming volcanic activity.[69] Low solar activity coupled with volcanic activity can significantly reduce global temperatures.

During the Late Antique Little Ice Age, aridification continued across North Africa, Procopius of Gaza reported. Ptolemais, a city in Cyrenaica, was almost deserted due to the scarcity of water.[70] The great city of Lepcis Magna, "which in ancient times was large and populous, had been deserted and largely buried in sand."[71] Procopius, writing in the sixth century, described scorching drought in Palestine: "The springs had become dry and salty, and Zeus no longer sent rain."[72] The desert expanded and the once fertile wine country of Gaza was put beyond the reach of even irrigation.[73] The Late Antique Little Ice Age also impacted the Americans. University of North Carolina anthropologist Joel Gunn points to the cold climate and droughts that led to the abandonment of the city Teotihuacan in Mesoamerica about 30 years after the start of the Late Antique Little Ice Age.[74]

The population in Europe fell to a low that was not reached again in any subsequent period.[75] According to Kyle Harper, "Climate change and disease exhausted the remnants of the Roman imperial order. The Rome of Pope Gregory the Great (540 CE–604 CE) may have been home to as few as 10,000 to 20,000 souls, huddled inside its walls; they would have barely filled the corner of the Colosseum."[76] In the year 784, a third of the population of Europe was said to have died. The cold climate not only impacted harvests, but the weight of pigs and cattle was lower than during the Roman Climatic Optimum.[77]

It also became cold in the Dark Ages in China, which is believed to have contributed to the fall of the Tang Dynasty (618 CE to 907 CE). Bao Yang et al. used a combination of multiple paleoclimate proxy records and historical documents and found a cool period in China from 240 CE to 800.[78] Ming Tan et al. analyzed a stalagmite from the Shihua Cave near Beijing. They found the temperature in northern China dropped by more than 1°C during the Tang Dynasty.[79] Pingzhong Zhang et al. analyzed a record from Wanxiang Cave China to reconstruct the Asian monsoon history over the past 1,810 years. They conclude: "The summer monsoon correlates with solar variability, Northern Hemisphere and Chinese temperature, Alpine glacial retreat, and Chinese cultural changes."[80] They state the monsoon was generally weak during the final decades of the Tang Dynasty.[81] They identified a period of low temperatures and weak monsoon precipitation between 850 CE to 930 CE, which they describe as the Late Tang Weak Monsoon Period. This is the same time as a major solar minimum that reached its lowest point at about 900 CE (see Figure 41 in Chapter 7).[82] Zhang et al. state that the 9th century dry period has been invoked as contributing to the decline of the Tang Dynasty.[83] GFZ Helmholtz Centre for Geosciences (Germany) paleoclimatologist Gergana Yancheva and coauthors analyzed cave stalagmites and sediments of Lake Huguang Maar to reconstruct climate in coastal southeast China. Their results and conclusions are similar to Zhang et al. They conclude that climate change and weaker summer monsoon rain contributed to the decline of the Tang Dynasty in China.[84]

Famines due to arid conditions were also common in America during the Dark Ages, and the Classical Maya Civilization collapsed between 750 CE to 900 CE. Studies cite wars, harvest failures, famine, and epidemics as significant causes for the Maya collapse.[85] Yancheva et al. conclude that climate change and weaker summer rain contributed to the decline of the Classic Maya Civilization in Central America.[86] University of Georgia geologist James W. Webster and coauthors reviewed paleoenvironmental data from a stalagmite from western Belize to analyze the droughts that impacted the Maya civilizations. They conclude that the stalagmite record indicates a series of droughts between 750 CE and 900 CE, which collectively led to the fall of the Classic Mayans.[87] University of Florida paleoclimatologist David A. Hodell and coauthors analyzed lake sediment cores from the Yucatan Peninsula, Mexico to reconstruct the climate history of the region. Oxygen isotope records revealed a recurrent pattern of drought with a periodicity of 208 years, which matches solar cycles and cosmic ray flux as measured by Carbon-14 and Beryllium-10 isotopes. They conclude that a significant component of the century-scale variability in Yucatan droughts is explained by solar forcing, which corresponds to the decline of Mayan culture.[88]

Medieval Warm Period

Following the cold Dark Ages, solar activity once again began to surge. The increase began about 1050 and peaked around 1150, but did not substantially decline until after 1300.[89] During this period the world warmed again into the Medieval Optimum. The Medieval Warm Period was a prosperous time for Europe, and paintings from this period tell the story of bountiful harvests.[90] May frosts, always a hazard for warm-loving crops, were virtually unknown between 1100 and 1300.[91] The increase in food supplies due to the warmer temperatures led to the substantial increase in populations during the High Medieval Ages of 1000 CE to 1300 CE. Populations began to expand in the ninth century, and between the years 1050 to 1300, the population of Europe increased from 46 million to 76 million.[92] This was a level never attained in any previous time. In the late 11th century England's population was about 1.4 million but rose to 5 million by the year 1300.[93] The population of France grew from 6.2 million to 17.6 million during this same time period.[94]

This was a period of great town formation. In German-speaking areas, towns grew from the few hundred that had survived the Cold Dark Ages to more than three thousand.[95] Farmland expanded even in mountainous regions such as high Alpine regions and fjords. High mountain pastures were opened as the warming allowed livestock to graze there for longer periods.[96] In China, cultivational limits for some crops were several hundred kilometers farther north during the Medieval Warm Period than they were at the end of the twentieth century, implying higher temperatures by 0.9°C to 1.0°C.[97] The growing population and abundant food supplies led to increased industrial and economic activity.

This was another time of great prosperity, as witnessed by the economic expansion and great cathedrals built in Europe and the prosperity and growing population in China. According to UC Santa Barbara Emeritus Professor of Anthropology, Brian Fagan, "The 12th and 13th centuries were golden years of architects, masons, and carpenters who moved from cathedral to cathedral." Such great cathedrals as Notre Dame in Paris, Canterbury Cathedral, Sainte-Chapelle in Paris, and Chartres Cathedral were all constructed at this time.[98]

Warmer summers and mild winters allowed small communities to grow crops on marginal soils and at higher altitudes than ever before.[99] The warmth allowed grain to be grown in high latitudes, and the countries of Scandinavia prospered. The favorable climate fostered the creation of Nordic nations that remain in existence today.[100] With the warmth, Iceland became an attractive land to settle. By the year 930, Iceland had a population of 60,000, and by the eleventh century, Iceland had a population of 80,000.[101] The warmth was so great that Vikings, led by Eric the Red, took settlers to Greenland in the year 985. Grain was grown in Greenland, and recent excavations of Osterbygd have uncovered about 450 farms.[102] The warm climate allowed the Vikings to explore farther west, and evidence from excavations have confirmed migrations to Newfoundland.

It was also warmer in China during the Medieval Warm Period. Bao Yang et al. used used a combination of multiple paleoclimate proxy records and historical documents and found the Medieval Warm Period in China between 800 CE to 1100 CE.[103] According to Quansheng Ge et al., "During the warming period of 900s–1300s, the maximum temperatures in some regions exceeded the highest level of the 20th century."[104] During this period, the Song Dynasty (960 CE to 1127 CE) flourished. The Song Dynasty is considered a peak of pre-modern Chinese prosperity. Agricultural productivity resulted in population growth from about 100 million to approximately 200 million. The Song developed a market economy with paper money, significant iron production (~125,000 tons annually), and robust maritime trade. It saw advances in moveable type printing, gunpowder, compass navigation, and Neo-Confucian philosophy. The Song expanded the Grand Canal connecting Beijing in northern China to the Hangzhou port city in southern China. This population shipped grain, salt, silk and textiles, iron, and manufactured goods.

Pingzhong Zhang et al. analyzed a record from Wanxiang Cave China to reconstruct the Asian Monsoon history over the past 1,810 years. They conclude: "The summer monsoon ... was generally strong during Europe's Medieval Warm Period. It was strong during the first several decades of the Northern Song Dynasty, a period of increased rice cultivation and dramatic population increase."[105] Zhang et al. identified a strong monsoon between 960 to 1020 CE, which they refer to as the Northern Song Strong Monsoon Period. They claim the "Northern Song Strong Monsoon Period may have contributed to the rapid increase in rice cultivation, the dramatic increase in population, and the general stability at the beginning of the Northern Song Dynasty."[106]

Zhang et. al. also point out that the Swiss record of Alpine glaciation corresponds broadly to the Medieval Warm Period and the strong monsoon precipitation in China between 950 CE and 1340 CE. They state: "These correlations demonstrate that the link between North Atlantic/European and East Asian climate, established deeper in time, continues into the late Holocene."[107] Ming Tan et. al. analyzed the chronology of a stalagmite record in Shihua Cave to reconstruct temperatures in China from 665 BC to 1985 CE. They found temperatures rose during the Medieval Warm Period and the Song Dynasty by over 1.5°C from the end of the Tang Dynasty at the end of the Dark Ages and stayed warm throughout the Song Dynasty's reign.[108]

The warm and moist climate of the Medieval Warm Period also benefited civilizations in America. A study of pollen content from a bog in the eastern Ecuadorian Andes by Institute of Evolutionary

Science of Montpellier paleoclimatologist Marie-Pierre Ledru and coauthors showed that "the Medieval Climate Anomaly interval was warm and moist."[109] French Institute of Andean Studies paleoclimatologist Alexander J. Chepstow-Lusty and coauthors point out that the rapid expansion of the Inca civilization from the highland area of Peru became the largest empire in the New World. They argue that it would not have been possible without increased crop productivity, which was linked to more favorable climatic conditions after 1100 CE, as confirmed from multi-proxy, high-resolution sediment record from Marcacocha in the heartland of the Inca Empire. In this region of Peru, the warmer climate of the Medieval Warm Period allowed the Incas to increase crop productivity by farming at higher altitudes on agricultural terraces.[110]

The Little Ice Age

After the solar maximum of the Medieval Warm Period, a series of solar minimums occurred including the Wolf Solar Minimum 1250-1340; the Spörer Solar Minimum 1400-1550; the deep Maunder Solar Minimum 1645-1715; and the smaller Dalton Solar Minimum 1810-1830.[111] Accordingly, by the fourteenth century, the warming ended, and the world was plunged into the Little Ice Age. The drop in temperatures in the Little Ice Age had devastating impacts. Crops failed, food supplies were limited, and agrarian economies collapsed. Living standards plummeted, life expectancy declined, famine and disease increased, war ensued, and societies imploded. With the cold weather of the Little Ice Age, droughts were common. A paleoclimate study of pollen deposits in a water spring in Syria by Paul Sabatier University paleoclimatologist David Kaniewski and coauthors reports: "The Little Ice Age is not only cooler, but also much drier than the Medieval Climate Anomaly and the present-day climate."[112]

The Great Famine, the largest famine ever recorded in Europe, occurred between 1315 and 1317 during the Wolf Solar Minimum.[113]

The Spörer Solar Minimum of the 1430s brought a run of exceptionally harsh winters in France leading to the well-documented food crisis from 1432 through 1439.[114] In France during the fifteenth century, the countryside was depopulated by famine, plague, and war, which led to the abandonment of as many as 3,000 villages.[115]

Historical records describe the "Years of Hunger" from 1450 to 1454 in the Aztec civilization as severe famines caused by crop failure due to poor weather. To cope with the crisis, desperate measures were taken including human sacrifice, cannibalism, and selling family members into servitude for food to address the shortages.[116] The second French Cartier expedition up the St. Lawerence River of Canada in 1535 recorded exceptionally cold weather. The whole St. Lawrence had frozen with ice up to 12 feet (3.66 meters) thick.[117] It was recorded that half of the company of Jacques Cartier died of cold. [118] The Spanish expedition of the Rio Grande basin by Coronado in 1540-1541 recorded unusually cold weather with the Rio Grand, not far from present-day Albuquerque, frozen over for more than one month, which allowed loaded horsed to cross over the top of the ice.[119]

Bao Yang et al. used a combination of multiple paleoclimate proxy records and historical documents and found the cool Little Ice Age in China from 1400 CE to 1920 CE.[120] It has been concluded from the freezing of major lakes in China during the Little Ice Age that the average temperature between 1470 and 1850 must have been one degree colder than in the late twentieth century.[121] Periods of spectacular cold were reported everywhere in China during the Little Ice Age. Although people died of cold, it was the aridity associated with cooling that brought Chinese agriculture to the point of collapse.[122] The effects of the Little Ice Age in China were many droughts and famines between the 1300s and 1700s.[123] Harvest failures in China led to increased mortality and susceptibility to disease, and population levels declined.[124] The Yuan Dynasty (1271 CE to 1368 CE) suffered from cold and dry weather that coincided with the Wolf Solar

Minimum. During this period, Northern China suffered famines, plagues, and local revolts. The starving turned to banditry that the government could not control. The Yuan Dynasty came to an end in 1368 and the emperor, Toghon Temür, fled Beijing, only to die in exile two years later. Pingzhong Zhang et al. refers to this period as the Late Yuan Weak Monsoon Period as the mid-14th century monsoon weakened with the colder temperatures. They concluded that climate played a key role in the demise of the Yuan Dynasty.[125]

After recovery from the cold induced by the Spörer Solar Minimum 1400-1550, the temperature once again began to drop. The year 1617 marked another beginning of a decline in solar activity.[126] Astronomers began to observe sunspot activity in the early 1610s after the invention of the telescope. Scientists Galileo Galilei, Thomas Harriot, and Christop Scheiner were among the first to observe and record sunspots. By the 1630s, Johannes Hevelius conducted regular observations of sunspot activity. In 1645, a dramatic reduction in sunspot activity was reported. During the 70-year period from 1645 to 1715, known as the Maunder Minimum, 50 sunspots were recorded. This compares with 40,000 to 50,000 sunspots observed in the 70-year period between 1954 and 2024.

Sunspots are a measure of the strength of the Sun. Fewer sunspots indicate a decline in solar activity including lower solar radiation and weaker solar magnetic field strength. The weaker solar magnetic field allows a higher flux of cosmic rays to enter the solar system, and cosmic rays catalyze the nucleation of low clouds, which reflect solar energy back out to space, leading to lower temperatures. The low solar activity during this period had a notable impact on the Earth's climate. Accordingly, the seventeenth century was one of the cloudiest and coldest centuries in recorded history with devastating impacts on the world's civilizations. In 1621, England was described by the Anglican bishop Robert Burton (1577–1640) as one of "endless days when dark clouds obscure the sunlight."[127]

During the Maunder Solar Minimum in the seventeenth century, there were more rebellions and civil wars than any other century. In 1640 wars and rebellions ravaged the globe. According to anthropologist B. B. Wagner: "As seen with many empires throughout the world, once a limited food supply is established, a series of calamities tend to follow."[128]

Renowned Ohio State, St. Andrews, Yale, and Cambridge professor and historian Geoffrey Parker has written a book titled *Global Crisis, War, Climate Change and Catastrophe in the Seventeenth Century.*[129] This book portrays the connection between the worldwide turmoil of the mid-seventeenth century and the climate change of the Little Ice Age. Dominic Sandbrook of *The Sunday Times* describes the book as "intellectually dazzling," Lisa Jardin of *The Financial Times* calls this work "a monumental new book," and Kenneth Pomeranz of *Historically Speaking* calls the book "a landmark book for environmental history and for world history."

Parker recounts: "In Germany, a Swedish diplomat expressed alarm in 1648 at a new bout of revolts by the people against their rulers everywhere in the world, for example in France, England, German, Poland, Muscovy, and the Ottoman empire."[130] Parker continued: "Seventeenth-century China also suffered. First, a combination of droughts and disastrous harvests; rising tax demands and drastic cutbacks in government programs releasing a wave of banditry and chaos. Few areas of the world survived the mid-seventeenth century unscathed. North America and West Africa both experienced famines and savage wars. In India, drought followed by floods killed over a million people in Gujarat between 1627 and 1630. In Japan, following several poor harvests in 1637 [and] 1638, the largest rural rebellion in modern Japanese history broke out."[131]

Parker points out that these events took place against a backdrop of extreme weather events. It was intensely cold in Europe and North America, and droughts afflicted the Middle East, Africa, Mexico,

Virginia, India, Japan, and China. The years 1628 and, again in 1675, were each known as a "year without a summer" with temperatures so low that many crops never ripened.[132] In 1642, John Winthrop, Governor of Massachusetts, wrote: "The frost was so great and continual this winter that all the Bay [Boston Harbor] was frozen over, so much and so long, as the like, by the Indians' relation, had not been so these forty years... To the southward also the frost was as great and the snow as deep, and at Virginia itself the great [Chesapeake] Bay was much of it frozen over, and all of their great rivers."[133] Parker wrote: "In the northern hemisphere, nine of the 14 summers between 1666 and 1679 were either cool or exceptionally cool—harvests in western Europe ripened later in 1675 than in any other year between 1484 and 1879."[134]

Droughts were common in the Greek Dark Ages, Dark Ages, and Little Ice Age because colder climates hold less moisture in the atmosphere. Storms are also generally more severe in cold times because they are driven by an increasing Meridional Temperature Gradient, higher winds, and a steeper lapse rate, all of which increase in a cold climate (see Chapter 3). During cold times, there is less Arctic warming (Arctic Amplification) and the contrast in temperature between the Arctic and the Tropics is greater. Known as the Meridional Temperature Gradient, this stronger temperature gradient produces more eddy kinetic energy (EKE)[135] and severe storms since extreme storms are driven by EKE and form when cold polar air collides with warm most tropical air. The stronger Meridional Temperature Gradient also produces stronger winds[136] which feed moist air into storm tracks, thereby strengthening the storms. Colder periods also produce less warming at the top of the troposphere from the condensation of water vapor. Known as the Lapse Rate or Vertical Temperature Gradient, this increase in the Lapse Rate during colder times produces more Convective Available Potential Energy (CAPE).[137] CAPE drives storms. **Accordingly, droughts and severe storms were com-**

mon during the Little Ice Age, and these weather events wreaked havoc on agriculture in the seventeenth century.

Historical records confirm weather was more severe during the Little Ice Age. Climate historians Dagomar Degroot and Sam White both show how storms were more severe during the Little Ice Age. University of Helsinki professor and climate historian Sam White states, "...the mid-1500s to early 1600s actually witnessed the highest average levels of hurricane activity for at least the past 500 years. In particular, counts of all storms, storms identified as hurricanes, total shipwrecks, and the rates of shipwrecks found in the Spanish colonial records all peaked during the late 1500s."[138] Georgetown University associate professor and climate historian Dagomar Degroot discusses how "a simultaneous rise in the frequency and severity of storms" occurred during the cold period of the Maunder Solar Minimum in the Little Ice Age.[139] The British Royal Navy kept meticulous records of weather on its ships, recording hourly wind speeds and weather. An examination of these logbooks reveals a markedly enhanced gale frequency during the period from 1685 to 1700, which was during the Maunder Solar Minimum (1645-1715), one of the coldest periods of the Little Ice Age.[140] In the 1730s, it became warmer just after the end of the Maunder Solar Minimum, which coincides with a significant decline in gale activity. Studies have confirmed the enhanced storminess during the Little Ice Age in Greenland,[141] Denmark,[142] the British Isles,[143] Sweden,[144] the Netherlands,[145] Scotland,[146] northern France,[147] southwestern France,[148] western Iberia Peninsula (Portugal and western Spain)[149], the French Mediterranean coast,[150] the western North Pacific,[151] the Guangxi Province in China,[152] and the Bahamian Archipelago in the Americas.[153]

India suffered a catastrophic drought in 1628.[154] Virtually no rain fell in the Valley of Mexico in 1640, 1641, and 1642.[155] The Indonesian rice harvest failed in both 1641 and 1642.[156] Data from Indonesia and Thailand show famines and epidemics in the early to mid-seventeenth century.[157] Java experienced the longest drought recorded in the past four centuries between 1643 and 1671.[158] In Northern China, numerous Gazetteers reported drought in 1640, and the following year, the Grand Canal, which brought food to Beijing, dried up for lack of rain.[159]

In the Mediterranean, Catalonia endured a drought in the Spring of 1640 that was so severe authorities fostered a pilgrimage to a local shrine to pray for water, one of only four such occasions in the past five centuries.[160] In 1641 the Nile River fell to the lowest level ever recorded.[161] In Chile the drought in the 1630s was so severe that no taxes could be collected due to the drought.[162] In Sub-Sahara Africa, a severe drought affected both Senegambia and the Upper Niger between 1640 and 1644.[163] In the border towns of Languedoc and Roussillon between France and Spain, residents saw the longest recorded drought in 1651.[164] The Aegean Sea and Black Sea regions experienced the worst drought of the last millennium in 1659.[165] A military engineer complained that Romania had not had a single drop of rain between 1683 and 1686, lakes and rivers dried up, and he doubted that there had ever another example of such a terrible and lasting drought.[166]

In 1641, the first winter snow fell on November 28 in Tokyo, nearly the earliest date on record. The average date is February 5.[167] It was uncommonly wet and cold weather between 1638 and 1641 in Hungary, and summer frosts in those years devastated crops in Bohemia (Czech Republic).[168] In Eastern France, each grape harvest between 1640 and 1643 began a full month later than usual, and grain prices surged, indicating a poor harvest.[169] The year 1641 remains the coldest ever recorded in Scandinavia.[170] In China the winter of 1649 to 1650 seems to have been the coldest on record.[171] In the winter of 1657 to 1658 Massachusetts Bay and the Delaware River froze over.[172] That same year, the people in Vienna rode their horses over the frozen Danube River.[173] Canals in The Netherlands froze over that same winter, and the canals between Haarlem and Leiden remained frozen for 63 days.[174] In February 1658, the Swedish army, with all its artillery, marched 20 miles over the Danis Sound from Jutland to Copenhagen on the ice.[175]

In 1644, 1666, and 1667, Poland experienced frost on several summer days. It experienced 109 days with frost in 1666 compared with an average of 63 days today.[176] Proxy data shows Russian winters between 1650 and 1680 were some of the coldest on record.[177] Winters between 1650 and 1680 were the coldest spell recorded in the Yangzi and Yellow River Valleys over the last two millennia.[178] A traveler in the 1670s to Egypt recorded that no one in Egypt knew about wearing furs. But now "we have severe winters and we have started wearing fur because of the cold."[179] The wooden backs of Antonio Stradivari violins in Northen Italy display narrow growth rings, reflecting the unique succession of cold summers in the mid-seventeenth century.[180]

Cold, drought, and severe weather during the seventeenth century led to crop failures and famines. In Germany in 1622 crop yields were cut nearly in half and much of Württemberg staved.[181] In the 1640s in the Northern Hemisphere, there was a drop of about 2°C, which reduced grain harvest yields by between 30 and 50 percent.[182] In Sicily, extreme weather in the 1640s drove crop yield ratios down on some lands by 80 percent, the lowest recorded in early modern history.[183] In Scotland, the cold and wet summers of the 1640s brought disaster. In the Lammermuir Hills area, three-quarters of the farms were abandoned and in the Mull of Kintyre, four-fifths of all townships were abandoned because the grain did not ripen.[184] In 1641 and 1642, cool temperatures destroyed the rice harvest throughout South China. Perhaps 500,000 people starved to death and public order collapsed.[185]

In the seventeenth century, food accounted for up to half the total expenditure of most families, so any increase in food prices caused hardship.[186] A 30 percent reduction in the grain harvest often doubled the price of bread, while a 50 percent reduction quintupled it.[187] A famine diet usually lacks protein and vitamins, resulting in a stunting of the long bones in the legs and arms of children. Human remains from the Little Ice Age show this unmistakable evidence.[188] The French Army kept records of the height of each soldier. Soldiers born in the second half of the seventeenth century were on average about one inch shorter than those born after 1700.[189] Those soldiers born in 1675, the year without a summer, or in the years of cold and famine of the early 1690s, were on average only 63 inches tall (5-foot-3), the lowest ever recorded.[190] The weakened condition cause by malnutrition led to the rapid spread of disease. Increased fatalities from famine and disease coupled with the growing cost of food often lead to public unrest and rebellion.

The prolonged imbalance between the production and consumption of food in the seventeenth century produced famine, disruption, and revolt. In many states the shortage and high price of bread caused unrest among many people.[191] Consequently, the mid-seventeenth century witnessed more civil wars than any previous or subsequent period.[192] Revolts broke out in China, Japan, the Ottoman Empire, England, the Holy Roman Empire, France, Spain, Russia, and Mexico.

Tree ring proxies for East Asia show 1643 and 1644 as the coldest years in the entire millennium between 800 and 1800 CE, and the winter monsoons brought little rain.[193] A study of Chinese glaciers suggests a late seventeenth century climate on average more than 1°C colder in the west and more than 2°C colder in the northwest than today.[194] A paleoclimate study by Ming Tan et al. using stalactite analysis shows temperatures in northern China dropped to their lowest level in 2,650 years during the Ming Dynasty (1368 to 1644).[195] Another paleoclimate study by Quansheng Ge et al. shows temperatures dropped to their lowest level in the past 2,000 years in northern China between 1620 and 1710.[196] Pingzhong Zhang et al. analyzed a record from Wanxiang Cave China to reconstruct the Asian monsoon history over the past 1,810 years. They conclude: "The summer monsoon correlates with solar variability, Northern Hemisphere and Chinese temperature, Alpine glacial retreat, and Chinese cultural changes."[197] They state that the Chinese monsoon was generally strong during Europe's Medieval Warm period and weak during Europe's Little Ice Age, including the final decades of the Ming Dynasty, a time characterized by popular unrest.[198]

In China, poor harvests due to the cold and dry climate culminated in high rice prices beginning in 1620 and, according to one Machus chronicle, "Thereafter there were no years in which rice was not expensive."[199] Poor harvests in China produced starvation, migration, and disorder leading to the erosion of Ming dynasty power. Roving bandits became common, and during a prolonged drought in Shaanxi in 1628, starving farmers abandoned their land and joined the outlaws.[200] In the Guanxi and Guangdong provinces, heavy snowfall was experienced in 1633 and 1634, and cold weather in 1636—all leading to a further plunge in crop yields.[201]

Global cooling impacted Manchuria more severely than more temperate areas. Cannibalism was reported as a means of survival during severe famines in 1636, 1637, and 1639.[202] In 1640, Northern China experienced the worst drought recorded during the last five centuries, and Jiangnan experienced severe frost and heavy snow.[203] In July of that year, the Grand Canal dried up in Shandon province, cutting off the supply of rice to the imperial capital in Beijing.[204] In 1642, even subtropical Lingnan reported snow and ice, and this prevented the gathering of two annual rice harvests as was customary.[205] Public granaries were emptied, and it was impossible to replenish them.[206]

The problem with roving bandits in China intensified in the 1640s. One government official observed, "When people are suffering, they can either risk death by flouting the law and becoming bandits, or face starvation if they do not; the weak starve and the strong become bandits."[207] One town in Shandong reported that 30 percent of the population had died of starvation, with another 40 percent so poor they could only survive by joining the bandits.[208] These outlaws further weakened the power of the Ming Dynasty in Beijing. As one Chinese government official in the mid-eighteenth century said, "The Demise of the Ming was due to banditry; the rise of banditry was due to famine."[209]

With the Ming government in this weakened condition, a Qing army attacked the capital of the Ming Empire. Although defended by great walls, inside the walls food supplies were running low and disaffected defenders opened one of the outer gates. The great Ming Empire came to an end in 1645 as the Emperor hanged himself to avoid being captured, humiliated, and executed.[210] **Many scholars believe the Ming Dynasty may have remained in power had the climate not become colder.**[211]

Japan also was impacted by the Little Ice Age. Cold weather led to several famines. In 1637, food prices doubled from 1633 prices, and by 1642, they were four times higher.[212] During the winter of 1641 to 1642, the first snow fell in Tokyo six weeks earlier than usual, and the prolonged cold weather caused the Kan'ei Famine.[213] In Osaka, food became so scarce that many people died of hunger leading to food riots from crowds congregating before the house of the city governor.[214] During the Kan'ei Famine it is estimated that 50,000 to 100,000 people starved to death in Japan.[215] These famines caused 26 revolts by vassals of the governing Tokugawa Shogunate in the years 1631 to 1650. The Tokugawa government took decisive action including opening granaries and food kitchens to the people and executing granary officials who were hording grain and withholding it from the people. During this time of adversity, taxes were also reduced, from 23 percent of production in 1636 to 6 percent in 1642.[216] These prudent actions prevented the overthrow of the government.

A poor climate during the Little Ice Age did not spare the Ottoman Empire. Early in the seventeenth century, climate-related famine conditions caused widespread hardship, especially in urban areas such as the capital in Istanbul. The high cost of bread and other staples triggered discontent among the populous. Osman II, sultan of the Empire from 1618 to 1622, tried to consolidate power and disband the Janissaries, an elite Ottoman military force. He also attempted to reduce the influence of the religious leaders, known as the Ulema. The hardships of the famines made it more difficult for Osman II to implement his reforms, as the suffering population did not support his reign, and inflation and lower tax collections induced from poor harvests created instability. In 1622, Osman II was deposed and executed. Although the cause of his overthrow was more related to his conflict with the Janissaries and Ulema, the poor conditions caused by the colder weather contributed to the unrest that resulted in his overthrow.

The climate continued to deteriorate. Egypt experienced drought and famine in 1630, 1631, and 1632 when the Nile River fell below the level required to irrigate fields. Drinking water ran short in the capital city of the Ottoman Empire, Istanbul.[217] In 1632, protests in Istanbul became violent, and perhaps 20,000 perished in the disorder.[218] Although the sultan, Murad IV, survived the rebellion, the empire was weakened by the famines.

The Nile reached its lowest level of the century from 1641 to 1643 and again in 1650.[219] Torrential rains in 1646 and drought in 1647 destroyed harvest surpluses on which Istanbul depended.[220] In 1648 high food prices resulted in riots in Istanbul. Crop failures resulted in widespread famine across the empire. Starvation and malnutrition became rampant, especially in Anatolia and the Balkans. Just as the poor harvests led to exorbitant food prices and lower tax collections, the sultan, Ibrahim I, lived an excessive lifestyle and indulged in extravagant spending. The military also suffered defeats which showed his poor leadership ability. The religious leaders (Ulema) viewed the poor weather as a sign of divine displeasure. The widespread suffering caused by famines and high taxes fueled unrest among the populations. Urban riots and rural revolts became more common. The population, and military and religious leaders, all experienced growing dissatisfaction with Ibrahim's rule. In August 1648 a coalition of military leaders, religious authorities, and court officials deposed Ibrahim. Shortly after, he was executed.

Misery and decline continued for the Ottoman Empire. From 1650 to 1660, the lands around the Aegean and Black seas experienced the worst drought in a millennium.[221] In Transylvania, meager harvests caused widespread starvation.[222] The Little Ice Age seems to have been most severe in the land around the eastern Mediterranean. Most lands in the Ottoman Empire in the eastern Mediterranean suffered drought and plague in the 1640s, 1650s, and 1670s.[223] The once great Ottoman Empire declined in the late seventeenth century through a combination of climatic adversity and military defeats. It surrendered most of Hungary to the Habsburg Monarchy in Vienna through the Treaty of Karlowitz in 1699. The Habsburgs also took over Ottoman territory in the Balkans in 1718 under the Treaty of Passarowitz. In his book, *The Climate of Rebellion in the Early Modern Ottoman Empire*, author Sam White argues that climate variability – especially cold, drought, and extreme weather during the Little Ice Age of the sixteenth to eighteenth centuries interacted with political, economic, and social pressures to destabilize the Ottoman Empire.[224] He finds that climate variability played a critical and underappreciated role in exacerbating rebellion, migration, and state crisis in the Ottoman Empire.

Climate also impacted England, under the Stuart monarchy, in the seventeenth century. Severe weather ruined harvests in 1621.[225] Food prices were driven up in the late 1620s due to continued poor harvests. Unwisely, in these times of need, King Charles declared war on Spain and soon thereafter on France. These wars increased spending and diverted food supplies to feed the troops. Parliament objected to the spending on "the king's wars," and Charles retaliated by dissolving Parliament three times between 1626 and 1629. By 1630, King Charles made peace with Spain and France, but his debts to fund the war totaled four times his annual revenues, which were paid through taxing his subjects. The weather continued to deteriorate. "[The year] 1630 saw widespread harvest failure; the summer of 1632 was the coldest that any then living ever knew; spring 1633 was wet, cold, and windy; summer 1634 brought drought; and the following winter saw such very intense cold that the entire Thames froze over. Then came two summer droughts, that of 1636 so excessive that everyone declares that there is no memory of such misfortune in England."[226] Food riots and unrest became more common, and these further destabilized Charles's rule.

King Charles's plan to unify England and Scotland included policies which confiscated church and crown lands to the king, and return them back under less favorable terms. His policies also imposed distasteful religious practices on the Scottish people. His timing of these actions was unfortunate as it was a time of suffering caused by the climate of the Little Ice Age. In June 1637 the Privy Council in Edinburgh issued emergency legislation to deal with a plague epidemic and a "universal scarcity of victuals" because of the poor harvest.[227] The economic strain of crop failures intensified Scottish resistance to the reforms of King Charles. The result was a Scottish rebellion from 1637 to 1640 that forced Charles to call out 35,000 troops to quell the rebellion.

Ireland also suffered poor harvests, and the population in Ireland may have fallen in the 1640s and 1650s by one-fifth.[228] The famines and suffering in Ireland deteriorated support for the monarchy. During this time of suffering and dissatisfaction Charles implemented policies to unite England and Ireland. The Irish rebelled against Charles's unification plan, which required the king to deploy further troops in Ireland.

Meanwhile, England was suffering its own crisis. Bad weather ruined the harvests of grain and hay for five years starting in 1646.[229] The price of grain soared and was predicted to increase to "such a price as was never known in England."[230] As with other wars, the people suffered as food supplies were diverted to feed the troops. Many blamed the king for their starvation and food price inflation. They believed Charles was too focused on his wars instead of addressing the suffering of his people. The cost of the military campaigns in Scotland and Ireland strained England's economy and heightened tensions between the king and Parliament. To fund his military campaigns, Charles imposed controversial taxes without Parliament's consent.

Although King Charles was Anglican, his marriage to the Catholic princess from France gave concern to Protestants. Charles implemented religious reforms which included Catholic-like rituals that were offensive to Puritans and other Protestants. These religious reforms caused resentment to his rule and concern in the Parliament, whose members were predominately Protestant with a significant number of Puritans. His actions to ignore, dissolve, and suppress Parliament led to a conflict between royalists, those supporting the king and parliamentarians, and supporters of parliamentary sovereignty, escalating into warfare. The First English Civil War of 1642 to 1646 and the Second English Civil War of 1648 led to the capture, trial, and execution of King Charles I on January 30, 1649.

Although the execution of the king was primarily driven by political and religious conflicts, the climate of the Little Ice Age played an indirect but significant role. Cooler temperatures and extreme weather led to harvest failures, famine, disease, and economic hardship that heightened discontent among the people. The suffering from cold in England did not end with King Charles I. In January through February of 1684 the cold was severe, and the price of coal tripled. [231] "People died throughout the land, as did beasts, birds and fish. Burials were suspended because the ground was too hard to dig up. Trees split apart and plants perished. Along the coasts of England, France and Holland the sea froze, ice reportedly extending more than three miles from shore in some places. Consequently, several ships were lost and international maritime trade halted throughout northern Europe."[232]

During the seventeenth century, Central Europe was under the governance of the Holy Roman Empire, and it included 20 million people in what is now Germany, Austria, the Czech Republic, Slovenia, western Poland, and eastern France. The Empire was ruled by an elected official, Emperor Ferdinand, an Archduke of the Habsburg Dynasty in Austria. During 1618 Bohemia Protestant nobles rebelled against the Catholic Habsburg emperor over religious freedom. The Holy Roman Empire used this incident to consolidate its authority under the Catholic Habsburg Dynasty, but many Protestants resisted. Armed resistance spread, and this was to balloon into the devastating Thirty Years War. Other countries joined the fight including France, Spain, Sweden, and Denmark.

Unusually cold and wet conditions ruined the harvests in 1617 and 1618 throughout Central Europe, just as the war was beginning.[233] Poor harvests caused by the cold weather and late springs led to shortages of food for both civilians and soldiers alike. The year 1628 was known as one without a summer, and some Alpine villages experienced snowfalls every month. In many areas, neither grain nor grapes ever ripened.[234] Food shortages led to widespread hunger, driving many peasants to banditry or revolution. The situation was exasperated as soldiers drained needed food supplies away from the population. Soldiers often resorted to looting villages to survive, and this worsened the suffering of civilians.

Malnutrition contributed to plagues and diseases. The combination of war, climate-induced famine, and disease devastated the region. Between 1618 and 1648, the populations in some areas of Germany and Bohemia declined by more than half from famine, disease, and the Thirty Years War. Bohemia lost 30 percent to 40 percent of its population, and Germany and Austria lost 15 percent to 30 percent of their entire population during these years.

In the northeast area of Spain, known as the Catalans, a drought hit the area in 1640 that was so severe that the authorities declared a holiday to allow the entire population to make a pilgrimage to a local shrine to pray for rain.[235] In 1647, the weather turned very cold, and frost killed the ears of grain resulting in the worst harvest of the century. The price of bread tripled.[236] Poor harvests continued, and by April 1652, a loaf of bread cost more than a working man could earn all week.[237] In May 1652 rioters attacked houses of those suspected of hoarding grain. Rebellion spread to at least 20 towns and cities across the region including Seville, Córdoba, Granada, and Málaga.

Harvest failures continued in 1676-1683 and 1685-1688.[238] Phillip IV also kept a standing army of 10,000 men in Lombardy in northern Italy. In 1628 and 1629 torrential rains ruined two harvests, and perhaps one-third of the population died between 1628 and 1631.[239] The soldiers of Phillip IV extracted food from the local Lombardy population, and this created widespread hardship and resentment. Due to the famines and disease enhanced by malnutrition, Philip IV of Spain ended his reign in 1665 with far fewer subjects than when it started in 1621.[240]

With about 20 million subjects, France, under the Louis monarchy, was the largest country in Europe during the seventeenth century. Urban uprisings in France in the 1630s and the Fronde Revolution in France from 1648 to 1653 were the most widespread rebellions in mid-seventeenth century Europe.[241] About one million French men and women died during these revolts.[242] Between 1625 and 1631 cold and severe weather decimated crops, leading to famines and disease.

France's second largest city, Lyon, lost half its population, and hundreds of thousands died in the countryside.[243] In Normandy grain prices increased by 50 percent in 1631. With the economy destroyed by famine and disease, the tax burden was increased for those who could pay. **More than six urban revolts were reported in 1631 as a combination of famine, disease, price increases, and crushing taxes, and all this moved the people to action**. In 1637 King Louis ordered royal judges to sequester grain for French troops. This instigated the largest popular uprising in French history in southwestern France.[244]

The Fronde rebellion was a series of civil wars between the monarchy and nobles, magistrates, and common people that began in 1648. The Little Ice Age climate had a profound impact on the Fronde, exacerbating the economic and social conditions that fueled rebellion. The Fronde began by the Parliament of Paris opposing increased taxation. Nobels and urban elites joined to challenge the crown. The winter of 1648 to 1649 lasted almost six months with intense cold and severe weather.[245] In Paris, starving crowds surrounded the Palace of Justice and shouted, "Give us bread or give us peace." The monarchy in France were concerned as the Fronde rebellion occurred soon after news was received of the execution of King Charles I in London and of the Sultan Ibrahim in Istanbul.

For five consecutive years, from 1647 to 1651, agriculture was victimized by bad weather, with the most disastrous harvests being those of 1649 and 1651.[246] In 1651, the combination of plague and the highest grain prices of the century provoked nearly 70 popular revolts in Provence.[247] Another disastrous winter and spring of 1661 brought another famine, and the price of bread tripled in Paris. [248] At least 300 revolts broke out in 1661 and 1662, and cold weather and poor harvests continued through 1694. Unlike King Charles I and Sultan Ibrahim, King Louis XIV was to survive the Fronde rebellion. He took actions to ameliorate the adverse effects of climate change, buying grain in Aquitaine, Brittany and the Baltic that he had shipped to Paris. He was able to consolidate power and reigned as the "Sun King."

The Little Ice Age continued through the eighteenth century, and some historians cite continued cold temperatures as the underlying cause of food shortages and the Bread Riots in France in 1789. **The French Revolution of 1789 to 1799 was driven by multiple political, social, and economic causes. However, the cold weather and resulting bread shortages played a key role in exasperating public unrest**. The 1780s saw especially harsh winters in France, and grain crops and vineyards were destroyed. France experienced a harsh winter in 1788-1789 and a renewed drought in 1789.[249] In the bitterly cold winter of 1788-1789, heavy snowfall blocked roads, major rivers froze over, and commerce came to a standstill.[250] Hailstorms in 1788 devastated much of the French wheat harvest, leading to severe shortages in 1789. The price of bread became so high that most common people could no longer afford food to eat.

Starving peasants and urban workers began to riot and loot grain stores. Louis XVI's extravagant spending and government debt payments for costs supporting the American Revolution fostered resentment from the people as the government did nothing to provide relief from their suffering. In October 1789, women protesters, angry about food shortages and high bread prices, marched from Paris to Versailles. This march forced the king and queen to move to Paris. Bread shortages became a rallying cry of the Revolution, and Bread Riots and protests increased in France in 1789. Grain prices reached their highest level on July 14, 1789, the day of the storming of the Bastille.[251] The monarchy was overthrown by revolutionaries storming the Bastille and arresting King Louis XVI and Queen Marie Antoinette. They were tried, convicted, and executed by guillotine on January 21, 1793.

America was not spared from the extreme cold and associated droughts of the seventeenth century. Timing for the colonization of Jamestown was tragic as the years 1607 to 1612 saw the most prolonged drought registered in the region in eight centuries.[252] During the winter of 1607/1608, one of the Jamestown settlers in Virginia wrote in a letter, that the winter grew "so intense that one night the [James] river at our fort froze almost all the way across."[253] Sam White, in his book, *A Cold Welcome*, describes in detail the death toll, starvation, and misery of the Jamestown settlers from the cold climate and how they only survived because of some limited help from Native Americans and provisions delivered from England on rescue ships.[254] The winter of 1609 was particularly brutal. According to White, "... roughly 240 people must have been in Jamestown at the start of winter in late 1609. By the following summer, just 60 were left alive." [255]

Evidence suggests the Jamestown settlers turned in desperation to eating cats, dogs, rats, and mice and even cannibalism of corpses of deceased settlers.[256] "According to George Percy's account and the testimony of survivors, colonists began to dig up corpses and eat the dead," which is confirmed by forensic examination of skeletal remains.[257]

We remember Jamestown because it survived the cold, although just barely. Most have not heard of the Popham Colony, established by the English in 1607, contemporaneous with Jamestown. The colony was located at the mouth of the Kennebec River in present-day Maine. The reason Popham Colony is forgotten is that it was abandoned within a year due to an unusually cold winter and harsh storms. Over the winter, the colonists suffered from the cold severe weather. Many of the colonists died from the bitter winter conditions, including the colony's leader, George Popham. According to White, "Most contemporaries blamed the colony's failure on the extreme cold..."[258]

A permanent settlement in New England would have to wait until the Pilgrams landed at Plymouth Rock in 1620. New England settlement grew from these humble beginnings. However, poor harvests in 1637 forced starving Connecticut settlers to return to the Bay Colony in Massachusetts to find food.[259] Colder winters and cooler summers were experienced in America during the 1640s and 1660s with almost all surviving harvest records showing a dearth during these decades.[260] In the winter of 1641 to 1642, the entire Chesapeake Bay froze over, and in the winter of 1657 to 1658 the Delaware River was "frozen so hard that a deer could run over it … an extraordinary case, which the oldest Indians had never known."[261] The year without a summer, 1675, remains the second coldest recorded in North America during the last six centuries.[262]

Impacts of the Little Ice Age were devastating to civilizations throughout the world. Cold, drought, and severe weather caused famines, and famines caused starvation and malnutrition. The weakened condition caused by malnutrition leads to the rapid spread of disease. Plague, smallpox, typhus, and measles epidemics all correlate closely with harvest yields.[263] Many cities in Northern Europe suffered deaths from plague in the mid-seventeenth century. Moscow was hit particularly hard with an epidemic in 1654. In 1663 and 1664, an epidemic killed 50,000 in Amsterdam and more than 100,000 in London. Severe plague breakouts were also reported in China in 1642.[264]

In contrast to growing populations during the Medieval Warm Period, the cold Little Ice Age brought a dramatic decline of the populace in many countries. In 1656 the abbess of the convent of Port-Royal near Paris lamented that a third of the world had died.[265] The Chinese emperor asserted that during the transition from the Ming to the Qing dynasties, more than half of China's population perished.[266] In Europe, the seventeenth century was a time of famine, war, and destruction. In Germany, parts of the north, including Hessen and the Palatinate, lost two-thirds of their population between 1618 and 1648, while the population in Württemberg declined from 450,000 in 1618 to 100,000 by 1639.[267]

The population of Finland declined by one half, and Scotland's population fell by 15 percent. The life expectancy of the poor in England plunged to 30 years, and the English Civil War erupted and ended with the execution of King Charles I. Russian populations declined by 20 percent in the seventeenth century due to famines, disease, and wars. In France, the Fronde revolution (1648 to 1653) was the most widespread rebellion in all of Europe during the mid-seventeenth century. Over one million French men and women died in this period.

Disease, including the Black Death plague in 1348, ravaged the world during the beginning of the Little Ice Age, taking many lives. An evaluation of dental and skeletal remains of many victims shows indicators of malnutrition.[268] This weakened condition may have made much of the population vulnerable to dysentery, typhus, tuberculosis, and plagues. Malnutrition on a large scale during the Little Ice Age was a result of poor agricultural harvests in the cold and stormy weather of the time. Greenland was abandoned during this period as it was too cold for inhabitants to survive.

Just as uninformed and misinformed climate alarmists of today blame humankind for climate change, bad weather during the Little Ice Age was deemed to be caused by man. The anthropogenic cause of severe weather events during the Little Ice Age was attributed first to the Jews, then to witchcraft, and finally to rulers. Witches were directly blamed for weather and were the scapegoats people needed to explain the disastrous conditions.[269] During this period, 50,000 victims accused as witches were prosecuted legally through the courts and executed to control what was seen as anthropogenic climate change.[270] In Southern Germany, a hailstorm in May 1626 followed by cold temperatures led to the arrest, torture, conviction, and execution of 900 men and women suspected of producing the adverse weather through witchcraft.[271] According to historian Sam White, "Accusations of weather magic, such as conjuring hailstorms and crop blights, were among the most common features of European witchcraft trials."[272]

In 1484, the book *Malleus Maleficarum* was published which became the "*consensus*" text outlining the powers and proper punishments of witches. **Just as so-called "*settled science*" today disseminates the myth that fossil fuel companies are totally to blame for climate change, so the *Malleus Maleficarum* provided the authoritative basis for it to be "*settled*" that witches were the cause of climate change during the Little Ice Age.** In the book's introduction Pope Innocent VIII recognizes the power of witches in the destruction of crops.[273] The book contained a chapter detailing the power of witches to control weather. To ensure everyone knew that the cause of severe weather was "*settled*" the chapter ends, "Therefore it is reasonable to conclude that, just as easily as they [witches] raise hailstorms, so can they cause lightning and storms at sea; and so, *no doubt at all* remains on these points." [274]

Witch hunts during the Little Ice Age were conducted with a religious zeal that characterizes the extreme climate activists' movement of today. Fanatics of today once again discount natural causes of weather and climate change, blame humankind, spread fear, and push detrimental policies. Rather than consider the contribution of heating of the oceans by lower cloud cover during the current Modern Solar Maximum, climate crisis fanatics of today blame fossil fuels for all of the recent warming and seek remedy through the courts for damages from climate change. **The cities of San Francisco, New York, and Baltimore have all filed lawsuits seeking damages from fossil fuel companies for costs relating to severe weather and sea level rise.**[275]

If the persecution of Jews and witches failed to feed the hungry, the crowds blamed their governments.[276] The pain and suffering caused by the Little Ice Age eventually led to popular uprisings against rulers. The total number of food riots in England rose from 12 from 1600 to 1620, to 36 between 1621 and 1631, and 14 more between 1647 and 1649.[277] In France, popular revolts peaked in the middle of the decades of the seventeenth century, and in Russia, urban rebellions in 1648 and 1649 shook the government to its foundations.[278] Japan had at least 40 rural uprisings between 1590 and 1642, and in China, the number of major armed uprisings rose to more than 70 in the 1620s and more than 80 in the 1630s, with over one million people taking part in revolts against the Ming Dynasty.[279] By the end of the Little Ice Age, Emperors of the Yuan and Ming Dynasty in China, two Sultans of the Ottoman Empire, and monarchs in England and France were all removed from power and met untimely deaths.

Modern Warming

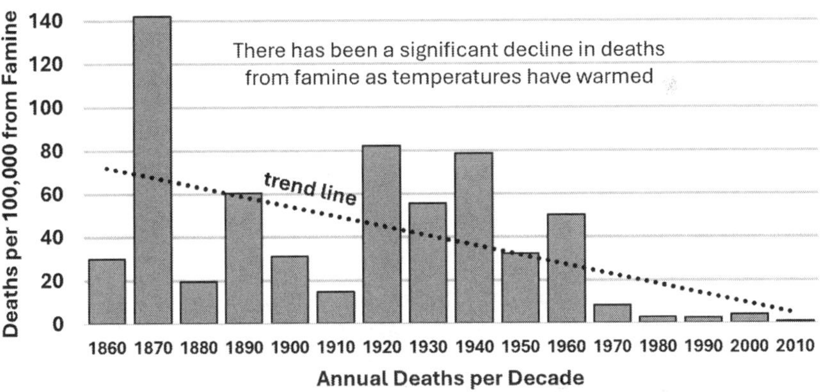

Figure 37 – Famine Deaths Are Declining as Temperatures Rise. *Since the 1860s, temperatures have increased, and the trend line of deaths from famines have dropped more than 100-fold from 70 per 100,000 to 0.5 per 100,000.* Source: *Our World in Data*, "Famines," https://ourworldindata.org/famines.

Beginning in the early years of the twentieth century, solar activity increased as we entered the Modern Solar Maximum (see Figure 39 in Chapter 7). The Little Ice Age ended in the nineteenth century, and the world has once again moved into a period of warmth and prosperity. The lessons of history teach us that warmth expands agricultural productivity and reduces the adverse toll on humans from famines and disease. Cold is the enemy of humankind as it has always brought droughts, famines, epidemics, misery, and civilization decline. Famines often lead to political unrest, wars, disease, declining populations, and economic hardship. Recent modern warming has led to a dramatic decline in deaths from famines.

Since the 1970s, we have been in a period of warming and declining deaths from famines. In the 1960s, 50 in 100,000 people in the world died of famine. After 2010, this number fell by 100-fold to less than 0.5 in 100,000, despite the growing world population (see Figure 37). History repeats itself; once again warmth has proven to be good for humanity, and we should welcome warming and not fear it. The tremendous growth in agriculture is largely due to mechanization powered by fossil fuels and nitrogen fertilizers produced from the Haber-Bosch process, that would have not be possible without the use of natural gas as the primary feedstock. The contribution of warmth plus increased CO_2 fertilization represents 78% of the enhanced greening of the Earth, which has also contributed to unprecedented agricultural productivity in the Modern Warming Period[280].

History teaches us to expect climate cycles with alternating warm and cool periods of approximately 500 years each in duration. These warm and cold cycles coincide with the solar maximums and solar minimums of Eddy solar cycles. Warmth has always been good for humanity, and the current Modern Warm Period, which is at the height of the warm period of the 1,000-year Eddy Cycle, has witnessed great prosperity, just as the world experienced in the Minoan, Roman, and Medieval warm periods. Modern warming should be celebrated, not feared.

In this chapter, we have discussed abundant harvests and their impact on population growth during the Roman Warm Period and Medieval Warm Period. Overall, European population tripled during the Medieval Warm period between the years 1000 to 1348.[281] Population growth during Modern Warming has been even greater. The benefits of warming coupled with fossil fuel powered mechanization of agriculture, nitrogen fertilizers, modern medicine, and improved sanitary conditions has led to the greatest population boom in history. In 1850, the global population stood at an estimated 1.2 billion.[282] By 2025, the world's population is assumed to be 8.2 billion.[283] This is nearly a seven-fold increase in population during Modern Warming. None of this growth would have been possible without modern advances in agricultural productivity. It is undeniable that warming and CO_2 fertilization has been a part of this agricultural miracle.

As George Santayana said, "Those who cannot remember the past are condemned to repeat it." There are several important, undeniable lessons history teaches us about climate. Cooler climates have always been disastrous for humankind, and warmer climates have always been hugely beneficial. There is a stark contrast between the prosperous Minoan, Roman, and Medieval Warm periods and the starvation, disease, and misery of the Greek Dark Ages, the Dark Ages, and the Little Ice Age. **Speaking about climate alarmism from global warming, climatologist Judith Curry has said, "The weakest part of the argument has always been that warming is dangerous."**[284] History teaches the blessings of warmth and the curse of colder climates. A desire to go back to the cold climate of the Little Ice Age shows a complete ignorance of history. Yet many climate alarmists use the end of the Little Ice Age in 1850 as a starting point when climate change began to lead to what they term "*disastrous*" warming. The Paris Climate Agreement has a goal to limit warming to 1.5°C from the temperature of 1850. However, anyone who studies history will know that this is an irrational and counterproductive goal as warmth has always been a blessing to humankind and civilizations.

We learn from history that climate change is cyclical. In addition, major historical climate cycles have always coincided with Milankovitch cycles, solar maxima, and solar minima. In Chapter 7 we will explore scientific evidence regarding the impact of the Sun and solar cycles on climate.

Chapter 7:
Significance of the Sun

Milankovitch Cycles

Because 99.9% of the heat received by the Earth is from the Sun, even small changes in solar energy can have major impacts on the climate. Climate reconstructions across the past 400,000 years have been undertaken from Antarctica ice cores (see Figure 51 in Chapter 8). These reconstructions show climate cycles every 100,000 years, 41,000 years, and 23,000 years. Serbian mathematician Milutin Milankovitch calculated that these climate cycles are due to changes in solar energy based upon the distance and angle of the Earth from the Sun. More specifically, these climate cycles occur because of changes in the Earth's orbital shape, axial tilt, and rotational progression in the seasonal orientation of the Earth's axis.

Eccentricity: Over 100,000 years, the Earth's orbital shape around the Sun cycles from one of nearly circular to more elliptical to less elliptical. Gravitational interactions with other planets, primarily Jupiter and Saturn, cause the orbit to stretch and contract. The degree of eccentricity value ranges from near 0 (perfect circle) to about 0.06 (more elliptical). When the orbit is more elliptical, the Earth approaches closer to and farther from the Sun than in a circular orbit.

During a circular orbit, the Earth is approximately (~) 149.6 million kilometers (km) from the Sun. During an elliptical orbit, the Earth is closest to the Sun in its orbit during perihelion (~147 million km) and farthest from the Sun in its orbit during aphelion (~152 million km). If the Earth is in perihelion during summer, the increased solar radiation from a more elliptical orbit results in hotter summers, and these melt ice sheets. Growing or shrinking ice sheets can have further impacts on the climate from the increase or decrease in reflectivity—or its albedo—and the reflection of shortwave solar radiation back out to space. They may also influence ocean and atmospheric circulation, which can modify climate patterns. The Northern Hemisphere dominates eccentricity effects because its large landmasses support extensive ice sheets. Precession, as described below, will determine when the tilt of the Earth's axis will result in the Northern Hemisphere being closest to the Sun (perihelion) in summer.

Obliquity: The tilt of the Earth changes every 41,000 years from 22.1 degrees to 24.5 degrees. Known as obliquity, this cycle is the angle of the Earth's rotational axis relative to a vertical orbital plane. Gravitational forces from the Sun, Moon, and other planets cause the tilt to oscillate. A higher tilt increases the contrast between summer and winter temperatures, particularly at high latitudes, which amplifies seasonal temperature differences. A greater tilt causes solar radiation at the poles to be higher during summer, and this melts ice cover. Weaker tilt weakens seasonality and can reduce summer ice sheet melt, thus promoting ice accumulation. The impact on growing or melting ice sheets to the Earth's albedo, and ocean and atmospheric currents, impact climate.

Precession: The third cycle impacting the climate, known as precession, is a rotation of the Earth's axis in a circular motion, which changes relative to the Earth's position in its orbit. Precession changes the seasonal timing of when the Earth is in perihelion (closest to the Sun) and aphelion (farthest from the Sun). For example, in our current day, perihelion occurs in January on all calendars, so the Southern Hemisphere is closest to the Sun in summer, and the Northern Hemisphere is closest to the Sun in winter. Precession cycles every 23,000 years, but this oscillation can vary from 19,000 years to 29,000 years. Therefore, in about 11,500 years, perihelion will occur in July, making the Southern Hemisphere closest to the Sun in winter and the Northern Hemisphere closest to the Sun in summer. Gravitational influences of the Sun, Moon, and planets drive precession. Precession amplifies the impact of eccentricity and obliquity. Because the impact of the Milankovitch cycles relates to summer ice melt and its impacts on climate, and the Northern Hemisphere has the most land mass to support ice sheets, precession is key to the climate impact of both eccentricity and obliquity.

We are currently in a near circular orbit (~0.017 eccentricity). This compares with eccentricity of ~0.06 during an elliptical orbit. We are in that part of the eccentricity cycle where the orbit of the Earth is getting more circular and will approach minimum eccentricity in 27,000 to 30,000 years from now. In 80,000 to 100,000 years from now eccentricity will be high again, and if precession aligns summer in the Northern Hemisphere with aphelion (farthest from the Sun) in this period, reduced summer temperatures could trigger ice accumulation and initiate glaciation.

Obliquity peaked at ~24.5 degrees about 9,000 years ago. This cycle contributed to summer warming, melting of the ice sheets, and the end of the last glaciation, which had peaked around 20,000 years ago. The delayed inertia of warming and ice melt caused the maximum temperatures to occur around 6,000 to 7,000 years ago. Obliquity will force continued cooling of the Earth as the next obliquity minimum of ~22.1 degrees is not expected until around 13,000 to 14,000 years from now. A reconstruction of the Holocene climate reveals temperatures have been declining for the past 9,000 years due to changes in obliquity (see Figure 38). The warmest period of the Holocene, known as the Holocene Climatic Optimum, occurred between 6000 BC to 7000 BC when obliquity was 24.2 degrees. Obliquity today stands at about 23.5 degrees, and it continues to decline, suggesting we are slowly moving into a cooler climate (see Figure 38). Glaciers and other temperature reconstructions follow this same pattern (see Figures 26, 29, and 35 in Chapter 5 and Figure 50 in Chapter 8).

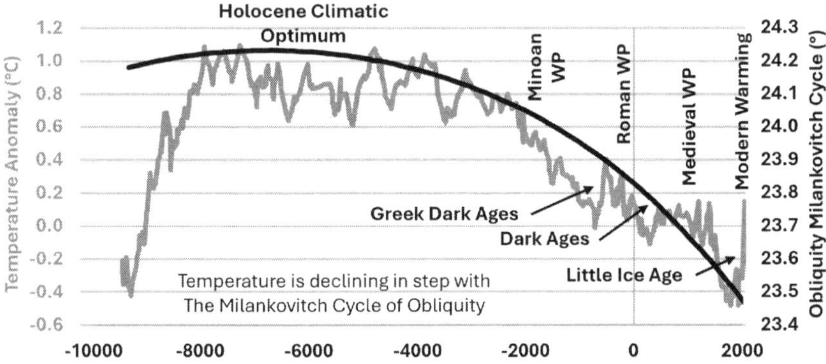

Figure 38 – Holocene Climate Reconstruction and Obliquity Cycle. *The figure above provides a visualization of the 41,000-year Milankovitch obliquity cycle. The obliquity cycle is overlaid with temperature reconstructions from Marcott et al. (2013). Temperatures have declined in step with the obliquity cycle since the Holocene Climatic Optimum around 7500 BC.* Sources: (Obliquity) Vinós, Javier, "Periodicities in solar variability and climate change: A simple model," Energy Matters, Energy, Environment and Policy, posted by Evan Mearns, 11 May 2016. https://euanmearns.com/periodicities-in-solar-variability-and-climate-change-a-simple-model/; (Temperature Anomaly) Marcott, Shaun A. et al., "A reconstruction of the regional and global temperature for the past 11,300 years," *Science*, Volume 339, Issue 6124, pp.1198-1201, March 2013. https://doi.org/10.1126/science.1228026.

The warmth of the Holocene Climatic Optimum (HCO) is confirmed with proxy reconstructions of temperature by Marcott[1] (Figure 38), Humlum[2] (Figure 26 in Chapter 5), Rosenthal[3] (Figure

35 in Chapter 5), and Kaufman[4] (Figure 50 in Chapter 8). Warm conditions of the Holocene Climatic Optimum are also established in the history of alpine glaciers, which were nearly ice-free during the Holocene Climatic Optimum but have not experienced such low ice levels since.[5] Studies also show the sea level was higher during the HCO than today.[6]

Solar Cycles

Other than Milankovitch cycles, perhaps the largest of the cycles impacting climate are solar cycles. Scientists have classified several solar cycles, ranging from 11 years to 2,500 years. The shortest solar cycle is the Schwab Solar Cycle, which averages 11 years, but it can range from 8 to 15 years per cycle. Additional solar cycles have been classified when the collective impact of a series of Schwab Solar Cycles are either strong or weak. The 87-year Gleissberg Solar Cycle has been identified and can vary from 87 to 88 years. The 208-year de Vries Solar Cycle can vary from 206 to 210 years. The 976-year Eddy Solar Cycle is one of the most constant solar periodicities reported and varies between 923 and 1,029 years.[7] Identified in 1968 by J. Roger Bray, the Bray Cycle (also known as Halstatt Solar Cycle) averages 2,500 years in length, but it can vary between 2,007 and 2,515 years in length.

When a series of consecutive strong Schwab Solar Cycles are experienced, we have a solar maximum. When consecutive weak Schwab Solar Cycles occur, that is referred to as a solar minimum (see Figure 39). Solar cycles impact both low cloud cover, stratospheric warming, and changes in solar irradiance. All of these factors impact climate. We have just come off the peak of the largest grand solar maximum in 10,000 years, known as the Modern Solar Maximum.[8]

Periods of strong solar activity are accompanied by strong solar magnetic fields, increased sunspots, higher solar irradiance, and fewer cosmic rays. Galactic cosmic rays are high energy particles that originate outside of our solar system, usually from the explosion of dying stars, known as supernovae. When the solar magnetic field is strong, the Earth is protected from cosmic rays, and fewer reach Earth. The magnetic field of the Sun primarily protects our solar system from charged particles such as protons by deflecting them. Because 85% of cosmic rays are positively charged protons, the magnetic field of the Sun deflects the path of cosmic rays away from our solar system and the Earth. Cosmic rays aid the formation of low clouds, so periods of low solar activity and high cosmic ray flux are times of more low cloud cover and colder temperatures.

Cosmic ray intensity can be estimated in ice core, tree rings, sediment, shell fossils, and stalactite samples based upon the mass spectrometry quantitation of Carbon-14 and Beryllium-10 isotopes in these samples. When cosmic rays collide with atoms in the Earth's atmosphere, they induce nuclear reactions which convert nitrogen and oxygen atoms into Carbon-14 and Beryllium-10 isotopes. When high-energy cosmic rays collide with oxygen and nitrogen nuclei in the atmosphere, they induce spallation reactions, breaking apart the nuclei into smaller fragments. In this reaction, Oxygen-16 spallation results in the creation of Beryllium-10, Lithium-6, and other particles. Nitrogen spallation from cosmic rays produces Beryllium-10 and Helium nuclei. Nitrogen spallation from cosmic rays also produces Carbon-14 and a free proton.

Once Beryllium-10 is formed, it becomes part of aerosols and eventually attaches to atmospheric particles. These particles are precipitated by rain and snow to the Earth's surface. Beryllium-10 has a half-life of 1.39 million years, so it can be used as a proxy for cosmic rays and solar cycles going back millions of years; its precision for a few thousand years, however, is not as good as Carbon-14. Carbon-14 atoms quickly oxidize to form carbon dioxide. $^{14}CO_2$ is taken up in plants during photosynthesis where they enter the food chain and become part of all living organisms. Once an organism dies, it stops absorbing Carbon-14, and the existing Carbon-14 in its tissues begins to decay. The half-life of Carbon-14 is about 5,730 years, which makes it useful for dating organic material up to about 50,000 years. This is the basis of radiocarbon dating.

Both Carbon-14 and Beryllium-10 isotopes are more abundant during periods of increased cosmic rays, so the levels of Beryllium-10 and Carbon-14 isotopes in sediment, stalactite, ice core, fossils and other samples can be used to estimate the strength of the solar cycle when the sample was deposited. We have accurate records of sunspots going back to the early 1600s, but scientists need to rely on estimated cosmic ray flux from isotope ratio analysis to infer solar activity during periods which predate the year 1600.

Temperature changes have a clear correlation to sunspot numbers, cosmic ray flux, and solar cycles. Over the past few thousand years

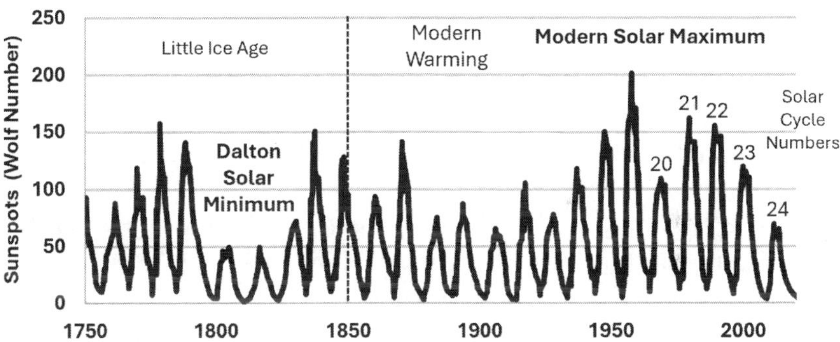

Figure 39 – Solar Cycles, 1749 to 2020. *The chart above provides solar activity based upon sunspot numbers (the Wolf Number). Schwab Solar Cycles occur about every 11 years. When consecutive strong Schwab Solar Cycles are experienced, this is a solar maximum. We are just coming out of the Modern Solar Maximum, the strongest grand solar maximum in 10,000 years. The last grand solar maximum was 1,000 years ago, during the Medieval Warm Period. When consecutive weak solar cycles occur, this is referred to as a solar minimum. The Dalton Solar Minimum was a time of three consecutive low Schwab solar cycles, low temperatures and the infamous year without a summer.* Source: AL FIN, "Global Cooling: A Return to the Age of a Frozen Thames?" 30 January 2012. https://alfin2100.blogspot.com/2012/01/global-cooling-return-to-age-of-frozen.html.

we have seen the Minoan Warm Period, which was followed by the cold Greek Dark Ages, then the Roman Warm Period, followed by the cold Dark Ages, followed by the Medieval Warm Period, followed by the Little Ice Age, and finally the current warm period. One factor remains constant: warm periods occur during grand solar maximums, and associated low cosmic ray flux and cold periods occur during grand solar minimums and associated high cosmic ray flux.

One of the most comprehensive studies of past solar cycles using $\Delta_{14}C$ and ^{10}Be isotope proxies during the Holocene is by Freidhelm

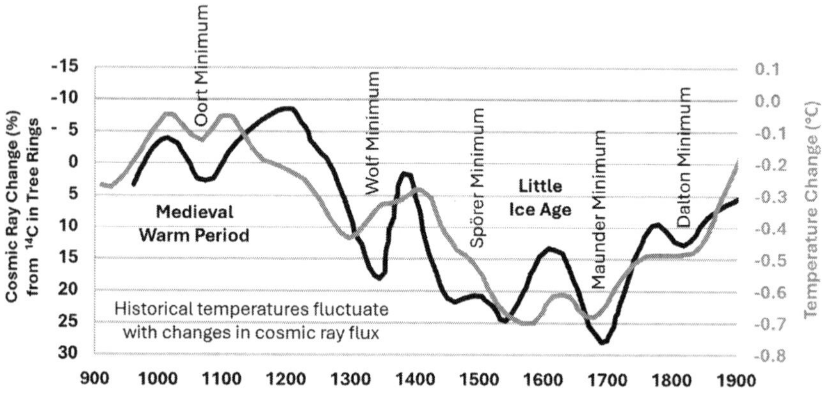

Figure 40 – Cosmic Ray Flux Matches Temperature Cycles of the Past 1,000 Years. *Warm and cool periods have followed solar maxima and minima cycles and the associated low levels of cosmic rays during a solar maximum, and elevated levels of cosmic rays during a solar minimum. The chart shows a reconstruction of temperature and cosmic rays (a proxy for solar cycles). During a strong solar cycle, the Sun's stronger magnetic field shields the Earth from cosmic rays, and fewer reach the Earth. During a weak solar cycle, more cosmic rays descend to the Earth, producing increased levels of Carbon-14. In the chart above, the cosmic ray change (%) above is inverted to show the close anti-correlation of cosmic ray flux to temperature changes.* Sources: (*Cosmic Ray Change*) Kirkby, Jasper, "Cosmic Rays and Climate," *Surveys in Geophysics*, Volume 28, pp. 333-375, 28 February 2007. https://doi.org/10.1007/s10712-008-9030-6 (*Temperature Change*) Moberg, A. et al., "Highly variable Northern Hemisphere temperature reconstructions from low-and high-resolution proxy data." *Nature*, 433 613-617 (2005). https://doi.org/10.1038/nature03265.

Steinhilber, et. al.[9] Findings from the Steinhilber paper reveal a correlation of low cosmic ray flux, due to solar maximums, during the Holocene Optimum, Minoan Warm Period, Roman Warm Period, Medieval Warm Period and Modern Warming. They also confirm high cosmic ray flux due to solar minimums during the Greek Dark Ages, Dark Ages, and Little Ice Age. Using sunspot numbers and

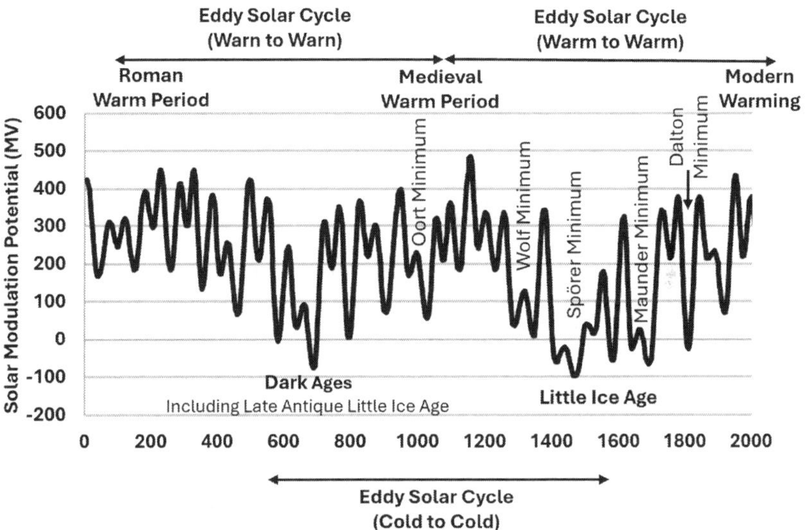

Figure 41 – Solar Activity Over the Past 2,000 Years. *Using solar isotope proxies from Steinhilber et al. provide a history of solar activity over the past 2,000 years. The data shows high solar activity during the warm climates of the Roman Warm Period, Medieval Warm Period, and Modern Warming. Solar minima occurred during the Dark Ages and Little Ice Age. Specific solar minima including the Oort Minimum, Wolf Minimum, Spörer Minimum, Maunder Minimum, and Dalton Minimum are clearly seen. The 1,000-year Eddy Solar Cycles are also evident in the graph.* Sources: Vinós, Javier, "Periodicities in solar variability and climate change: A simple model," *Energy Matters,* Energy, Environment and Policy, posted by Evan Mearns, 11 May 2016. https://euanmearns.com/periodicities-in-solar-variability-and-climate-change-a-simple-model/; See also, Steinhilber, F. et al., "9,400 years of cosmic radiation and solar activity from ice cores and tree rings." *Proceedings of the National Academy of Sciences,* Volume 109, Issue 16, pp. 5967-5971, 2 April 2012. https://doi.org/10.1073/pnas.1118965109.

solar activity reconstructions from Steinhilber et al. provides a solar activity chart over the past 2,000 years (see Figure 41).[10] By overlaying Eddy Solar Cycle periods, the impact on climate from the 1,000-year Eddy Solar Cycles is clearly seen.

The Roman Warm Period, Medieval Warm Period, and Modern Warming all occurred during high solar activity during grand solar maximums of Eddy Solar Cycles. The cold Greek Dark Ages, Dark Ages, and Little Ice Age all occur during times of low solar activity during grand solar minimums of the Eddy Cycles. The coldest of these solar minimums is during the Little Ice Age, and this aligns with the cold periods of both the 1,000-year Eddy Solar Cycle and the 2,500-year Bray Solar Cycle. Between 1680 and 1730, the coldest cycle of the Little Ice Age temperatures plummeted and the growing season in England was about five weeks shorter than during the twentieth century's warmest decades.

As shown in Figure 41, 1,000-year Eddy Solar Cycles are clearly seen in the Steinhilber et al. data. Due to shorter solar cycles such as the Gleissberg and de Vries Solar Cycles, the actual trend is jagged and not smooth like a sinusoidal curve. For example, recovery from the low solar activity of the Little Ice Age should have been underway by the 1800s, but the Dalton Solar Minimum occurred in the early 1800s, and this further drew the temperatures into decline. In the year 1816, the Dalton Solar Minimum, coupled with the eruption of the Mount Tambora Volcano, caused what was known in North America and Europe as the "Year without a Summer." According to the Eddy Solar Cycle, the next cold period should not be for a few hundred years from now. **However, just like the Dalton Solar Minimum, solar physicist Valentina Zharkova is predicting a significant solar minimum in Schwab Solar Cycle 26, which begins in about 2031.**

Examination of the historical record of solar activity reveals that the Roman Warm Period, Medieval Warm Period, and current Modern Warming are in sync with the solar maximums of the

1,000-year Eddy Solar Cycle and correlate directly with increases in temperature and decreases in cosmic rays. The cold periods of the Greek Dark Ages, Dark Ages, and the Little Ice Age also correlate with solar minimums as reconstructed from isotope ratios of Carbon and Beryllium. The coldest times during the Little Ice Age, 1645 to 1715, was a time when solar activity was low since aurora borealis events were rare, and observed sunspot activity almost ceased.[11] *No such correlation can be seen in CO_2 levels to explain these 1,000-year historical warm and cold temperature cycles.*

Solar Cycles and Total Solar Irradiance: The Sun is more active during a solar maximum and less active during a solar minimum. Because ^{10}Be flux and Δ^{14}C levels fluctuate in sync with temperature proxies such as δ^{18}O, several past climate studies assumed that past temperature cycles were primarily caused by changes in Total Solar Irradiance (TSI) between solar minimum and solar maximum.[12] Such studies assumed TSI was the cause of climate change, and that cosmic rays were a marker of TSI, but not the cause of warming or cooling. Recent research has suggested that these assumptions were premature.

Several studies have shown a temperature difference of nearly 1°C in the Northern Hemisphere between the Little Ice Age and both the Medieval Warm Period and Modern Warming.[13] Recent satellite records have given us accurate information to measure the change in solar irradiance between the peak and trough of recent 11-year Schwab Solar Cycles. Surprisingly, the total solar irradiance variations at the top of the atmosphere, as measured by satellites, is only ~0.2 W/m^2 to ~0.3 W/m^2.[14] This would imply climate sensitivity to 11-year Schwab Solar Cycle TSI is less than 0.1°C, which is one-tenth of the warming found from proxy records of the 1,000-year Eddy Solar Cycle. Because of the low-measured TSI by satellites of recent 11-year Schwab Solar Cycles, the IPCC and its climate models

(CMIP6 Global Climate Models) assume both TSI and all solar cycles are insignificant drivers of climate. In its *Sixth Assessment Report*, the IPCC states: "Changes in solar and volcanic activity are assessed to have together contributed a small change of –0.02 [–0.06 to +0.02] °C since 1750 (medium confidence)."[15]

Some controversy exists over Total Solar Irradiance (TSI) satellite measurements. There are two satellite data sets of TSI including the Active Cavity Radiometer Irradiance Monitor (ACRIM) and the Physikalisch-Meteorologishes Observatorium Dovos (PMOD). These two datasets showed a multidecadal TSI divergent trend throughout solar cycles 21 to 23. The PMOD TSI satellite composite shows a slight TSI decrease, while the ACRIM shows a moderate increase from 1980 to 2000.[16] The IPCC climate models (CMIP6 Global Climate Models) are derived from scientific literature that have low TSI impact, implying a preference for the PMOD composite index that shows a recent decline in TSI.[17] However, both of these satellite datasets show low TSI variance of only 0.07% (or ~0.2 W/m^2 to ~0.3 W/m^2) over recent 11-year Schwab Solar Cycles. The IPCC would appear to be correct in discounting TSI during 11-year Schwab Solar Cycles as a driver of recent climate. However, **evidence, to be covered later in this chapter, reveals that it may not be appropriate to extrapolate 11-year Schwab cycle TSI to longer solar cycles.** Furthermore, solar cycles impact more climate drivers than just TSI. Therefore, discounting the impact of solar cycles on climate by the IPCC seems presumptive and contrary to historical and paleoclimate evidence.

Not all scientists agree that the current satellite record of recent 11-year Schwab Solar Cycles can be extrapolated to past solar minimum and maximum that extend over decadal, centennial, or millennial periods. The minima and maxima of decadal Gleissberg Solar Cycles, centennial Vries and Eddy Solar Cycles, and millennial Bray Solar Cycles are a consecutive series of extreme 11-year Schwab Solar Cycles which would have larger variances than individual Schwab

Solar Cycles. Astrophysicist, Professor Adjunct at ETH Zürich, and former Director for the Physical-Meteorological Observatory Davos/ World Radiation Center, Werner K. Schmutz, argues that, over centennial time scales, the Sun might vary its radiance significantly more than during the last 40 years of space TSI measurements.[18] National Institute for Space Research (Brazil) solar physicist Luis Eduardo Antunes Viera and coauthors published a study titled, "Evolution of the solar irradiance during the Holocene" which reconstructed solar irradiance over the past 3,000 years using the physics-based relationship between TSI and the Sun's magnetic flux as measured by sunspots and $\Delta^{14}C$ isotopes. They concluded, "We present the first physics-based reconstruction of the total solar irradiance over the Holocene, which will be of interest for studies of climate change over the last 11,500 years. The reconstruction indicates that the decadal averaged total solar irradiance ranges over approximately 1.5 W/m^2 from grand maxima to grand minima."[19] Using the Stefan Boltzmann Law, 1.5 W/m^2 is just under 0.5°C. This estimation is about five times larger than the observed Total Solar Irradiance variance between the peak and trough of 11-year Schwab Solar Cycles. An estimate is provided by astronomer and solar physicist Natalie A. Krivova and coauthors who reconstructed TSI going back to 1610, and "suggest a value of about 1-1.5 W/m^2 for the increase in the cycle averaged TSI since the end of the Maunder Minimum."[20] Dr. Natalie Krivova heads the Solar Variability and Climate research group at the Max Planck Institute for Solar Systems Research in Göttingen, Germany.

Atmospheric levels of CO^2 and other anthropogenic greenhouse gases varied by less than 10 ppm between the Medieval Warm Period and the Little Ice Age (see Figure 48).[21] The drop in CO^2 from 285 ppm to 275 ppm from the Medieval Warm Period to the Little Ice Age equates to only -0.15 W/m^2 [4.33 x ln (275/285) = -0.15 W/m^2] and a temperature change of less than 0.05°C. Therefore, **greenhouse gases cannot explain the warming of the Medieval Warm Period and the**

cooling of the Little Ice Age. If TSI between the Medieval Warm Period maxima to the Little Ice Age minima changed only by 0.2 to 0.3 W/m^2 as seen in 11-year Schwab Solar Cycles in recent years, the total of greenhouse gas and TSI together would be less than 0.45 W/m^2, or about 0.13°C, which is nearly eight times smaller than the drop in global temperatures of about 1° C between the Medieval Warm Period and the Little Ice Age. A study by Stockholm University paleoclimatologist and climate historian Fredrik Ljungqvist and coauthors systematically compared proxy-based reconstructions and climate model simulations of temperature variability of past climate cycles. These climate models assume TSI over all solar cycles is similar to the recent satellite measurements of the 11-year Schwab Solar Cycles. They note that climate simulations using these models fail to reproduce peak medieval warmth around 900-1100 CE.[22] It would appear that decadal, centennial, and millennial TSI variance has been underestimated and solar-climate mechanisms complementary to TSI forcing are ignored in present-day global climate models.[23]

Although decadal solar cycles (Gleissberg), centennial solar cycles (Vries and Eddy), and millennial solar cycles (Bray) likely have a larger impact on TSI than is found in 11-year Schwab Solar Cycles, evidence suggests other mechanisms, outside of TSI, account for much of the impact of solar cycles on climate. Israeli solar physicist Nir Shaviv analyzed the signature of 11-year solar cycles in ocean heat flux, sea-level change rate, and sea-surface temperature variations and concluded that the climate sensitivity to solar cycles could be about 5-7 times larger than associated TSI variations alone.[24] Italian physicist Nicola Scafetta estimates the climate sensitivity to solar cycles is 4-7 times greater than radiative forcing alone.[25]

Historical climate temperature cycles have an excellent correlation to solar cycles.[26] Yet the correlation cannot fully be explained by TSI alone. Evidence suggests that there must be solar-cycle mechanisms, other than TSI, that are modulating temperature. Three hypotheses have been proposed, including 1) UV and ozone increases in the stratosphere; 2) heat transport to polar regions; and 3) cosmic rays and low cloud cover. These mechanisms are not mutually exclusive, and evidence suggests they are all valid and additive to TSI in driving temperature from solar cycles.

UV and Stratospheric Ozone: Total Solar Irradiance includes all spectra of ultraviolet, visible, and infrared radiation received by the Earth from the Sun. Over these light spectra, total irradiation changes over an 11-year Schwab Solar Cycle is only 0.7%. Solar cycles have a much greater impact on total ultraviolet (UV) light irradiation. Between the 200 nm to 320 nm UV wavelength, *total UV irradiance varies between 4% to 6% over the 11-year Schwab Solar Cycle.*[27] In the troposphere, the temperature declines as you increase in altitude until you reach the tropopause. This is because the troposphere is heated by the Earth's surface. The stratosphere exists above the tropopause, and the temperature begins to increase as you rise in altitude because the heating is from above. The ozone layer is concentrated roughly at 15 km to 35 km altitude. Ozone (O^3) absorbs UV-B and UV-C radiation from the Sun, which converts UV energy into heat, warming the surrounding air. Because the ozone concentration is highest in the upper stratosphere, that is where most of the UV absorption and heating occur. As a result, temperature increases with height in the stratosphere.

UV-C radiation (below 240 nm wavelength) produces ozone in the stratosphere. UV-C splits oxygen (O^2) molecules, and the free oxygen atoms react with other O^2 molecules to form ozone (O^3). This process releases heat which warms the stratosphere. Satellite and models confirm that ozone levels increase in the stratosphere during a solar maximum.[28] Ozone is a greenhouse gas and therefore traps infrared radiation from escaping to space, thereby changing the Earth's Energy Balance and producing heating, although far less heating than UV interactions with ozone. While UV-C creates ozone, UV-B (280 to 320 nm wavelength) destroys ozone and releases heat in the process. This destruction of ozone by UV-B provides another mechanism to release heat into the stratosphere. According to Duke University atmospheric physicist, Drew Shindell and coauthors, "Increases in ozone abundance and incoming UV radiation lead to greater UV absorption and hence greater solar heating during solar maximum."[29]

When total UV irradiation heats the stratosphere, it creates a top-down heating effect on the troposphere.[30] *When UV irradiation decreases during a solar minimum, this produces less heat, which cools the stratosphere and leads to top-down cooling of the troposphere.* Perhaps more significant, changes in stratospheric temperatures affect atmospheric circulation, which impacts climate.[31] For example, a warmer stratosphere during a solar maximum is observed to produce a more stable jet stream, and a stronger, less wavy polar vortex, which transports less heat to the poles where it could more easily escape out to space, thus increasing the Earth's Energy Budget, which increases temperatures.

Heat Transport to the Poles: Spanish scientist Javier Vinós has postulated a solar cycle mechanism which explains much of the climate change impact of UV heating of the stratosphere during solar maxima. Vinós has presented evidence that a major contributor to climate change is the movement of heat from the Tropics to the poles. Water vapor in the atmosphere retains vastly more heat than any other greenhouse gases. Because the poles are cold and have limited water vapor, heat sent to the poles more easily radiates out to space. Polar regions are often referred to as the Earth's radiator, as heat from the Tropics is moved to the poles and sent out to space from polar regions. **If heat is not efficiently transported to the radiator of a car, the engine will overheat. In a similar manner, if the transport of heat to the poles is blocked, the Earth warms.** When more heat is transported to the poles, more of the Earth's energy is lost. Conversely, when less heat is transported, more energy is conserved, and this changes the Earth's Energy Balance. Vinós has shown that this loss of heat is especially strong during winter when it is colder and there is lower water vapor in the polar skies. Known as the "Winter Gatekeeper Hypothesis," Vinós has set forth evidence for this phenomenon in his book, *Solving the Climate Puzzle*.[32]

We have discussed how solar cycles impact UV radiation in the stratosphere, which warms or cools the stratosphere. The temperature of the stratosphere drives the strength of the jet stream, waviness of the jet stream, strength of the polar vortex, trapping of polar air, expansion of the Hadley Cell, and polar movement of subtropical jets. Scientific literature has well documented these dynamic changes in atmospheric circulation that are tied to solar cycles.[33] All these changes impact the transport of heat to the poles. In summer, heat is transferred to the poles in the oceans and atmosphere, but in winter, when the polar regions are covered by ice, the transport of heat is primarily from the atmosphere. For atmospheric heat to radiate out through polar regions in winter, it must cross the stratospheric polar vortex. This vortex is a wall of high winds that can reach up to 160 km/hour (100 miles/hour). This wall of wind of the vortex acts as a formidable barrier, blocking the entry of warmer air into polar regions.[34] During a solar maxima, the stratosphere heats up, the winds of the polar vortex are strengthened, and less hot air is transported to the poles. During a solar minima, the stratosphere cools down, the polar vortex is weakened, and more warm air is transported to the poles. Since the poles are the Earth's primary radiators, where much of the Earth's heat is sent out to space, any change of heat transport to the poles upsets the Earth's Energy Balance leading to warming or cooling temperatures.

Javier Vinós writes that paleoclimate evidence suggest that this transport of heat to the poles and its impact from solar cycles is very important and probably the main driver of climate change.[35] For example, during the Medieval Warm Period, a solar maximum would have increased UV heating of the stratosphere. This warm stratosphere induced a strong polar vortex preventing heat from being transported to the poles and subsequently out to space, thus leading to an excess of heat, which warmed the Earth. During the Little Ice Age, the Wolf, Spörer, Maunder, and Dalton solar minimums would have reduced UV heating of the stratosphere, weakened the polar vortex, and released heat into the polar regions that was radiated out to space. Temperatures dropped significantly during these solar minima, especially in the winter months as predicted by the Winter Gatekeeper Hypothesis.

Consistent with the Winter Gatekeeper Hypothesis, a NASA study suggests the lower ultraviolet radiation during the Maunder Solar Minimum caused a drop in temperature and ozone in the stratosphere, which weakened the polar vortex and resulted in chilly temperatures.[36] The Winter Gatekeeper Hypothesis also explains why cooling from the 70-year Maunder Minimum was much more severe than the TSI change in an 11-year Schwab Solar Cycle. The transport of heat to the poles and the radiation of that heat out to space is not an instantaneous process. A multi-year solar minimum such as the Wolf, Spörer, Maunder, and Dalton solar minimums of the Little Ice Age would have provided sufficient time to transport additive heat, year after year, to polar regions and out to space. This theory also explains why climate change has been more about warming winters than hotter summers in the Extra Tropics (see Figure 23 in Chapter 4), since the transfer of heat from the Extra Tropics to polar regions in ocean currents is hindered in winter due to sea ice.

Cosmic Rays and Low Clouds: Moderate Resolution Imaging Spectroradiometer (MODIS) and Earth's Radiant Energy System (CERES) satellites have shown us that warming since the year 2000 has been primarily due to less clouds (see Chapters 10 and 11).[37] Cloud cover is a powerful driver of climate, so it is therefore reasonable to look at cloud cover and its relationship to past solar cycles. A drop of 1° C in temperature (or 3.4 W/m^2), as experienced during the Little Ice Age, can easily be explained by greater cloud coverage. In fact, a 1% change in albedo forces 3.4 W/m^2 of warming or cooling, which is equal to 1° C.[38] Isotope records of ^{10}Be flux, $\Delta^{14}C$, and $\delta^{18}O$ confirm the temperature dropped and cosmic rays increased during the Wolf, Spörer, Maunder, and Dalton solar minimums during the Little Ice Age. A connection between cloudiness and solar minimums would explain past cold cycles and their correlation to solar cycles.

Unfortunately, there is no direct record of cloud cover from the Little Ice Age. We have already seen that some described this time as a period of cloudiness. The Anglican bishop Robert Burton (1577 to 1640) described the times in the British Isles as one of "endless days when dark clouds obscure the sunlight."[39] During the Little Ice Age, the Reverend Daniel Schaller, of Stendal in the Prussian Alps, wrote, "There is no real constant sunshine, neither a steady winter nor summer; the earth's crops and produce do not ripen, are no longer as healthy as they were in bygone years."[40] A weather diary kept by Tycho Brahe in Denmark in the 16th century noted prolonged cloudy periods.[41] The French explorer Cartier recorded the summer of 1605 in Massachusetts, stating, "During the time we were there it blew a gale from the northeast, with the sky so overcast that the sun was hardly visible at all."[42]

Meteorologist and professor Hans Neuberger from the Department of Meteorology of Penn State University analyzed 12,000 paintings in museums in the United States and eight European countries to look at cloud cover in these historical paintings between 1400 to 1967. Neuberger's analysis suggest that artists during the Little Ice Age frequently depicted cloudier skies, reflecting the climate conditions they observed.[43] He found that paintings between 1550 to 1849 had the darkest, cloudiest skies.[44]

Agricultural records also seem to support a connection between cloudiness and solar cycles. During the Roman Warm Period, there was high and stable solar activity between 200 BC to 100 CE.[45] During this period, Romans expanded viticulture (grape cultivation) into new territories. Growing grapes requires significant sunlight. During the high solar activity of the Roman Warm Period, there was a noteworthy expansion of viticulture to regions in Britain and Northern Gaul, where vineyards are less viable today. Archaeological evidence of Roman vineyards in Britain near Hadrian's Wall suggest abundant sunshine and fewer clouds in that era. Solar activity declined around 260 CE, and the Great Aletsch Glacier began to grow around 272 CE.[46] Solar activity dropped further in the middle of the fifth century,[47] and by the sixth century the Lower Grindelwald Glacier and Zmut Glacier both advanced.[48] During the coldest period of the Dark Ages, known as the Late Antique Little Ice Age (536 CE to 660 CE), viticulture contracted throughout the Roman Empire, suggesting a period of increased cloud cover and less sunshine. No archaeological vineyards or wine production facilities from the 5th to 10th centuries have been identified in Britain, suggesting that grape growing diminished or ceased.[49]

Solar activity increased once again into the Medieval Warm Period, and vineyards were established in England, Scandinavia, and Scotland, including areas where grape cultivation is marginal today.[50] Barley cultivation also requires significant sunlight, and archaeological evidence from Greenland shows barley cultivation during the Medieval Warm Period suggesting another period of low cloud cover and sunny days.[51] In 1280, the Wolf Solar Minimum began to reduce solar activity initiating the Little Ice Age. Tax records show the decline in viticulture in England and southern Scandinavia during the Little Ice Age, suggesting a return to cloudier skies.[52] English manorial records such as the Winchester Pipe Rolls show reduced barley yields during the Little Ice Age implying cloudier skies and limited sunlight.[53]

Consistent with satellite data, IPCC models capture only ~0.2 W/m^2 to ~0.3 W/m^2 of Total Solar Irradiance variation from 11-year Schwab Solar Cycles. However, these models do not include the cooling impact of clouds. The IPCC states, "The net global mean cloud radiative effect of approximately –20 W/m–^{2} implies a net cooling effect of clouds on the current climate."[54] Counter to these facts, the IPCC models use an average number of 0.6 W/m^2 of net warming from clouds.[55] IPCC climate models greatly underestimate the cooling during the Little Ice Age as reconstructed from paleoclimate proxies, glacier growth, tree line elevations, and sea level rise. IPCC models estimate temperature changes from the Maunder Minimum resulted in only about -0.3° C to -0.4°C.[56] It would appear that their improper treatment of clouds in the models may be the underlining cause of their underestimation.

During solar maxima, the strong magnetic field of the Sun limits the number of cosmic rays that reach the Earth. Conversely, during a solar minimum, the weak magnetic field allows more cosmic rays to hit the Earth's surface. During recent 11-year Schwab Solar Cycles, cosmic ray flux has varied from 10% to 20% between the peak and trough of each solar cycle.[57] Danish astrophysicist Henrik Svensmark and Danish space/solar-terrestrial physicist Eigil Friis-Christensen observed 3% to 4% variation of global cloud cover during a recent solar cycle as measured by satellite. The cloud cover was strongly correlated with cosmic ray flux and, in turn, inversely correlated with the solar cycle. They postulated that cosmic rays help create cloud condensation nuclei which help form low clouds and thus lead to cooling from energy being reflected out to space off the clouds.[58] They proposed the solar influence through cosmic ray flux and cloud cover totaled 1.4 W/m^2 between 1901 to 1995. This would account for about half of all warming during that period, with greenhouse gases accounting for the other half of warming in the 20th century.

The theory makes sense, as cosmic rays create ions in the atmosphere which stabilize atmospheric sulfate and ammonia molecules so they can cluster into cloud condensation nuclei (see Chapters 11 and 12). The impact of low clouds on temperature is known to be significant, reducing solar radiation by 476 W/m^2 under the cloud cover at the equator. You can feel this difference on a hot, sunny day, when the Sun is covered by a low passing cloud; the temperature drops significantly.

Cosmic rays create positively charged nitrogen and oxygen ions in the atmosphere by stripping off electrons forming ion pairs of positively charged nitrogen and oxygen ions ("cations") and negatively charged free electrons. These cations react with water in the atmosphere and produce additional ions including H^+ and $H30^+$ cations. They also react with ammonia (NH_3) in the atmosphere and produce NH_4^+ cations. Positively charged ion clusters are then formed from these cations. The ionosphere is positively charged relative to the ground.[59] The strength of the electric field in the atmosphere, often referred to as the fair-weather electric field, is typically about 120V/m near the surface.[60] Since the Earth's surface has a negative electrical charge, the extremely small mass and negative charge of free electrons move these negatively charged particles away from the Earth's surface. This is because negative charges repel each other, and the gravitational force is weak due to the low mass of electrons. On the other hand, positively charged cation clusters, with their heavier mass and active charge, are attracted to the Earth from gravity and their positive electrical charge. These positive ion clusters would thus travel down through the layers of the atmosphere to where low clouds are formed.

Negatively charged ions (anions) are also produced in the atmosphere including O^-, O^{2-}, and HO_4^- anions. Cosmic rays produce OH which reacts with nitrogen oxides in the atmosphere to produce nitrate anions, and OH radicals react with volatile organic compounds in the atmosphere to produce carboxylic anions. Negatively charged anion clusters are also formed.

The number of high-energy cosmic rays entering the atmosphere is relatively small, estimated to be only 1,000 particles per second per square meter.[61] However, they have a significant impact in creating large numbers of ion pairs which include positively charged nitrogen and oxygen cations, the cascade of other cations, anions, and free electrons. When a high-energy cosmic ray enters the atmosphere, it produces a shower of secondary particles including electrons, positrons, and muons. Through interactions with atmospheric molecules, primarily nitrogen and oxygen, these secondary particles contribute significantly to ionization. It is estimated that cosmic rays produce on the order of one million to ten million ion pairs per second per cubic meter in the lower atmosphere.[62] Millions of cations per cubic meter are attracted through the atmosphere toward the Earth, where they interact with anions contained in ionic molecules including cloud nucleating aerosol, which catalyze the formation of low clouds. Anions also play a role in cloud formation as they interact with the cations contained in ionic molecules that become cloud nucleating nuclei.

As the cations fall to the earth, they will pass by and react with negatively charged particles including free electrons, and they may recombine to form neutral molecules, reducing the flux of cations. Since the secondary particles created by cosmic ray continuously produce ion pairs, including free electrons, there are ample free electrons to neutralize cations formed high in the atmosphere. Solar UV radiation also ionizes nitrogen and oxygen molecules in upper layers of the atmosphere. However, there are still abundant cation clusters at lower altitudes because the high energy of cosmic rays allows them to penetrate through the atmosphere and form cations at the height of low clouds. Cosmic rays are likely the most important contributor to ion-pair production from ~3 km to ~50 km in the atmosphere.[63] They are thus the primary source of positive ion clusters in the low atmosphere where cumulus and stratus clouds are formed. As described above, they also contribute to the formation of anions.

Ions are the third factor in cloud formation along with water vapor and aerosols in the atmosphere. Sulfate cloud condensation nuclei are ionic molecule clusters in that they contain cations and anions. The molecules are neutral because the positive charge of the cations in the molecule neutralizes the negative charge of the anions in the molecule. For example, the most common sulfate cloud condensation nuclei, ammonium sulfate ($(NH_4)_2SO_4$), contains two NH4+ cations that are neutralized by SO_4^{-2} anions. Neutral molecules, such as ammonium sulfate, are less likely to form clusters than when ions are introduced. Free ions of opposite charges in the atmosphere electrostatically attract NH_4^+ cations and SO_4^{-2} anions in ammonium sulfate molecules. The electrostatic charge provided by these ions attracts additional sulfate molecules to join the cluster, thus forming small sulfate aerosols of 1 nm to 2 nm in size. To become cloud condensation nuclei these aerosols must grow to the sufficient size of about 50 nm. Due to the large surface area of small 1 nm to 2 nm particles, many of the neutral ammonium sulfate molecules are lost to evaporation. The electrostatic charge provided by free ions helps hold the sulfate molecules in the cluster, thus reducing losses to evaporation. Because the ions are attracted to other sulfate molecules, they facilitate growth of the cluster to cloud condensation nuclei size. Thus, atmospheric ions help initiate aerosol formation and enhance the clustering of sulfate aerosols to sufficient size to become cloud condensation nuclei, which are the seeds around which water condenses to form low clouds. Martin Enghoff and Henrik Svensmark from the Technical University of Denmark as well as coauthors from other universities conducted experiments that confirm ion-induced nucleation. Using an ion beam to nucleate sulfate aerosols, they found a clear contribution from ion-induced nucleation.[64]

Svensmark and his research group built a cloud chamber containing sulfate aerosol precursors and an ionization beam and proved that cosmic rays and the ionization created by cosmic rays produce small 1 nm to 2 nm aerosols from gas molecules (sulfate vapors). These results were independently confirmed in a cloud chamber operated by the European Organization for Nuclear Research CERN group in Meyrin, Switzerland. Cosmics Leaving Outdoor Droplets (CLOUD) experiments at CERN showed that ions accelerated aerosol formation by 2 to 10 times.[65] It has long been known that low clouds form when water vapor condenses on aerosol particles in the air. These aerosol particles are considered essential to the formation of clouds. To become cloud condensation nuclei, aerosols need to grow to sufficient size (~50 nm). Colorado State University atmospheric scientist Jeffrey Pierce performed computer simulations that showed that cosmic rays do not impact cloud formation on the basis that aerosols created by cosmic rays are too small to become cloud condensation nuclei. The computer simulation showed that these particles were lost as they were absorbed into larger particles; consequently, they concluded that the hypothesized effect is too small to play a significant role in current climate change.[66]

Because the cosmic ray theory provides an alternative to the greenhouse gas theory, the climate alarmist media had a field day stating that the cosmic ray climate theory was dead. This was reinforced by scientists who echoed the media message that cosmic rays do not impact the climate. However, as Mark Twain said, "The reports of my death have been greatly exaggerated." Despite the criticism, Professor Svensmark and his research team persevered. Svensmark and his team did additional work in their cloud chamber and showed that the small aerosols are lost in normal conditions, as predicted by the computer models, but when exposed to ionization, as created by cosmic rays, they cluster to become larger aerosols *in about five days*, large enough to nucleate clouds.

With ionization, these small particles are not lost to the larger particles. The Svensmark team even conducted their experiment in a mine over a mile deep in the Earth to ensure no contamination of their results from cosmic rays at the Earth's surface. **Svensmark's discovery may be the greatest breakthrough in climate science in our generation**, yet after he and his research group completed a paper on their discoveries, and because it provided an alternate to the greenhouse gas-driven climate crisis, it took one and one-half years to find a journal that would publish the results. **Svensmark said one scientific journal after another refused to publish his findings, yet they provided no scientific reason not to accept his paper for publication**. Finally, it was published in 2017.[67]

The *five-day delay* in aerosol and cloud formation in Svensmark's experiments proved to be a pinnacle detail. Previously, in 2009, Svensmark realized that nature had provided the perfect experiment to confirm his theory that cosmic rays influence cloud formation. Solar flares generate large magnetic waves for a period of about 10 days. It has been observed that this spike in the magnetic field of the Sun reduces the cosmic ray flux reaching the Earth. Such events are known as Forbush Decreases, and are named for Scott Forbush, an astronomer and physicist who discovered the Forbush Decrease effect in 1937. Svensmark and his team used four separate satellite data sets, since 1982, to evaluate the impact of these 10-day solar flares on cosmic rays, aerosols, and cloud formation.[68]

The correlation was perfect. The data showed a significant decline in cosmic rays during the Forbush Decrease. ***Five days after* the start of the Forbush event, there was a significant drop in aerosols as measured by the AERONET satellite**. The SSM/1, MODIS, and ISCCP satellites all showed lower cloud formation and lower water content in clouds with measurements of liquid water, liquid cloud

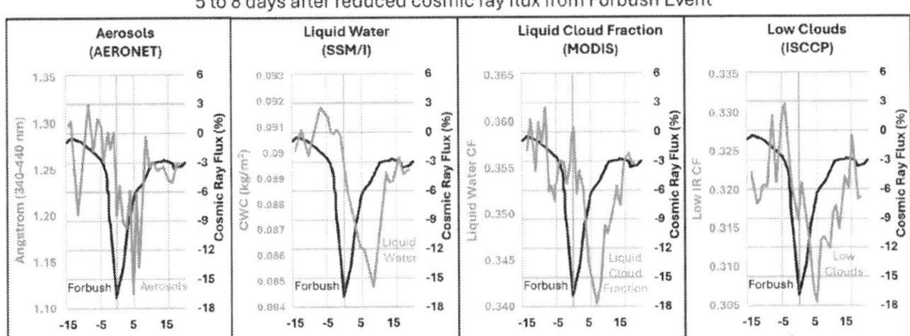

Figure 42 – Cloud Formation Matches Solar Flares and Cosmic Ray Flux. *Solar flares from the Sun, known as Forbush Decreases, result in a substantial increase in solar magnetic waves, which reduce the number of cosmic rays reaching the Earth and lower cosmic ray-induced ionization in the atmosphere. Observations show cloud nucleating aerosols, cloud formation, and cloud water content declines* ***after about five days****, consistent with Svensmark's cloud chamber experiments. Data in the figures above summarizes the five largest Forbush event since 1982. The black line depicts the decline in cosmic rays and atmospheric ionization during the solar flare induced Forbush Decreases. Four separate satellite measurements are shown in gray lines, including aerosols (AERONET satellite data), liquid water in clouds declines (SSM/1 and MODIS satellites) and cloud formation (ISCCP satellite).* Source: Svensmark et al., "Cosmic Ray Decreases Effect Atmosphere Aerosols and Clouds," *Geophysical Research Letters,* 1 August 2009. https://doi.org/10.1029/2009GL038429.

fraction. *The declines were all in step with the change in cosmic ray flux after a five-day delay* (see Figure 42). Interestingly, the MODIS satellite did not show this same correlation for ice cloud fractions, which are the higher cirrus clouds. The data reveals that cosmic rays contribute primarily to the formation of low clouds, and not high clouds. It is the low clouds that have a cooling effect on the climate by reflecting more net radiation back out into space, while high clouds have a warming effect by emitting more net radiation back *to* Earth. Both experimental and observational data support the Svensmark theory of the role of cosmic rays in low cloud formation.

In 2021, Svensmark and other scientific collaborators published another paper on the five strongest week-long Forbush events since the year 2000 and the accompanying declines in atmospheric ionization from reduced cosmic rays.[69] These Forbush events showed natural decreases in ionization of 10% to 20% for each Forbush Decrease. Using data from the CERES satellite, they showed the change in shortwave radiation leaving the Earth from before and during these Forbush Decreases. The results were significant. The data shows that shortwave radiation changed by an average of 1.7 W/m^2 over a large part of the Earth during these Forbush events, which average nine

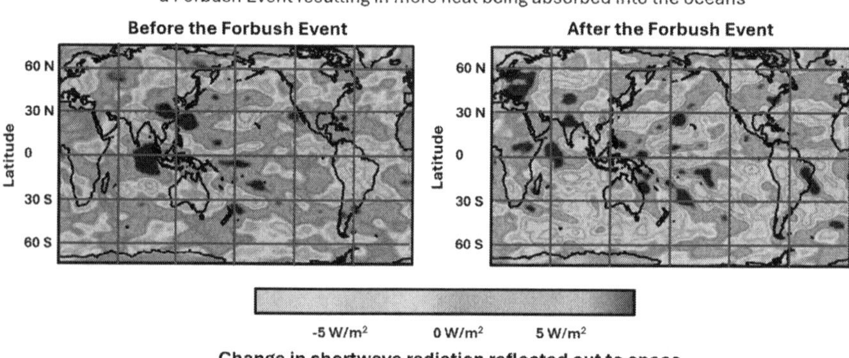

Figure 43 – Ocean Warming and Cosmic Ray Flux Correlation Confirmed in Measurements*. Using data from* the *Earth's Radiant Energy System (CERES) satellite to measure shortwave light radiating out to space during the five largest Forbush Decreases since 2000 reveals a meaningful change in heating of the oceans when cosmic ray flux changes. The chart above shows the significant decrease in light radiated out to space, when cosmic ray flux declines as depicted in lighter shades of gray. This results in a 1.7 W/m^2 increase in ocean heat over the nine-day period of each Forbush Decrease. Most of this impact is over the oceans. Using these figures, Nir Shaviv and Henrik Svensmark have calculated that cosmic ray flux and associated cloud cover accounts for 50% to 66% of all global warming during the 20th century.* Source: Svensmark, Henrik et al., "Atmospheric Ionization and Cloud Radiative Forcing," *Nature Scientific Reports*, Volume 11, Article Number: 19668, 11 October 2021. https://doi.org/10.1038/s41598-021-99033-1.

days in duration. The other major finding was that the changes for each Forbush were almost entirely over the oceans.

This observational data proves that changes in the solar magnetic field and the associated changes in cosmic rays and atmospheric ionization alter low cloud cover, which primarily impacts heat in the oceans. To put 1.7 W/m^2 in nine days in context, to achieve radiative forcing of about 1.7 W/m^2 from CO_2 radiative forcing would require increasing CO_2 concentrations from today's 427 ppm to 635 ppm. Such an increase in CO_2 levels would take many decades to reach. Astrophysicists Nir Shaviv and Henrik Svensmark have used this data to calculate the total warming in the 20th century was 50% to 66% driven by the low cosmic ray flux and lower cloud cover during the Modern Solar Maximum (see Figure 43).[70]

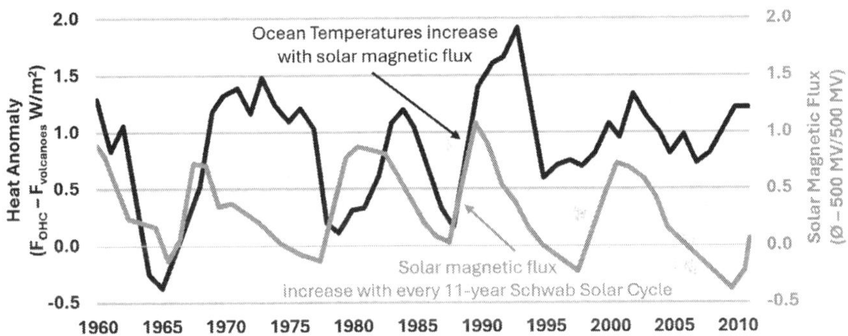

Figure 44 – Ocean Temperatures Follow Cosmic Ray Flux, 1960 to 2010. *The chart above tracks ocean temperatures in black and solar cycle magnetic flux in gray. Eleven-year Schwab Solar Cycles produce a fluctuation in the strength of magnetic flux which can clearly be seen in the gray line in the chart above. When solar magnetic flux is high, there are fewer cosmic rays, less low clouds, and the oceans heat up. The chart above shows how the heat in the oceans follows the decline in cosmic rays due to high solar flux from each solar cycle.* Source: Astrophysicist Nir Shaviv charting ocean temperatures from NOAA, adjusted for volcanic forcing from NASA GISS, and solar modulation from Matthes, K. et al., "Solar forcing for CMIP5 (v3.2)," *Geoscientific Model Development*, Volume 10, Issue 6, pp. 2247-2302, 22 June 2012. https://doi.org/10.5194/gmd-10-2247-2017.

Observational data over every time scale confirm the warming and cooling of ocean temperatures with cosmic ray flux. Hebrew University astrophysicist Nir Shaviv constructed a comparison of ocean temperatures, adjusted for volcanic forcing and the strength of the solar magnetic flux, which modulates cosmic rays. A correlation is clearly seen between the 11-year Schwab Solar Cycles and ocean temperatures during the years of 1960 to 2011 (see Figure 44).

Confirming Svensmark's theory, University of Oulu (Finland) professor, astrophysicist, and head ot the Oulu Cosmic Ray station, Ilya G. Usoskin, and coauthors published a study on the latitudinal dependence of low cloud amount on cosmic ray-induced ionization, which demonstrates a close correlation to low cloud cover in the Tropics and low cosmic ray concentrations in satellite data from 1985 to 2000.[71] This phenomenon coincides precisely with the main warming phase of the late 20th century and a period with a strong solar magnetic field. Astrophysicist at AIC (Spain), Enric Pallé and coauthors published a paper in *Science* on the changes in Earth's reflection during the timeline between 1984 and 2000.[72] The paper shows a decline in albedo (reflectivity of a surface), which would result from less cloud cover, equal to an increase of 2 W/m^2 to 6 W/m^2, just as Svensmark's theory predicts.

Cosmic Rays and Climate Observations: Every eleven years there is a Schwab Solar Cycle where solar activity, the solar magnetic field, and cosmic rays increase and decrease. Modern sea temperature records reflect these 11-year solar cycles. The 11-year solar cycles are also reflected in sea level rise and fall as measured from both satellite measurements since 1980 and tide gauges since 1920.[73] As the solar cycle peaks the strong solar magnetic field shields the Earth from cosmic rays, and the oceans heat as more shortwave radiation penetrates the water. When solar cycles decline to the valley of the curve, the solar magnetic field declines allowing more cosmic rays

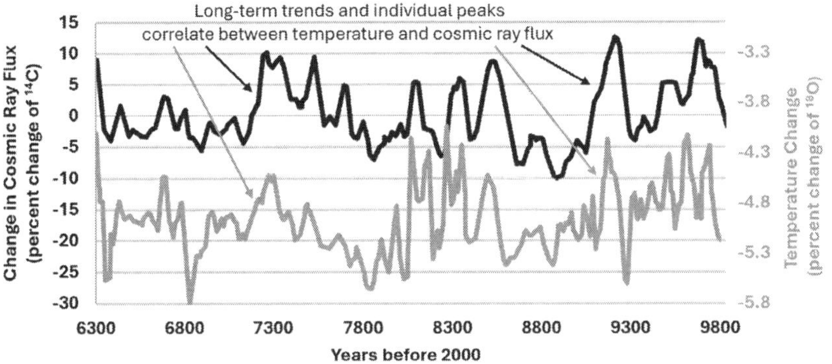

Figure 45 - Oman Holocene Temperatures Match Cosmic Ray Flux*. A reconstruction of Holocene temperature from $\delta^{18}O$ isotopes from stalactite layers taken from caves in Oman (in gray). Comparing these temperatures with cosmic ray reconstructions from $\Delta^{14}C$ isotopes (in black) reveals a strong correlation between cosmic ray flux and temperature.* Source: Neff, U. et al., "Strong coherence between solar variability and monsoon in Oman between 9 and 6 kyr ago," *Nature*, Volume 411, pp. 290-293, 1 May 2001. https://doi.org/10.1038/35077048.

into the atmosphere, which results in increased cloud formation and more radiation reflected out to space leaving less shortwave radiation to warm the oceans. When the oceans warm, the water expands, and sea levels rise. When oceans cool, the water becomes denser, and sea levels decline.

Ulrich Neff from the Heidelberg Academy of Sciences and coauthors analyzed $\delta^{18}O$ isotopes from layers of stalactite from caves in Oman. They were able to construct temperatures for the Indian Ocean during the Holocene. When compared with $\Delta^{14}C$ isotopes a correlation can be found between cosmic rays, solar cycles, and temperature. These results show a strong correlation between cosmic rays as measured by $\Delta^{14}C$ isotopes and temperature as measured by $\delta^{18}O$ isotopes in the stalactites at various layers (see Figure 45).[74]

Working independently from Svensmark, isotope chemist and Distinguished Professor at the University of Ottawa, Jan Veizer, analyzed isotopes in 24,000 calcium carbonate fossil shells from various sediment levels to reconstruct temperatures of the Earth over the past 500 million years, using ^{18}O to ^{16}O isotope ratios, depicted as $\delta^{18}O$. He was surprised to see a strong cycle of temperature swings of up to 10°C every 145 ± 7 million years. For years he searched for the reason for these temperature swings but could find no answers. There was no correlation between temperature and CO_2 on these time scales. During the coldest period, 450 million years ago, Veizer discovered CO_2 was over 5,000 ppm, and in the period 180 million years ago, temperatures were colder than now, yet CO_2 levels were nearly 1,500 ppm as compared to 427 ppm today.

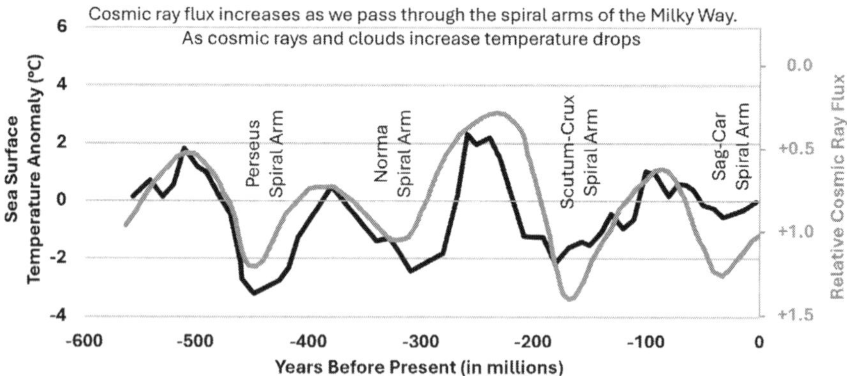

Figure 46 - Temperature Correlates to Cosmic Ray Flux Over 500 Million Years. *The chart above shows the work of Jan Veizer in reconstructing the temperature by analyzing isotopes in 24,000 shell fossils from various sediment layers (black line), and the work of Nir Shaviv in the reconstruction of cosmic rays reaching the Earth from analyzing isotopes in meteorite samples (gray line). Both correlate well and show how temperatures decrease as cosmic rays increase about every 140 million years as the Earth passes through the spiral arms of the Milky Way, which have higher levels of supernovae, the source of cosmic rays.* Source: Shaviv, Nir J., and Veizer, Ján, "Celestial driver of Phanerozoic climate." *GSA TODAY*, July 2003. https://www.atmos.washington.edu/academics/classes/2003Q4/211/articles_optional/CelestialDriver.pdf.

Astrophysicist Nir Shaviv studied isotopes in meteorites to reconstruct historical cosmic ray flux in our solar system and found cosmic rays increased every 143 ± 10 million years as our solar system passed through the spiral arm of the Milky Way, which had many supernovae, and the resulting elevated levels of cosmic rays. Supernovae are exploding stars that are the primary source of galactic cosmic rays. Astronomical data reveals that supernovae are concentrated in the spiral arms of the Milky Way, and that our solar system passes through these spiral arms every 135 ± 25 million years,[75] consistent with Shaviv's cosmic ray flux reconstructions.

Shaviv was aware of Svensmark's work on cosmic rays and climate, and he approached Jan Veizer with a solution to his quandary. Figure 46 shows how the temperature decreases about every 140 million years, each time our solar system passes through spiral arms of the Milky Way. Each of these Milky Way spiral arms—including the Perseus, Norma, Scutum-Crux, and Sag-Car arms—are areas of numerous supernovae and a resulting abundance of cosmic rays.[76] This data is independent of solar cycles and is yet another confirmation of the impact of cosmic rays on climate.

The correlation between global temperature and cosmic ray flux has been confirmed over millions of years, thousands of years, hundreds of years, and even over recent decades. Yet despite all the scientific evidence, the IPCC does not account for cosmic rays in its climate models and assigns a measly 0.02°C to account for all solar cycle and volcanic drivers of climate since 1750.[77] Since the IPCC considers CO_2 the control knob of climate, they cannot explain the warm and cold periods in the past such as the Holocene Climatic Optimum, Minoan, Roman, and Medieval warm periods, and the cold Greek Dark Ages, Dark Ages, and Little Ice Age since these were all before significant increases in CO_2 levels. The scientific method demands confirmation from experimental and observational data. Assigning nearly all warming to CO_2 concentrations fails this test, while the influence of solar cycles and cosmic rays on climate meet this test with overwhelming evidence.

The link between cosmic rays and climate is one of the most important new areas of modern climate research. However, the climate community finds it difficult to accept Svensmark evidence because the cosmic ray influence on climate detracts from the narrative of the anthropogenic greenhouse gas hypothesis. Sadly, as Dr. Svensmark has commented, "Funding for this research is nearly impossible to obtain."[78]

Skeptical Views on Cosmic Rays' Impact on Clouds: The hypothesis that cosmic rays influence cloud condensation nuclei, cloudiness, and climate is a highly debated topic in climate science. In the end, this is probably a positive development as the scientific method demands that theories be challenged and tested against experimental and observational data to verify or disprove the underlying theory. Furthermore, experiments need to be repeatable. There is a need for the anthropogenic greenhouse gas theory to face similar scrutiny. However, the experience of scientists who have challenged the hypothesis that greenhouse gases are the dominant factor of climate change or have provided alternative climate change mechanisms make it difficult to publish and receive funding as experienced by Henrik Svensmark and others.

A good example of how criticism has led to more thorough analysis is found in the critique of the Shaviv and Veizer work on the increased cosmic ray flux every 143 ± 10 million years as our solar system passes through the spiral arms of the Milky Way, and in the ice age epochs periods that coincide. Astrophysicist and astronomer from the Max Planck Institute, Knud Jahnke, criticized the statistical method used by Shaviv in determining the cosmic ray flux from isotopes in meteorites. Jahnke attempted to repeat Shaviv's previous analysis and claimed there was no significant periodicity in cosmic ray flux variability as shown by Shaviv. Shaviv responded to this critique by pointing out the faults of Jahnke's reanalysis, which used poorly dated meteorites that had a grave effect on signal-to-noise and impacted the statistical significance of Jahnke's reanalysis of periodic cosmic ray flux.[79] Shaviv did acknowledge weaknesses in the Kolmogorov-Smirnov statistical method he employed as brought to his attention by Jahnke. As a result, Shaviv reanalyzed the meteorite data using another statistical tool known as Rayleigh Analysis.[80] The Rayleigh Analysis method is far better suited to the analysis of cosmic ray flux periodicity from meteorites as it was designed to establish periodic deviations from the uniform occurrence of discrete events. Using the Rayleigh Analysis, Shaviv found a periodicity of cosmic ray flux of 145 million years with high statistical confidence.[81] Furthermore, he mentions that, using purely astronomical data, the cosmic ray flux varies every 135 ± 25 million years, which is consistent with the meteorite analysis.[82] Based on this astronomical data he observes, "The fact that a signal is present should not be surprising at all. On the contrary, it would have been a great surprise if no signal would have been observed in the meteorites!"[83]

Papers by Max Planck Institute astrophysicists Sami K. Solanki & Natalie A. Krivova and Purdue University atmospheric scientist Ernest Agee and coauthors are often cited to show the correlation between cosmic ray flux and the temperature anomaly breaks down in recent years.[84] The Solanki & Krivova paper shows a strong correlation between cosmic ray flux and the temperature anomaly from 1870 to the mid-1980s, and divergence after that time.[85] The Agee et al. paper states that the high cosmic ray flux in 2008 to 2010 did not result in an increase in cloud cover.[86] These observations appear to be correct. The International Satellite Cloud Climatology Project (ISCCP) shows a 6.5% reduction of the global cloud-cover fraction between 1986 and 2018.[87] This was a time when the trend of cosmic ray flux generally increased, which, according to Svensmark's hypothesis, should have increased cloud cover.

The answer to this divergence from expectations may have been provided in a paper by University of Oslo (Norway) astrophysicist Benjamin A. Laken. and coauthors. This study provided what the authors termed "perhaps the most compelling evidence presented thus far of a Galactic Cosmic Ray-climate relationship." They concluded that cloud condensation nuclei creation is governed by both Galactic Cosmic Rays and internal atmospheric precursor conditions.[88] The ions produced from cosmic rays can only assist the growth of cloud condensation nuclei if precursor aerosol material such as sulfates and ammonia are present in the atmosphere. It does not matter how strong the cosmic ray flux is if there are no precursor aerosols from which to build cloud condensation nuclei. Therefore, the level of aerosol precursors in the atmosphere is perhaps more important than cosmic ray flux in forming clouds. If this is true, the CLAW hypothesis, which proposes a cooling feedback from increased sulfate aerosols produced by phytoplankton growth as the climate warms, might be more powerful than thought (see Chapter 12). Experiments run in the CERN CLOUD chamber seem to confirm there is a large effect on cloud condensation nuclei from variations in aerosol emissions.[89]

Svensmark's experiments showed the impact of cosmic rays is primarily with sulfate aerosols. Since 1979 industrial emissions of SO_2 have declined by about 50% due to air pollution regulations and the implementation of pollution controls.[90] The decline in China of sulfate pollution was significant after 2006.[91] This is the period of the Agee et al. study. In 2020, SO_2 was further reduced by new regulations that require an 85% decline in SO_2 emissions from maritime shipping (see Chapter 12). A study by University of Maryland atmospheric scientist Tianle Yuan and coauthors states that the warming effect from the reduction in maritime sulfur emissions "is equivalent in magnitude to 80% of the measured increase in planetary heat uptake since 2020."[92] Evidence suggests the decline in clouds and increase in temperature after the mid-1980s may have been driven by lower sulfate aerosols in the atmosphere, which higher cosmic ray flux could not impact.

Critics of the connection between cloud cover and cosmic ray flux continue to cite the paper by Pierce & Adams from results of their computer simulation that showed aerosols created by cosmic rays are too small to become cloud condensation nuclei. The authors conclude: "Any changes in the cosmic ray flux during the 20th century have likely had little impact on present-day climate."[93] This paper was based on computer simulations, not experiments or real-world data. As we have already discussed, Svensmark and his collaborators proved the computer simulations to be wrong from his cloud chamber experiments and Forbush Decrease observations that both confirm cosmic rays grow sulfate aerosols of sufficient size to be cloud condensation nuclei in about five days.[94]

In 2008, University of Oslo (Norway) atmospheric scientist Jon E. Krisjánsson and coauthors looked at the correlation between Forbush Decrease events and cloud formation. They found that only in the Forbush events with the highest amplitude was there a positive correlation to cloud impact, and many of the correlations were weak. They also found introducing the time lag of a few days for cloud response to cosmic ray signal become weaker, counter to the Svensmark et al. five-day lag findings.[95] This challenges the reproducibility of Svensmark's observations. Yet others who ran this same experiment found a statistically significant correlation. Professor of Climate Change at the University of Sussex (UK), Dominic R. Kniveton, ran the same experiment in 2004 and found cloud formation declined during Forbush events, with statistical significance and a lag of several days after the start of the Forbush event, similar to the Svensmark et al. results.[96] Kniveton ran the experiment again with University of Sussex Professor of Environmental Science, Martin Todd, and similar cloud anomalies and lag were found as in his first study.[97] In addition, a study of the cosmic ray ionization of clouds from the Lerwick Observatory in Shetland during a Forbush event by University of Reading (UK)professors and atmospheric physicists, R. Giles Harrison and Maarten H. P. Ambaum, concluded there is a rapid timescale cloud response to cosmic ray changes.[98] **It is also of note that Svensmark ran the Forbush experiment twice, once in 2009 and again in 2021, both with similar results**. Perhaps more important, Svensmark's collaborators were different for his 2009 and 2021 studies. For the 2009 study his collaborators included atmospheric physics researcher Torsten Bondo, while in his 2021 study collaborators included atmospheric/space physics researcher Martin Bødker Enghoff and Israeli astrophysicist Nir Shaviv. Given that Svensmark was able to reproduce the results again in 2021 on five separate Forbush Decrease events with new collaborators, it appears he passed the test of demonstrating a repeatable experiment.

Besides solar cycles, the Earth's magnetic field shields the Earth from cosmic rays. The Earth periodically undergoes a magnetic field shift. During that shift there is a low magnetic field protecting the Earth, and the flux of cosmic rays increases significantly. This provides another event in which to test the impact of cosmic rays on cloud formation and climate change. Penn State paleoclimate scientist Richard Alley cites an instance of extremely high cosmic ray flux between 41,000 to 42,000 years ago from a magnetic field shift, which resulted in only a modest decline in temperature, thus putting a shadow on the impact of cosmic rays on climate.[99] According to ice cores studied by the Neils Bohr Institute, the decline in temperature was not modest. Temperatures were 3°C colder than today, and they plunged 4°C more after the increase in cosmic rays around 41,000 to 42,000 years ago, which is 7°C colder than today. Temperatures after this high cosmic ray flux 41,000 to 42,000 years ago dropped to one of the coldest temperatures of the 120,000-year interglacial period.[100] The most recent Ice Age glaciation of North America and Europe occurred between 70,000 to 20,000 years ago. This ice age has been attributed to the Milankovitch cycles of eccentricity, the Earth's orbit around the Sun, and obliquity and precession of the tilt of the Earth's axis. The geomagnetic shift occurred during the middle of this glaciation period; glaciers covered Northern Europe, Canada, and several northern states in America. The great Laurentide Ice Sheet covered nearly all of Canada and reached as far south as Chicago and St. Louis. This glacier was up to two miles thick in Nuavik, Quebec.

Alley argues that there were other times during past Ice Ages, with lower cosmic ray flux, that were as cold as temperatures during the geomagnetic shift. This fact is also true but may be due to Earth's Thermostat that keeps the world from getting too cold (see Chapter 11). As the temperature cools, water vapor in the atmosphere dramatically declines. According to the CLAW hypothesis, colder waters result in less phytoplankton growth, which is a major contributor to sulfate aerosols. Since cloud formation requires water vapor, aerosols, and positive ions (produced from cosmic rays), the decline in water vapor and sulfate aerosols in extreme cold limits the formation of low clouds, despite the presence of increased cosmic rays (see Chapter 12). This example given by Richard Alley does not disprove the impact of cosmic ray flux on cloud formation at today's warmer temperatures. It only shows that the impact on climate of cosmic rays may be limited in very cold temperatures due to the low water vapor content of the atmosphere and reduced sulfate aerosols when temperatures decline. It also demonstrates that you cannot isolate one variable in a complex climate system, which has multiple interdependent variables.

Alley's conclusions are in stark contrast to a paper in *Science* on this topic. Prominent University of Adelaide (Austraila) Professor of Evolutionary Biology, Alan Cooper and coauthors, created a precise dating of the Laschamps geomagnetic reversal about 41,000 to 42,000 years ago using radiocarbon records from swamp trees in New Zealand. They found a substantial increase in carbon-14 indicating a high cosmic ray flux. They found a major global climate shift causing marked environmental changes, extinction events, and transformations in the archaeological record.[101] Other studies have found a connection between the Earth's geomagnetic field and climate change. French Academy of Sciences geophysicist, Vincent Courtillot and coauthors reviewed evidence for connections between climate and the Earth's magnetic field variations going back one million years. Their study identified geomagnetism through its effect on cloud cover as an additional climate driver.[102]

Complutense University of Madrid (Spain) geophysicist Saioa A. Campuzano and coauthors conducted a study of the link between past climate changes and the geomagnetic field. The authors conclude that they found a connection between the Earth's magnetic field and climate with a confidence level of about 90%. They state that the most plausible explanation is the flux of galactic cosmic rays being modulated by the Earth's magnetic field and their role in cloud formation.[103] Paris Institute of Earth Physics geophysicist Yves Gallet and coauthors obtained archeological geomagnetic intensity data to document the occurrence of changes in the geomagnetic field in Western Europe over the past three millennia. The intensity variations show several maxima which coincide in time with cooling events documented in the region. "This coincidence suggests a causal link between enhanced secular variation of the geomagnetic field and climate change over centennial time scales, challenging the role of solar forcing as the sole factor provoking these climatic variations."[104]

Climate alarmists ignore the poor historical correlation between climate and CO_2 concentrations but cannot seem to accept the overwhelming evidence in all ages of the remarkable correlation between cosmic ray flux, ocean temperatures, and climate. I believe most competent scientists would agree there is a link between cosmic rays, atmospheric ionization, formation of cloud condensation nuclei, cloud cover, and climate change. The theoretical, experimental, and observational evidence supporting these facts is overwhelming. The question debated remains: how much is the impact of cosmic rays on climate? As described in Chapters 8 and 9, anthropogenic greenhouse gases cannot explain all the warming since 1850 or during past climate cycles. Unlike CO_2, the correlation between solar cycles and climate is consistently high. **What we do not know is how much of the warming from solar cycles is from TSI, UV heating of the stratosphere, the transport of heat to polar regions, or the impact on cloud cover from cosmic ray flux.** More research is needed to understand such nuances.

The World Will Be Getting Colder

Northumbria University (UK) professor and solar physicist and mathematician Valentina Zharkova has published several scientific papers on the Sun's magnetic field.[105] The magnetic field of the Sun is composed of two components, including the poloidal magnetic field and the toroidal magnetic field. The poloidal magnetic field runs north-south, with field lines along meridians, extending from the solar North Pole to the South Pole. The toroidal magnetic field wraps east-west, around the Sun's equator along lines of latitude. Toroidal fields are created by the shearing of poloidal fields by the Coriolis force due to the rotation of the Sun. Toroidal fields produce sunspots and poloidal fields are regenerated by the decay of sunspots through the Babcock-Leighton mechanism.

Using two years of continuous helioseismology observations from the Solar Dynamics Observatory, Stanford University helioseismologist Junwei Zhao and coauthors discovered a second meridional circulation cell in the poloidal magnetic field.[106] Building off the work of Zhoa et al. Zharkova and her research team applied Principal Component Analysis to magnetograms from the Wilcox Solar Observatory. They discovered poloidal and toroidal magnetic fields have two principal components with magnetic waves of opposite polarities.

Both the poloidal and toroidal magnetic fields are involved in producing the 11-year Schwab Solar Cycles, but it is the toroidal magnetic field that has the greatest impact on the strength of these cycles. The toroidal magnetic field produces sunspots, aurora, solar flares, and coronal mass ejections. It also enhances the Sun's ability to shield the solar system and the Earth from galactic cosmic rays by extending the heliospheric magnetic field far beyond the Sun. The toroidal magnetic field regulates the propagation of grand solar maxima and grand solar minima.

Zharkova and her team demonstrated that the total magnetic field of the Sun is modulated by two dynamo magnetic fields within the poloidal and toroidal fields. When these duo dynamo waves are in resonance, they create a strong magnetic field. When the waves are in anti-phase, they cancel each other out (using the same principle as sound cancelling headphones), resulting in a weak magnetic field.

Using principal component analysis of the duo dynamo waves of the toroidal field, Dr. Zharkova has been able to predict past solar maximums and solar minimums with great accuracy. Her models of solar cycles, using this magnetic field proxy, correlate remarkably with observed sunspots, cosmic rays, and temperatures as reconstructed from Carbon-14, Beryllium-10, and Oxygen-18 isotopes in ice cores, sediment samples, and stalactites. She has been able to retroactively predict the Dalton Minimum (1810-1830), Maunder Minimum (1645-1715), Wolf Grand Minimum (1200), Oort Grand Minimum (110-1050), and Homer Grand Minimum (800-900 BC) as well as the Medieval Warm Period (900-1200) and Roman Warm Period (400-10 BC).

University of Oulu (Finland) astrophysicist Ilya G. Usoskin and Russian Academy of Sciences astrophysicist Gennady A. Kovaltsov criticized criticized the findings of Zharkova and team stating that the past predictions of solar maxima and solar minima over the past 800 years using the duo dynamo model did not provide precise dating of these events. Furthermore, the Spörer solar minimum (1400-1550) was not produced at all in the Zharkova duo dynamo model.[107] Zharkova responded to this criticism with updated duo-dynamo model results that do provide precise dating of these past solar cycles with the exception of the Spörer Minimum, which is not produced by the model.[108] Zharkova explains the Spörer Minimum by attributing the increase in galactic cosmic rays due to supernovae (the source of cosmic rays) rather than a drop in the heliospheric magnetic field

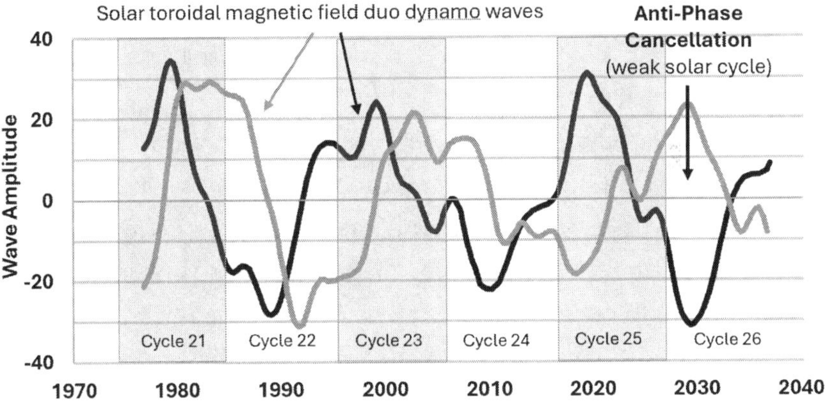

Figure 47 – Solar Minimum Predicted for Solar Cycle 26*. Principle component analysis of the duo dynamo fields of the toroidal magnetic field of the Sun reveal above that when the two dynamo fields are in resonance the solar cycles are strong as seen on the top (solar cycle 21 in 1980 and solar cycle 22 in 1990—the end of the Modern Solar Maximum). When they are in anti-phase, they cancel each other out, and we have weak solar cycles as we have had since 2010 with solar cycle 24. We entered solar cycle 25 in about 2020, and the growing anti-phase between the duo dynamo waves of the toroidal magnetic solar field should result in very weak solar cycles after 2031. Dr. Zharkova predicts solar cycle 25 will be similar to solar cycle 24 but solar cycle 26 will be very weak, like the Maunder Minimum that was experienced during the Little Ice Age.* Source: Zharkova, Valentina, "Heartbeat of the Sun from Principal Component Analysis and prediction of solar activity on a millennium timescale," *Nature Scientific Reports*, Volume 5, Article Number 15689, 29 October 2015. https://doi.org/10.1038/srep15689.

from a solar minimum.[109] The supernova Vela Junior, Tycho's supernova and the Kepler's supernova show that the galactic core of the Milky Way was very active during the time of the Spörer Minimum. Since isotope studies confirm temperatures dropped and cosmic ray flux spiked during the Spörer Minimum (see Figures 26, 40, and 41 in Chapter 5 and Figure 58 in Chapter 10), if the Spörer event was caused by supernovae rather than a solar minimum, this suggests the impact on clouds from cosmic rays may be the primary driver of this cold climate period during the Little Ice Age rather than a solar minimum that would reduce Total Solar Irradiance and cool the stratosphere, resulting in a weak polar vortex and more heat being transferred to polar regions where it easily goes out to space.

Schwab Solar Cycles occur every 11 years, and we are currently in the middle stages of solar cycle 25 (2020 to 2031). Dr. Zharkova accurately forecasted the decline of solar activity in cycle 24 (2009 to 2020), and the dual dynamo of magnetic waves reveals cycle 25 will be similar to cycle 24. This analysis of magnetic waves also foresees cycle 26 (2031 to 2042) as being exceptionally low due to the predicted anti-phase of the two solar dynamo waves. Dr. Zharkova warns cycle 26 should be like the Maunder Minimum of the Little Ice Age, and temperatures will begin to decline after 2031 (see Figure 47).[110] If Zharkova, Svensmark, and Vinós are correct, we will soon have the convergence of lower solar irradiance, higher cosmic rays, increased low clouds, a cooler stratosphere, and more heat passing out to space from polar regions due to a weaker polar vortex. This will also coincide with the beginning of natural cold phases of the oceans from both the Atlantic Multidecadal Oscillation and Pacific Decadal Oscillation. **The future looks cold after 2031. Fortunately, warming from anthropogenic greenhouse gases provides us with a cushion of warmth so that we will likely not face devastation from extreme cold temperatures as was experienced during the Maunder Solar Minimum in the seventeenth century**. Mark my words: future generations will be thankful for the modest warming provided by our current fossil fuel emissions.

The evidence that past climate cycles were significantly impacted by Milankovitch and solar cycles is overwhelming. In Chapter 8 we shall see that the evidence for CO_2 as a driver of past climate cycles is greatly lacking.

Chapter 8: Carbon Dioxide

CO2 Does Not Explain Past Climate Cycles

A recent paper in the peer-reviewed journal, *International Journal of Geosciences* by independent researhers Michael Nelson and David B. Nelson examined the correlation between CO_2 and global temperatures over several historical periods. They examined a 500-million-year period, a 50-million-year period, and the past one million years. The results showed that 70% of the time there was no correlation, or a negative correlation, between CO_2 concentrations in the atmosphere and global temperature. For the one-million-year period, there was no correlation, or negative correlation, 87% of the time. CO_2 concentrations in the atmosphere and temperatures moved in the opposite direction 35% of the time during the 500-million-year period, 42% of the time during the 50-million-year period, and 27% in the one-million-year period.[1] *This study shows that CO_2 is clearly not the primary driver of climate.*

We have already covered warming during the Holocene Climatic Optimum, Minoan Warm Period, Roman Warm Period, and the Medieval Warm Period. Historical, archeological, and paleoclimate evidence of global warming in these periods is beyond doubt (see Chapters 5 and 6). Historical CO_2 levels can be determined by measuring air bubbles trapped in ice cores. An examination of such ice cores clearly shows that elevated CO_2 levels have only been experienced since the industrial revolution, beginning in the nineteenth century (see Figure 48).

Since CO_2 levels cannot explain the Holocene Climatic Optimum, Minoan Warm Period, Roman Optimum, Medieval Warm Period, or cold Greek Dark Ages, cold Dark Ages, and Little Ice Age, such warming and cooling must have come from natural causes, such as Milankovitch cycles and the 1,000-year Eddy Solar Cycle, which matches these climate periods and the current Modern Warming exactly (see Chapters 5, 6, and 7). Other than minor blips in concentration levels, CO_2 shows no major correlation with the Holocene Climatic Optimum, Minoan Warm Period, Roman Warm Period, or Medieval Warm Period in which temperatures were as warm or

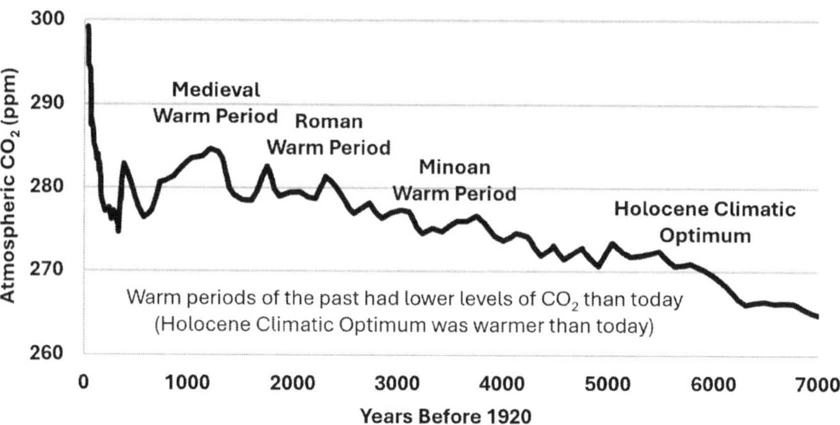

Figure 48 – CO_2 Concentrations Over Seven Millennia Do Not Correlate with Climate Cycles. The chart above is a reconstruction of CO_2 concentrations in the atmosphere over the past 7,000 years. The Holocene Climatic Optimum 7,000 years ago was about 2°C warmer than today in the Pacific Ocean and in Greenland, yet CO_2 levels were much lower than today. Source: Indermühle A. et al., "Holocene carbon-cycle dynamics based on CO_2 trapped in ice at Taylor Dome, Antarctica," *Nature*, Volume 398, pp. 121–126, 11 March 1999. https://doi.org/10.1038/18158.

warmer than today. Since these climate cycles have been established beyond any doubt, this data is convincing evidence that natural variation is a major driver of climate. Yet despite this evidence, the IPCC states that it is very likely that well-mixed greenhouse gases were the main driver of tropospheric warming in recent years.[2] This assumption is not supported by data. There is no doubt that increased levels of CO_2 warm the atmosphere. However, there is also no doubt that a sizeable portion of Modern Warming is from natural climate cycles.

A recent paper by statistician John K. Dagsvik and information technologist Sigmond H. Moen from Statistics Norway finds a weak correlation between CO_2 emissions and global warming. The paper states, "Using theoretical arguments and statistical tests we find that the effect of human-caused CO_2 emissions does not appear to be strong enough to cause systematic changes in the temperature fluctuations during the last 200 years."[3]

Between 1944 and 1976 the burning of fossil fuels exploded, resulting in the annual emissions of CO_2 growing from around one gigaton in 1944 to nearly five gigatons by 1977. Yet the temperature declined during this 33-year period despite a five-fold increase in CO_2 emissions (see Figure 49). Many scientists thought the temperature decline would continue, and they sounded alarm of "The Big Freeze" and a coming ice age (see Figure 76 in Chapter 13). The five-fold increase in CO_2 emissions would have driven higher radiative forcing from CO_2, but such CO_2-induced warming was more than offset by natural cooling, including the convergence of the cold ocean cycles of the Atlantic Multidecadal Oscillation and Pacific Decadal Oscillation. IPCC reports offer no explanation for this cooling period from 1944 to 1976.[4] This period shows that oceans, which are warmed by the Sun, can have more impact on the climate than greenhouse gases.

There has been extensive climate research on the Holocene. There are hundreds of proxy records analyzed by University of Wisconsin paleoclimatologist Shaun Marcott in 2013.[5] More recently the

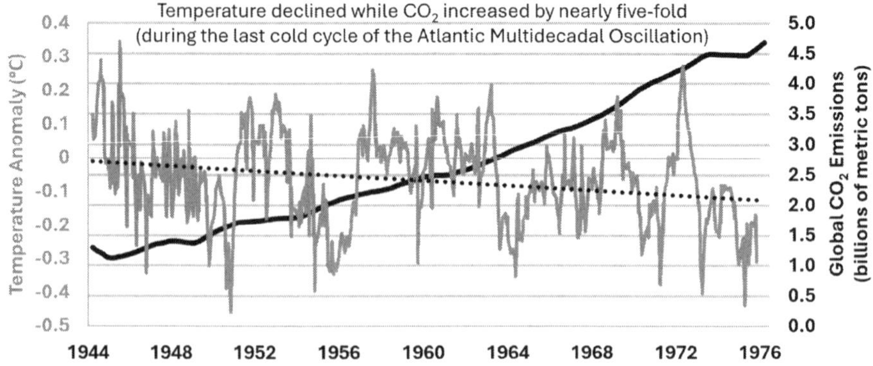

Figure 49 – Cooling During the 1940s through 1970s Does Not Correlate with Increases in CO_2 Concentrations. *Between 1944 and 1976 the temperature dropped despite an increase in CO_2 emissions from about one gigaton in 1944 to nearly five gigatons by 1976.* Sources: Wrightstone, Gregory, "Fact #27, CO_2 rose after the Second World War, but temperature fell," CO_2 Coalition. https://co2coalition.org/facts/ ; (Temperature Anomaly) from Hard CRUT4 (1977) and CO_2; (Global CO_2 Emissions) from Boden et al., "Global CO_2 emissions from Fossil-Fuel Burning and Cement Manufacture and Gas Flaring 1751–2013" (2016), CDIC, Oak Ridge National Laboratory, U.S. Department of Energy.

Northern Arizona University paleoclimatologist Darrell Kaufman paper with 92 other authors uses 1,319 proxies including lake sediments, marine sediments, peat, glacier ice, and other natural records to establish regional and global temperature trends.[6] We are currently in the Holocene, which started about 11,000 years ago. According to these proxy records, the Holocene Climatic Optimum (HCO), around 6,000 to 7,000 years ago, was a time when temperatures were warmer than today. This is confirmed with the glacier record of the Alps. These glaciers have remained ice covered for the past 5,900 years. However, a study by Austrian Academy of Sciences glaceologist Pascal Bohleber and coauthors shows nearly ice-free conditions during the HCO.[7] According to Spanish scientist Javier Vinós, "The Holocene Climatic Optimum was the period in the last 100,000 years when glaciers were at their smallest, while the Little Ice Age was the period in the last 7,000 years when glaciers were at their largest."[8] Furthermore, numerous studies have shown tree lines during the HCO in the Italian Alps, Swiss Central Alps, Pyrenees, Swedish Scandinavia, and British Columbia were much higher than modern times, indicating the HCO was warmer than today.[9]

This warming impacted most of the globe, with the exception of Antarctica. In a post by Renee Hannon, levels of CO_2 and temperature reconstructions of the Holocene are discussed. Antarctica temperatures seem to follow CO_2 levels, but the rest of the globe does

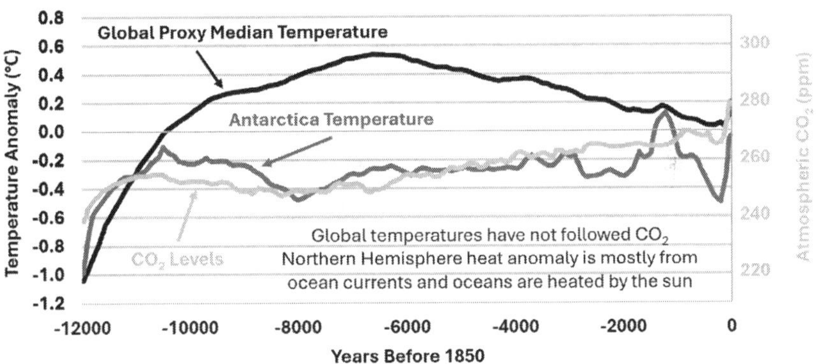

Figure 50 – Holocene Temperature and CO_2 Reconstruction. *Hundreds of proxy records of the Holocene allow for a credible reconstruction of global and regional temperature and CO_2 levels. Antarctica temperatures seem to follow CO_2 levels, but the remainder of the globe does not. This is especially true in the well-documented Holocene Climatic Optimum (HCO) around 6,000 to 7,000 years ago. Global temperatures were higher than today, but CO_2 levels were lower than at present. This lack of correlation suggests there are natural causes of global warming far more powerful than CO_2 levels.* Sources: (Atmospheric CO_2) data from: Monnin, Eric et al., "Evidence for substantial accumulation rate variability in Antarctica during the Holocene, through synchronization of CO_2 in the Taylor Dome, Dome C and DML ice cores," *Earth and Planetary Science Letters,* Volume 224, Issues 1-2, ppl 45-54, 30 July 2004. https://doi.org/10.1016/j.epsl.2004.05.007; (Temperature Anomaly) from Kaufman, D. McKay, N., Routson, C. et al., "A global database of Holocene paleotemperature records," *Scientific Data* 7, 115, 12 August 2020. https://doi.org/10.1038/s41597-020-0445-3; See also, Hannon, Renee, "The Holocene CO_2 Dilemma," *Andy May Petrophysicist,* May 26, 2023. https://andymaypetrophysicist.com/2023/05/26/the-holocene-co2-dilemma/.

not.[10] This is likely due to ocean currents surrounding Antarctica that do not allow warm water to penetrate to the shores of that continent and transfer the heat to the atmosphere (see Chapter 13). The same situation holds today. Satellite records show the warming at the North Pole is 25 times (x) that of the South Pole (see Figure 68 in Chapter 13).[11] During the Holocene Climatic Optimum and Modern Warming, ocean warming from the Sun and the transfer of this heat to the Northern Hemisphere and Arctic dominate regional warming.

During the HCO, temperatures were warmer globally than today, yet CO_2 levels were about 260 ppm. Today, CO_2 concentrations are about 427 ppm (see Figure 50). The lack of correlation between CO_2 levels and the global temperatures during the HCO suggests there are natural causes that are far more important than CO_2 levels in forcing climate change. Since climate models rely on CO_2 as the primary driver of temperature, they cannot reproduce the well-documented Holocene Climatic Optimum.

Temperature Leads and CO_2 Follows

In studying paleo reconstructions of the Earth's past temperatures, one fact is clear: CO_2 has never been the primary driver of climate as declared by climate alarmists. Temperature reconstructions from Antarctica ice cores have provided a history of climate for more than 400,000 years (see Figure 51). The ice cores reveal a correlation between increased temperature and CO_2 levels, but in this case, correlation is not causative. In the movie *An Inconvenient Truth*, former U.S. Vice President Al Gore presents Antarctica ice core temperature reconstructions to make the claim that CO_2 is the driver of warming temperatures during each interglacial period, which occurs every 100,000 years. Citing CO_2 as the primary driver of these past interglacial temperatures recorded in Antarctica ice cores is ridiculous, which any climate scientist should know. Yet you can still find countless lectures online that perpetuate this false claim.

The reasons this claim has no merit are as follows:

Milankovitch Cycles: First, it has been well-established that these 100,000-year glacial periods are caused not by CO_2 but are attributed to the Milankovitch cycles of eccentricity, of the Earth's orbit around the Sun. Over 100,000 years, the Earth's orbital shape cycles from more elliptical around the Sun to less elliptical. Known as eccentricity, this orbital change is understood to be a major factor in generating and ending ice ages as it moves the Earth farther from or closer to the Sun and lowers or increases solar shortwave radiation from the Earth's Energy Budget on a 100,000-year cycle.

Minimal CO_2 Radiative Forcing: Second, you will notice the temperature varies by over 11°C, but CO_2 concentrations vary by only 120 ppm from a low of 180 ppm to a high of 300 ppm. Using the ERF adjusted Myhre radiative forcing equation for CO_2,[12] increasing from 180 ppm of CO_2 to 300 ppm of CO_2 increases radiative forcing by 2.2 W/m², which is only 0.7°C (see Tables 7 and 8; see also, Chapter 9). There is no physical way for a 120 ppm increase in CO_2 from 180 ppm to 300 ppm to increase temperature by 11°C. Such a claim is pseudoscience.

Table 7 Effective Radiative Forcing of CO_2
180 ppm to 300 ppm of CO_2
K × ln(C/Co),
K = 4.33
C = 300 ppm
Co = 180 ppm
300 ppm/180 ppm = 1.667
ln(1.667) = 0.5108
4.33 × 0.5108 = 2.21 W/m²
235.54 W/m² + 2.21 = 237.75 W/m²

Table 8 Stefan-Boltzmann Conversion of Heat to Temperature	
180 ppm of CO_2	**300 ppm of CO_2**
$m^2 = 1$	$m^2 = 1$
Effective Emissivity of the Earth = 0.6127	Effective Emissivity of the Earth = 0.6127
Radiative Forcing = 235.5 W/m^2	Radiative Forcing = 235.54 + 2.21 = 237.75 W/m^2
Temperature = 286.94 K	Temperature = 287.62 K
	Temperature Change = 0.68°C

Table 8 Source: *Stefan-Boltzmann Law Calculator, BYJU'S. https://byjus.com/stefan-boltzmann-law-calculator/.*

Temperature Leads CO_2: Third, CO_2 rises in hotter times and falls in colder times, but it follows, not proceeds, the heating and cooling, usually by 200 to 1,000 years. This can be clearly seen in Antarctica ice cores.[13] In fact, the temperature begins to fall just before CO_2 levels peak in each cycle. According to Institute of Oceanography (Russia) oceanographer, Nadezda Vakulenko and coauthors, "On the whole, the temperature variations turned out to be 800 ± 200 yrs. ahead of the Greenhouse Gas Concentration (GGC) variations."[14] The authors also state, "Thus, one can conclude that temperature variations always preceded GGC variations during the four main glacial cycles recorded in the Vostok ice core. Of particular importance is the fact that the temperature began to decrease after reaching a very high value, although the GGC values continued to increase."[15] If CO_2 were the control knob of temperature, the temperature would not decline as CO_2 levels continued to rise.

Several studies confirm the fact that temperature precedes changes in CO_2 levels. National Center for Scientific Research (France) paleoclimatologist and glaciologist Nicolas Caillon and coauthors studied the Antarctica Vostok ice sheet over the past 240,000 years.

They conclude the series of events during this period "suggests that CO_2 increase lagged Antarctic deglacial warming by 800 ± 200 years … "[16] One of the world's leading ice core researchers from the University of Bern, Hubertus Fischer, and coauthors studied the relationship between greenhouse gases and climate in the past by examining air trapped in bubbles in polar ice cores. They conclude: "High-resolution records from Antarctic ice cores show that carbon dioxide concentrations increased by 80 to 100 parts per million by volume 600 ± 400 years after the warming of the last three deglaciations."[17] University of Bern (Switzerland) climate physicist and ice core expert, Eric Monnin, and coauthors studied ice cores from the Dome Concorda (Dome C) Antarctica covering a period from 22,000 years before present to 9,000 years before present. "We found that the start of the CO_2 increase thus lagged the start of the δD increase by 800 ± 600 years, taking the uncertainties of the gas-ice age differ-

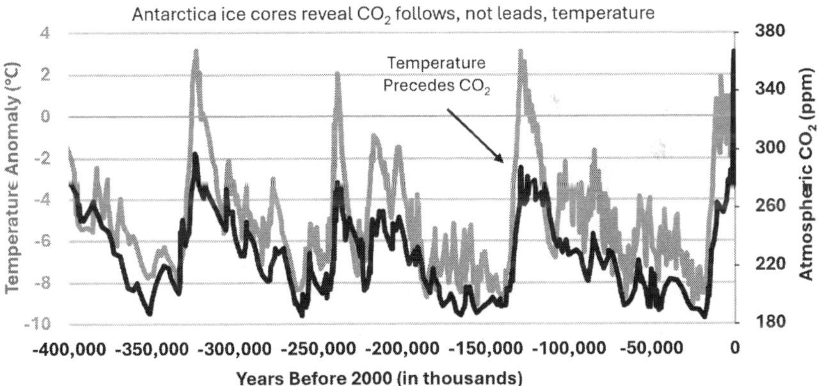

Figure 51 - Temperature Leads CO_2 Concentrations. *Antarctic ice cores show that temperature precedes CO_2 levels by 200 to 1,000 years. CO_2 is not the primary driver of temperature. CO_2 increases as the oceans outgas CO_2 at higher temperatures. The oceans absorb CO_2 at lower temperatures.* Source: Pettit, J. R. et al., "Climate and atmospheric history of the past 420,000 years from the Vostok ice core, Antarctica," *Nature*, Volume 399, pp. 429-436, 3 June 1999. https://doi.org/10.1038/20859.

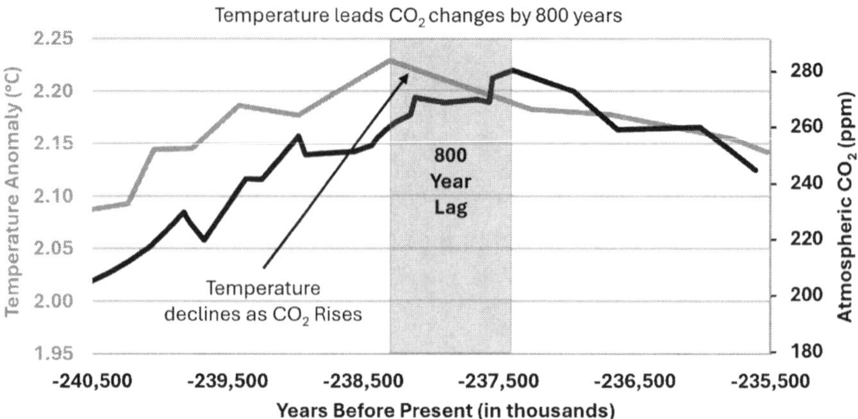

Figure 52 – CO_2 Lags Temperature by About 800 Years. *A closeup of temperature and CO_2 records from an Antarctica ice core clearly shows the delay of 800 years between temperature levels and subsequent CO_2 levels. The chart also shows how CO_2 levels continue to rise while temperatures are falling.* Source: Ian Clark, University of Ottawa, Antarctica ice core data from Vostok. Clark, Ian, "CO_2 lags temperature rises in ice core data," Skeptical views of CO_2 driven global warming. http://hyperphysics.phy-astr.gsu.edu/hbase/thermo/icecore.html.

ence and the determination of the increases into account."[18] ("δD" is a proxy for surface air temperature based upon the deuterium abundance of the ice.) University of Tasmania (Australia) paleoclimatologist Joel B. Pedro and coauthors studied Antarctic ice cores over the last deglaciation between 19,000 years before present to 11,000 years before present. They reported: "Utilizing a recently developed proxy for regional Antarctic temperature, derived from five near-coastal ice cores and two ice core CO_2 records with high dating precision, we show that the increase in CO_2 likely lagged the increase in regional Antarctic temperature by less than 400 years and that even a short lead of CO_2 over temperature cannot be excluded."[19]

Outgassing of CO_2 From the Oceans: The change in CO_2 is explained by the oceans' absorption and release of CO_2. The oceans contain 60 times more CO_2 than the atmosphere.[20] Just as a Coca-Cola or beer remains carbonated in the refrigerator, but goes flat when it is warmed, so the oceans absorb and release CO_2 subsequent to temperature changes, pursuant to Henry's Law. Henry's Law states that the amount of a gas dissolved in a liquid is directly proportional to its partial pressure above the liquid and inversely related to temperature. According to Henry's Law, at constant pressure, the solubility of CO_2 decreases by 4% for each 1°C increase in temperature leading to outgassing.[21] Using Henry's Law to calculate the outgassing of CO_2 for each degree Celsius of warming, CO_2 concentrations in the atmosphere may increase by approximately 6 ppm from the top 100 meters of the ocean to 10 ppm if the entire ocean is considered.[22]

This is only part of the story. Internationally recognized ocean and climate scientists, Princeton professor Jorge L. Sarmiento and ETH Zurich professor Nicholas Grüber, discuss how other factors enhance outgassing.[23] When temperatures rise, there is more deep water mixing of ocean waters bringing CO_2 up from deep carbon reservoirs, and this releases more stored CO_2 into the atmosphere. Isotope analysis in benthic foraminifera from deep ocean sediments show deep ocean carbon accumulations during ice ages and subsequent release in warmer times.[24] Sea ice blocks gas exchange between the ocean and atmosphere. Warmer weather melts ice, which not only releases CO_2 trapped in ice bubbles in the ice, but it allows greater ocean surface area to outgas directly from the ocean surface into the atmosphere. Paleo-sea-ice reconstructions show links between ice edge retreat and CO_2 spikes.[25] Warming and CO_2 release lowers ocean pH which releases more CO_2 from deep-sea shells, which are made of carbonate ($CaCO_3$). Marine sediment cores show decreased carbonate preservation during deglaciation, consistent with this buffering process.[26] Warmer temperatures also melt permafrost, and this releases CO_2 from aerobic decay and methane from anaerobic decay. The methane released from permafrost converts into CO_2 and water vapor within nine to 12 years through oxidation.

All of these processes result in the release of about 10 ppm or more for each degree Celsius in temperature increases. Therefore, it is more than reasonable to assume that 11°C of warming would outgas CO_2 in the atmosphere by 120 ppm as seen in the Antarctic ice core data. Temperature absolutely drives CO_2 concentrations more than CO_2 drives temperature.

The conclusion that temperature drives CO_2 is also found in observations over the past 60 years. A paper published in *Science* in 2023 by highly respected National Technical University of Athens (Greece) professor and hydrologist Dementris Koutsoyiannis and coauthors confirms temperature as the driver of CO_2 through stochastic evaluation of UAH satellite temperature data and CO_2 measurements at Mauna Loa, Hawaii. The authors of this paper found that all evidence resulting from the analysis establishes a causal link with temperature as the cause and CO_2 as the effect. Their conclusion is that CO_2 concentrations do not drive temperature, but temperature drives CO_2 concentrations. This direction of causality holds for the entire 60-year period covered by the observations, and the causal link applies to all timescales of the available data from monthly to decadal.[27]

The bottom line from the Antarctica ice core studies and the Koutsoyiannis et al. study is not that CO_2 has no impact on warming, but that its impact is tepid compared to the considerable impact temperature has on CO_2 absorption and outgassing from the oceans. The Antarctica ice core studies show us that an 11°C increase in temperature changes CO_2 concentrations from 180 ppm to 300 ppm, which is a 67% increase. At the same time, the 120 ppm CO_2 increase (from 180 ppm to 300 ppm) results in a small temperature increase, from 286.9 K to 287.6 K, which is only a 0.2% increase. From this perspective, temperature drives CO_2 concentrations in both historical ice core records and modern measurements of temperature and CO_2. In comparison, CO_2 has little impact on temperature.

Studies have consistently shown that CO_2 has not been the control knob of climate over the past 500 million years, 50 million years, one million years, 400,000 years, 11,000 years, 200 years, and even 60 years. The scientific reasons will be explained in Chapters 9 and 10. In Chapter 9, we will explore the science of greenhouse gases and see how the radiative forcing of human CO_2 emissions will not and cannot lead to extreme temperatures as claimed by some climate alarmists.

Chapter 9:

Science of Greenhouse Gases

CO2 Warming Is Not Significant

The Sun warms the Earth primarily from shortwave radiation of visible light and ultraviolet radiation. The Earth radiates this heat back into space in the form of longwave infrared radiation. When the amount of heat received by the Earth equals the amount of heat radiated to space from the Earth, the system is in balance and the temperature remains constant. Greenhouse gases and clouds absorb some of the infrared radiation emitted from the Earth and emit some of this radiation back to Earth, and this disturbs the Earth's Energy Budget by reducing infrared radiation sent out to space. This decrease in longwave infrared radiation going out to space warms the atmospheric temperature. This process is known as the greenhouse effect. Table 9 sets forth the estimated greenhouse effect of each greenhouse effect substance. If there were no greenhouse effect, the Earth would be about 33°C colder than it is.

Table 9 Greenhouse Effect		
Greenhouse Substance	**Warming Impact**	**Percent of Total**
Water Vapor	16.5°C	50%
Clouds	8.3°C	25%
Carbon Dioxide	6.3°C	19%
Nitrous Oxide	0.8°C	2%
Methane	0.6°C	2%
Ozone	0.5°C	2%
Halogenated Gases	<0.1°C	<1%
TOTAL	~33°C	100%

Increases in CO_2 do raise the Earth's temperature by reducing the amount of longwave radiation going out to space. **However, counter to the popular narrative, satellite observations have demonstrated that CO_2 greenhouse gas warming is not the primary driver of recent global warming.** If CO_2 greenhouse gas warming were the major source of recent warming, then the amount of outgoing long-wave radiation at the top of the atmosphere would be decreasing as CO_2 traps increasing amounts of infrared heat.[1] However, Clouds and the Earth's Radiant Energy System (CERES) satellite measurements since 2001 have recorded an increase of about 0.6 W/m^2, rather than a decrease, in outgoing longwave radiation.[2] This may be explained due to a decline of low-cloud coverage since 2001, since clouds have a stronger greenhouse effect than CO_2. During this same period, incoming shortwave solar radiation has increased, consistent with lower cloud cover.[3] University of Wisconsin professor and atmo-spheric scientist Tristan L'Ecuyer and coauthors estimate that clouds have a greenhouse warming effect of 26.3 ± 3.8 W/m^2 and a cooling impact of -51.1 ± 7.8 W/m^2 due to their white color, which reflects incoming solar shortwave radiation back out to space.[4]

USDA physical scientist Ned Nikolov and USDA atmospheric scientist Karl Zeller state, "The assumption that Earth's albedo has been decreasing for the past 37 years (and thus driving climate change) is based on the latest version of the ISCCP-FH radiative-profile flux product by the International Satellite Cloud Climatology Project (ISCCP), which shows a 6.5% reduction of the global cloud-cover fraction between 1986 and 2018."[5] Using this rate of decline, over a 22-year period from 2001 to 2023, the decline in cloud cover would be about 4.5%. This number is consistent with decline of cloud cover of 1.5% to 3.0% per decade over the past 24 years, as measured by the Moderate Resolution Imaging Spectroradiometer (MODIS) M3 Terra Total Cloud Cover satellite.[6]

From 2001 to 2023 there has been a 14% increase in concentrations of CO_2 (370 ppm to 420 ppm). Using the accepted ERF adjusted Myhre radiative forcing equations, this 14% increase equates to about 0.5 W/m^2 less longwave radiation being sent out into space.[7] When you include all anthropogenic greenhouse gases (GHG), the trapped longwave radiation increases to about 0.8 W/m^2. Therefore, if cloud cover did not change, CERES satellites would have measured a decrease in outgoing longwave radiation of -0.8 W/m^2. However, CERES satellites measured +0.6 W/m^2 of longwave radiation going out to space during this period (see Figure 53).[8] This means the reduction in cloud cover of 4.5% between 2001 to 2023 must have resulted in 1.4 W/m^2 more longwave radiation escaping out to space since -0.8 (GHG) + 1.4 (Clouds) = +0.6 W/m^2 (CERES net measurement). Since the greenhouse effect of clouds is -26.3 ± 3.8 W/m^2, a decrease in cloud cover of 4.5% results in -4.5% of -26.3 ± 3.8 W/m^2 or +1.2 ± 0.2 W/m^2 more longwave infrared heat radiated out to space. Since +1.4 W/m^2 is within this range, all the net +0.6 W/m^2 of observed outgoing longwave radiation increase between 2001 to 2023 can be explained from lower cloud cover offsetting the radiative forcing from anthropogenic greenhouse gases.

As cited by Tristan L'Ecuyer et al. as a global average, the cooling effect of clouds is −51.1 ± 7.8 W/m^2 by reflecting shortwave radiation out to space off the white tops of clouds.[9] A decrease of 4.5% in cloud cover would increase incoming shortwave radiation by 4.5% of -51.1 ± 7.8 W/m^2 or +2.3 W/m^2 ± 0.4 W/m^2. The increased amount of incoming shortwave radiation measured by CERES satellites between 2001 and 2023 is +2.0 W/m^2.[10] These measurements align with the estimate of increased incoming shortwave radiation of 2.3 W/m^2 ± 0.4 W/m^2. **The data suggests warming since the year 2001 is not primarily from increases in CO_2 and other anthropogenic greenhouse gases, but from a 4.5% decrease in cloud cover.** This also illustrates

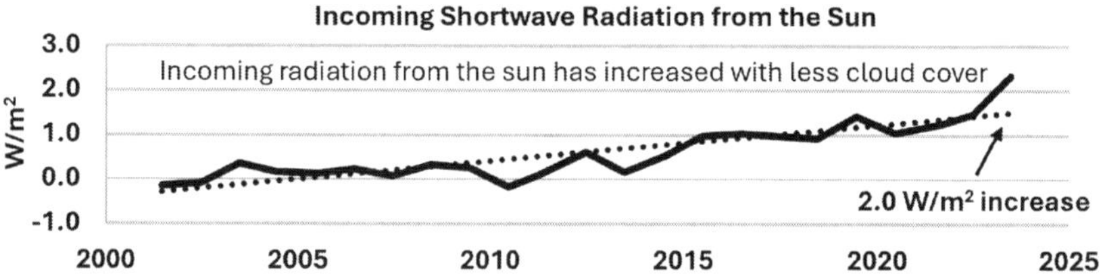

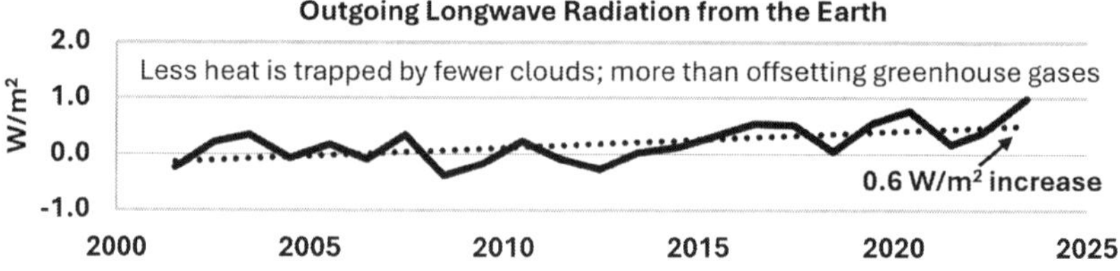

Figure 53 – Increase in Both Incoming Solar Shortwave Radiation and Outgoing Longwave Radiation. *The graphs above chart CERES satellite data to show the increase of both longwave radiation from the Earth to space and an increase of incoming shortwave radiation from the Sun to the Earth between 2001 to 2023. Measurements show a 2.0 W/m^2 increase in incoming shortwave radiation from the Sun to the Earth and a 0.6 W/m^2 increase in outgoing longwave radiation from the Earth to space.* Data Source: Qualglia, Ilaria and Visioni, Daniele, "Modeling 2020 regulatory changes in international shipping changes in international shipping emissions helps explain anomalous 2023 warming," *EDS Letters,* Volume 15, Issue 6, pp. 1527-1441, 28 November 2024. https://doi.org/10.5194/esd-15-1527-2024.

that the climate is much more sensitive to changes in cloud cover than to changes in anthropogenic greenhouse gases.

CO_2 is limited in its ability to warm as it can only absorb and emit warming from within a very narrow spectrum of infrared radiation based upon quantum mechanics. The full spectrum of infrared radiation is from wavelengths of 0.7 to 100 micrometers, but CO_2 can only absorb radiation in the ranges of 2 to 4 micrometer and 13 to 17 micrometer wavelengths. The Earth radiates very little infrared radiation at the 2 to 4 micrometer wavelengths, so there is little radiation in this spectrum for CO_2 to absorb. As a result, the 13-to-17-micron wavelength is the only meaningful absorption spectrum of CO_2.

Water vapor, which is a powerful greenhouse gas, absorbs over half of the infrared radiation at these 13 - to 17-micrometer wavelengths (see Figure 54). Therefore, the amount of radiation CO_2 can absorb and emit back to the Earth is limited because most of the energy in its radiative absorption spectrum is already absorbed by water vapor. This is especially true on the Earth's surface, where there is ten times more water vapor than CO_2 in the atmosphere. At current levels of water vapor and CO_2, 99.4% of all radiation in the 13 - to 17-micrometer wavelength is absorbed between the ground and the first 10 meters of the atmosphere. Therefore, increased levels of CO_2 cannot absorb any meaningful additional heat near the surface of the Earth because there is so little leftover heat to absorb. This phenomenon is known as *infrared absorption saturation or greenhouse gas saturation.*

Because of infrared absorption saturation, most of the CO_2 greenhouse warming takes place at higher altitudes in the troposphere, where there is less water vapor and where infrared radiation absorption is less saturated. Such warming in the higher altitudes of the troposphere does raise the surface temperature by changing the amount of heat radiated out to space, which alters the Earth's Energy Budget. This warming high in the atmosphere expands the altitude of the troposphere, resulting in an increase in the tropopause height and a temperature increase on the Earth's surface.

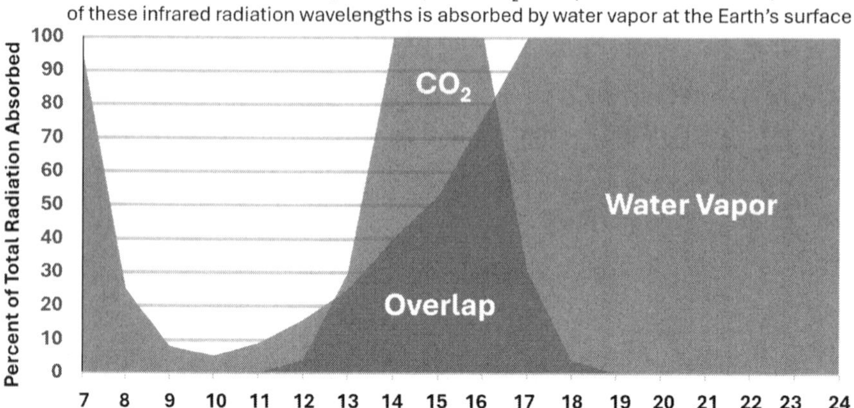

Figure 54 – Overlap of Water Vapor and CO_2 Infrared Radiation Absorption Bands. *Greenhouse gases absorb some of the heat radiating from the Earth, which lowers the amount of heat radiated from the Earth out to space. This changes the Earth's Energy Budget and warms the Earth. Each greenhouse gas can only absorb radiation from a limited wavelength spectrum unique to the quantum mechanics of the molecule. As seen in the chart, water vapor is the strongest greenhouse gas absorbing radiation from a broad spectrum of radiation wavelengths. The only effective radiation band of CO_2 is in the 13 to 17 micrometer wavelengths. However, from the chart above, you can see that water vapor absorbs over half of the radiation in the 13 to 17 micrometer wavelengths, leaving less radiation for CO_2 to absorb.* Data Source: Wright, Jason, "Name That Function," *Astro Wright, Astronomy and Meta-Astronomy* by Jason Wright. https://sites.psu.edu/astrowright/2012/06/04/name-that-function/.

Logarithmic Decline of CO_2 Marginal Warming

Perhaps the most important scientific principle which goes against the climate alarmist narrative is the fact that the power of CO_2 to warm (known as radiative forcing) declines dramatically as concentration is increased. For example, an increase of 400 ppm of CO_2, from today's levels, produces approximately 0.9°C increase in temperature. An additional 800 ppm increase is required for another 0.9°C increase in temperature, an additional 1,600 ppm for another 0.9°C increase in temperature, and an additional 3,200 ppm is required for yet another 0.9°C increase in temperature. The equation for calculating the change in radiative forcing for increasing concentrations of CO_2 is:

Change in Radiative Forcing = K × ln(C/Co)

Change in radiative forcing is measured in watts per square meter (W/m^2)

"K" is a constant (4.3 for CO_2)

"ln" is the natural log

"C" is the new concentration in parts per million (ppm) for CO_2

"Co" is the original concentration in parts per million (ppm) for CO_2

This equation was formalized in a 1998 paper by University of Oslo climate scientist Gunnar Myhre and coauthors.[11] It was derived from detailed radiative transfer calculations through radiative transfer simulations and spectroscopic data. The unique equations to calculate radiative forcing of CO_2, methane, nitrous oxide, and halogenated gases are set forth in Table 3 of the Myhre et al. paper. These equations have since been validated and widely applied in studies including those by the IPCC.[12]

Known as the radiative forcing coefficient for CO_2, the constant K varies for each greenhouse gas, but for CO_2 the constant is 5.35.[13] This is the value of the constant under fixed conditions known as Instantaneous Radiative Forcing (IRF). Under IRF, nothing in the atmosphere is allowed to adjust. However, in the real world, the stratosphere, tropospheric temperature structure, water vapor, and clouds all adjust rapidly to the increased radiative forcing. Adjusting for these real-world responses is known as Effective Radiative Forcing

(ERF). If you fit ERF into a Myhre-like logarithmic equation and consider the impact of overlapping greenhouse gases, you get a lower constant between about 4.2 to 4.4 instead of 5.35. Improved spectroscopy from the updated High-resolution Transmission Molecular Absorption database (HITRAN) has improved handling of the overlap and atmospheric adjustments for ERF. ERF correlates better with observed temperature response and thus is more representative of the real-world climate. "ERF gives a more complete picture of the overall expected energy budget change."[14]

In 2015, Lawrence Berkley National Laboratory climate scientist Daniel R. Feldman and coauthors observed through measurements the radiative forcing for increasing CO_2 from 368 ppm in 2000 to 390 ppm in 2010 to be 0.2 W/m^2, which infers an ERF constant K for CO_2 of 3.45 ($0.2 \div \ln(390/368) = 3.45$).[15] In 2019, York University (Canada) physicist William van Wijngaarden and Princeton University physicist William Happer estimated the radiative forcing for doubling CO_2 from 400 ppm to 800 ppm at 3.0 W/m^2, which implies a ERF coefficient for CO_2 constant K to be 4.33 ($3.0 \div \ln(800/400) = 4.33$).[16] Happer and Wijngaarden appear to perhaps be the most accurate, as they conducted a detailed spectral line-by-line analysis of one-third of a million infrared absorption intensities from the HITRAN database, and they validated their model using satellite observational data.[17] For my calculations, I use a radiative forcing coefficient constant K of 4.33, which is consistent with both the detailed analysis of HITRAN data performed by van Wijngaarden and Happer (3 W/m^2) and CO_2 radiative forcing estimated by MODTRAN radiation transfer software (3 W/m^2), which we will cover later in this chapter.[18]

As shown in Table 10 below, you need to double CO_2 concentrations successively to achieve the corresponding one unit of increases of heating. To increase heat by 5-fold requires a 32-fold increase in CO_2 concentrations. In other words, to force a 4.5°C temperature increase from CO_2 emissions would require increasing CO_2 levels from today's

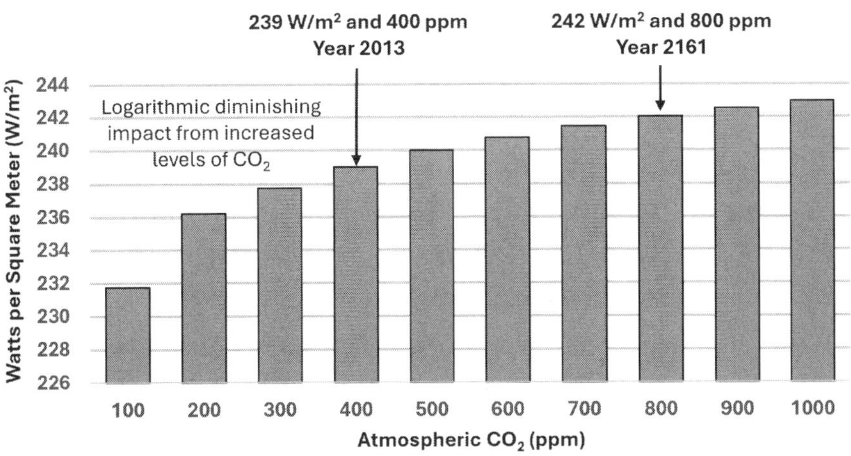

Figure 55 – Logarithmic Decline of CO_2 Marginal Warming Effect with Increasing Concentration. *Successive doubling of CO_2 concentrations is required to achieve successive one-unit increases to heating. As a result, each additional ppm increase in CO_2 concentration has a logarithmically declining impact on warming as seen in the logarithmic decay curve in the chart above. Moving from 100 ppm to 200 ppm of CO_2 has much more impact on temperature than moving from 400 ppm to 500 ppm. Doubling CO_2 from 400 ppm in 2013 to 800 ppm increases radiative forcing by only 3 W/m² and, assuming the current annual increase of ~2.7 ppm per year, it would take nearly 150 years. Assumptions in Chart: Change in CO_2 Radiative Forcing = 4.33 × natural log / (concentration original concentration).*

level of 427 ppm to 13,860 ppm, which at today's increase of ~2.7 ppm per year[19] would take nearly 5,000 years!

Table 10 Logarithmic Decline of the Natural Log	
Natural Log of Successive Doubling	**Equals Only Unit Increases ($32\times CO_2 = 5\times$ Temperature)**
ln(2) = 0.693	0.693 = 1 × 0.693
ln(4) = 1.386	1.386 = 2 × 0.693
ln(8) = 2.079	2.079 = 3 × 0.693
ln(16) = 2.722	2.722 = 4 × 0.693
ln(32) = 3.465	3.46 = 5 × 0.693

Because "ln" is the natural logarithmic function, the greenhouse warming of CO_2 is not linear to CO_2 concentrations but decreases logarithmically, so increasing levels of CO_2 have only a modest and rapidly declining impact on temperature. See Figures 55 and 56 for visualizations of the logarithmic decline of marginal radiative forcing (marginal warming) from CO_2 as concentrations of CO_2 increase.

Another factor which reduces the power of greenhouse gases to increase temperature is the Stefan-Boltzmann Law. The Stefan-Boltzmann Law is used to convert heat (in watts per square meter) of a black body, such as from radiative forcing on the Earth, into temperature (in Kelvin). This law states that the total energy radiated per unit surface area of a black body is directly proportional to the fourth power of its absolute temperature. Temperature is thus the fourth root of the energy divided by the Stefan-Boltzmann constant. Therefore, it takes exponentially more energy to increase temperature, to the power of 4!

This law means that as temperature rises, it takes exponentially more energy to increase temperature by each degree. The hottest temperature recorded on Earth was 134 degrees Fahrenheit (56.7°C), measured on July 10, 1913, in Death Valley, California. **The reason it has been so difficult to exceed this record is the extra energy required to heat over 134 degrees is exponentially higher than the temperature increase.** According to physics teacher Michel van Biezen, the Stefan-Boltzmann Law is another thermostat of the Earth, which acts to limit extreme temperatures.[20]

Temperature By Doubling CO_2 Concentration

MODTRAN software and physicist William Happer and William van Wijngaarden agree with my calculations that doubling CO_2 from present levels of about 400 ppm to 800 ppm would result in an increase in radiative forcing (warming) by about 3 W/m^2.[21] At 400 ppm of CO_2, radiation emitted back to space from the Earth is 239

W/m^2. As set forth in Table 11, doubling CO_2 from 400 ppm to 800 ppm would change this number to 242 W/m^2, which is a net change of 3 W/m^2.[22]

You may hear the number of 3.7 W/m^2 for doubling CO_2, but that is the number for doubling CO_2 using the Myhre CO_2 radiative forcing coefficient of 5.35 that relies on the Instantaneous Radiative Forcing method that is less representative of real-world mechanisms. An estimate of 3.0 W/m^2 reflects the Effective Radiative Forcing method, which incorporates more advanced radiative transfer models that account for adjustments to the stratosphere, tropospheric temperature structure, water vapor, clouds, and overlapping infrared radiation absorption of other greenhouse gases.

Confirmation of 3.0 W/m^2 for doubling CO_2 concentrations from 400 ppm to 800 ppm, without feedbacks, is provided by using MODerate resolution atmospheric TRANsmission (MODTRAN) software. MODTRAN is a widely used computer program designed to model atmospheric propagation of electromagnetic radiation across a spectrum range from 0.2 to 100 micrometers. It is used to model how radiation is absorbed and scattered by atmospheric components. MODTRAN was developed and is maintained by Spectral Sciences, Inc. and validated by the US Air Force. Derrek Wilson and Julio Gea-Banacloche from the physics department of the University of Arkansas used MODTRAN software, which solves radiative transfer equations in scenario analysis, to determine the radiative forcing for doubling CO_2 concentrations from 390 ppm to 780 ppm. They found that when CO_2 is isolated, the radiative forcing is 3.5 W/m^2.[23] However, when CO_2 is combined with other greenhouse gases, including water vapor, methane, and ozone, the number goes down to 3 W/m^2 due to the impact of overlapping infrared absorption bands (see Figure 54 as an example).[24] Since the real atmosphere is a mixture of all these gases, 3.0 W/m^2 is more realistic than 3.5 W/m^2. Therefore, the preferred Effective Radiative Forcing method for doubling from CO_2 levels in

2013 of 400 ppm to 800 ppm results in radiative forcing from CO_2 of about 3 W/m^2. Since 3 W/m^2 is provided by both the MODTRAN software and the ERF adjusted Myhre radiative forcing equation, with a CO_2 radiative forcing constant of 4.33, this is a good independent validation of the accuracy of both methods and the constant.

Table 11 **Effective Radiative Forcing of CO_2**
400 ppm to 800 ppm of CO_2
K × ln(C/Co),
K = 4.33
C = 800 ppm
Co = 400 ppm
800 ppm/400 ppm = 2
ln(2) = 0.6931
4.33 × 0.6931 = 3.0 W/m^2

Watts per square meter can be translated into temperature using the Stefan-Boltzmann Law. In order to make use of the Stefan-Boltzmann equation you need to know the Effective Emissivity of the Earth. *Emissivity* is a measure of how effectively a surface emits thermal radiation compared to an ideal black body at the same temperature. The CERES satellite reports the Earth radiates 239 W/m^2 of heat out to space, and satellites confirm the current surface temperature of the Earth is 288 K. Using this data and the Stefan-Boltzmann Law, the Effective Emissivity of the Earth is calculated to be 0.6127.

Using the Stefan-Boltzmann Law, an increase of 3 W/m^2 translates to an increase of 0.9 Kelvin (K) or degrees Celsius (C). Using a Stefan-Boltzmann Law calculator, you can enter the figures in Table 12 to determine that an increase of 3 W/m^2 produces a temperature increase of 0.90°C.[25] At the current rate of fossil fuel use, it will take about 150 years to increase CO_2 from 427 ppm today to 800 ppm. Due to the declining logarithmic function of CO_2 marginal warming, CO_2 levels would need to increase by another 800 ppm to produce another 3 watts/square meter of heating, and by 1600 ppm for yet another 3 watts/square meter of heating. **Therefore, to increase temperature by 2.7°C (3 × 0.9°C) requires a CO_2 level of over 3,200 ppm (400 + 400 + 800 + 1,600), almost 8 times the current level!** At the current rate of increase of CO_2 concentrations of ~2.7 ppm per year,[26] it would take more than 1,000 years to reach 3,200 ppm and 2.7°C in warming. However, 3,200 ppm of CO_2 could never occur due solely to the burning of fossil fuels since it exceeds the amount of CO_2 available in the entire Earth's known fossil fuel reserves.

Table 12 Stefan-Boltzmann Conversion of Heat to Temperature	
400 ppm of CO_2	**800 ppm of CO_2**
$m^2 = 1$	$m^2 = 1$
Effective Emissivity of the Earth = 0.6127	Effective Emissivity of the Earth = 0.6127
Radiative Forcing = 239 W/m^2	Radiative Forcing = 239 + 3 = 242 W/m^2
Temperature = 288 K	Temperature = 288.90 K
	Temperature Change = 0.90°C

Table 12: *Table 12 Source: Stefan-Boltzmann Law Calculator, BYJU'S. https://byjus. com/stefan-boltzmann-law-calculator/.*

The physics behind these calculations is not disputed. Even using the overstated Myhre equation IRF constant K for CO_2 of 5.35, the warmth for doubling CO_2 from 400 ppm to 800 ppm would be only 3.7 W/m^2 or 1.1°C. Physicists who are climate alarmists, such as Brad Marston of Brown University[27] and climate crisis skeptic William Happer, from Princeton University,[28] both calculate the number to be approximately 1°C or less by doubling CO_2 concentrations without feedbacks. Using the MODTRAN radiation transfer code also yields temperature increase of 0.9°C for an increase in CO_2 from 400 ppm to 800 ppm, without feedbacks.[29]

Temperature increases from rising CO_2 concentrations can also be estimated using the lapse rate, or effective emission height method. The lapse rate is the rate at which temperature decreases with altitude in the troposphere. The lapse rate varies at different altitudes. A decline in temperature of 6.5°C per kilometer rise in altitude is a global weighted average, derived from thermodynamic principles.[30] Thus, the weighted average lapse rate is 0.0065°C per meter. Model simulations estimate the expansion of the troposphere from doubling CO_2 concentrations from 400 ppm to 800 ppm, without feedbacks, is about 150 meters.[31] Using a value of 150 meters multiplied by 0.0065°C per meter yields a temperature increase of 0.975°C. However, this is only an approximation, the true radiative forcing is more complicated than this simple calculation. Using the ERF adjusted Myhre equation (with a CO_2 radiative forcing constant of 4.33), or using scenario analysis with MODTRAN software, both yield a temperature increase of 0.9°C for doubling CO_2 concentrations from 400 ppm to 800 ppm.

Warming Since 1850

The IPCC reported the following in a press release on August 9, 2021: "Emissions of greenhouse gases from human activities are responsible for approximately 1.1°C of warming since 1850-1900."[32] The year 1850 is considered the start of the industrial revolution for climate studies. CO_2 levels in the atmosphere in 1850 were 285 ppm.[33] Using the ERF adjusted Myhre radiative forcing equation for CO_2 yields 1.65 W/m^2 and a temperature increase of 0.5°C to increase CO_2 from 285 ppm to 2021's level of 417 ppm (see Tables 13 and 14).[34] Princeton physics professor William Happer also calculates 0.5°C of warming from CO_2 over the past two centuries.[35] Even using the overstated Myhre equation IRF constant K for CO_2 of 5.35, the warmth between 1850 to 2021 would be only 2.0 W/m^2 or 0.6°C. This suggests greenhouse gas radiative forcing from CO_2 represents about half of the warming since 1850. Most of the remainder of warming must be from climate feedbacks, other greenhouse gases, or natural causes, such as changes in solar activity and cloud cover.

Table 13 Effective Radiative Forcing of CO_2
417 ppm to 285 ppm of CO_2
K × ln(C/Co),
K = 4.33
C = 285 ppm
Co = 417 ppm
285 ppm/417 ppm = 0.6834
ln(0.6834) = -0.3806
4.33 × -0.3806 = -1.65 W/m^2
239.18 W/m^2 – 1.65 = 237.53 W/m^2

Table 14 Stefan-Boltzmann Conversion of Heat to Temperature	
285 ppm of CO_2	**417 ppm of CO_2**
m^2 = 1	m^2 = 1
Effective Emissivity of Earth = 0.6127	Effective Emissivity of Earth = 0.6127
Radiative Forcing = 239.18 – 1.65 = 237.53 W/m^2	Radiative Forcing = 239.18 W/m^2
Temperature = 287.556 K	Temperature = 288.054 K
	Temperature Change = 0.49^{o}C

Table 14 Source: Stefan-Boltzmann Law Calculator, *BYJU'S*. https://byjus.com/stefan-boltzmann-law-calculator/.

The End of Fossil Fuels

According to the Energy Institute (EI) and its "2024 Statistical Review," the estimated global reserves of fossil fuels are 1,730 billion barrels of oil, about 200 trillion cubic meters of natural gas, and 1.1 trillion tons of coal.[36] If burned, this represents 235 gigatons of CO_2 from oil (0.136 tons of CO_2 per barrel × 1,730 billion barrels), 550 gigatons of CO_2 from natural gas (2.75 tons of CO_2 per 1,000 cubic meters × 200 billion cubic meters), and 3,146 gigatons of CO_2 from coal (2.86 tons of CO_2 per ton of coal × 1.1 trillion tons)—or a total of 3,931 gigatons of CO_2 (see Table 15). This aligns closely with the Carbon Tracker estimate that suggests total emissions potential is 3,500 gigatons of CO_2.[37] Since one part per million (ppm) of CO_2 in the atmosphere represents about 7.81 gigatons of CO_2, 3,931 gigatons of total fossil fuel reserves divided by 7.91 gigatons equals a 503 ppm increase in CO_2 in the atmosphere.

Table 15
Total ppm Change of CO_2 if All Known Reserves of Fossil Fuels Are Burned

Fossil Fuel	Known Reserves	CO_2 Tons Per Unit	CO_2 Total	CO_2/ppm	ppm of CO_2
Oil	1,730 B barrels	0.136/barrel	235 gigatons	7.81 gigatons	30
Natural Gas	200 T m3	2.75/K m3	550 gigatons	7.81 gigatons	70
Coal	1.1 T tons	2.86/ton	3,145 gigatons	7.81 gigatons	403
Total			3.931 gigatons	7.81 gigatons	503 ppm

If all oil, natural gas, and coal reserves were burned, and none of the resulting CO_2 was absorbed in the oceans, it would produce an increase of 503 ppm of CO_2. Adding that to the current CO_2 level of 427 ppm would lead to a total CO_2 concentration of 930 ppm (503 + 427). This estimate is higher than would ever be experienced, however, as the ocean would certainly absorb substantial amounts of CO_2 from the atmosphere. At the current level of ~2.7 ppm per year[38] increase in CO_2, it would take more than 175 years to reach 930 ppm, and even longer if the ocean's absorption of CO_2 is considered. Furthermore, it is not expected that we will burn all fossil fuel reserves, as some will be used for lubricants, asphalt, plastics, and other petrochemicals (which will see growing demand), and it is expected that alternative forms of energy and transportation, including nuclear energy, hydrogen fuel, and electric vehicles (EVs) will be deployed long before the Earth's carbon reserves are exhausted.

Using the radiative forcing equation for the use of all fossil fuel reserves scenario, increasing CO_2 levels from today's 427 ppm to 930 ppm increases radiative forcing by 3.37 watts/meter squared ($R = 4.33 \times \ln(930 \text{ ppm}/427 \text{ ppm}) = 3.37 \text{ W/m}^2$, Table 16). Using the Stefan-Boltzmann calculator to convert to temperature, we see an increase of temperature of 1.01°C in 175 years (see Table 17). Even using the overstated Myhre equation IRF constant K for CO_2 of 5.35, the warmth for burning all known fossil fuel reserves would be only 4.16 W/m^2 or 1.2°C. Given the absorption of CO_2 into the oceans over the next 175 years, however, a temperature increase of 1.01°C or 1.2°C, from CO_2 emissions may be overstated.

According to economic studies, there is a net benefit to warming up to 0.6°C, and there are no significant net negative impacts economically from climate change until temperatures exceed 2.4°C in warming (see Figure 24 in Chapter 4).[39] Therefore, burning all fossil fuel reserves will likely have no net negative effect through the end of this century and only minor negative economic impacts thereafter.

Table 16 Effective Radiative Forcing of CO_2
427 ppm to 930 ppm of CO_2
K × ln(C/Co),
K = 4.33
C = 930 ppm
Co = 427 ppm
930 ppm/427 ppm = 2.178
ln(2.178) = 0.778
4.33 × 0.778 = 3.37 W/m²

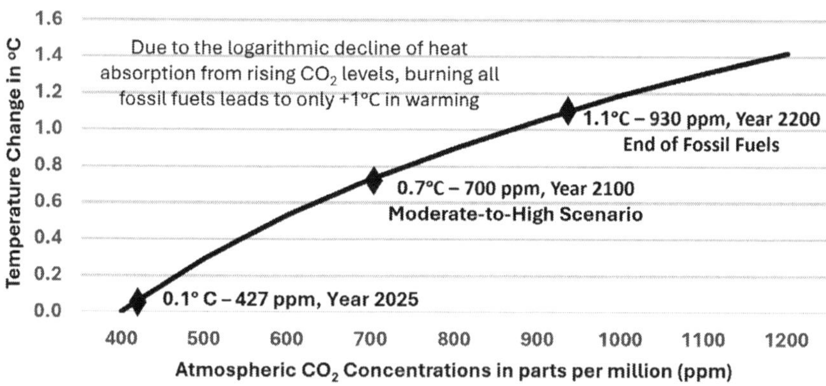

Figure 56 – End of Fossil Fuels - The Marginal Radiative Forcing of CO_2 in Degrees Celsius Since 2013 (400 ppm). *The marginal radiative forcing (warming) of CO_2 concentrations in ppm. The marginal radiative forcing of CO_2 would produce warming of about 0.6°C between 2025 to 2100 if we do little or nothing to curb fossil fuel use. All remaining fossil fuel reserves are estimated to produce a total of 930 ppm of CO_2 in the atmosphere—if they were all burned, and no CO_2 was absorbed into the oceans (very unlikely). This unlikely worst-case scenario would result in a temperature increase of 1.01°C from 2025 and would take about 175 years, assuming we continue to add the current rate of ~2.7 ppm of CO_2 per year to the atmosphere. Assumptions: CO_2 base level 400ppm in 2013; Forcing=4.33 × ln(New Concentration/400); Stefan-Boltzmann Law Calculator to convert watts/square meter to temperature; average Earth emissivity of 0.6127; 2013 temperature of 288 K; and 239 W/m².* Stefan-Boltzmann Law Calculator, *BYJU'S.* https://byjus.com/stefan-boltzmann-law-calculator/.

Table 17 Stefan-Boltzmann Conversion of Heat to Temperature	
427 ppm of CO_2	**930 ppm of CO_2**
$m^2 = 1$	$m^2 = 1$
Effective Emissivity of Earth = 0.6127	Effective Emissivity of Earth = 0.6127
Radiative Forcing = 239.28 W/m^2	Radiative Forcing = 239.28 + 3.37 =242.65 W/m^2
Temperature = 288.08^{o}K	Temperature = 289.09^{o}K
	Temperature Change = 1.01^{o}C

Table 17 Source: *Stefan-Boltzmann Law Calculator, BYJU'S. https://byjus.com/stefan-boltzmann-law-calculator/.*

Other Greenhouse Gases

Increases in anthropogenic greenhouse gases methane, nitrous oxide, tropospheric ozone, and halogenated compounds—chlorofluorocarbons (CFCs) and hydrofluorocarbons (HFCs)—have far less impact on warming than CO_2. Each molecule of methane (NH_4), nitrous oxide (N_2O), and halogenated greenhouse gases—CFCs and HFCs—have a much higher warming capacity per molecule than one molecule of CO_2, but because the amount of these gases released into the atmosphere is so tiny (parts per billion for methane and nitrous oxide and parts per trillion for CFCs and HFCs) relative to the amount of CO_2 emissions (parts per million), the contribution of methane, nitrous oxide, CFCs, and HFCs to warming is far less than that caused by CO_2.

CO_2 is increasing in the atmosphere at a rate of ~2.7 ppm per year[40] or 2,700 parts per billion (ppb) per year, nitrous oxide at a rate of 11 ppb per year, nitrous oxide at a rate of 1.3 ppb, and CFCs and HFCs collectively at a rate of 19 parts per trillion (ppt), although CFC and HFC emissions are declining dramatically due to regulations. A crude estimate of the warming power of these other greenhouse gases can be approximated by comparing the annual CO_2 increase in concentration and power to warm per molecule with the annual increase in concentration of each greenhouse gas and the power of each molecule to warm.[41]

The radiative forcing per molecule of methane is about 30 times larger than for CO_2 and increasing by about 11 ppb. However, because the rate of increase per year of methane in the atmosphere is about 245 times less than CO_2 (2,700 ppb/11 ppb). The increase in radiative forcing is about 30/245 or about 1/8 that of CO_2. Methane decomposes and has an atmospheric lifetime of only 12 years. The radiative forcing of nitrous oxide is 273 times greater than CO_2, but the rate of annual increase of nitrous oxide is very small at 1.3 ppb. Thus, nitrous oxide is increasing in the atmosphere at a rate of 2,077 times less than CO_2 (2,700 ppb/1.3 ppb = 2,077). Nitrous oxide thus warms the atmosphere at a rate that is 273/2,077, or about 1/8 that of CO_2. Nitrous oxide lasts longer in the atmosphere and has an atmospheric lifetime of approximately 114 years.

Ozone (O_3) in the stratosphere is naturally produced by the interaction of solar ultraviolet radiation with molecular oxygen. Almost all the warming from ozone molecules occurs from this natural source in the stratosphere. Low levels of ozone in the troposphere that are by-products of human activities have a negligible impact on global warming. Ozone has a short atmospheric lifetime because it's a powerful oxidant and is highly reactive and removed from the atmosphere from chemical reactions. It is formed in the troposphere where sunlight drives photo-chemical reactions with nitrogen oxides (NO_x) and volatile organic compounds (VOCs). Its lifetime is a couple of hours to a few weeks. Not only is it created from NO_x and VOCs, it also reacts with these same compounds, so in areas where it is produced it has a short lifetime, typically a couple of hours to a few days. Tropospheric ozone can impact temperature in a local area where ozone is produced and bands of wind currents emanate from these sources. However, it has a short enough atmospheric life to prevent it from accumulating extensively on a global scale. Although it partially disperses in the troposphere, it does not pass through the tropopause and therefore does not enter the stratosphere to add to the ozone stratospheric warming. Furthermore, the infrared radiation absorption bands of ozone overlap with CO_2 and water vapor, and this also limits its impact on global temperatures in the troposphere.

Halogenated gases are primarily used as refrigerants for air conditioning and refrigeration, as propellants in aerosol spray cans, and for halogen fire extinguishers. Halogenated greenhouse gas emissions of CFCs and HFCs are measured in parts per trillion. They are powerful greenhouse gases with a weighted average warming power of 3,000 times greater than CO_2 per molecule. Despite this powerful warming capability, they have an almost insignificant impact on warming since they are increasing at such a low level: about 19 parts per trillion per year (ppt). NOAA estimates they contribute only 1.3% of global warming from anthropogenic greenhouse gases.[42] They are increasing at levels that are 142,000 times lower than CO_2 (2,700,000 ppt/19 ppt). Therefore, their impact on global temperature is negligible: less than 3,000/142,000, or about 1/50th the warming of CO_2. **A rapid increase in halogenated compounds would be alarming due to their powerful greenhouse effect per molecule. However, halogenated gas emissions and concentrations are now and will continue to decline dramatically due to global regulations.** CFCs were effectively banned in 2010 under the Montreal Protocol. The observed radiative forcing from CFCs ceased increasing in 2000 and has continued to decline ever since.[43] The concentration of CFCs will continue to decline slowly. The dominant CFCs are CFC-11, with an atmospheric life of 52 years, and CFC-12, with a 100-year atmospheric life.

Signed in January 2019, the Kigali Amendment to the Montreal Protocol will phase down long-lived HFCs by more than 80%, along with a reduction of greenhouse warming of 0.4°C by 2100.[44] This suggests HFC warming will be limited to 0.1°C by 2100. Of the 19 ppt increase per year in halogenated gases in the atmosphere each year, nearly 13 ppt per year is from HFC-125 and HFC-32.[45] Another 5 ppt is from HFC-134a.[46]

It is estimated that China produces 70% of the world's HFC products, and that nation will phase out 97.5% of long-lived HFCs by 2030.[47] HFC-125 has an atmospheric life of 29 years, HFC-32 has an atmospheric life of 5.2 years, and HFC-134a has an atmospheric life of 14 years. The Kigali Amendment is moving industry to shift to halogenated gases with shorter atmospheric lifetimes, such as HF-152a, which has a 1.6-year atmospheric lifetime, and Hydrofluoroolefins (HFOs), with an atmospheric lifetime of days to weeks. In 50 to 100 years, halogen gases should have virtually disappeared from the atmosphere since all long-lived CFCs and HFCs will have deteriorated by then, and the remaining short-lived halogen gases will disappear within a relatively short period after they are emitted.

Using these estimates, the approximate radiative forcing of all anthropogenic greenhouse gases is as follows during the time that CO_2 concentrations double: CO_2 (3.0 W/m^2) + Methane (0.4 W/m^2) + Nitrous Oxide (0.4 W/m^2) + Halogenated Gases (0.1 W/m^2). Thus, other anthropogenic greenhouse gases add about 30% more warming than CO_2 alone. However, the estimates above are only approximations as the saturation to infrared absorption bands of each molecule needs to be considered to determine its precise radiative forcing. A more accurate determination of the warming of each of these other greenhouse molecules is possible by understanding the specific radiative forcing of each molecule. Such analysis was used in determining the radiative forcing equations set forth in Myhre et al.[48] IPCC *Assessment Report 5 (AR5)* includes the use of the Myhre et al. equations to determine the radiative forcing of anthropogenic greenhouse gases since 1750.[49]

Using the IRF Myhre et al. radiative forcing equations over the next 165 years (with unadjusted IRF warming coefficients), in which time CO_2 concentrations double, provides the following: CO_2 (3.7 W/m^2) + Methane (0.6 W/m^2) + Nitrous Oxide (0.6 W/m^2) + HFCs (0.1 W/m^2). Together, all these other greenhouse gases contribute 35% more warming than CO_2 alone, using the Myhre et al. equations. When the radiative forcing of all greenhouse gases are ERF adjusted, the warming of other greenhouse gases is nearly 57% of the warming of CO_2 alone (1.7 W/m^2 vs. 3.0 W/m^2 for CO_2 alone, see Table 18). I have used this figure in my calculations. This is similar to historical figures as the IPCC *AR5* estimates these greenhouse gases represent about 55% of the warming of CO_2 between 1750 to 2011 (1.01 W/m^2 vs. 1.82 W/m^2 for CO_2).[50]

Moving away from the Instantaneous Radiation Forcing method, of the Myhre CO_2 radiative constant of 5.35 to a more realistic 4.33 constant, which reflects the Effective Radiative Forcing (ERF) method, is more representative of the real-world climate. Using updated HITRAN data and the ERF method show that the CO_2 radiative warming coefficient was overstated by 24% (5.35 vs. 4.33) in the Myhre et al. equations because they did not account for the impact of ERF rapid adjustments to the stratosphere, tropospheric temperature structure, water vapor, clouds, and overlapping infrared absorption bands of multiple greenhouse gases. To get a more accurate impact on the energy balance from other greenhouse gases you need to make ERF adjustments to the IRF Myhre equation result to determine the Effective Radiation Forcing. Because each greenhouse gas has a different impact on the environment, the ERF adjustment for each greenhouse gas varies. For CO_2, the adjustment is a reduction in warming, for methane the warming is enhanced and for nitrous oxide and halogenated gases, the value is similar between IRF and ERF, so no adjustment is needed. Table 18 makes ERF adjustments for each greenhouse gas.

Table 18 Effective Radiative Forcing Increase by 2177 (152 years) from Each Anthropogenic Greenhouse Gas (uses IRF Myhre equations adjusted for ERF)					
Greenhouse Gas	Current Concentration	Increase per year	Concentration by 2177	Radiative Forcing Increase	Warming From Increase
CO_2	427 ppm	2.7 ppm	840 ppm	3.0 W/m^2	0.90°C
C_{H4}	1950 ppb	11ppb	3765 ppb	1.0 W/m^2	0.30°C
N_2O	336 ppb	1.3 ppb	550 ppb	0.6 W/m^2	0.18°C
HFCs	375 ppt	2.2 ppt	727 ppt	0.1 W/m^2	0.03C
TOTAL				**~4.7 W/m^2**	**~1.41°C**

Sources for Table 18. *(Radiative Forcing Increase)* Myhre, Gunnar et al., "New estimates of radiative forcing due to well mixed greenhouse gases," *Geophysical Research Letters,* Volume 25, Issue 14, pp. 2715-2718, 15 July 1998. https://doi.org/10.1029/98GL01908 ERF adjustments have been made for all greenhouse gases to determine the Effective Radiation Forcing. *(Warming from Increase)* Stefan-Boltzmann Law Calculator, *BYJU'S.* https://byjus.com/stefan-boltzmann-law-calculator/.

The IPCC RPCP6.0 scenario (moderate-to-high emissions) forecasts a CO_2 level of 700 ppm by the end of this century.[51] A change from the current level of CO_2 of 427 ppm to 700 ppm produces an increase of 2.1 W/m^2 in radiative forcing ($R = 4.33 \times \ln(700\ \text{ppm}/427\ \text{ppm}) = 2.1\ \text{W/m}^2$), or a rise in temperature of 0.6°C. Including ERF values for CO_2, methane, nitrous oxide, CFC, and HFC emissions to the year 2100, the temperature will increase by about 0.94°C, in this RPCP6.0 scenario. Hardly a crisis.

Such mild warming of 0.94°C, in this RPCP6.0 scenario, is not a crisis; it would have only small, or no, negative net impacts on the economy (see Figure 24 in Chapter 4). Furthermore, the increase in CO_2 would certainly boost agricultural productivity. So why do some insist we are having a climate crisis? The entire climate crisis hysteria is based on an unsubstantiated supposition that warming of 0.94°C will be amplified 2.2 times—to 2.1°C—by water vapor, melting ice caps, and cloud feedbacks, which are not offset by negative feedbacks.

The climate alarmist narrative stands or falls on this one conjecture. In Chapter 10, we shall see that feedbacks of this magnitude disagree with observations and the scientific literature and are therefore wrong.

Chapter 10:
Climate Feedbacks

IPCC Climate Models Exaggerate Water Vapor and Other Feedbacks

Climate Models Assume 2.2x Warming Amplification: As discussed in Chapter 9, calculations of the greenhouse effect of increasing CO_2 to 700 ppm by the year 2100 results in a temperature increase of 0.6°C. You may increase the number to about 0.9°C, if you also include increases in methane, nitrous oxide, HFCs, and CFCs. The United Nations Emissions Gap Report states, "Even if countries meet commitments made under the 2015 Paris Agreement, the world is heading for a 3.2 degrees Celsius global temperature rise over pre-industrial levels ..."[1] We have already seen 1.1°C in warming between preindustrial times to 2021.[2] This leaves a forecasted increase in temperature of 2.1°C between now and the year 2100. **This increase of 0.9°C of warming from anthropogenic greenhouse gases to a projected total increase in temperature of 2.1°C from now to the end of this century is a *2.2-fold amplification.*** Climate models upon which these forecasts are based assume the impact of the water vapor climate feedback, clouds, and polar ice melt as a *2.2× (times)* *warming multiplier* despite the fact that the IPCC and other scientific studies discussed below support only about 1/2× warming feedback.

IPCC climate models assume temperatures will be amplified by 2.2× due to long-term climate feedbacks, not included in rapid ERF adjustments, including added water vapor, melting ice caps, and increased cloud cover, with water vapor being the dominant force. As previously stated, Richard Feynman summarized the scientific method as follows: "... we say compare to experiment or experience, compare it directly with observations to see if it works. If it disagrees with experiment, it's wrong."[3] As we shall see, the hypothesis that water vapor, clouds, and ice melt has a 2.2× warming multiplier effect, which is the basis of climate alarmism, disagrees with observations and is therefore wrong. Climate models, which incorporate these feedbacks, run about 2.4-fold warmer than actual temperature measurements (see Figure 50). Renowned MIT Atmospheric Science Professor Emeritus Richard Lindzen said the positive feedbacks in the climate models "are assumed—not derived or observed."[4]

The alarm about global warming comes, not from CO_2, methane, nitrous oxide, and other anthropogenic greenhouse gas warming, but from this unsubstantiated conjecture of a positive climate feedback that increased temperature will raise humidity, melt polar ice caps, and amplify warming through the greenhouse effect of water vapor and clouds and reduced albedo of ice melt, resulting in temperatures which are magnified by 2.2×. But this positive climate feedback supposition is counter to how nature works, which is that it rebalances itself through negative feedbacks, and as a result, the world has never seen oceans boil from positive temperature feedbacks, despite the fact that we have had periods on this Earth with far higher levels of CO_2 (about 427 ppm today versus over 7,000 ppm in the Cambrian era[5]). For billions of years, the Earth has varied in temperature within a relatively narrow range.

Known as the Earth's thermostat, negative (cooling) feedbacks rebalance temperatures. Low cloud cover is the major thermostatic feedback as low clouds reflect 70% to 95% of sunlight back to space and prevent this solar energy from ever reaching the Earth.[6] This is a negative feedback of at least 476 W/m^2 of solar energy under low clouds at the equator. If it gets too warm, more low clouds are formed from the increased water vapor in the atmosphere, and an increase in cloud nucleating sulfate aerosols from accelerated phytoplankton growth in the oceans. These low clouds cool the temperature. If it gets too cold, less low clouds are formed due to decreased water vapor in the atmosphere and the decline of cloud nucleating sulfate aerosols from decelerating phytoplankton growth in the oceans. Fewer low clouds allows more solar energy to heat the Earth's surface. The Stefan-Boltzmann Law also makes it difficult for the Earth's temperature to increase to extremes, since it requires an exponentially larger amount of energy, to the fourth power, to increase temperature linearly. Physics teacher Michel van Biezen says the Stefan-Boltzmann Law is yet another natural thermostat which prevents the Earth from overheating.[7]

Scientific evidence demonstrates that net climate feedbacks are not positive (warming), as speculated by climate alarmists, but slightly negative (cooling), as would be expected by the natural thermostat of the Earth. Changes in cloud albedo and aerosols can have significant cooling impacts on radiative forcing. Despite the presence of well-established negative (cooling) feedbacks in nature which offset warming, IPCC climate models have ignored the net negative (cooling) feedback of clouds and aerosols and greatly exaggerate the positive feedback (warming) from water vapor, melting ice caps, and clouds to *exaggerate the temperature impact of CO_2 and other anthropogenic greenhouse gas warming by 2.2×.*

Water Vapor and Ice Sheet Melt Together Have a 1/2x Warming Amplification

The 2.2× warming amplification in climate models is from water vapor melting ice caps and increased clouds. The IPCC has concluded that clouds have both positive and negative feedbacks to temperature and the net impact of clouds is to lower temperature.[8] Internationally recognized Caltech atmospheric scientist Graeme L. Stephens and coauthors show how clouds have a net cooling impact of -21 W/m^2, mostly by reflection of sunlight from clouds in the mid-latitude summer hemisphere.[9] In its *AR5 Report,* the IPCC estimated the feedback of clouds to be one of net cooling of -20 W/m^2.[10] A more recent study by prominent atmospheric scientist from the University of Wisconsin, Tristan S. L'Ecuyer, and coauthors, "...estimate the annual mean net cloud radiative effect at the surface to be −24.8 ± 8.7 W/m−^{2} (−51.1 ± 7.8 W/m−^{2} in the shortwave and 26.3 ± 3.8 W/m−^{2} in the longwave)."[11] The estimated net cooling effect of -24.8 W/m^2 from the L'Ecuyer et al. paper is likely the most accurate as it uses updated datasets from the CloudSat and CALIPSO satellites (see Chapter 11). Observations confirm the net impact of cloud cover is one of cooling (see Chart 63 in Chapter 11). Therefore, the positive warming feedback of clouds is more than offset by the negative cooling effect of clouds as set forth in the Stephens et al., IPCC *AR5 Report*, and L'Ecuyer et al. paper (see Chapter 11 for additional details).

The net cooling impact of clouds will be addressed later, but this section will now focus on the remaining major warming amplification feedbacks, which are water vapor and melting ice caps. Methane release from melting permafrost and ocean sediments is another positive (warming) feedback, but its impact is so small as to be insignificant[12] (see Chapter 4, Tipping Points, for more details). Increased water vapor warms because it is a greenhouse gas. It amplifies temperature because, as the Earth warms, more water evaporates out of the oceans. Melting ice caps warm because they reflect solar shortwave radiation back out to space due to their white reflective surface. As ice is reduced, less solar shortwave radiation is reflected out to space—and the Earth warms. Melting ice amplifies temperature because, as temperature warms, reflective snow and ice-covered areas shrink in size.

Internationally recognized Caltech atmospheric scientist Graeme L. Stephens and coauthors show how that most evidence supports "a magnitude of 1.5 W/m^2-C to 2.0 W/m^2-C" from water vapor feedback.[13] A study published in the *Journal of Climate* by University of Miami professor and atmospheric scientist Brian J. Soden and NOAA meteorologist Isaac M. Held examines water vapor feedback and other positive climate feedbacks and finds similar numbers.[14] The authors look at various climate feedback mechanisms to determine their impact. Soden and Held conclude water vapor is the largest positive feedback in the climate. They look at the feedback of water vapor and other factors based on a doubling of CO_2 from levels found in the year 2000. The result they report is an amplification from water vapor of 1.8 W/m^2-C and changes to surface albedo as ice caps melt of 0.26 W/m^2-C.[15] The IPCC cited the Dressler and Sherwood, and Soden and Held, papers in its *AR5* report in 2013. In its *AR6* report in 2021, it stated that new information has model estimates of the water vapor feedback of 1.77 W/m^2 of global mean radiative feedback in response to a uniform 1°C warming, and 1.85 W/m^2-C as measured by satellite.[16] The *AR6* report points out that the actual feedback is less since water vapor warming occurs at the Top of the Atmosphere (TOA), and because some of this heat in the TOA is emitted out to space, *the effective water vapor feedback to the Earth at the TOA (combined water vapor feedback and lapse rate feedback) is reduced to 1.3* W/m^2*-C,* and they cite a range of 1.2 W/m^2-C to 1.4 W/m^2-C with high confidence.[17]

The IPCC has stated: "The surface albedo feedback, primarily due to changes in snow and sea ice cover, is positive (warming) and contributes approximately 0.1 to 0.8 W/m^2-C to global mean radiative feedback, with the best estimate around 0.3 W/m^2-C."[18] This figure is close to the Soden and Held estimate of the Earth's albedo from melting ice caps of 0.26 W/m^2-C. The warming from polar ice cap melt could be even less than 0.3 W/m^2-C in recent years. Antarctica sea ice has seen net growth over the past 40 years.[19] Furthermore, the annual minimum sea ice extent in the Arctic has increased from 1.31 million square miles in 2012 to 1.65 million square miles in 2024, a 26% increase![20] **Therefore, the impact of sea ice albedo between 2012 and 2024 should be to lower temperature, not increase temperature.** Therefore, a warming amplification of 0.3W/m^2-C is likely conservative.

Adding 0.3 W/m^2-C from melting ice caps to water vapor feedback of 1.3 W/m^2-C equals 1.6 W/m^2-C. Therefore, the combined warming amplification of water vapor and melting ice caps is about 1.6 W/m^2 for each 1°C in warming at the top of the atmosphere. Using the Stefan-Boltzmann Law with the Earth's Effective Emissivity of 0.6127, the amount of radiative forcing to increase temperature by 1°C is 3.4 W/m^2. In the IPCC *AR5* report, the authors explain how you can calculate the feedback per degree Celsius of temperature increase. *AR5* states that the feedback of water vapor "may be compared with the basic 'black body' response of 3.4 W/m^2/°C."[21] Since 1°C warming equals 3.4 W/m^2, and 1.6 W/m^2 is less than half of 3.4 W/m^2, *the water vapor and melting ice cap positive climate feedback (warming), as estimated by the IPCC, is less than 1/2×, <u>not</u> 2.2× as found in their climate models.*

Water Vapor Feedback Is Overstated Because Relative Humidity Is Declining: The effective water vapor amplification is thus about 1.3 W/m^2 for each 1°C in warming. It may be less than this since these calculations are based largely on models that assume stable relative humidity in the upper troposphere, which is counter to weather balloon measurements. According to the Clausius-Clapeyron Principle, in dry air and constant pressure, a 1°C increase in temperature would increase humidity by 7% to remain at relative humidity; this is confirmed in laboratories. IPCC models rely on the Clausius-Clapeyron Principle to estimate the positive feedback of water vapor. IPCC *AR6* states: "The Clausius–Clapeyron equation determines that low-altitude specific humidity increases by about 7% for each degree Celsius of warming, assuming that relative humidity remains constant, which is approximately true at a global scale but not necessarily valid regionally."[22] IPCC climate models base their calculations on the Clausius-Clapeyron Principle and generally assume relative humidity will remain constant in the upper troposphere as temperature increases. These models attribute a 2.2× increase in temperature, mostly from water vapor, for every 1× increase in temperature from other factors, including CO_2, methane, nitrous oxide, and halogenated greenhouse gas warming, as well as natural causes. However, this 2.2× amplification is much too large.

According to Dressler and Sherwood: "The water vapor feedback mainly results from changes in humidity in the topical upper troposphere, where temperatures are far below that of the surface and the vapor is above most of the cloud cover."[23] They also explain that the water vapor feedback amplifies climate change "because of the expectation that the *atmosphere's relative humidity would remain roughly constant*—meaning that the specific humidity would increase at the rate of the equilibrium vapor pressure, which rises rapidly with temperature." Thus, constant relative humidity in the upper troposphere is an essential factor in current estimates of water vapor amplification. Therefore, questions arise as to the accuracy of the water vapor amplification forecasts because relative humidity has declined in the upper troposphere as the temperature has increased (see Figure 57).[24]

Model calculations of global relative humidity are extremely complex as they are influenced by local weather conditions, climate patterns, wind speeds, and cloud formation, all of which are in a constant state of flux. Higher rates of condensation will lower humidity, and this is another dynamic having multiple variables. Higher winds significantly increase evaporation, which you experience when you place a fan in front of a wet towel. With the moderating temperatures in the Arctic, there is a smaller temperature difference between the Arctic and Tropics, which induces lower wind speeds and results in lower evaporation rates.[25]

Measured data from weather balloons since 1948 shows relative humidity has remained constant on the Earth's surface and has *declined* in the mid-troposphere and upper-troposphere (see Figure 57).[26] Although relative humidity has remained relatively constant on the Earth's surface, water vapor amplification only matters high in the troposphere. This is significant, as global warming from greenhouse gases occurs in the upper troposphere because greenhouse gases are saturated at lower altitudes, which means they have almost no capacity to absorb additional heat near the surface. Since the calculation of the water vapor temperature amplification is so complex with many dynamic variables, actual relative humidity measurements (Figure 57) provide the best clarity as to what is happening. Climate models assume constant relative humidity in the upper troposphere, which assumption is not supported by observations.

The fact that relative humidity is constant at the surface and declines in the upper troposphere can be explained by condensation of water vapor in the troposphere. Figure 57 shows this phenomenon is pronounced between the ground and 4.2 kilometers. This suggests a significant amount of water vapor is being condensed to form low clouds before reaching a height of 4.2 kilometers. Low clouds have a net negative cooling feedback, lowering temperature by reflecting sunlight back out into space. This cloud feedback offsets the positive warming water vapor feedback.

Andy May has written an article titled "Atmospheric Water Vapor (TPW) and Climate Change." In this article, he examines specific humidity, or Total Atmospheric Water Vapor (TPW), from weather balloon measurements going back to 1948.[27] May cites studies by Benestad (2016), Miskolczi (2014), and the NOAA R1 datasets, all of which show declines in TPW from 1979 to 2011, a period of rapid global warming. From 1985 to 2008, TPW declines sharply as temperature increases. May cites from Paltridge, Arking, and Pook (2009): "Negative trends in 'q' [TPW] as found in the NCEP data

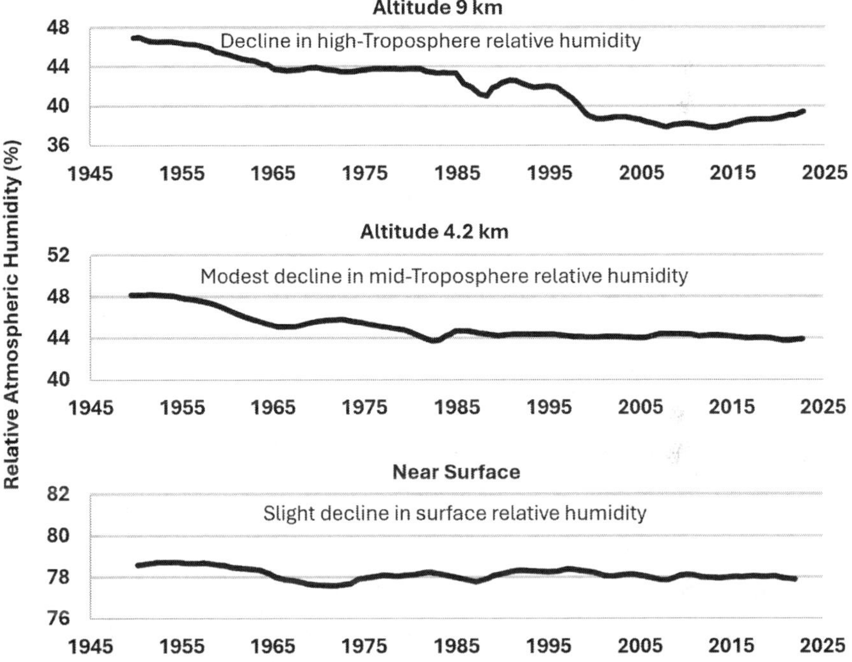

Figure 57 – Global Relative Humidity, 1948 to 2023. *Measurements by the National Oceanic and Atmospheric Administration (NOAA) show that relative humidity in the upper troposphere has declined and relative humidity on the Earth's surface has been stable.* Data: NOAA Earth System Research Library. Source: *Climate4you,* "Relative Atmospheric Humidity," Greenhouse Gases, Atmospheric Water. https://www.climate4you.com/GreenhouseGasses.htm#Atmospheric%20water%20vapor.

would imply that long-term water vapor feedback is negative—that it would reduce rather than amplify the response of the climate system to external forcing such as that from increasing atmospheric CO_2." May concludes: "It seems likely that the Clausius-Clapeyron relation is not the only factor affecting TPW. This casts considerable doubt on the CMIP6 [IPCC] model results, which rely only on Clausius-Clapeyron, human activities, and sporadic volcanism."[28]

Increased Cloud Cover Cools By 1/2×

Esteemed Professor Emeritus of Physics at Princeton University William Happer has said, "The reason the models get these huge numbers is because of the water vapor feedback. The reason those feedbacks are overestimated is that they don't treat clouds correctly at the same time. You cannot increase water vapor without affecting clouds. Just a few percentage change in cloudiness completely overwhelms any change from CO_2 over water vapor. We are all familiar with the fact that if it is a hot summer day and a cloud comes by, it cools down. If you take that on a global scale, it turns out there is exquisite sensitivity to cloudiness."[29]

Using the assumption from the Clausius-Clapeyron Principle, and observations (see Figure 63 in Chapter 11), that water vapor increases by 7% at the Earth's surface for each 1°C in warming, cloud cover will also increase by about 7% for each 1°C in warming if clouds increase at the same rate as water vapor. Climate models and traditional meteorological theory predict that global clouds and precipitation will increase at a slower rate than water vapor. However, a paper by prominent earth scientist and CEO of Remote Sensing, Frank J. Wentz, and coauthors states, "A recent analysis of satellite observations does not support this prediction of a muted response of precipitation to global warming. Rather, the observations suggest that precipitation and total atmospheric water have increased at about the same rate over the past two decades."[30] For precipitation to scale at the same rate as water vapor, clouds would also need to scale at a similar rate.

The 2019 paper by L'Ecuyer et al., estimates the net cooling effect of clouds of -24.8 W/m^2 (see Chapter 11).[31] Seven percent of -24.8 W/m^2 equals -1.7 W/m^2. Therefore, the net cooling of clouds is about -1.7 W/m^2 for each one degree C increase in temperature or cooling of 0.5°C. This figure is confirmed by observations. Observations of clouds by the International Satellite Cloud Climatology Project have shown that an increase in cloud cover of 7% lowers temperature by 0.5°C (see Chart 63 in Chapter 11). This is the net impact of clouds and includes both negative (cooling) and positive (warming) feedbacks.

The Wentz et al. paper suggests a 1°C increase in temperature will increase clouds by 7% due to increased humidity. Assuming a 7% increase in clouds, the cooling impact of clouds would be -0.5°C or -1.7 W/m^2 for each 1°C in warming (see Table 19), which is a negative cooling feedback of 1/2×. **Thus, the net cloud feedback of cooling more than offsets the water vapor and ice melt feedback of warming.**

Table 19 Stefan-Boltzmann Conversion of Heat to Temperature
-0.5°C for each 1°C in warming
m^2 = 1
Effective Emissivity of the Earth = 0.6127
Temperature = 287.5 K (288 K - 0.5K)
Radiative Forcing = 237.3 W/m^2
Radiative Forcing Change = -1.7 W/m^2 (237.3 W/m^2 - 239 W/m^2)

Table 19 Source: *Stefan-Boltzmann Law Calculator, BYJU'S. https://byjus.com/stefan-boltzmann-law-calculator/.*

Temperature amplification of 1.6 W/m^2-C from water vapor and melting ice caps is easily counteracted by negative cooling feedbacks including increased cloud albedo from increased humidity and increased sulfate aerosols. The CLAW Hypothesis proposes that increased temperatures lead to increased phytoplankton algae blooms and a corresponding increase in dimethyl sulfide (DMS) sulfate aerosols, which cool the Earth's surface (see Chapter 12). A paper in *Geophysical Research* found that a 1°C increase in the sea surface temperature of the Indian Ocean was sufficient to increase sulfate aerosol concentrations in the atmosphere by as much as 50%.[32] Extrapolating the findings of another paper by University of East Anglia environmental scientist, M. A. Thomas, and coauthors, suggests a cooling of -0.6 W/m^2-K from the CLAW Hypothesis effect (see Chapter 12, Table 22).[33] The combined cooling effect of increased cloud cover from higher humidity of -1.7 W/m^2-C plus -0.6 W/m^2-C from DMS aerosols equals -2.3 W/m^2-C, which more than offsets the 1.6 W/m^2-C of warming from the water vapor and melting ice cap climate amplification. Exactly as would be expected from the "Earth's Thermostat," which rebalances itself. The net change of all feedbacks is one of cooling of -0.7 W/m^2. The warming of +1.6 W/m^2 from water vapor and ice melt combined with -2.3 W/m^2 of cooling from clouds and DMS aerosols results in -0.7 W/m^2 or -0.2°C of cooling feedback for each 1.0°C of warming.

Facts do not support positive feedbacks from water vapor, melting ice caps, and clouds as the primary driver of climate. The United Nations claims temperatures will increase by 3.2°C by the end of the twenty-first century from pre-industrial times (1850/1900).[34] They claim we have already seen 1.1°C in temperature increase,[35] leaving 2.1°C in increase to happen from now to 2100. Of this 2.1°C in warming, we have seen that only 0.6°C can be attributed to radiative forcing of CO_2 and another 0.3°C from other anthropogenic greenhouse gases in that period. This means the IPCC believes 1.2°C of this warming will be due to positive feedbacks. The IPCC *AR6* report and the paper by Soden and Held demonstrate that water vapor and melting ice caps are the dominant positive (warming) feedback mechanisms, but these would only account for less than 1/2× of 0.9°C, or about 0.4°C, in warming by the end of the century—not 1.2°C. Furthermore, the water vapor, ice cap melting, and cloud positive feedbacks of 0.4°C would be more than offset by the -0.6°C of cooling feedback of increased low cloud cover due to increased water vapor in the atmosphere and the CLAW effect of increased DMS sulfate aerosols from phytoplankton. *Therefore, the expected net temperature increase by 2100, including all greenhouse gases, all warming feedbacks, and all cooling feedbacks, is about 0.7°C, not 2.1°C.* ***The IPCC forecast is inflated by nearly 1.4°C.***

History Shows No Net Feedback Warming

Since anthropogenic greenhouse gases did not begin to rise substantially until after 1850, historical temperatures provide a guide to understanding the temperature contribution of natural variation. As dendrochronologist and professor at Johannes Gutenberg University (German), Jan Esper, and coauthors observed, "One way to disentangle the influence of the Sun on the climate is to look for its influence on climate in pre-industrial times, when the influence of man-made greenhouse gases can be neglected and the natural variations can be isolated."[36] Using 30 paleoclimate proxies, Professor and paleoclimatologist Fredrik Ljungqvist from Stockholm University (Sweden) reconstructed temperatures in the Northern Hemisphere for the past two millennia.[37] This reconstruction places the nineteenth century in context. **It is important to see that the total warming since 1850 is about 1°C as reconstructed from proxy data.** Professor Ljungqvist shows the temperature change between 1850 to 2000 is 0.6°C. There has been an additional 0.35°C in warming since 2000. So the temperature increase from 1850 to 2021 would be slightly less than 1°C. This compares closely with the 1.1°C figure quoted by the IPCC for heating from 1850/1900 through 2021.[38] In Chapter 9 we discussed how the radiative forcing of CO_2 can only account for 0.5°C of warming, which is less than one half of the 1.1°C of warming from 1850/1900 to 2021 (see Tables 13 and 14 in Chapter 9). Data reveals that natural variation is certainly responsible for most of the balance of warming since 1850. The Ljungqvist reconstructed temperatures in Figure 58 show how the Earth, in 1850, was recovering from the cold temperatures of depth of the Little Ice Age and was impacted by natural solar

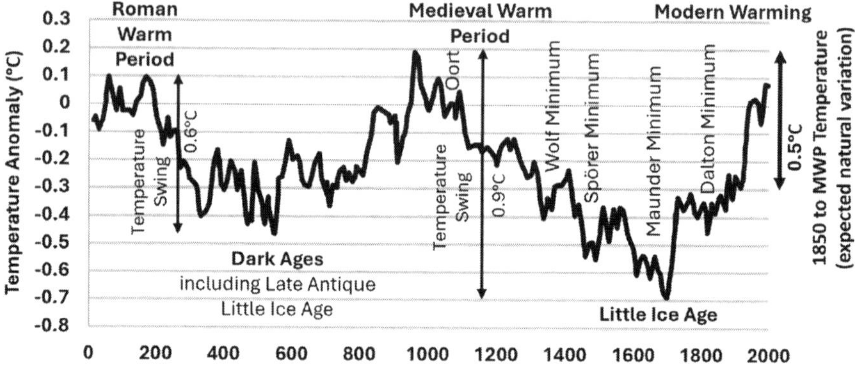

Figure 58 – Natural Climate Variation from Historical Climate Cycles. *Using 30 paleoclimate proxies, Professor Fredrik Ljungqvist reconstructed temperatures for the past two millennia. From this graph, you can see that temperatures varied from between 0.6°C and 0.9°C between the peak of warm and cold cycles in preindustrial times. Since CO_2 levels were stable during preindustrial times, these temperature variations show the impact of natural climate forcings. Since temperatures during the Medieval Warm Period reached 0.5°C warmer than experienced in 1850, we should expect about 0.5°C from natural variation during Modern Warming since 1850. Ljungqvist data also clearly shows the impact on temperature from the Oort, Wolf, Spörer, Maunder, and Dalton solar minima as well as the Late Antique Little Ice Age.* Source: Ljungqvist, Fredrik, "A new reconstruction of temperature variability in the extra-tropical Northern Hemisphere during the last two millennia," *Geografiska Annaler: Series A, Physical Geography,* Volume 92, Issue 3, pp. 339-351, 6 September 2010. https://doi.org/10.1111/j.1468-0459.2010.00399.x See also, "The Hockey Schtick, Paper: Roman & Medieval Warming Period Temps Reached or Exceeded the 20th Century," 25 September 2010. https://hockeyschtick.blogspot.com/2010/09/paper-roman-medieval-warming-period.html.

variation, including the Oort, Wolf, Spörer, Maunder, and Dalton solar minima.

In the twentieth century, the Earth moved into the Modern Solar Maximum, the largest solar maximum in 10,000 years.[39] As explained in Chapter 7, the increased solar irradiance, reduced cosmic rays, stratospheric warming, and lower cloud cover of the twentieth century had a major impact on increasing temperatures, just as these forces warmed the Earth in the Roman Optimum and Medieval Optimum. The Earth experiences grand Eddy Solar Cycles about every 1,000 years. These solar cycles correlate to the Ljungqvist temperature reconstructions in Figure 58, including Modern Warming. As can be seen in the graph, the Roman Warm Period was 0.6°C warmer than the Dark Ages; the Dark Ages was 0.7°C colder than the Medieval Warm Period; the Medieval Warm Period was 0.9°C warmer than the low of the Little Ice Age; and 1850 was 0.4°C warmer than the coldest part of the Little Ice age in the seventeenth century. Since there were virtually no anthropogenic gas emissions in Roman and Medieval times, these temperature cycles must be from natural causes. Independent paleo temperature reconstructions of the Northern Hemisphere by Moberg et al.[40] (see Figure 27 in Chapter 5) and Jan Esper[41] show climate cycles and temperature variances similar to those provided by Ljungqvist.

Paleo temperature reconstructions from Ljungqvist, Moberg, and Esper teach us to expect a natural variation of temperature between the peak to trough of Eddy Solar Cycles to be about 0.6°C to 0.9°C, and for Modern Warming, to match temperatures of the Medieval Warm Period, natural variation should drive about 0.5°C more warming since 1850 (see Figure 58). This warming does not account for greenhouse gases. When you add ~0.5°C from CO_2 warming (see Tables 13 and 14), plus ~0.3°C for other greenhouse gases (~57% of CO_2 radiative forcing), plus ~0.5°C from natural variation, this number totals ~1.3°C. The actual measured temperature increase since 1850/1900 is 1.1°C. The net climate feedbacks from clouds, the CLAW Hypothesis, water vapor, and polar ice melt is about -0.2°C (cooling) resulting in a net temperature change of 1.1°C.

The observed temperature of 1.1°C since 1850/1900 accounts for the impact of solar cycle-related natural variation (~0.5°C), radiative forcing of greenhouse gases (~0.8°C), and net climate feedbacks (~-0.2°C). Nearly 50% of this warming is from natural causes, and just over 50% is from CO_2 and other anthropogenic greenhouse gas radiative forcing. Natural variation may be higher than 0.5°C, and more than 50% of recent warming, since the Modern Solar Maximum has been stronger than the solar maximum during the Medieval Warm Period. The estimate of ~0.5oC of solar-cycle variation is estimated from the temperature difference between 1850 to the peak temperature of the Medieval Warm Period, so more solar variation during the Modern Solar Maximum would make that differential higher. In fact, we have just come off the peak of the Modern Solar Maximum, the largest grand solar maximum in 10,000 years.[42] Interestingly, internationally known solar physicist from the Max Planck Institute, Sami K. Solanki, and solar physicist and head of the the Max Planck Institute Solar Variability and Climate research group, Natalie A. Krivova, analyzed solar data between 1856 to 1999 and concluded that most of the warming between 1856 to 1970 is from solar variation and up to 30%–50% of warming between 1970 to 1999 is from solar variation.[43] Well known German scientist, Fritz Vahrenholt, and coauthor Sebastion Lüning cite eight peer-reviewed scientific papers estimating 40% to 70% of global atmospheric warming in recent decades is caused by the Sun.[44]

Climate Models Run Too Warm

Climate models predict that evidence of the water vapor amplification of the atmosphere should be seen by a hotspot 10 kilometers (km) up in the Tropics. After water vapor evaporates at the surface, in the Tropics it rises in the atmosphere and then condenses at

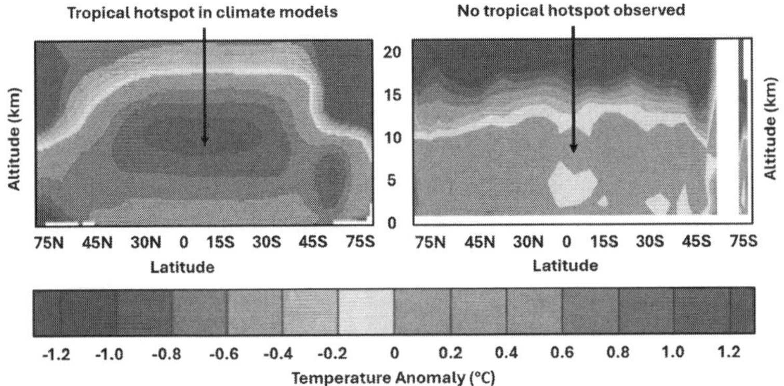

Figure 59 – Tropical Hotspot from Climate Models is Not Seen in Observations. *Climate models assume global warming will increase water vapor high in the troposphere, which will amplify warming. Climate models predict a hotspot at an altitude of about 10 kilometers in the Tropics from the latent heat released by condensing water vapor. However, this predicted tropical hotspot has not been found in observations by weather balloons or satellites. The figure on the left shows the predictions of the tropical hotspot in typical climate models (this example is from the Canadian IPCC climate model), while the figure on the right shows actual observation with no tropical hotspot.* Source: Singer, Fred S., "Lack of Consistency Between Modeled and Observed Temperature Trends," *Energy & Environment*, Volume 22, Issue 4, 1 June 2011, https://doi.org/10.1260/0958-305X.22.4.375. See also, Shaviv, Nir, "What role has the Sun played in climate change? What does this mean for us?" European Climate and Energy Institute (EIKE), 8 December 2022. https://youtu.be/hRFIzVB4Qss?si=NBxs5eu1qxHwl4Li.

about 10 km from the surface. When the water condenses, it releases stored latent heat, and this produces a hotspot. It should be primarily seen in the Tropics as this is the area of greatest water evaporation and humidity. The IPCC's *4th Assessment Report* included charts to depict this theoretical tropical hotspot. However, observational data from weather balloons and satellites has not been able to confirm the presence of this hotspot as predicted in models. **Dr. Steven M. Japar, atmospheric chemist with Ford Motor Company Research**

Laboratories, participated in the 1995 and 2001 IPCC assessment reports, and he resigned over the absence of this hotspot anomaly.[45] Dr. Jasper commented: "Temperature measurements show that the hot zone, that is predicted by the models in the mid-troposphere, is non-existent. This is more than sufficient to invalidate global climate models and projections made by them."[46]

Meteorologist and head of the UAH satellite record, Dr. Roy Spencer, wrote: "I believe the missing hotspot is indirect evidence that upper tropospheric water vapor is not increasing, and so upper tropospheric water vapor (the most important layer for water vapor feedback) is not amplifying warming from increasing CO_2."[47] The IPCC has removed the hotspot from more recent assessment reports since observational data does not confirm its existence. The absence

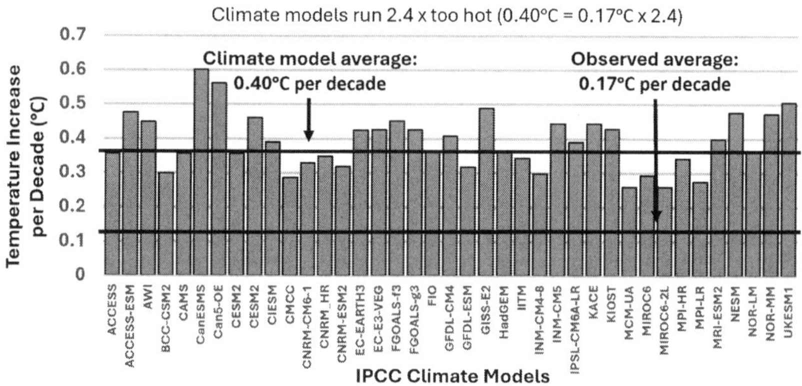

Figure 60 – Climate Models Run Too Hot: Tropical Mid-Tropospheric Temperature Variations, Models versus Weather Balloon and Satellite Observations, 1979 to 2019. *The IPCC climate models run 2.4× (times) warmer than measured observations. In the chart above, temperature increase predictions of 39 IPCC CMIP6 climate models are displayed, which average 0.40°C per decade. The average actual temperature increase that weather balloons and satellites have measured is 0.17°C per decade between 1979 and 2019 and 0.40 ÷ 0.17 = 2.4×.* Source: McKitrick, Ross and Christy, John, "Pervasive Warming Bias in CMIP6 Tropospheric Layers," *Earth and Space Science*, September 2020, Volume 7, Issue 9, 15 July 2020. https://doi.org/10.1029/2020EA001281.

of a consistent tropical hotspot confirms the fact that water vapor temperature amplification has been highly exaggerated in IPCC climate models.

Actual temperature measurements from satellites and highly calibrated weather balloons show that there has been no amplification of warming, as is predicted in the climate models.[48] Figure 60 is from Professor of Atmospheric Sciences, John Christy, at the University of Alabama, Huntsville and Ross McKitrick, of the University of Guelph, Ontario, Canada. McKitrick and Christy appropriately look at mid-tropospheric temperatures, as this is the altitude where the climate models predict substantial CO_2-induced greenhouse gas warming occurs. Because almost all CO_2 radiative forcing is saturated at the surface of the Earth (99.4% in the first 10 meters from the ground), most CO_2-induced warming occurs in the mid to high troposphere. They compare results from 39 IPCC CMIP6 climate models versus actual satellite and weather ballon observations. The Coupled Model Intercomparison Project Phase 6 (CMIP6) climate models are currently the most up-to-date set of global climate models and are used for climate projections by the IPCC. McKitrick and Christy found that *CMIP6 climate model predictions are 2.4 times larger than observed measurements.*[49]

McKitrick and Christy show that between 1979 and 2019, climate models have predicted a 0.40°C increase in temperature per decade. However, satellite and balloon observations have measured only a 0.17°C increase per decade. Warming in the future is likely to be less than 0.17°C per decade since several warming events have occurred since 1979, including warm ocean oscillations of the Atlantic Multidecadal Oscillation (AMO), Pacific Decadal Oscillation (PDO), and El Niño-Southern Oscillation (ENSO) and unusually low levels of sulfate aerosols (see Chapter 12) and low levels of cloud nucleating cosmic rays (see Chapter 7). Both the AMO and PDO will be in cold phases after 2030, and this should force temperatures down, like in the

1970s. Furthermore, the rapid logarithmic-declining power of CO_2 to warm, as concentrations increase, will lessen future greenhouse gas warming. However, even in the unlikely event that temperatures continue to increase at the current rate of 0.17°C per decade, the temperature by the year 2100 would add up to only 1.3°C warmer than today (0.17°C × 7.5 decades), which is far less than the average prediction of climate models of 3.0°C (0.4 × 7.5 decades).

Some have cited a paper by Berkeley Earth climate scientist Zeke Hausfather and coauthors to suggest that climate models are accurate.[50] However, the Hausfather et al. paper only examines 17 old climate models, mostly from the 1970s, 1980s, and 1990s.[51] These climate models are not used by the IPCC to forecast future temperatures. The IPCC currently uses the CMIP6 models that are the subject of the McKitrick and Christy paper, which shows these current IPCC climate models overestimate warming by 2.4 times. McKitrick and Christy are not the only scientists who suggest CMIP6 models used by the IPCC to predict future climate scenarios run too warm. McKitrick and Christy point out that "Our findings mirror recent evidence from inspection of CMIP6 ECS's (Voosen, 2019)[52] and paleoclimate simulations (Zhu et al., 2020)[53] which also reveal a systematic warm bias in the latest generation of climate models."[54]

Climate alarmists base their fears on IPCC climate model predictions, yet it is clear from observational measurements that these climate models run 2.4× too warm. We have already seen that IPCC climate model projections for the year 2100 require feedbacks that are 2.2× larger than the warming from greenhouse gases. When the net cooling of all feedbacks, including clouds and DMS aerosols, are included, an overestimation of warming by about 2.4× is to be expected.

Atmospheric Sciences Professor John Christy has pointed out that climate models include the water vapor feedback, ignore negative cooling feedbacks, and predict 1.4 watts per meter of heat radiating from Earth out to space for each 1°C of warming. Actual satellite measurements record 2.6 W/m^2 of energy radiates out into space for each 1°C of warming.[55] This is clear evidence that the climate models overestimate the trapping of radiation from water vapor and clouds. *When compared with actual measurements, we must conclude the water vapor amplification hypothesis is exaggerated and is more than offset by negative cooling feedbacks, which reflect or radiate heat out into space.*

The supposition that water vaper, ice cap melting, and clouds will amplify greenhouse gas warming by 2.2× is the entire basis of climate alarm. The true warming from these feedbacks is likely about 0.5×, and this is more than offset by the negative cooling feedback of -0.7× from low cloud cover and increased sulfate aerosols from phytoplankton. Water vapor feedback assumptions are by far the primary reason that climate models run too hot and are not able to predict the future. Such inadequate models should not be used for policy decisions. **The positive water vapor, melting ice caps, and cloud feedback of 2.2× is not supported by experiment, the historical record, or the ignored cooling impact of clouds. Ignoring the cooling of clouds is the reason physicist and Nobel laureate John Clauser said, "The popular narrative about climate change reflects a dangerous corruption of science."**[56]

Dr. Clauser properly identifies clouds as a key driver of climate change. Clouds not only counteract the warming feedbacks of water vapor and melting ice caps, they can also be an independent driver of climate change. In Chapter 11, we will explore, in detail, the formation and impact of clouds on our climate.

Chapter 11:
The Impact of Clouds

Warming and Cooling Impact of Clouds

Although there are several subcategories of clouds—these include cumulonimbus, stratocumulus, and nimbostratus—the three major meteorological classifications of clouds are *cumulus, stratus,* and *cirrus*. Cumulus clouds are puffy, white, with cotton-like billowy shapes. Stratus clouds appear as uniform grayish clouds that cover the sky like a blanket, although the top of stratus clouds appear bright white as viewed from an airplane. Cirrus clouds are high-altitude clouds composed of ice crystals which appear thin and wispy. Of these three categories, cumulus and stratus clouds are considered low clouds and cirrus clouds are classified as high clouds.

Established meteorological theory assumes increased water vapor leads to increased cloud cover. Water vapor content is known to drive low cloud formation, while there is little correlation between high cloud formation and water vapor. Clouds drive both positive (warming) and negative (cooling) climate feedbacks. High cirrus clouds, composed primarily of ice crystals, are mostly transparent to incoming solar radiation, and are known to act as a banket, keeping radiation from escaping out to space. Since they are at high altitudes and thin, their impact on temperature is far less than low clouds. Low clouds have a high albedo (reflectivity) to incoming sunlight and reflect much of the Sun's radiation back out into space. Both high cirrus clouds and low clouds have a greenhouse effect of radiating heat back to Earth, as can be felt on a chilly winter night, which is warmer when overcast and cooler during clear skies.

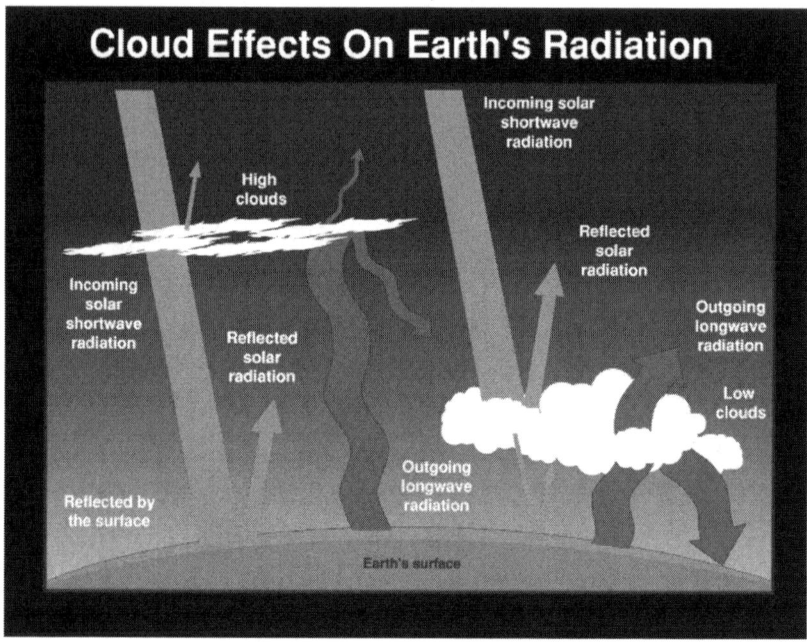

Figure 61 – Cloud Effects on Earth's Radiation. *Clouds have a major impact on the Earth's temperature. Low cumulus and stratus clouds are white on top with high albedo, and they cool the Earth by reflecting more heat out into space than they radiate to the Earth. They are known to reflect 70% to 95% of sunlight back to space. High clouds are somewhat transparent to incoming solar radiation and emit more heat back to the Earth than is reflected out to space. Both high and low clouds have a greenhouse effect and emit infrared or longwave radiation back to the Earth. However, measurements and IPCC statements agree that the net impact of total cloud cover is one of cooling.* Chart Source: Courtesy of NASA. NASA, "Analyzing Cloud Effects on Earth's Energy Budget," My NASA data. https://mynasadata.larc.nasa.gov/mini-lessonactivity/analyzing-cloud-effects-earths-energy-budget.

Three Essential Elements of Cloud Formation

Scientific experiments have revealed that there are *three factors which drive low cloud cover: 1) water vapor; 2) aerosols; and 3) electrically charged ions.* Since cloud formation requires these three elements, an increase in any of these factors can increase low cloud formation. Furthermore, lack of any of these factors results in less cloud formation. As relative humidity rises in the atmosphere, increased low cloud cover is expected. Less water vapor results in less clouds. It has also long been known that clouds are formed as water vapor condenses on cloud nucleating aerosols in the atmosphere. Therefore, increases in aerosols also increase cloud cover, and decreases in aerosols decrease cloud cover. Because electrical charge is needed to produce and grow cloud-nucleating aerosols, an increase of ions in the atmosphere also increases cloud cover, and fewer ions reduces cloud condensation nuclei and cloud cover.

Water Vapor and Cloud Formation: When relative humidity in the atmosphere is sufficiently high, water vapor begins to condense on hydrophilic (water loving) particles, known as cloud condensation nuclei. This forms tiny liquid water droplets around hydrophilic aerosol particles, and these can grow into cloud droplets. The affinity of these hydrophilic aerosols for water molecules facilitates the condensation of water vapor onto the aerosol particles, given sufficient water vapor in the air. Hydrophilic aerosols undergo hygroscopic (water attracting) growth. They absorb water vapor and grow as relative humidity increases. Due to this process, increases in relative humidity lead to more low cloud formation, and decreases lead to less low cloud formation.

Leipzig University (Germany) professor and atmospheric scientist Johannes Quaas and coauthors evaluated data using three satellite datasets to understand cloud formation. They concluded:

"Observations show that cloud cover increases with relative humidity, with a sharp rise in cloud fraction as relative humidity approaches 80-100%, reflecting the enhanced condensation of water vapor onto Cloud Condensation Nuclei under humid conditions."[1] This principle is well established. In their classic book, *Microphysics of Clouds and Precipitation*, which remains the foundational book on cloud physics, Johannes Guttenberg University atmospheric scientist Hans R. Pruppacher and University of New Mexico/Los Alamos Scientific Laboratory atmospheric scientist James D. Klett state, "The formation of cloud droplets requires that the ambient relative humidity exceeds the saturation ratio, which is more readily achieved in regions of high humidity where water vapor concentrations are elevated."[2] Water vapor is also an essential element in the chemical reactions that form cloud-nucleating aerosols, so increased water vapor facilitates the creation of more cloud condensation nuclei, which also leads to more cloud cover.

Aerosols Key Role in Cloud Formation: When humid air rises and cools, the water vapor in the air condenses onto the surface of Cloud Condensation Nuclei (CCN) aerosols. This condensation process forms tiny cloud droplets around each nucleus leading to the development of clouds. As more water vapor condenses onto these droplets, they grow and coalesce into larger droplets forming visible clouds. Oceans are a major collector of solar energy and a dominant player in controlling the climate (see Chapter 13). Oceans cover 71% of the Earth's surface and absorb 90% of the global heat of the Earth from the Sun.[3] Therefore, clouds over the oceans are the most important modulator of solar energy and climate.

Due to their abundance and hydrophilic properties, the primary CCN aerosols over the oceans are sulfates and sea salt particles. These two aerosols represent 80% to 90% of all CCN aerosols over the oceans. A part of sea salt aerosols are negatively charged chlorine ions

(Cl^-). A part of sulfate aerosols are the double negatively charged sulfate ions (SO_4^{2-}), which play a significant role in cloud formation over the oceans by acting as CCN. Negative ions, known as anions, particularly those contained in sulfate and chloride aerosols, are crucial in cloud formation over the ocean. These anions are attracted to hydrogen molecules in water, thus making them hydrophilic (water attracting). The sulfate anions attract and organize water molecule around them, forming a hydration shell. The hydrogen atoms in water orient themselves toward the sulfate anion due to the electrostatic attraction. These interactions make sulfate ions very hydrophilic. Their hydrophilic nature is critical in atmospheric chemistry, where sulfate aerosols play a significant role in cloud formation. The hydrophilic properties of these sulfate and chlorine CCN attract water vapor, and this vapor initiates the formation of cloud droplets. The abundance and hygroscopic nature of these ions ensure that they effectively contribute to the formation and properties of marine clouds, playing a vital role in weather and climate processes over the oceans.

While both salt aerosols and sulfate aerosols are the primary CCN over the oceans, their impact varies by location. Salt aerosols are generated from the ocean surface through the action of wind and waves, and these produce sea spray. As a result, salt aerosols dominate in mid-latitudes and coastal regions, and these are areas of strong wind and wave activity. Sulfate aerosols are produced from phytoplankton and are the dominant aerosol over the Tropics, because this region is a biologically productive area. The Tropics are characterized by warm temperatures, nutrient rich waters, and abundant sunlight which provides ideal conditions for the growth of phytoplankton. Phytoplankton produces dimethyl sulfide (DMS), which is released into the atmosphere and eventually forms sulfate aerosols. The familiar sulphury smell of the ocean is the presence of sulfates produced by phytoplankton. As a result, sulfate aerosols are the primary CCN over the open oceans in the Tropics. Sulfate aerosols containing the anion

(SO_4^{2-}) are especially effective CCN because they are very hygroscopic due to their double negative electric charge. Globally, sea spray (sea salt) aerosols contribute less than 30% to the CCN population while cloud condensation nuclei population between latitudes 70°S and 80°N is composed primarily of non-sea salt sulfate aerosols.[4]

Observations of large phytoplankton algae blooms in the South China Sea showed that these blooms contributed a substantial amount of sulfate aerosols into the atmosphere resulting in observed increased cloud formation.[5] The process of creating sulfate aerosols containing (SO_4^{2-}) anions is complex. DMS is oxidized in the atmosphere by OH radicals and oxygen to form sulfur dioxide (SO_2). Sulfur dioxide is further oxidized in the presence of water vapor in the atmosphere to become sulfuric acid (H_2SO_4). This process is accelerated by cations and iodine oxoacids. A study by the CERN CLOUD project shows that the formation rate of sulfuric acid particles increased by 10- to 10,000-fold in the presence of iodine oxoacids which are present in marine and polar regions.[6] Sulfuric acid reacts with naturally occurring ammonia (NH_3) to form ammonium sulfate ($(NH_4)_2SO_4$) and ammonium bisulfate (HN_4HSO_4). Sulfuric acid also reacts with sodium and calcium to form sodium sulfate (Na_2SO_4) and calcium sulfate ($CaSO_4$). All of these sulfate molecules contain the hydrophilic SO_4^{2-} anions and are the building blocks of sulfate CCN.

Ammonium sulfate aerosols are by far the most abundant CCN. They are extremely hydrophilic because they contain not only the hydrophilic SO_4^{2-} anion that attracts hydrogen molecules in water, but they also contain the hydrophilic NH_4^+ cation that attracts oxygen molecules in water. Ammonia (NH_3) is a biologically produced chemical found in the atmosphere. Ammonium sulfate is formed when sulfuric acid (H_2SO_4) reacts with NH_3 to form ammonium sulfate ($(NH_4)_2SO_4$).

Besides phytoplankton, other sources of sulfate aerosols include volcanoes and industrial sulfur dioxide emissions. Volcanic and industrial emissions of sulfur dioxide are significant. They can be transported over long distances by wind and atmospheric currents. The primary source of sulfates in the atmosphere over oceans is the oxidation of DMS produced by marine phytoplankton. This natural process is the most significant contributor to sulfate aerosols in these regions. Nevertheless, volcanic and man-made emissions from shipping and long-range transport of industrial pollutants do contribute to sulfate levels over the oceans.

Because of the angle of the Sun, the region between 23.5 degrees north and 23.5 degrees south account for approximately 60% to 70% of total solar radiation received by the Earth (see Figure 69 in Chapter 13). **Since low cloud cover can reflect up to 90% of this radiation back out into space, cloud formation from sulfate aerosols over the oceans in the Tropics is one of the most important drivers of climate.** Sulfate CCN are small relative to other aerosols such as dust. Clouds formed over oceans with a high concentration of sulfate aerosols tend to have smaller, more numerous droplets, which increase the cloud's reflectivity and can affect the local climate by reflecting 10% to 20% more sunlight back out to space (see the "Twomey Effect" in Chapter 12[7]).

Sulfate aerosols primarily influence the formation of low clouds. Sulfate aerosols are highly effective in producing low cloud CCN due to their hygroscopic nature. Sulfate aerosols are crucial for the formation of stratus and cumulus clouds. As the air rises and cools, water vapor condenses onto sulfate aerosols, forming cloud droplets and leading to the development of stratus and cumulus clouds. On the other hand, cirrus clouds form in the cold upper troposphere, typically 20,000 feet in altitude or higher, and are composed primarily of ice crystals rather than liquid water droplets. The primary ice nuclei for cirrus clouds are typically different particles, such as mineral dust or biological particles, which are more effective at nucleating ice crystals. The direct role of sulfate aerosols in cirrus cloud formation is limited. This is significant for climate since stratus and cumulus clouds have a net cooling effect and cirrus clouds have a net warming effect. Therefore, sulfate aerosols have a large impact on cooling, especially over the Tropics where the Earth receives most of all solar heating.

Ions Play a Critical Role in Cloud Formation: Ions are electrically charged particles. Ions play a crucial role in the formation of sulfate CCN aerosols containing sulfate ions (SO_4^{2-}). The formation of aerosols containing SO_4^{2-} ions in the atmosphere are caused by a complex cascade of reactions. When galactic cosmic rays enter the atmosphere, they strip electrons from nitrogen and oxygen molecules forming nitrogen cations (N_2^+) and free electrons (e^-) as well as oxygen cations (O_2^+) and free electrons (e^-). The O_2^+ cations react with water in the atmosphere and produce H^+, H_3O^+, and OH^-. Cosmic rays are the primary source of OH in the troposphere, and OH is critical to producing sulfate CNN. H_3O^+ cations produced from cosmic rays react with natural ammonia (NH_3) in the atmosphere to produce NH_4^+ cations. Cosmic rays also produce anions such as O^-, O_2^- and HSO_4^-. The OH produced by cosmic rays in the troposphere reacts with nitrogen oxides to produce nitrate anions (NO_3^-), and it reacts with volatile organic compounds in the atmosphere to create carboxylic anions ($HCOO^-$ and CH_3COO^-). All of these ions facilitate sulfate CNN production.

Phytoplankton produces DMS (C_2H_6S) which is emitted into the atmosphere. DMS is oxidized by OH through a series of steps that produces sulfur dioxide (SO_2). Sulfur dioxide is oxidized from OH radicals and oxygen to form SO_3, which reacts with water vapor to produce sulfuric acid (H_2SO_4). Sulfuric acid is not itself a sulfate cloud-nucleating aerosol, but it is a critical precursor that leads to the formation of aerosols containing SO_4^{2-} anions. Sulfuric acid reacts with ammonia, sodium, and calcium to form ionic molecules including ammonium sulfate, ammonium bisulfate, sodium sulfate, and calcium sulfate, all of which contain the hydrophilic SO_4^{2-} anion and cations which neutralize the charge of the molecule. These processes are all enhanced by the presence of ions. Furthermore, free ions help these CCN molecules form into 1-2 nanometers (nm) clusters because the electrostatic charge of ions attracts the oppositely charged ions in ionic molecules, and they facilitate the clustering of these neutral molecules. CLOUD experiments at CERN in Switzerland demonstrated that ammonia increases the nucleation rate of sulfate particles by 100- to 1,000-fold, and that the presence of ions increases this rate by another 2- to 10-fold.[8] East China Normal University environmental scientist Gehui Wang and coauthors studied the aerosol formation impacting air pollution in China. They found that "sulfate aerosols formation was enhanced by four orders of magnitude [or 10,000 times] larger than the traditionally held value due to the occurrence of ions, which remarkably accelerates sulfate formation."[9]

Sulfate aerosols are initially only 1-2 nm in diameter but need to grow to at least 50 nm in diameter to be sufficiently large to become CCN aerosols. According to internationally recognized atmospheric scientist and professor at the University of Helsinki (Finland), Mikael Ehn, and coauthors, "The growth of sulfate aerosols is accelerated by the presence of cations such as NH_4^+, which facilitate the condensation of sulfuric acid and organic vapors onto existing particles, increasing their size from a few nanometers to the Aitken [10-100 nm-diameters] and accumulation [100-1000-nm diameters] modes."[10] Professor Ehn is known for applying advanced mass spectrometric measurements to study gas-phase chemistry and particle formation. Positive ions, known as cations, and negative ions, known as anions, help sulfate aerosols to cluster to become of sufficient size to initiate cloud droplet formation. The aggregation of sulfate aerosols with cations such as N_2^+, H_3O^+, NH_4^+, and anions such as O^-, O_2^-, HSO_4^-, NO_3^-, $HCCO^-$, and CH_3COO^- can stabilize sulfate aerosols, which helps facilitate particle formation.

Free ions are attracted to neutral ionic molecules such as ammonium sulfate, ammonium bisulfate, sodium sulfate, and calcium sulfate via charge-dipole interactions, making it harder for the aerosol molecules to escape the aerosol cluster. The electrostatic charge provided by ions holds the aerosol clusters together so that aerosol clusters of ionic molecules with ions are more stable than a cluster of ionic molecules alone. The ions hold the clusters together and make it less likely that the ionic molecules will leave the cluster through evaporation. Clusters of ionic molecules alone (neutral clusters) often see molecules evaporate out of the cluster due to high surface area of small-diameter clusters. Therefore, aerosol clusters with ions allow more particles to survive the volatile 1 nm to 10 nm particle size growth range. Due to the electrostatic attraction of ions with ionic molecules, charged clusters also attract additional ionic molecules more efficiently than neutral clusters. This accelerates the addition of molecules, increasing cluster size. Because of the higher nucleation rates, more clusters survive and grow, and this speeds up the growth to CCN size. Ion-induced nucleation has been confirmed through experiments. Technical University of Denmark atmospheric scientist Martin Enghoff, astrophysicist Henrik Svensmark, and coauthors used an ion beam to nucleate sulfate aerosols. They found clear evidence of ion-induced nucleation as a source of aerosol production.[11] Thus, ions play a crucial role in helping to both create sulfate aerosols from precursor molecules and in assisting sulfate aerosols to grow in size to become CCN aerosols, thus facilitating cloud formation.

By stabilizing aerosol clusters at the 1-2 nm size and enhancing growth rates, ions increase the fraction of such clusters that survive to 10 nm in size and beyond. Once the particles reach ~10 nm, they are more stable and can grow to CCN size. Ion-induced nucleation is especially important in clean air environments where there are low levels of precursor aerosol molecules. In clean environments, such as over the oceans, ion-induced nucleation is a major pathway for sulfate CCN production. Ion-induced nucleation of CCN is thus particularly important in cloud formation over the oceans. As will be discussed in Chapter 13, the oceans play a key role in the climate because they act as a massive solar collector, absorbing about 90% of all solar heat received by the Earth.[12] **Therefore, ions play a key role in cloud formation over the oceans, which impacts the heat content of the oceans and the transfer of such heat to the atmosphere.**

Cloud formation requires CCN, and the production of CCN is enhanced by ions. The primary source of ions in the atmosphere at the level of low cloud formation is from cosmic rays.[13] When high-energy cosmic rays interact with other molecules in the atmosphere, they induce a cascade of cation and anion formation, creating a variety of ions in the troposphere. These ions are key elements in aerosol formation and cloud nucleation. As described in Chapter 7, cosmic rays increase low cloud cover. Recent global warming has occurred during a time of low cosmic rays and low cloud cover. NASA and others are predicting we are just now moving into a period of higher levels of cosmic rays. "During the next solar cycle, we could see cosmic ray dose rates increase by as much as 75 percent," said lead author Fatemeh Rahmanifard of the University of New Hampshire's Space Science Center.[14] **The expected future increase in cosmic ray flux should seed the formation of more low clouds which should lower temperatures.**

The Net Cooling Effect of Clouds

Clouds have a significant cooling effect because they act as a parasol shading the Earth and reflect incoming solar shortwave radiation out to space from their reflective white color. Even the darkest clouds appear white on the top when viewed in sunlight from space. In fact, the reason dark clouds look gray or nearly black from Earth is because they block much of the sunlight passing through them. Darker-looking clouds from below are typically thicker and denser, meaning they block more sunlight from getting through to the Earth. The reflectivity of such clouds is so strong at the top that they appear brilliantly white, even more so than other clouds, unless they are shaded by clouds above them. In full sunlight clouds that are darkest from below are generally brightest from above.

According to section 1.1.2 of the IPCC climate assessment, *Natural Forcing of the Climate System*, "the net average effect of the Earth's cloud cover in the present climate is a slight cooling: the reflection of radiation more than compensates for the greenhouse effect of clouds."[15] This is a gross understatement. According a 1989 paper by renowned atmospheric scientist and Distinguished Professor of the Scripps Institute of Oceanography, Veerabhadran Ramanathan and coauthors, clouds have a greenhouse warming effect of 31.3 W/m^2 and a cooling impact of -44.5 W/m^2 due to their white color, which reflects incoming solar shortwave radiation back out to space.[16] Therefore the net impact of clouds is a net cooling effect of -13.2 W/m^2. This is an early estimate, and better measurements by satellite have shown the net cooling impact of clouds is *higher*. Caltech atmospheric scientist Graeme Stephens and coauthors cite the following: "The so-called albedo effect of clouds enhances the reflected solar flux by 47.5±3 $Wm–^2$, whereas clouds reduce the outgoing long-wave flux relative to clear skies by approximately 26.4±4 $Wm–^2$ (a measure of their greenhouse effect). This gives a net loss of radiation from Earth by clouds of 21.1±5 Wm^{-2}, mostly by reflection of sunlight from clouds in the mid-latitude summer hemisphere."[17] A more recent study by University of Wisconsin professor and atmospheric scientist Tristan L'Ecuyer and coauthors states: "… estimate the annual mean net cloud radiative effect at the surface to be −24.8 ± 8.7 W m^{-2} (−51.1 ± 7.8 W m^{-2} in the shortwave and 26.3 ± 3.8 W m^{-2} in the longwave)."[18]

In its ***AR5*** **Report, the IPCC estimated the feedback of clouds to be one of cooling**, stating the following: "By enhancing the planetary albedo, cloudy conditions exert a global and annual shortwave cloud radiative effect (SWCRE) of approximately –50 W m^{-2} and, by contributing to the greenhouse effect, exert a mean longwave effect (LWCRE) of approximately +30 W m^{-2}, with a range of 10% or less between published satellite estimates (Loeb et al., 2009). Some of the apparent LWCRE comes from the enhanced water vapour coinciding with the natural cloud fluctuations used to measure the effect, so the true cloud LWCRE is about 10% smaller (Sohn et al., 2010). **The net global mean cloud radiative effect of approximately –20 W m^{-2} implies a net cooling effect of clouds on the current climate.**"[19] The more current paper by L'Ecuyer et al. estimate of a net cooling from clouds of -24.8 W/m^2 is likely the most accurate estimate since it uses the most modern datasets from CloudSat and CALIPSO satellites.

Despite these facts, according to the IPCC, "All global models continue to produce a near-zero to moderately strong positive net cloud feedback."[20] In other words, these climate models assume the net impact of clouds is *warming*. The models use an average number of 0.6 W/m^2 of net warming from clouds.[21] The IPCC attributes most of this positive impact of clouds due to the "fixed anvil-temperature" mechanism that assumes the infrared emissions of clouds will change with height as the lapse rate heightens the top of the troposphere. However, the IPCC admits that there is weak support from observations for this fixed anvil-temperature mechanism and "observed cloud height trends do not appear sufficiently reliable to test this cloud-height feedback mechanism."[22] *Nevertheless, rather than use the IPCC's stated -20 W/m^2 net cooling impact of clouds in climate models, IPCC climate models use a net warming effect of +0.6 W/m^2*. **This blatant contradiction reveals the desperate and deceptive measures the IPCC allows to perpetuate a climate crisis narrative**. It also further explains why climate models run warmer than measured temperatures (see Figure 60 in Chapter 10).[23]

There is observational evidence to confirm clouds provide a net negative cooling feedback to the climate. It is estimated that a 1%

increase in albedo reduces incoming solar shortwave radiation by about 3.4 W/m^2.[24] USDA physical scientist Ned Nikolov and USDA atmospheric scientist Karl Zeller state: "The assumption that Earth's albedo has been decreasing for the past 37 years (and thus driving climate change) is based on the latest version of the ISCCP-FH radiative-profile flux product by the International Satellite Cloud Climatology Project (ISCCP), which shows a 6.5% reduction of the global cloud-cover fraction between 1986 and 2018."[25] Using this rate of decline of 6.5% over 32 years (1986 to 2018), over a 24-year period from 2000 to 2024, the decline in cloud albedo would be between 4% to 5%. Measurements from the Clouds and the Earth's Radiant Energy System (CERES) satellites between 2001 to 2023 have shown a decrease in the reflectivity (albedo) of the Earth of 0.6 (see Figure 62).[26] This decline in low cloud cover has reduced the amount of solar shortwave radiation reflected out to space by ~2.0 W/m^2 from 2000 to 2024, as measured by the CERES satellite (see Figure 62).[27] These observations are consistent with the estimated change of 3.4 W/m^2 for each 1% change in albedo since 2 W/m^2 divided by an albedo of 0.6 equals about 3.4 W/m^2 (2.0 ÷ 0.6 = ~3.4).

Observations suggest recent warming is primarily driven by an increase of incoming shortwave radiation from the Sun that is caused by decreased reflection of sunlight from lower albedo.[28] Between the years 2000 and 2024, albedo has declined by ~0.6%, and incoming shortwave radiation has increased by ~2.0 W/m^2 (see Figure 62).[29] Low clouds are white on top and extremely reflective with a high albedo. **These satellite observations show that the decrease in low cloud cover over the past 24 years had a more powerful effect of warming the Earth than increased levels of CO_2 and other anthropogenic greenhouse gases.**

CERES satellite measurements also show an increase in outgoing longwave radiation of ~0.6 W/m^2 between 2001 to 2023 (see Figure 53 in Chapter 9).[30] If the warming since 2001 was primarily from increases in anthropogenic greenhouse gases, the longwave radiation emitted out to space would have *declined* as infrared radiation is trapped in the atmosphere by such greenhouse gases. The *observed increase* in longwave radiation passing out to space is consistent with a decline in cloud cover, since clouds also trap longwave radiation and prevent it from escaping out to space. According to Tristan L'Ecuyer et al. clouds have a greenhouse effect of 26.3 W/m^2 which is nearly nine times larger than the 3.0 W/m^2 for doubling CO_2 concentrations.[31] Between 2000 and 2024, CO_2 and other anthropogenic greenhouse gases have increased by about 15%. A decline of 4% to 5% in low cloud cover sends more than twice as much longwave radiation out to space as is trapped by an increase of 15% in anthropogenic greenhouse gases (see Chapter 9). This has resulted in a net loss of -0.6 W/m^2 of longwave radiation emitted out to space as is observed in the CERES satellite measurements (see Figure 53 in Chapter 9). **Unbelievably, despite this clear observational evidence from CERES satellites, the IPCC states that it is very likely that well-mixed greenhouse gases were the *main* driver of tropospheric warming since 1979.**[32]

Because cloud cover has declined in recent years during a time of increasing temperatures, Helmholtz Center for Polar and Marine Research (Germany) atmospheric scientist Helge F. Goessling and coauthors have speculated that global warming leads to less cloud cover as a possible explanation.[33] This is a surprising hypothesis as it is counter to common sense and the established consensus view that warmer temperatures increase water vapor flux in the atmosphere, thus increasing cloud cover.[34] Futhermore satellite measurements have confirmed precipitation (and thus rain clouds) scale up with warming.[35] In all fairness, the Goessling et al. paper suggests global warming as only one possible explanation of reduced cloud cover and they also mention other potential causes, including reduced aerosols.

Although the moderating lapse rate due to global warming reduces vertical convective power and limits the formation of cumulous clouds, the atmospheric stability of a lower lapse rate enhances the formation of horizontal stratus clouds. This is especially true in the Tropics where direct sunlight and higher temperatures sustain humidity. Global warming can change the type and regionality of cloud formation. However, if there are sufficient CNN, the formation of low clouds increases with warming due to increased evaporation and humidity, especially in the Tropics, where clouds have the most impact on climate. A more plausible explanation for lower cloud cover in recent years is the decrease in clouds due to the dramatic decline in sulfate-aerosol-based cloud cover. A study by University of Wyoming atmospheric scientist Daniel T. McCoy and coauthors suggest that a decline in cloud coverage caused by aerosols can be confused for evidence of negative cloud feedback from warming, and separating the two effects has been difficult.[36]

The decline in cloud cover and the Earth's albedo since 2000 is likely mostly driven by reduced industrial emissions of sulfur dioxide during this period. Sulfur dioxide is a major source of sulfate CCN. Not only does more SO_2 in the atmosphere drive more cloud cover, it also creates smaller cloud droplets, which has a higher albedo (reflectivity) than clouds formed from other CCN (see the "Twomey Effect" in Chapter 12[37]). SO_2 molecules in the atmosphere also increase albedo by scattering incoming shortwave radiation which limits the amount of radiation reaching the Earth's surface. Since 1979 industrial emissions of SO_2 have declined by about 50% due to air pollution regulations and the implementation of pollution controls. In 2020, SO_2 was further reduced by new regulations that require an 85% decline in SO_2 emissions from maritime shipping (see Chapter 12). This reduction in sulfur dioxide emissions has thus lowered albedo and raised temperature by allowing more solar incoming shortwave radiation to strike and heat the Earth. A 2024 study by University of Maryland atmospheric scientist Tianle Yuan and coauthors states that the warming effect from the reduction in maritime sulfur emissions "is equivalent in magnitude to 80% of the measured increase in planetary heat uptake since 2020."[38]

If 80% of the increase in heat uptake by the Earth between 2020 and 2024 was due to a reduction in sulfate emissions, as set forth in the Yuan et al. paper, only about 20% or less remains for anthropogenic greenhouse gases. Similar numbers have been observed since 2001. The CERES satellite record between 2001 to 2023 shows an increase in solar shortwave radiation of 2.0 W/m^2.[39] From 2001 to 2023 there was a 14% increase in concentrations of CO_2 (370 ppm to 420 ppm). Using the accepted ERF adjusted Myhre radiative forcing equations,

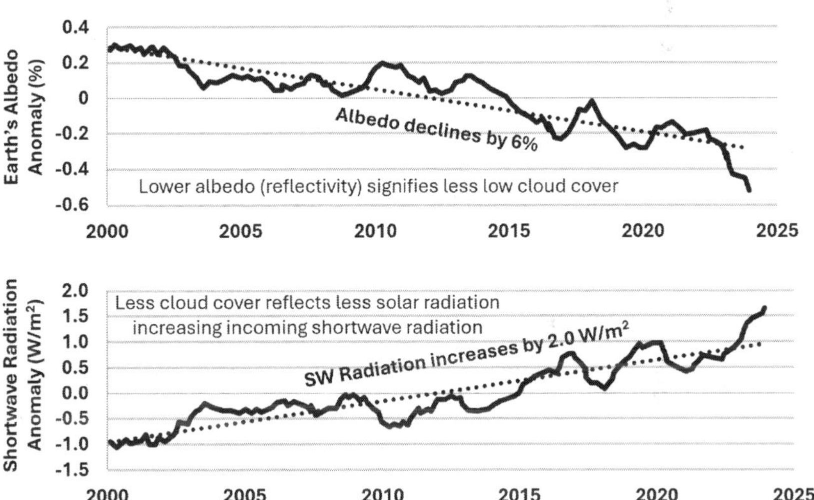

Figure 62 – Decline in Earth's Albedo and Increase of Incoming Shortwave Radiation. *The charts above use CERES satellite data to show the decline in albedo (reflectivity) of the Earth (top graph), and the increase in incoming shortwave radiation from the Sun to the Earth (bottom graph), between 2000 and 2024. Measurements show a 2.0 W/m^2 increase in incoming solar radiation which is expected from the observed 0.6% decline in albedo.* Source: Nikolov, Ned and Zeller, Karl F., "Role of Earth's Albedo Variations and Top-of-the-Atmosphere Energy Imbalance in Recent Warming: New Insights from Satellite and Surface Observations," *MDPI Geomatics*, Volume 4, Issue 3, pp. 311-341, 20 August 2024. https://doi.org/10.3390/geomatics4030017.

this 14% increase equates to about 0.5 W/m^2 less longwave radiation being sent out into space.[40] When you include all anthropogenic greenhouse gases (GHG), the trapped longwave radiation increases to about 0.78 W/m^2. When you consider both of these warming impacts they total 2.78 W/m^2. Therefore, the impact from more incoming solar shortwave radiation, due to reduced sulfate aerosols, represents about 72% of this warming (2.0 ÷ 2.78 = 72%) and the impact of all anthropogenic greenhouse gases represents about 28% (0.78 ÷ 2.78). **The evidence is clear, greenhouse gases have *NOT BEEN* the *main* driver of tropospheric warming in recent years as claimed by the IPCC.**

The high albedo of low clouds is a major source of reflecting heat out into space. Using the assumption of the Clausius-Clapeyron relationship that water vapor increases by 7% for each 1°C in warming, this additional water vapor in the atmosphere will condense as it rises so cloud cover may also increase by 7%, if sufficient cloud-nucleating aerosols are present. Climate models and traditional meteorological theories have predicted that clouds and precipitation will not scale with water vapor increases. However, a study by earth scientist and CEO of Remote Sensing, Frank J. Wentz, and coauthors points out that "a recent analysis of satellite observations does not support this prediction of a muted response to precipitation to global warming. Rather, the observations suggest that precipitation and total atmospheric water have increased at about the same rate over the past two decades."[41] Using the 2019 L'Ecuyer et al. estimate of the net cooling of clouds[42] of -24.8 W/m^2 x 7% = -1.7W/m^2. Satellite observations confirm this number. Using measured data from the International Cloud Climatology Project (ISCCP) satellite network confirms temperatures decline by 0.5°C for every 7% increase in total cloud cover (see Figure 63).[43] The 0.5°C in cooling for a 7% increase in cloud cover therefore means that clouds cool 1.7 W/m^2 (see Table 20) for each 1°C in warming, given the Clausius-Clapeyron relationship. The cooling of clouds thus offsets all and more of the estimated warming of 1.3 W/m^2 from water vapor positive feedback for each 1°C in warming (see Chapter 10).

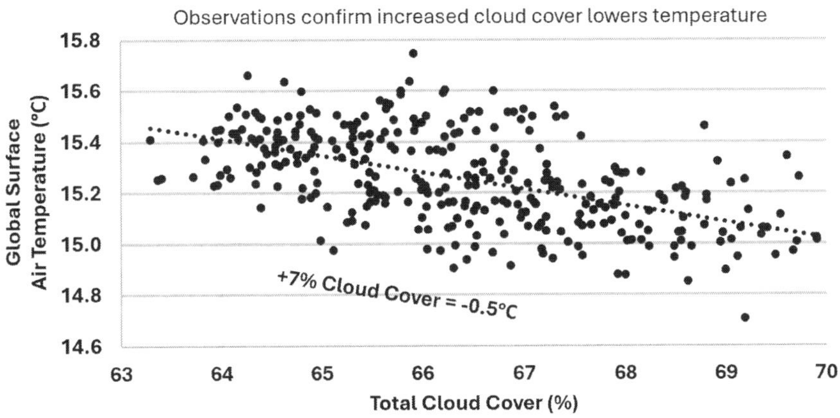

Figure 63 – Global Cloud Cover and Temperature. *The scatter diagram above charts monthly total cloud cover versus global surface temperatures since 1983. The decline in temperature as cloud cover increases demonstrates the net cooling effect of clouds. A linear fit model reveals a 0.5°C drop in temperature for every 7% increase in global cloud cover.* Source: hardCRUT3 and The International Satellite Cloud Climatology Project from *Climate4you.com,* "Total cloud cover vs. global surface temperature," "Climate and Clouds, Cloud cover effect on climate," https://climate4you.com/ClimateAndClouds.htm#TotalCloudCoverVersusGlobalTemperature.

Table 20 Stefan-Boltzmann Conversion of Temperature to Heat	
Current Temperature plus 1°C	0.5°C Decline in Temperature
$m^2 = 1$	$m^2 = 1$
Effective Emissivity of the Earth = 0.6127	Effective Emissivity of the Earth = 0.6127
Radiative Forcing = 242.7 W/m²	Radiative Forcing = 241.0 W/m²
Temperature = 288.1 K + 1°C = 289.1 K	Temperature = 289.1 K – 0.5°C = 288.6 K
	Radiative Forcing Change = -1.7 W/m²

Table 20 Source: Stefan-Boltzmann Law Calculator, *BYJU'S*. https://byjus.com/stefan-boltzmann-law-calculator/.

Cloud feedback is dynamic and changes with the amount and type of cloud cover since low clouds reduce temperature by reflecting more energy out to space than they absorb and emit, and high clouds act as a blanket to increase temperature and are not effective at reflecting radiation. Clouds are not evenly distributed around the globe. Measurements report 10% to 15% more cloud coverage over oceans than over land. About 72% of the oceans are covered by clouds and therefore changes in cloud cover have a significant impact on the heating of the oceans. Since the year 2000, cloud cover has decreased by between 4% to 5% (see Chapter 9),[44] reflecting less solar energy out to space, thereby producing a 2 W/m^2 increase in incoming solar shortwave radiation.[45] Ocean temperatures over this period mirror these changes since oceans are primarily warmed by the Sun (see Figure 64). Measurements of global ocean temperature and incom-

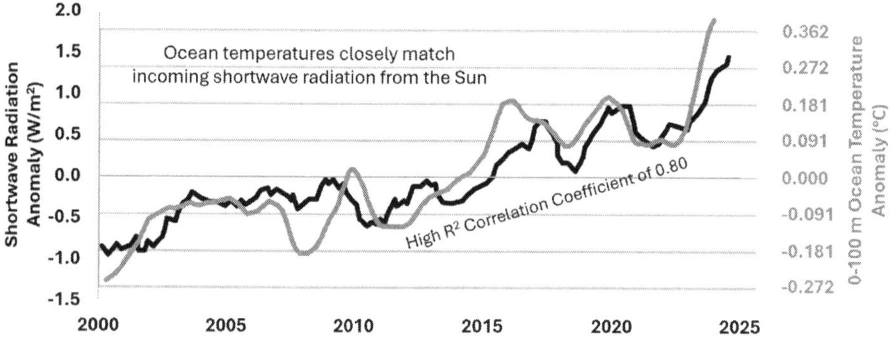

Figure 64 – Incoming Shortwave Radiation and Ocean Temperatures, 2000 to 2024. *Since the year 2000, there has been a reduction in cloud cover, between 4% to 5%. Less solar energy is reflected out to space with lower cloud cover and more shortwave radiation is absorbed in the oceans from the Sun. Observed temperature anomalies in the upper 100 meters of the global ocean layer reflect the increase of incoming solar radiation.* Source: Nikolov, Ned and Zeller, Karl F., "Role of Earth's Albedo Variations and Top-of-the-Atmosphere Energy Imbalance in Recent Warming: New Insights from Satellite and Surface Observations," *MDPI Geomatics,* Volume 4, Issue 3, pp. 311-341, 20 August 2024. https://doi.org/10.3390/geomatics4030017.

ing shortwave radiation from the Sun over the past 24 years have a high 80% R^2 correlation coefficient as expected. **This CERES satellite data confirms that it is the Sun, and not anthropogenic greenhouse gases, that explains recent warming of the oceans.** Ocean currents move this warmth to the Arctic, causing accelerated Arctic warming, known as Arctic Amplification, and the associated moderation of the Meridional Temperature Gradient (see Chapter 13 for more details).

Respected MIT atmospheric physicist Richard Lindzen and his research group published a paper which describe how changes in cloud cover produce negative cooling climate feedbacks, which lower the temperature. This process deals with high cirrus clouds and is known as the "iris effect."[46] The iris effect proposes that every 1°C of sea surface warming reduces high cirrus cloud cover by about 20%, because higher sea surface temperatures lead to stronger, more concentrated updrafts that produce fewer cirrus clouds. Fewer cirrus clouds allow longwave infrared radiation from the Earth to more easily pass out into space. According to Lindzen, this induces a negative cooling feedback which lowers temperature enough to offset warming due to water vapor positive feedback.[47] Lindzen's iris effect has been challenged by Gavin Schmidt (a well-known climate alarmist) of the NASA Goddard Institute for Space Studies. Nevertheless, observations seem to validate the iris effect, as satellite data from the Earth Radiation Budget Experiment (ERDE) shows energy passing out into space increases noticeably in the Tropics whenever sea temperature increases, as expected from the iris effect.[48] **In a blatant censorship of views which are counter to the climate alarmist narrative, the editors who allowed the iris effect paper to be published were fired.[49] Such suppression of ideas is detrimental to scientific research.**

The IPCC admits uncertainty by stating, "The role of thin cirrus clouds for cloud feedback is not known."[50] What is assumed is that the impact of high cirrus clouds is much smaller than the impact of low clouds on the climate. Compared to a net cooling of -24.8 W/

m^2 for all clouds, it has been estimated that cirrus clouds have a net warming impact of about 5.1 W/m^2. This is the net of 21.8 W/m^2 of warming from the greenhouse effect and -16.7 W/m^2 of cooling by reflecting shortwave radiation out to space.[51] Cirrus clouds are made up of ice crystals, which have a larger diameter than water droplets in low clouds, resulting in a lower albedo. Not only is the warming of cirrus clouds much lower than the cooling impact of low clouds, but low clouds dominate the sky with about four times more cloud coverage globally than high cirrus clouds. The net impact of all cloud cover is one of cooling of about -24.8 W/m^2. Therefore, low clouds are the dominant feedback in the climate system. Without cirrus clouds, the net cooling impact of low clouds would be -26.1 W/m^2 (-24.8 W/m^2 minus one fourth of 5.1 W/m^2).

Consequently, the most significant cloud feedback is how increased water vapor in the atmosphere leads to greater low cloud formation. The high albedo of low clouds reflects sunlight radiation back out into space, producing a negative feedback of at least 476 W/m^2 of irradiative energy away from the Earth, under cloud cover at the equator (see Chapter 13). This is a significant amount. Dr. John F. Clauser, recipient of the 2022 Nobel Prize in Physics, has stated that existing climate models greatly underestimate this negative cloud feedback, which provides "a very powerful dominant thermostatic control of the Earth's temperature." These white, low clouds reflect 70% to 95% of the sunlight which shines on them back out into space.[52] As seawater is heated and evaporates, it produces low clouds. According to Dr. Clauser, the radiative forcing from CO_2 is nearly two orders of magnitude (10_2, or 100-fold) smaller than the effective stabilization of the input-power provided by the low cloud-based thermostat.[53] **Because of this negative cloud feedback, Dr. Clauser says, "There is no real climate crisis."**

Lindzen and Clauser believe these negative cooling cloud feedbacks are stronger than the positive water vapor feedback, and this results in a net-negative cooling feedback from water vapor and clouds. **The IPCC admits the greatest uncertainty in climate models is clouds.** The IPCC stated: "The representation of cloud processes in climate models has been recognized for decades as a dominant source of uncertainty in our understanding of changes in the climate system."[54] Despite this uncertainty, they include positive (*warming*) water vapor and net positive (*warming*) cloud feedbacks in their climate models and ignore or minimize net-negative *cooling* cloud feedbacks, which could explain, in part, why their predictions of warming temperatures have been about 2.4 times higher than actual observational measurements (see Figure 60 in Chapter 10).[55]

Professor Lindzen was the lead author of Chapter 7, "Physical Climate Processes and Feedback" in the IPCC *Assessment Report 3*. He later resigned from the IPCC, in part because the IPCC refused to consider the negative cooling feedback of clouds in its climate models.[56] William Gray, professor of Atmospheric Science at Colorado State University, known as one of the greatest minds in hurricane research, agreed with Lindzen that the net feedback from increasing cloud cover was cooling, not warming. Among Dr. Gray's many accomplishments, he established the accumulated cyclone energy index, the accepted measure of hurricane intensity. Despite Dr. Gray's renown, his research funding and career were "cancelled" for expressing views counter to the climate crisis narrative.[57]

Clouds Act as the Earth's Thermostat

Dr. Clauser, Nobel laureate, has repeatedly stated that clouds are the Earth's thermostat.[58] When the Earth warms, more water vapor evaporates into the atmosphere forming low cloud cover, and this acts as a parasol to cool the Earth. In addition, warmer temperatures enhance phytoplankton algae blooms, which are a major source of sulfide CCN aerosols from phytoplankton release of DMS into the atmosphere (see The CLAW Hypothesis in Chapter 12). Since water vapor and aerosols are two of the three elements of cloud formation, as the earth warms, more low clouds are formed which reflect more solar energy back out to space—and the oceans begin to cool. When the earth cools, such as in an ice age, less water evaporates and phytoplankton algae blooms decline leading to lower water vapor in the atmosphere and fewer DMS aerosols, which result in fewer clouds and more ocean warming from the Sun. The third element of cloud formation is ions produced from cosmic rays. However, even if cosmic ray flux increases, cloud formation will not appreciably increase in cold temperatures as there is limited water vapor and DMS in the atmosphere for ions from cosmic rays to act upon.

Observational data supports these assumptions. Due to the cold temperatures in the polar regions, water vapor and sulfate aerosols are limited. However, cosmic ray flux at the poles is greater than elsewhere. The magnetic field of the Earth directs cosmic rays toward the poles. The cosmic ray flux at the poles can be significantly higher—often two to three times greater than at the equator—as measured by global neutron monitor networks.[59] Because there is less water vapor and DMS in polar regions, the higher cosmic ray flux does not create more sulfate CCN aerosols and clouds. According to Georgia Tech Arctic climate specialist, Judith Curry and her co-authors, "Cloud cover in the Arctic is highly seasonal, with a minimum in winter (often below 50%) due to the lack of moisture and stable atmospheric conditions, contrasting with higher cloud fractions in mid-latitudes and Tropics, where annual averages often exceed 60-70%."[60] The South Pole also experiences less cloud cover than the global average, with cloud cover ranging between 45% to 50% in the summer and 55% to 65% in winter.[61]

Stratus clouds have the highest reflectivity of all cloud types. They are composed of small water droplets that are densely packed and therefore have high reflectivity. They also blanket the sky with thick opaque optics. Their considerable thickness and widespread coverage enhance their ability to reflect incoming solar radiation back out to space. Cumulus clouds extend high into the sky. Their limited horizontal extent and larger water droplet sizes result in a lower overall albedo compared to stratus clouds. Cumulus clouds are formed by convective vertical updrafts and are more prevalent in unstable conditions. Stratus clouds form under stable conditions with lower convection that allow for horizontal spread of cloud formation. As discussed in Chapter 3, recent climate change has fostered lower convection and stable atmospheric conditions by lowering the lapse rate due to more rapid warming in the high troposphere. **Therefore, global warming is more conducive to the formation of stratus clouds, which reflect more sunlight back to space and provide more cooling than cumulus clouds. This is another way nature provides a natural thermostat to prevent the Earth from overheating.**

Clouds formed with water droplets have smaller particle diameters than clouds formed with ice crystals. Therefore, water-formed clouds have higher reflectivity than ice-formed clouds and have more of a cooling effect due to their higher albedo. This is why cirrus clouds (ice crystal-based) have a net warming effect, and stratus clouds and cumulus clouds (both water-based) have a net cooling impact. As the atmosphere warms, some of these ice crystal clouds will melt into smaller water droplets, thus increasing the reflectivity of sunlight back to space and cooling the Earth. Global warming should hasten this transformation of clouds from ice-based to water droplet-based. This is yet another way nature provides a natural thermostat.

Aside from the massive climate impact of clouds, other climate forcings include aerosols and volcanoes. These both play a part in climate change. Chapter 12 will cover the key role played by these factors, including the cause of recent warming and the interesting CLAW Hypothesis.

Chapter 12:

Other Climate Forcings

Aerosols

The IPCC climate models include a cooling effect from aerosols. Aerosols cool the Earth by scattering incoming solar radiation and nucleating low clouds. Scripps Institute of Oceanography atmospheric scientist Chul E. Chung and coauthors estimate the total cooling impact of aerosol light scattering of all anthropogenic aerosols combined is a total of -3.4 W/m^2 at the surface of the Earth.[1] The impact on climate from aerosol cloud nucleation is in addition to this number. Aerosols act as Cloud Condensation Nuclei (CCN) which are an essential part of cloud formation. According to a study by University of Wisconsin atmospheric scientist Tristan L'Ecuyer and coauthors clouds reflect 51.1 ± 7.8 W/m^2 of sunlight back out to space.[2] The impact on climate from cloud formation can therefore have a much larger impact on climate than the scattering of light properties of aerosols. A 10% increase in CCN and associated cloud cover equates to -5.1 W/m^2 of cooling, whereas a 10% increase in aerosols produces only -0.3 W/m^2 in cooling from light scattering. The primary aerosols that impact the climate are listed below in Table 21.

Table 21 Primary Cooling Aerosols		
Aerosol	**Sources**	**Mechanism**
Sulfate Aerosols	Phytoplankton, fossil fuel emissions, volcanoes	Scatter incoming light and nucleate clouds
Sea Salt Aerosols	Ocean spray	Scatter incoming light and nucleate clouds
Nitrate Aerosols	Combustion from vehicle exhaust and power plants	Scatter incoming light and nucleate clouds
Organic Carbon Aerosols	Vegetation, burning biomass, fossil fuels	Scatter incoming light and nucleate clouds
Volcanic Ash	Volcanoes	Scatter incoming light and nucleate clouds

As discussed in Chapter 11, sulfate cloud condensation nuclei (CCN) are essential in the formation of low clouds over the oceans in the Tropics. Sulfate ions are the single most important aerosol impacting the climate. This is because the most prevalent cloud formation in the Tropics is from sulfate CNN and most of all solar energy received by the Earth is in the Tropics. Furthermore, sulfate CCN produce fine cloud particles with high reflectivity. Sulphate aerosols are primarily produced from phytoplankton in the ocean releasing dimethyl sulfide into the atmosphere, which is converted into Sulfur Dioxide (SO_2) and eventually sulfate CCN (see Chapter 11). In addition, industrial emissions and volcanoes emit SO_2 (sulfur dioxide), which is converted into H_2SO_3 (sulfuric acid) and further converted into sulfate CCN.

The CLAW Hypothesis

In 1987, scientists Robert Jay **C**harlson, James **L**ovelock, Meinrat **A**ndreae, and Stephen G. **W**arren proposed the CLAW Hypothesis, named for their initials, as a climate-regulating feedback loop involving marine phytoplankton and atmospheric processes. The hypothesis suggests that the growth of phytoplankton increases with warmer temperatures and produces more sulfate CCN aerosols from higher dimethyl sulfide (DMS) emissions, resulting in more cloud cover, increased albedo of the Earth, and cooling on the Earth's surface. According to Charlson et al. this cooling could cancel out global warming caused from increases of CO_2. They claimed a 50% increase in DMS emissions would counteract a doubling of atmospheric CO_2 in the atmosphere.[3]

The CLAW Hypothesis has been controversial with a 2014 review paper by Scottish marine scientists Tamara K. Green and Angela D. Hatton suggesting it is an oversimplification, and that the effect might be much weaker than proposed.[4] However, Green and Hatton admit the CLAW Hypothesis is true in some respects. They cite nine years of data from Cape Grim that shows "increased DMS emissions during summer do in fact lead to an increased CCN population, which regulates the local climate on seasonal timescales."[5] Highly respected Max Plank Institute atmospheric scientist Meinrat O. Andreae and coauthors measured DMS and CCN over the South Atlantic and concluded: "Our results provide strong support for several aspects of the CLAW Hypothesis, which proposes the existence of a feedback loop linking DMS emission from marine plankton to sulfate aerosol and global climate."[6] Stanford University oceanographers Mathieu Ardyna and Professor Kevin Robert Arrigo discuss how the reduction in sea-ice extent in the Arctic Ocean over the last few decades has resulted in a longer phytoplankton growing season and increased open-water habitat for phytoplankton growth resulting in an annual net primary production increase in phytoplankton of 30% between 1998 and 2012 over the entire Arctic Ocean.[7] Studies have shown that there is a cooling effect from DMS aerosols which impact local cloud cover, especially in the Southern Hemisphere, where there is less land area and more ocean coverage. Approximately 61% of the Northern Hemisphere's surface is covered by oceans, while about 81% of the Southern Hemisphere is ocean-covered. A paper in *Geophysical*

Research found that a 1°C increase in the Indian Ocean sea-surface temperature was sufficient to increase sulfate aerosol concentrations in the atmosphere by as much as 50%.[8] This, however, is a local phenomenon and cannot be extrapolated on a global basis. Experiments run in the CERN CLOUD chamber seem to confirm there is a large effect on cloud condensation nuclei from variations in aerosol emissions.[9] This suggest the CLAW effect might be significant.

University of East Anglia (UK) atmospheric scientist M. A. Thomas and coauthors from the Max Planck Institute, Cornell University and other institutions studied DMS aerosol-cloud-climate interactions. They found that DMS produced CCN levels that reduced the radius size of cloud droplets by 15% to 18%.[10] These smaller droplet sizes are significant as smaller cloud particles have a higher albedo and reflect more solar shortwave radiation back out to space. Known as the "Twomey Effect," clouds formed around sulfate aerosols are about 10% to 20% more reflective than other clouds. The Twomey Effect is based on studies of aerosol-cloud interactions and their impact on albedo.[11] The Twomey Effect of an increase in reflectivity of clouds formed around sulfate CCN being 10% to 20% higher than other clouds is confirmed in other papers including a paper by University of Miami atmospheric scientist Bruce Albrecht.[12] A paper in *Science Advances* by McCoy et al. including renowned University of Washington scientist Dennis Hartmann, found that the smaller droplet size of clouds formed from DMS aerosols over parts of the Southern Ocean increases summertime mean reflected solar radiation of more than 10 W/m^2.[13]

Thomas et al. found the global mean of DMS aerosol cooling is -2.03 W/m^2 and DMS increases by 2.5% in summer over the southern oceans provide a DMS aerosol radiative forcing cooling that changes from -2.03 W/m^2 to -9.32 W/m^2, or a net cooling of -7.29 W/m^2 (-9.32 W/m^2 minus -2.03 W/m^2).[14] Since 81% of the Southern Hemisphere is covered by oceans, the impact on the Southern Hemisphere is -5.9

W/m^2 (-7.29 W/m^2 x 81%). In the Northern Hemisphere, 61% of the surface is covered by oceans, so the extrapolated impact is -4.4 W/m^2 (-7.29 W/m^2 x 61%).

To convert these numbers into the impact of DMS aerosols in lowering temperature for each 1°C increase in temperature requires us to understand the mean difference in temperature between summer and other seasons. This temperature difference varies greatly by latitude. In the Northern Hemisphere the average temperature difference between summer and other seasons is 2°C to 5°C in the Tropics (0° to 23.5° north), 10°C to 20°C in the Mid-Latitudes (23.5° to 66.5° north) and 20°C to 30°C in the High-Latitudes (66.5° to 90° north). In the Southern Hemisphere, temperatures are more moderate due to significant ocean coverage. The average temperature difference between summer and other seasons in the Southern Hemisphere is 2°C to 5°C in the Tropics (0° to 23.5° south), 4°C to 12°C in the Mid-Latitudes (23.5° to 66.5° south), and 10°C to 20°C in the High-Latitudes (66.5° to 90° south). Because the Earth's circumference is largest at the equator, 50% of the surface of the Earth is in the Tropics (0° to 23.5° north and south). This has a larger impact on the mean temperature difference. Taking this fact into account, the mean temperature difference between summer and the other seasons is about 11°C in the Northern Hemisphere and 7°C in the Southern Hemisphere.

Extrapolating the Thomas et al., cooling data from DMS aerosols, for each 1°C increase in temperature, there is a result in DMS aerosol cooling of about -0.8 W/m^2 (-5.9 W/m^2 ÷ 7°C) in the Southern Hemisphere and about -0.4 W/m^2 (-4.4 W/m^2 ÷ 11°C) in the Northern Hemisphere. The weighted average is thus -0.6 W/m^2 (see Table 22). This is far less than the 3.0 W/m^2 of doubling CO_2 concentrations as claimed by Charlson et al. Yet it does have an impact. When the DMS aerosol cooling effect of -0.6 W/m^2-K is added to the net -1.7 W/m^2-K cooling effect of clouds from increased humidity, the total cooling impact is about -2.3 W/m^2-K, which more than offsets the 1.6 W/

m^2-K of warming amplification of water vapor and melting ice caps (see Chapter 10).

Table 22 CLAW Hypothesis DMS cooling from phytoplankton for each 1°C increase in temperature	
DMS overall annual cooling	-2.03 W/m^2
DMS cooling in summer over oceans	-9.32 W/m^2
Net summer DMS cooling over oceans	-9.32 W/m^2 – (-2.03 W/m^2) = -7.29 W/m^2
Southern Hemisphere DMS cooling (81% ocean)	-7.29 W/m^2 x 81% = is -5.9 W/m^2
Northern Hemisphere DMS cooling (61% ocean)	-7.29 W/m^2 x 61% = -4.4 W/m^2
Southern Hemisphere summer warming	7°C more than other seasons
Northern Hemisphere summer warming	11°C more than other seasons
Southern Hemisphere DMS cooling per 1°C	-5.9 W/m^2 ÷ 7°C = -0.8 W/m^2-C
Northern Hemisphere DMS cooling per 1°C	-4.4 W/m^2 ÷ 11°C = -0.4 W/m^2-C
Global average DMS cooling per 1°C	(-0.8 W/m^2-C + -0.4 W/m^2-C)/2 = -0.6 W/m^2-C

Source for Table 22: (DMS overall annual cooling and DMS cooling in summer over oceans). Thomas, M. A. et al., "Quantification of DMS aerosol-cloud-climate interactions using the ECHAM5-HAMMOZ model in a current climate scenario," *Atmospheric Chemistry and Physics,* Volume 10, pp. 7425-7438, 10 August 2010, p. 7425. https://doi.org/10.5194/acp-10-7425-2010.

Anthropogenic Aerosols

Human emissions of sulfur dioxide also produce sulphate aerosols. The cooling influence of increasing levels of anthropogenic sulfate aerosols was in fact the concern that drove "The Big Freeze" scare of the 1970s (see Figure 70 in Chapter 13). Like DMS aerosols, industrial aerosols scatter incoming solar radiation and provide sulfate aerosols which nucleate the formation of low clouds and reflect solar shortwave radiation out to space. Anthropogenic sulfur emissions have a cooling effect globally, but primarily in the Northern Hemisphere where most of the emissions originate. In recent years, measurements have shown that industrial aerosols in the atmosphere have been declining, which has created warming.

Sulfur dioxide emissions have declined globally by almost 50% since 1979 following the implementation of environmental regulations, including the Clean Air Act in the United States and similar regulations in other countries, especially China. Sulfur dioxide emissions were first restricted under the Clean Air Act in 1971 but further restricted in 1990 with the addition of the Acid Rain Program, regulating SO_2 and NO_x from power plants to gradually reduce emissions. China began to regulate SO_2 emissions in 1998, but made little progress before the 11th Five-Year Plan was introduced in 2006. The year 2006 was a pivotal date. Ocean University of China ocean & atmospheric scientist Hai Wang and coauthors document a significant decline in aerosol emissions from China after this date.[15]

According to *Our World in Data*, global sulfur dioxide emissions peaked in 1979 at about 141 million metric tons, declined to about 106 million tons by the year 2000, fell to 80 million tons by 2020, and then 73 million tons by 2022.[16] The International Maritime Organization (IMO) issued regulations that capped sulfur emissions beginning in 2020. These regulations lowered the sulfur in maritime fuels to 0.5%, down from the previous limit of 3.5%.[17] It is estimated that these new regulations have lowered sulfur dioxide emissions from shipping by 77%, equivalent to an annual reduction of 8.5 million metric tons of SO_2, which represents more than a 10% global decline in sulfur dioxide emissions.[18]

A 2024 study in *Nature Communications Earth & Environment* states that the *warming effect from the reduction in maritime sulfur emissions "is equivalent in magnitude to 80% of the measured increase in planetary heat uptake since 2020."*[19] In other words, 80% of the additional shortwave solar radiation received by the Earth and longwave radiation from greenhouse gases from 2020 to 2024 can be attributed to the lower albedo of the Earth (less clouds and light scattering aerosols) caused by the 10% reduction in global SO_2 emissions from the new IMO regulations. CERES satellite measurements of the Earth's albedo show a decline in albedo of about -0.1 between 2020 through mid-2024 and an accompanying increase of incoming solar radiation of about 0.3 W/m^2 (see Figure 58 in Chapter 9). University of Maryland atmospheric scientist Tianle Yuan and coauthors estimate the increase in solar radiation of 0.2 $W/m^2 \pm 0.11$ W/m^2 due to the IMO regulations between 2020 and early 2024, representing 80% of the Earth's heat uptake.[20] Actual measurements of 0.3 W/m^2 x 80% = 0.24 W/m^2 are within this range and suggest that the decline in human SO_2 emissions could explain most of the warming found in the CERES satellite record since 2020, as well as explain a large amount of warming since 1979 when human SO_2 emissions started to decline.

The previously cited 2024 paper by Hai Wang et al. concludes that the warm temperatures in the Northeastern Pacific Ocean between 2010 and 2020 were primarily driven by the rapid aerosol abatement in China during this period.[21] This paper concludes that aerosols had a greater impact on Arctic warming than greenhouse gases in the period of 1955 to 1984.[22] This is not surprising since ocean currents bring warm water to the Arctic, and the oceans are primarily warmed by the Sun (see Chapter 13). They report: "Our findings highlight the asymmetric Arctic climate response to Greenhouse Gases and Aerosol forcings and show that clean air policies which have reduced aerosol emissions may have exacerbated the Arctic warming over the past few decades."[23] Such warming of the Arctic from lower cloud cover has led to melting Arctic sea ice, which further reduces the Earth's albedo. A 2024 paper by NCAR climate scientists Ilaria Quaglia and Cornell University assistant professor and climate scientist Daniele Visioni demonstrates that the regulatory changes to sulfate emissions from international shipping routes resulting in a significant reduction in sulfate particulate released have been a major contributor to the anomalous increase in temperatures in the summer of 2023.[24] Another 2024 paper by Pacific Northwest National Labs cloud physics scientist Andrew Gettleman and coauthors analyzed Northern Hemisphere surface temperature anomalies in 2022-2023 and correlated temperatures with marine cloud radiative forcing from the emission changes of sulfur and aerosols from 2020 shipping regulations.[25]

The decline of industrial and marine shipping SO_2 emissions has certainly been a major contributor to lowering the Earth's albedo by providing less sulfate cloud condensation nuclei, creating less low cloud formation, and producing less highly reflective clouds. This conclusion is consistent with experiments run in the CERN CLOUD chamber, which confirm there is a large effect on cloud condensation nuclei from variations in aerosol emissions.[26] As a result of lower SO_2 emissions in recent years, fewer clouds are formed, less sunlight is reflected out to space, and the Earth has warmed. Measurements from CERES satellites have shown a decline of the Earth's albedo by 0.6 and an increase in shortwave solar radiation by about 2 W/m^2 between the years 2001 and 2023 (see Figure 58 in Chapter 9). This decline in albedo can be explained from the observed decline in cloud formation, less light scattering aerosols, and the reduction of Arctic sea ice in recent years. This ice melt has been driven by Arctic warming, which has been primarily due to warming oceans from the decline of sulfate aerosols.

Volcanoes

Dust particles and sulfate aerosols from volcanoes also have a major short-term impact on incoming shortwave radiation, low clouds, and climate. Aerosols, particularly sulphate aerosols, cool the temperature by nucleating clouds and scattering incoming solar radiation, thereby reflecting some of that shortwave radiation out to space. One of the significant gases from volcanoes is sulfur dioxide. These sulfur dioxide aerosols are quickly transformed into sulfate aerosols, which scatter sunlight and nucleate low cloud formation. If the volcanic eruption occurs during a solar minimum when there is higher cosmic ray flux, the increased atmospheric ionization of the atmosphere from cosmic rays expedites the formation of cloud condensation nuclei further impacting climate. "The most important mechanism by which volcanic eruptions perturb climate is the injection of large amounts of SO_2 gas into the stratosphere."[27] Therefore, large eruptions which reach the stratosphere have the largest impact on climate.

In the year 1815, the eruption of Mount Tambora, in what is today Indonesia, spewed SO_2 and volcanic ash high into the atmosphere, and this blocked incoming solar radiation and seeded the formation of increased low cloud cover. The sulfate aerosols from this volcano, coupled with the increased positive ions created by cosmic ray flux induced from the Dalton Solar Minimum, resulted in reduced incoming solar shortwave radiation, increased low cloud cover, cold temperatures, crop failures, and forced migrations. Crop failures due to cold weather were reported in Europe and North America. In July and August, lake and river ice was observed as far south as northwestern Pennsylvania and frost was reported in Virginia on August 20 and 21.[28] The year 1816 is often known as "The Year Without a Summer" in both Europe and North America. A similar freeze occurred in 536 CE with a volcanic eruption that initiated the volcanic winter of the Late Antique Little Ice Age (see Chapter 6). The eruption of Mount Tambora was the largest known since 536 CE.[29] Volcanic eruptions are known to be a major force in short-term global cooling.

Volcanoes can also contribute to warming. When underwater volcanoes explode, they can eject massive amounts of water vapor into the upper atmosphere. Since water vapor is a greenhouse gas, such an increase in water vapor can boost world temperatures. The eruption of the submerged Hunga Tonga-Hunga Ha'apai Volcano on January 15, 2022, injected vast amounts of water vapor into the stratosphere. A paper in *Nature Climate Change* estimates an increase in water vapor in the stratosphere by 10% to 15%.[30] A study by University of South Wales (Australia) atmospheric scientist Martin Jucker and coauthors of the impact of water vapor in the stratosphere from the Tonga-Hunga Ha'apai Volcano states that the amount of water vapor injected into the stratosphere from the volcano is unprecedented. The paper goes on to say that stratospheric water vapor anomalies similar to those caused by the Tonga-Hunga Ha'apai Volcano led to strong and persistent warming of the Northern Hemisphere landmass and winter cooling over Australia. The paper concludes that this stratospheric water vapor spike will increase temperatures over large regions by over 1.5°C for several years and cause some areas to experience cooling of 1°C.[31] After the 2022 eruption of the Tonga-Hunga Ha'apai Volcano, global temperatures spiked by 0.8°C.

Ash and sulfate aerosols from volcanoes block the Sun and nucleate cloud cover, which has a short-term cooling impact. A study by Rutgers University distinguished professor and climate scientist Alan Robock states, "Large volcanic eruptions inject sulfur gases into the stratosphere, which convert to sulfate aerosols with an e-folding residence time of about 1 year. Large ash particles fall out much quicker."[32] Evidence suggests that the cooling influence on temperature from volcanoes is short-term, typically lasting one to three years, with minor cooling for at most four to five years.[33] The large eruption of Pinatubo in 1991 injected an estimated 20 megatons of SO_2 into the atmosphere, but this only led to a two-year cooling of temperature.[34] Ice cores can measure not only temperature proxies, but volcanic activity as well, based upon the quantification of sulfate deposition in the ice core sample. Such proxies reveal that even the largest volcanic eruptions have an impact on temperature for no more than five years.[35] **Because volcanoes have only a short-term effect on temperature, they are not considered a cause of long-term climate change.**

Papers by University of Wisconsin climate scientist Yafang Zhong and coauthors[36] and University of Colorado distinguished professor and paleoclimateologist and coauthors[37] suggest that the Little Ice Age (1300 to 1850) was primarily caused by volcanism. Zhong and Miller are coauthors on both papers. Because it is known that volcanoes only have a short-lived effect on climate, Zhong et al. proposed sea ice/ocean feedbacks amplified the volcanic effect extending the impact over years. The evidence for this hypothesis is not compelling. Both the Zhong and Miller papers base the longer-term impact on cooling from volcanoes from a conclusion of a paper by National Center for Atmospheric Research atmospheric scientist David P. Schneider and coauthors. conclude: "The impacts of large high-latitude eruptions are largely restricted to the short-term, even in the Arctic, whereas large and closely spaced tropical eruptions could have long-term impacts."[38] This may be a true statement, but the observation of closely spaced tropical eruptions is not found in the volcanic record during the Little Ice Age. According to University of Edinburgh climate scientists Professor Thomas J. Crowley and Matthew B. Unterman, except for volcanic activity around 1450, the period between 1300 to 1575 had very low volcanic activity.[39] During the 1600s volcanic activity was up, but most of the volcanic activity was spaced decades apart (see Figure 65).[40]

Schneider et al. used a climate model to analyze volcanic activity between 1250 to 1300, the most volcanically perturbed half-century of the last 2000 years. His conclusion that "closely spaced tropical eruptions *could have* long-term effects" is not definitive, and it is counter to the conclusions of the Robuck (2000)[41] and Hegerl (2023)[42] studies. Furthermore, there is no reason to believe these findings could be extrapolated to times of less volcanic activity after 1300, which contrasts with the extreme conditions between 1250 to 1300 that were the subject of the Schneider et al. study.

Zhong et al. proposed ice/ocean feedbacks from volcanic eruptions could amplify the cooling impact. Miller et al. relies on the Zhong paper to justify this feedback mechanism. They propose cooling from volcanoes will create more ice coverage which will both reflect more heat out to space from the increased albedo of ice and lower the salt content of oceans, thereby reducing the thermohaline transport of heat in the Meridional Overturning Circulation. They propose that such impacts would extend the climate impact of volcanic eruptions for years. These impacts are real, but likely minor. The IPCC states that ice amplifies temperature changes by only 0.3 W/m^2 for each 3.4 W/m^2 (1°C) change.[43] This is a modest impact since 0.3 W/m^2 is less than 9% of 3.4 W/m^2 of heating or cooling caused by a 1°C change in temperature, suggesting ice melt feedback cannot account for the severe cold during the Maunder Solar Minimum (1645 to 1715). Furthermore, MIT oceanographer Carl Wunsch concluded that the thermohaline (salt) impact on the Meridional Overturning Circulation current is small.[44] It is telling that the Zhong et al. study included four simulations of this sea ice/ocean feedback and two of the four simulations failed to substantiate this feedback mechanism.[45]

Most significantly, attributing volcanic activity as the cause of the Little Ice Age is not supported by the paleorecord. The previously cited study by T. J. Crowley et al. shows volcanic activity over the past 1200 years.[46] This study is widely accepted and was featured in the IPCC *Fourth Assessment Report*.[47] There were several significant volcanic eruptions during the Little Ice Age including around 1450, 1600, 1640, 1670, 1680, 1810 and 1815.[48] The Aztec civilization suffered severe famine from 1450 to 1454,[49] Chesapeake Bay and Boston Harbor froze over in 1642,[50] and 'Frost Fairs' were held on the Thames River in London in the 1680s.[51] The combined effect of volcanic activity and the Spörer Solar Minium in the 1450s likely caused a drop in temperature. The Maunder Solar Minimum and volcanic activity would have made the 1640s and 1680s colder just as the Dalton Minimum and volcanic activity in 1815 led to the "Year Without a Summer."

However, volcanic activity alone cannot explain the cold of the entire Little Ice Age. The largest famine recorded in Europe occurred during a cold period during the Wolf Solar Minium in 1315 to 1317,[52] yet volcanic activity remained low. It was exceptionally cold in the 1430s during the Spörer Minimum in Europe[53] yet volcanic activity was low. In the Americas, the Rio Grande River froze over in 1540 for over one month and afforded horse crossings on the ice[54] during the Spörer Solar Minimum and a period of exceptionally low volcanic activity. The early part of the Little Ice Age from 1350 to 1600 saw lower volcanic activity than the warm period of the Medieval Warm Period between 950 to 1050 (see Figure 65). Although volcanic activity increased from 1600 to 1700, the period from 1800 to 1900 saw

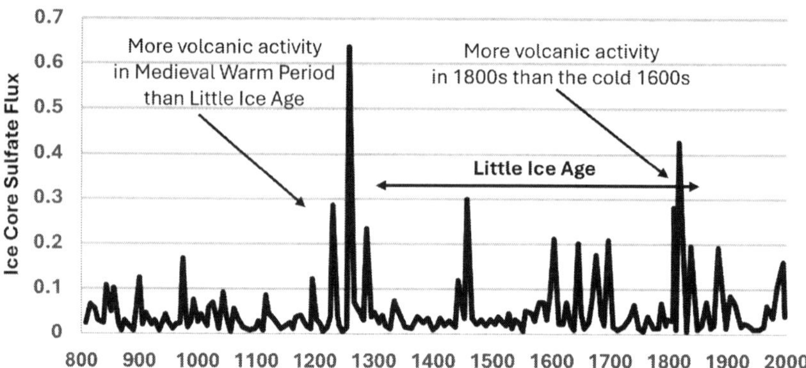

Figure 65 – Global Volcanic Activity, 800 CE to 2000. *Sulfate layers in ice cores from the Arctic and Antarctica have allowed a reconstruction of volcanic activity of the past. Volcanoes induce short-term temperature spikes lasting one to three years in duration. Source:* Crowley T. J. et al., "Technical details concerning development of a 1200 yr proxy index for global volcanism," *Earth System Science Data,* Volume 5, pp. 187-197, 23 May 2013. Figure 8., pg. 194. https://doi.org/10.5194/essd-5-187-2013.

higher volcanic activity than during the very cold Maunder Solar Minimum (1645 to 1715), yet temperatures were warmer between 1800 to 1900 than they were from 1645 to 1715 (see Figure 65).[55] There is excellent correlation of temperatures of the Little Ice Age and solar cycles (see Figures 27, 28, 29, and 31 in Chapter 5, Figure 58 in Chapter 10, and Japanese cherry blossom dates discussed in Chapter 5). However, the correlation between volcanic activity and temperature during the Little Ice Age between 1300 to 1850 does not always align (see Figure 65). This suggests that solar cycles played the more dominant role in the Little Ice Age with volcanic activity contributing shortterm cold spikes in temperature.

"Some oceanographers estimate that there may be as many as one million volcanoes on the Pacific Ocean floor alone – roughly 750 times the number on dry land."[56] Australian geologist Ian Plimer claims there are 3.4 million recognized submarine volcanoes globally.[57] Dr. Plimer believes these volcanoes heat the oceans. Geologist Arthur Viterito has demonstrated an excellent correlation of ocean temperatures with mid-ocean seismic activity from 1979 to 2022 after a two-year lag. Dr. Viterito claims mid-ocean seismic activity warms the oceans, and the thermohaline circulation (aka, the Meridional Overturning Circulation, or MOC) of the oceans carries this heat to northern latitudes. James Edward Kamis has suggested that ENSO ocean circulations may be influenced by seismic activity.[58] Submarine volcanoes may disrupt ocean circulation, however; more research in this area is needed to fully understand its impact on climate.

Ice melting may also be impacted by submarine volcanoes. The West Antarctic Ice Sheet sits atop the tectonically active West Antarctic Rift System, which hosts more than 100 known volcanoes beneath the ice sheet. These volcanoes contribute geothermal heat that melts ice from below. University of Rhode Island professor and oceanographer Brice Loose and coauthors observed that this volcanic source supplies ~2500 ± 1700 megawatts of heat beneath the ice sheet and may play a role in the melting and stability of the West Antarctic Ice Sheet.[59] It is therefore no surprise that the West Antarctic Ice Sheet is losing mass while the East Antarctic Ice Sheet has remained relatively stable and has even gained significant mass in recent years.[60]

Submarine volcanoes undoubtedly contribute to warming of the oceans, but the primary driver of ocean temperatures is the Sun, which is estimated to account for 99.9% of the heating of the oceans. Whereas the Sun is estimated to account for 173,000 TW of the Earth's heat, geothermal heat flow from radiogenic decay and primordial heat is estimated at 47 TW.[61] Therefore, submarine volcanoes likely play a minor role in climate change.

As opposed to the minor and short-term role of volcanoes in climate change, oceans are a major contributor. Physicist Steven Koonin said climate is driven by the oceans.[62] The final piece of the climate change puzzle is provided by the oceans.

In Chapter 13 we will dig into the significant impact on the climate from oceans and the impotence of greenhouse gases on ocean temperatures. We will also cover how the oceans explain climate cycles over the past 100 years, including the hot temperatures of the "Dustbowl" 1930s and 1940s, the cold of the "Big Freeze" during the 1970s, and the return to hot temperatures in recent years.

Chapter 13:
The Oceans' Impact on Climate

Oceans Drive Atmospheric Warming

Increasing levels of anthropogenic greenhouse gas emissions have contributed to global atmospheric warming. However, many credible scientists have attributed 40% to 87% of current global atmospheric warming to solar heat stored in the world's oceans. Astrophysicists Henrik Svensmark and Nir Shaviv attribute 50% to 66% of global atmospheric warming to solar heating of the oceans (see Figure 43).[1] Physicist Max Derakhashani says ENSO ocean temperatures account for 72% or more of atmospheric temperature variation since 1979.[2] Fritz Vahrenholt cites eight peer-reviewed scientific papers estimating 40% to 70% of global atmospheric warming in recent decades is caused by the Sun.[3] A recent paper in *Climate* by 37 international authors concludes that 70% to 87% of recent global atmospheric warming of the Northern Hemisphere can be explained by natural forcing from the Sun and volcanoes.[4]

The oceans are a massive collector of solar heat. Because the seas have a very low albedo (reflectivity) they readily absorb heat from the Sun. The reflective albedo of land masses is, on average, higher than the oceans, so, in general, the oceans absorb more heat from the

Sun than land. The oceans represent just over 70% of the area of the globe but absorb about 90% of the world's solar heat—again, due to their low albedo. For the Earth's Energy Budget to remain in balance (energy-in equals energy-out), solar heat in the oceans is transferred to the atmosphere in the process of radiating such heat out to space.

Incoming solar radiation is estimated at 173,000 TW, which constitutes over 99.9% of the energy input into the Earth's climate system.[5] The scientific reason solar radiation is the dominant source of ocean heat is because radiation from the Sun crosses a broad spectrum of shortwave ultraviolet light, visible light, and longwave infrared radiation. The shorter wavelength light spectrums carry more energy. Most important, however, is the fact that UV rays and sunlight can penetrate the water several meters, and some of these rays can do so up to a depth of about 1,000 meters. The deep penetration of sunlight and UV rays into the ocean is why this is so effective at heating the oceans. The vast thermal capacity and heat storage capability of the oceans play a crucial role in regulating Earth's climate. Oceans absorb a considerable amount of solar radiation, storing heat within their vast volumes of water. The process of transferring heat from the oceans to the atmosphere occurs primarily through evaporation and radiation. Evaporation from the ocean surface absorbs latent heat from the water's surface, transferring it into the atmosphere. As the oceans warm, they radiate infrared radiation out through the atmosphere and into space.

CO_2 and Ocean Temperature

Climate alarmists falsely claim recent ocean temperatures are rising at an alarming rate due primarily to increased CO_2 emissions in the atmosphere. Not only is this statement false, but these alarmists also have the relationship backwards. ***Oceans are a major driver of atmospheric warming, while the atmosphere cools the oceans.*** The Second Law of Thermodynamics is based on the observation that heat *always* moves from warmer objects to colder objects until they reach thermal equilibrium (the same temperature for both objects). *On average, the surface of the oceans are nearly* ***4°C*** *warmer than the atmosphere at sea level.* As of July 31, 2023, the daily global average Sea Surface Temperature (SST) was 20.96°C as measured in the Copernicus ERA5 dataset. Berkley Earth's average air temperature was 15.4°C, and the estimated air temperature at sea level in 2023 was 15.7°C to 15.9°C. A study in the *Journal of Marine Research* estimates the global mean ocean temperature is 3.7°C higher than atmospheric temperature above the ocean surface.[6]

Because the atmosphere is on average colder than the oceans, the atmosphere cannot directly warm the oceans overall on a global basis because the Second Law of Thermodynamics holds that heat always moves from the warmer object (oceans) to the colder object (atmosphere). Although there may be regional or local areas in which there is CO_2-induced heating of the oceans, there can be no global net CO_2-induced warming of the oceans as this would violate the thermodynamic second law. Respecting climate change, the global impact is what matters.

According to laws of thermodynamics, heat transfer can only be accomplished through radiation, conduction, and convection. CO_2 warming of the atmosphere is ineffective in transferring heat to the oceans because CO_2-induced longwave infrared radiation, conduction, and wind convection only warms the top surface skin of the oceans, and such surface skin warming is lost to evaporation and outgoing radiation. In contrast, solar energy penetrates deep into the oceans. Solar energy from shortwave radiation (ultraviolet and visible light) penetrates several meters into the water and is the primary source of heating the oceans.

CO_2 can only effectively absorb longwave radiation around the infrared 15-micron wavelength (13μ to 17μ spectrum) and emit such radiation back to the Earth at this same 13μ to 17μ absorption band (see Figure 54 in Chapter 9). However, longwave radiation around the 15-micron infrared radiation wavelength can only penetrate the very top surface of the water, so this heat is lost to evaporation and outgoing radiation. University of Miami scientists of marine and atmospheric sciences, Elizabeth Wong and Peter Minnett, stated: "This is because the penetration depth of infrared radiation in water is within submillimeter scales thereby implying that the incident longwave radiation does not directly heat the layers beyond the top submillimeter of the ocean surface."[7]

Infrared radiation around the 15-micron wavelength absorption band (13μ to 17μ spectrum) have relatively high absorption coefficients, which means such radiation is readily absorbed by water molecules. Remote sensing scientists, James H. Churnside of the of the National Oceanic and Atmospheric Administration's Environmental Technology Laboratory, and Montana State University Distinguished Professor Jospeh A. Shaw and other scientists, studied this 15-micron absorption band of CO_2 and its ability to penetrate ocean water. They concluded: "The skin depth of the ocean for this infrared band width is only 3 micrometers[,] or slightly more than one-fifth of the optical wavelength."[8] Therefore, the oceans absorb 100% of the radiative heat emitted from CO_2 in the first 3 microns of the water's surface. To put this into perspective, a human hair is 70 to 100 microns. **Because CO_2 radiation can only penetrate the very top surface of water, this heat is lost to radiation and evaporation**. Evaporation is a powerful means of transferring latent heat away from the surface skin of water. Radiation also removes heat from the surface skin because when only the skin surface layer is heated, that layer radiates an almost equal amount of energy back to the atmosphere immediately.[9] Therefore, CO_2 infrared radiation has no effect on warming the oceans below the ocean skin layer.

When CO_2 absorbs infrared radiation, the molecule begins to vibrate. As excited CO_2 molecules collide with nitrogen and oxygen, they increase kinetic energy in atmospheric O_2 and N_2 molecules. Since 99% of the atmosphere is composed of nitrogen (78%) and oxygen (21%), CO_2 warms the air. Because nitrogen and oxygen are not greenhouse gases, they are poor radiators in the spectral region relevant for atmospheric heat transfer. Heat in the air is thus primarily transferred to the oceans by conduction, as the sea surface touches the atmosphere, and from convection, as warm winds blow across the ocean surface. Both impact only on the ocean skin surface. Conduction is a poor means of warming the sea surface since air is such a poor conductor of heat. Furthermore, winds significantly increase evaporation and the loss of latent heat from the ocean surface. Because warmer air can only transfer heat to the ocean skin surface, such heat is lost to evaporation and radiation. According to scientist Javier Vinós: "The oceans transfer almost all of the energy [they] receive from the Sun to the atmosphere to maintain equilibrium. There is no net heat flux from the atmosphere to the ocean, and *the atmosphere has no net warming effect on the ocean.*"[10]

This loss of heat from the sea skin is well-known. According to a paper by globally recognized University of Exeter (UK) professor and ocean & atmospheric scientist Andrew J. Watson and coauthors in *Nature Communications*, the uppermost 1,000 microns of the ocean is cooler than the underlying water "because the *ocean surface is a net emitter of heat*, both via latent heat and longwave radiative fluxes, to the atmosphere."[11] Measurements of the surface temperature of the oceans reveal the fact that the surface of the ocean is generally cooler than just below the surface. Known as the "sea skin," this thin layer, typically only about a millimeter thick (1,000 microns) at the top of the oceans, is almost always cooler than just below the surface. This is true in cold regions of the Earth and most of the hotter regions of the Tropics. The temperature difference is generally 0.2°C cooler on the surface at night and 0.5°C cooler on the surface in the day (see Figure 66).[12] On a global average, the sea skin is 0.2°C to 0.3°C cooler than the ocean just below the sea skin. We know the statement by Watson et al. that the ocean surface is a "net emitter of heat" is true because if the net impact was to absorb heat from the atmosphere, the sea skin would be warmer than the ocean temperature just below the sea skin, yet it is colder.

Sea skin acts like a barrier to the absorption of heat into the ocean. In their paper "The Response of the Ocean Thermal Skin Layer to Variations in Incident Infrared Radiation," Elizabeth Wong and Peter Minnett write: "It is also not possible for the additional energy in the thermal skin layer (TSL) to be conducted into the bulk of the ocean (i.e., beneath the viscous skin layer) as that would require conduc-

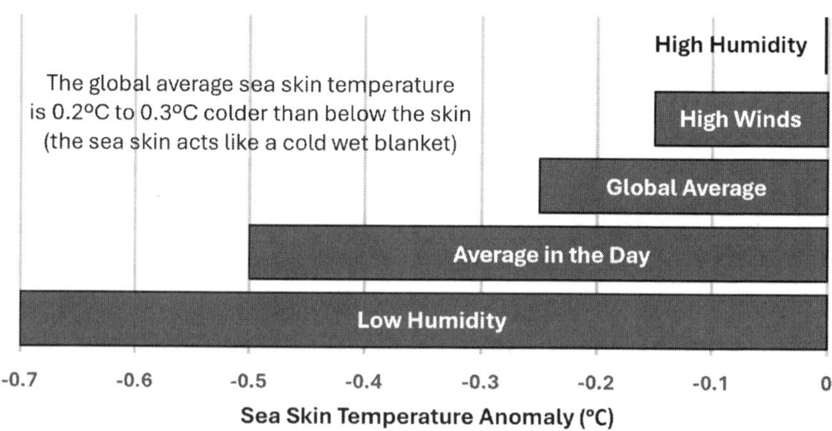

Figure 66 – Sea Skin Temperature (SST). *The top thin layer of the ocean is known as the sea skin. Measurements confirm the sea skin temperature is generally colder than water just below the surface. At night the sea skin surface is about 0.2°C colder than below the sea skin, and the sea skin temperature in the day is about 0.5°C colder in the first millimeter of surface waters.* Sources: Minnett P. J. et al., "Half a century of satellite remote sensing of sea-surface temperature, *Remote Sensing of Environment*, Volume 233, November 2019. https://doi.org/10.1016/j.rse.2019.111366 and Yan, Yunwei et al., "Tropical Cool-Skin and Warm-Layer effects and Their Impact on Surface Heat Fluxes," *Journal of Physical Oceanography*, pp. 45-62 January 2024, Figure 6. https://doi.org/10.1175/JPO-D-23-0103.1.

tion up a mean temperature gradient in the TSL."[13] In other words, for the ocean to warm, heat must penetrate below the sea skin layer. According to environmental scientist Jim Steele: "To date there has been no provable mechanism illustrating how heating from CO_2 can heat anything more than the ocean's skin surface."[14] Because atmospheric global warming from CO_2 does not penetrate the sea skin layer, it is not a cause of ocean warming. On the other hand, solar energy from shortwave radiation (ultraviolet and visible light) penetrates through the sea skin layer and down several meters into the water. Therefore, solar radiation is the dominant source of heating the oceans, not CO_2 radiative forcing or atmospheric temperature.

The narrative that CO_2 warming of the atmosphere warms the oceans is based on a hypothesis that warming the atmosphere will impede the cooling of the oceans by restricting the emission of heat out of the oceans. This hypothesis assumes that a warmer atmosphere will slow the transfer of ocean heat to the atmosphere. This would lead to a buildup of solar heat in the oceans which raises ocean temperatures. We all know that we remain warm under a warm blanket but lose heat under a cold, wet blanket. **You can think of the sea skin as a cold, wet blanket**. So long as you are covered by a cold, wet blanket, you will continue to lose heat. The theory of ocean warming from CO_2 emissions is based on an assumption that CO_2 infrared radiation and warmer air will lower the difference in temperature between the sea skin "wet blanket" and the ocean below the sea skin. If you heated your wet blanket so that it was not as cold, relative to your body temperature, you would slow down your body's loss of heat. However, observational measurements show the difference between the sea skin "wet blanket" temperature and the ocean temperature just below the sea skin does not change due to increased levels of CO_2 radiative forcing or atmospheric warming.

Since the sea skin is a net emitter of heat, all heat transferred from the atmosphere to the sea skin is lost to evaporation and outgoing infrared radiation. Although there are regional variations, on a global basis, the difference in temperature between the sea skin and the ocean below the sea skin layer is stable. *On a global scale, the cool-skin temperature difference—typically around 0.2°C to 0.3°C—tends to be relatively robust despite overall atmospheric warming.*

According to NOAA atmospheric physicist and known expert on air-sea interactions, Chris W. Fairall, and coauthors, the cool skin is caused by the combined cooling effects of the net longwave radiation, the sensible heat flux, and the latent heat flux and "is almost always present."[15] The stability of the sea skin temperature gradient arises because the cool-skin effect is primarily controlled by near-surface processes, such as radiative heat loss and evaporation, that tend to scale with the temperature difference itself. Moderate increases in air temperature, such as from CO_2 warming, do not significantly disrupt this balance because the heat fluxes adjust to maintain the gradient. A warmer ocean, from solar radiation, increases the temperature of the sea skin, but not the temperature gradient between the sea skin layer and just below the sea skin layer since evaporation and radiative cooling compensate to preserve the cool skin temperature gradient. The temperature of the atmosphere is evidently not a factor impacting the top to bottom sea skin temperature difference. Therefore, **the temperature gradient of the "wet blanket" of the ocean is not impacted by CO_2 and global warming of the atmosphere, so it does not slow the cooling of the oceans to any appreciable extent**.

Variables that impact sea skin temperature are wind speeds and humidity. Winds create turbulence on the ocean surface that brings warmer water to the surface, thereby heating the sea skin. In a 2024 paper by Hohai University (China) professor and oceanographer Yunwei Yan and coauthors in *Journal of Physical Oceanography*, the top to bottom *Sea Skin* temperature difference can change from 0.4°C to 0.7°C, without winds, to 0.15°C in high winds.[16] Because current global warming is moderating temperatures in the Arctic, there is a smaller temperature difference between the Arctic and Tropics, which has induced lower wind speeds. University of Albany distinguished professor and atmospheric scientist Aiguo Dai and Najing University of Information Science and Technology atmospheric scientist Jiechun Deng discuss this slowing of winds speeds due to Arctic Amplification.[17] Therefore, global warming of the atmosphere may have resulted in making the sea skin colder due to lower wind speeds.

The other major impact on sea skin temperature is humidity, since evaporation declines as relative humidity increases. The sea skin is not as cold in areas of high relative humidity.[18] Yan et al. show the top to bottom sea skin temperature difference can change from 0.7°C with low humidity to 0.0°C with high humidity.[19] Measurements between 1948 and 2023 reveal that global humidity at sea level has declined slightly from 79% to 78% relative humidity (see Figure 57). This slight decline would increase evaporation, which may have led to further cooling of the sea skin. Due to high local winds and/or local high humidity, CO_2-induced ocean warming does occur in some local regions. However, on a global scale, there is no CO_2-induced net ocean warming, and it is the global trend that matters for climate change.

In a paper by Wong and Minnett, the authors hypothesized that longwave radiation from greenhouse gases could slow the transfer of heat out of the oceans by changing the vertical temperature gradient within the sea skin layer.[20] They state: "It is, however, not clear how the greenhouse effect directly affects the oceans['] heat uptake in the upper 70m of ocean. This is because the penetration depth of infrared radiation in water is within submillimeter scales, thereby implying the incident longwave radiation does not directly heat the layers beyond the top submillimeter of the ocean surface." This is not surprising since CO_2-infrared radiation can only penetrate 3 microns into the 1,000-micron sea skin layer, and heat rises. Therefore, it is not clear how longwave radiation from CO_2 could impact the sea skin vertical temperature gradient. Wong and Minnett hypothesize that "any changes in the vertical temperature gradient of the thermal skin layer due to variations in the absorbed longwave radiation could provide an explanation of the mechanism as to how the ocean heat content increases through the retention of heat from the absorption of solar radiation within the bulk of the ocean."

Several facts discussed in this paper cast doubt on this hypothesis. First, the authors admit that attempts to directly relate the vertical temperature gradient within the sea skin layer "to changes in the incident infrared radiation did not produce a convincing dependence, at least on the time scales of our measurements. Revealing such a relationship will require more sensitive instruments than currently available." Second, they do not provide data on how much ocean cooling should be slowed from this hypothesis, and since instruments are not sensitive enough to measure such changes, if such slowing exists it is likely to be extremely small. Third, they admit the fact that turbulence in the ocean surface caused by wind erodes the sea skin layer, causing it to be thinner and warmer. They state: "The temperature difference ranges from ~0.1 K during high wind speeds of > 7 m/s to ~0.6 K during low wind conditions of <2. K m/s." They acknowledge that "breaking waves would disrupt the thermal skin layer at wind speeds of >7 m/s." Yet they admit that 52.87% of global ocean wind speeds are over 7 m/s. Because of the thinning and warming of the cold wet blanket of the sea skin due to wind, they had to limit their analysis to wind speeds of 2 m/s or less.

Changing the temperature of the "cold wet blanket" of the sea skin from ~0.6 K to ~0.1 K cooler than below the sea skin by higher winds is significant. Known as global terrestrial stilling, Arctic Amplification from global warming has reduced global wind speeds by an estimated 0.7 m/s over the past 50 years.[21] Given that a 5 m/s decrease in wind speeds (7 m/s to 2 m/s) decreases the sea skin temperature by ~0.5 K (0.6 K from 0.1 K), global terrestrial stilling has likely cooled the "cold wet blanket" of the sea skin by ~0.1 K over the past 50 years. Since a "cooler wet blanket" transfers more heat out of your body—or the ocean—climate change may have *increased* rather than decreased the cooling of the ocean, and the impact of wind is certainly much more impactful than a hypothetical change in the vertical temperature gradient of the sea skin layer that cannot be measured.

Heat is removed from the oceans by latent heat evaporation, radiation, convection, and conduction. Evaporation and radiation represent about 89% of this heat removal. Yet both evaporation and radiation of heat from the oceans are enhanced by warmer atmospheric temperatures. It is estimated that the oceans radiate 36% of their released heat to space through the atmosphere.[22] Known as Absorbed Solar Radiation (or ASR), the average global solar radiation absorbed by the Earth's surface is 240 W/m^2.[23] Since 36% of this energy is radiated out to space from the oceans, this represents 240 x 36% or 86 W/m^2. Such heat is radiated out of the oceans in the form of a full spectrum of infrared longwave radiation from wavelengths of 3 to 100 micrometers, with the peak emission around 10 micrometers.

Any object heated by the Sun, including a body of water, will radiate longwave infrared radiation back out to space. Such heat radiates only from its surface. Thermal radiation originates from the outermost microns of an object's surface because infrared photons cannot easily escape from deeper layers. The infrared emission depth of liquid water is typically about 10-20 microns. Because infrared from CO_2 radiative forcing can only penetrate 3 microns into the surface of the water, much of it is radiated back to space and it does not heat the layers beneath the infrared emission depth.

Warmer air typically does not impede the radiation of heat from the surface of the Earth. In fact, warmer air can facilitate the radiation of heat from the Earth's surface up through the atmosphere and into space. Warmer air molecules allow for more energetic collisions, and this can enhance convective heat transfer, moving heat upward through the atmosphere. In essence, warmer air does not block or impede the radiation of heat from the surface of the Earth. Instead, it contributes to the overall heat exchange process, facilitating the movement of heat energy away from the Earth's surface and up in the atmosphere and out to space. However, warmer air results in higher specific humidity, and this impedes radiation of heat to space as water vapor absorbs and emits infrared radiation. Nevertheless, such infrared radiation returned to the ocean surface from water vapor does not penetrate the sea skin layer.[24] Therefore, such heat is lost again to evaporation and radiation.

Evaporation, or latent heat, is the largest pathway for heat exchange from the oceans to the atmosphere. Globally, evaporation representants an estimated 52% of heat transfer away from the Earth's surface.[25] This would be even greater over the oceans since more evaporation occurs over the oceans than on land. Heat captured in evaporation is carried high into the atmosphere in water vapor. The heat remains in the water vapor until the water vapor is condensed high in the troposphere, where the latent heat from the oceans is released into the atmosphere. Evaporation is driven by heating the ocean surface; therefore the warming of the atmosphere by CO_2 radiative forcing and increased air temperature contributes to higher evaporation rates. Evaporation is an endothermic process; it requires energy to convert liquid water into water vapor. Increased heat transferred to the ocean's sea skin from CO_2 radiation and warm air provides energy to increase evaporation. Increased temperatures in the atmosphere also allow the atmosphere to hold more moisture, which also aids evaporation of the ocean surface.

Evaporation is a powerful mechanism of removing heat from the ocean surface, and such heat is transferred to the atmosphere. Since the average Absorbed Solar Radiation (ASR) at the Earth's surface, including cloud cover, is 240 W/m^2, and 53% of this energy is transferred out of the oceans by evaporation, this represents 240 x 53% or 127 W/m^2. This can be verified by calculating evaporation from the oceans.

The latent heat of evaporation for water is 2,260 joules per gram, so 2,260 joules of heat are absorbed to convert one gram of liquid water into vapor from the ocean surface. As a rough estimate, the average global evaporation rate for the world's oceans is generally considered to be in the range of approximately 1 to 8 millimeters per day. Using 4.5 millimeters as an average within this range could serve as a rough approximation. In one square meter of water, the surface depth of 4.5 millimeters includes 4,500 cubic centimeters of water, or 4,500 grams. Therefore, as a general average, 118 watts of heat is absorbed through evaporation in each square meter of ocean surface every day (see Table 23). This is close to the 127 W/m^2 figure representing 53% of the ASR estimate from NASA.

To put this into perspective, it would require 3 W/m^2 of radiation forcing from CO_2 to increase atmospheric temperature by about 0.9C, and this would take nearly 150 years of emissions to achieve at our current level of fossil fuel use (see Chapter 9). By contrast, it would take approximately 36.6 minutes to remove the equivalent energy of 3 W/m^2 from the ocean's surface through evaporation (see Table 23).

Table 23 Power of the Latent Heat of Evaporation	
Latent Heat of Evaporation	2,260 joules per gram of water
Average Global Ocean Evaporation Rate	4.5 milliliters or 0.45 centimeters per day
Average Global Evaporation Per m^2	1 m^2 = 10,000 cm^2 x 0.45 cm^2 = 4,500 cm2 or 4,500 grams
Joules of Heat Absorbed Per Day Per m^2	4,500 g x 2,260 joules/g = 10,170,000 joules
Watts Absorbed Per Day Per m^2	10,170,000 joules/24 hrs/60 min/60 sec = 118 W/m^2
Joules in 3 W/m^2	3 W/m^2 x 24 hrs. x 60 min x 60 sec = 259,200 joules
Time for Evaporation to Absorb 3 W/m^2	259,200 joules ÷ 118 watts = 2,197 sec ÷ 60 = 36.6 minutes

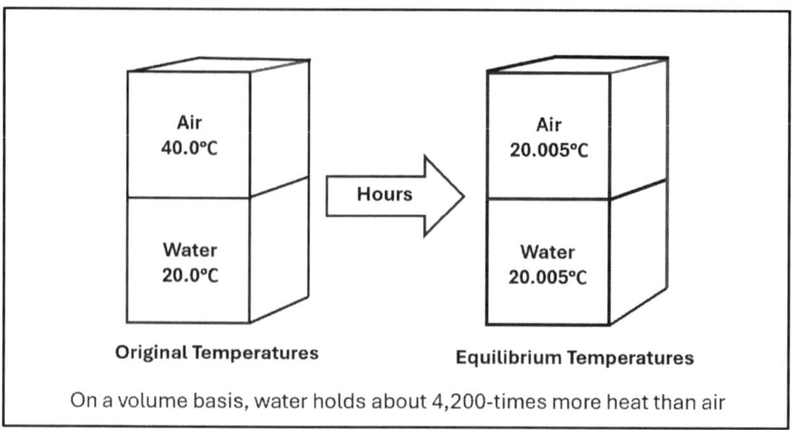

Figure 67 – Air Temperature Has Little Impact on Ocean Temperature. *Oceans have a notable impact on the climate, while the atmosphere has a minimal effect on ocean temperatures. If you place one cubic meter of air at 40°C next to one cubic meter of water at 20°C, after a few hours the air and water will come to equilibrium, both at a temperature of 20.005°C, or a net change in water temperature of only 0.005°C. By contrast, the Sun will heat one square meter of water by 0.005°C in only 20 seconds.* Source: Ott, Markus, "What Warms the Oceans?" *Tom Nelson Podcast*, 18 March 2023, https://www.youtube.com/watch?v=m9PCgCGo17w.

Heat Capacity of Water

As Professor Ian Plimer explained, if you fill a bathtub with warm water and close the bathroom door, the air temperature in the bathroom increases rapidly. However, if you fill a bathtub with cool water and heat air in the room, the temperature of the water in the bathtub hardly changes.[26] On a volume basis, the heat capacity of water is about 4,200 times more than the heat capacity of air. If one cubic meter of air at 40°C is next to one cubic meter of water at 20°C, the heat transfer to equilibrium results in the air cooling to 20.005°C and the water warming to a temperature of 20.005°C, which takes a few hours to complete (see Figure 67).[27] In a good conductor, such as silver, the process would be complete in minutes, but it takes hours to reach equilibrium due to the poor conductivity of air. In contrast to the hours needed to transfer heat from air to water, the Sun will heat one cubic meter of water by the same amount of 0.005°C in 20 seconds.[28]

Today, the average global ocean temperature is nearly 4°C warmer than air at sea level. The amount of heat you need to transfer 1°C from one cubic meter of water to one cubic meter of air would be 0.00024°C to warm the air by 1°C. If you were to heat a one square meter column of air one kilometer high by 1°C with the heat in one cubic meter of water, you would need to transfer only 0.24°C of heat from this one cubic meter of water. The oceans store a tremendous amount of heat, and, once again, they transfer heat to the atmosphere through evaporation, radiation, convection, and conduction. The weakness of CO_2 to heat the atmosphere pales in comparison to the power of the oceans to impact climate.

Direct heating of the oceans by the Sun's energy is the primary driver of ocean warming. Therefore, an increase or decrease in the solar heating of the oceans can have a major effect on the climate. *The oceans cover 71% of the Earth's surface and cloud cover, which plays a leading role in the amount of solar radiation that reaches and penetrates the oceans, and this makes it a major climate driver.*

Low cloud cover shades the Earth and reflects sunlight back out to space due to its high albedo of 0.70 to over 0.95 (1.0 is a perfect reflector, and 0.0 absorbs all light).[29] Therefore, on a cloudy day only 5% to 30% of total solar radiation reaches the oceans compared to the solar radiation reaching the ocean on a cloudless day. Known as the solar constant, on average the Sun delivers 1361 W/m^2 to the Earth. Because only one side of the Earth faces the Sun and the Earth is a sphere, on average, globally, only about 340 W/m^2 hits the Earth's surface. Of the ~340 W/m^2 of solar radiation received by the Earth, an average of ~48 W/m^2 is reflected back to space by clouds; ~29 W/m^2 is reflected by the atmosphere, water vapor, and aerosols; and ~23 W/

m^2 is reflected by the surface albedo, leaving ~240 W/m^2 of solar energy input to the Earth's Energy Budget.[30] But because only half of the Earth receives Sun at any given time, an average of about 680 W/m^2 is received by the Earth at the equator on the sunny side. Since clouds block 70% to 95% of this heat, that translates into at least 476 W/m^2 of heat being blocked under cloud cover at the equator. This energy is not available to heat the oceans. This is a massive amount compared to the 3 W/m^2 of radiative forcing required to increase atmospheric temperature by 0.9°C from doubling CO_2 concentrations (see Chapter 9). This is consistent with the statement from Nobel laureate John Clauser, who stated that the radiative forcing from CO_2 is nearly two orders of magnitude (1 x 10_2, or 100-fold) smaller than low cloud cover.[31]

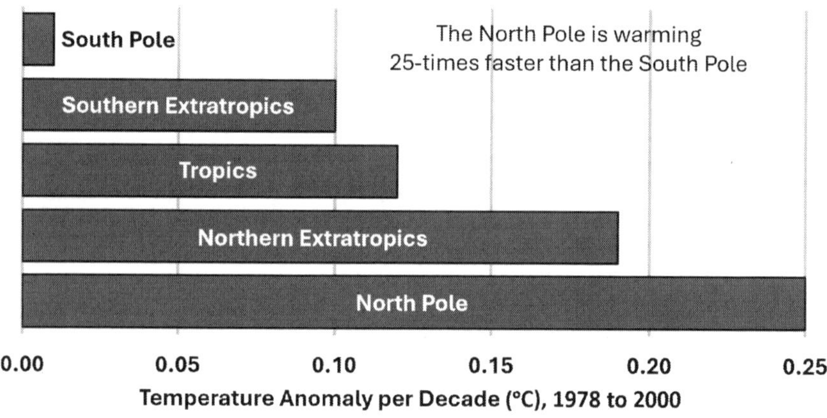

Figure 68 – Warming Is Mostly in the Arctic and Northern Latitudes, UAH Satellite Temperature Record, 1978 to 2022. *UAH Satellite temperature records from 1978 through 2022 reveal temperatures have increased much faster in northern latitudes, especially in the Arctic.* Data: Dr. Roy Spencer, University of Alabama-Huntsville. Source: Viterito, Arthur, "Viterito/Kamis/Yim/Catt: Impacts of Geothermal Energy on Climate," *Tom Nelson Podcast*, #181, 23 December 2023. https://www.youtube.com/watch?v=lbDbA32fNek.

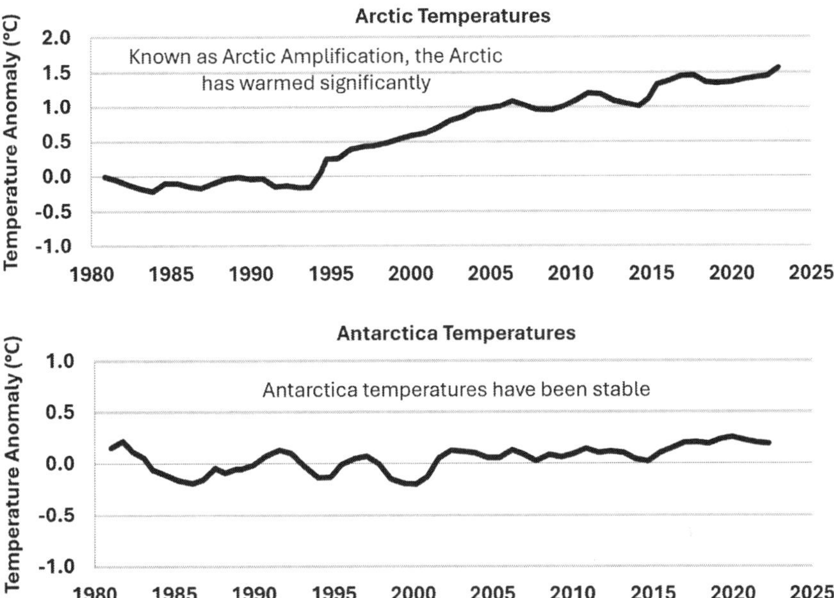

Figure 69 – TIROS-N Satellite Polar Temperature Measurements, 1978 to 2023. *Known as Arctic Amplification, the Arctic is warming faster than the rest of the world. NOAA TIROS-N Satellite temperature measurements mirror UAH Satellite temperature records from 1978 through 2023. During this period, NOAA TIROS-N Satellites show warming in the Arctic of about 1.5°C and virtually no warming in Antarctica.* Source: NOAA TIROS-N Satellite. Data interpreted by Dr. Carl Meise of the Remote Sensing System. *Climate4you*, "Temperature in Polar regions; Arctic and Antarctica; Polar regions as key regions for global climate change." https://climate4you.com.

Arctic Amplification

Modern global atmospheric warming is not global; it is primarily warming of the Northern Hemisphere and, most specifically, the Arctic. Between 1978 and 2022, University of Alabama in Huntsville (UAH) Satellites measured 1.1°C warming at the North Pole, 0.84°C warming in the Northern Extra Tropics, 0.53°C in the Tropics, 0.44°C in the Southern Hemisphere Extra Tropics, and 0.04°C at the South

Pole (see Figure 68).[32] Since 1978, the Television Infrared Observation Satellite-Next (TIROS-N) satellite measures an even greater disparity with the Arctic warming by about 1.5°C and virtually no warming measured in Antarctica (see Figure 69).[33] **Most atmospheric global warming has occurred in the Arctic. This fact is well known and is referred to as Arctic Amplification**. This has significant consequences. Climate alarmists continually point to scary scenarios of sea level rise due to the melting of Antarctica ice sheets. However, according to NASA climatologist Claire L. Parkinson an coauthors the actual data does not support such a claim. Despite periods of decline, "As a whole the Antarctica sea ice continues to have positive overall trends in yearly average ice extent."[34]

CO_2 radiative forcing cannot explain why the northern latitudes are warming faster than the southern latitudes since CO_2 readily disperses evenly throughout the atmosphere. **The concentration of CO_2 is on average similar between the North Pole and the South Pole, usually in the range of 400 to 420 ppm. Yet the North Pole is warming at a rate that is 25 times faster than the South Pole**. This would not be the case if CO_2 were the only driver of global warming. This temperature discrepancy is explained by the transport of heat to northern latitudes in ocean currents and the release of this heat into the atmosphere.

Observations confirm the role of ocean heat transfer in forcing Arctic Amplification. Figure 75 depicts Arctic temperatures since 1920, which cycle between warm and cool periods. These warm and cool cycles mirror the natural temperature changes of the Atlantic Ocean, known as the Atlantic Multidecadal Oscillation. If greenhouse gases were the sole factor of global warming, then the temperature would rise in step with CO_2 emissions, but observations reveal this is not the case.

Other hypotheses of Arctic warming include climate feedbacks, such as: 1) the loss of albedo as ice melts, which reflects less radiation back out to space; 2) weakening of the polar vortex, which transports warm air masses to the Arctic; and 3) the release of the greenhouse gas methane as permafrost melts. However, observational measurements of these feedbacks do not explain why the Arctic has warmed 25x (times) more than Antarctica. Furthermore, they cannot explain observed Arctic temperature oscillations.

Loss of Albedo as Ice Melts: A study published in the *Journal of Climate* by University of Miami professor and atmospheric scientist Brian J. Soden and NOAA meteorologist Isaac M. Held examines impact on temperature of the changes to surface albedo as ice caps melt.[35] They determined that the feedback was 0.26 W/m^2 for a 1°C increase in temperature (see Chapter 10). The IPCC places this number at 0.3 W/m^2 in their *AR6* report.[36] Since the global temperature has only increased by 0.75°C since 1978, the feedback from ice melt would be 0.75 x 0.3 or 0.23 W/m^2, which is a change of temperature of about 0.07°C, not even a noticeable fraction of the 1.1°C (UAH) to 1.5°C (TIROS-N) warming in the Arctic over Antarctica between 1978 to 2022. Melting ice can explain only a very small temperature change, and therefore melting ice caps are not a credible explanation of Arctic Amplification.

Weakening of the Polar Vortex: The polar vortex is a rotating mass of cold air consisting of strong westerly winds that circulates around the polar regions in the stratosphere. When the polar vortex is strong, it keeps cold polar air at the polar region. When it weakens, it becomes wavy, and it brings warm air into the polar regions and cold air into the mid-latitudes. These warm waves are known as Sudden Stratospheric Warming (SSW) events. The Antarctic polar vortex is more stable than the Arctic polar vortex. Some have postulated that Arctic Amplification is due to the warm air brought into the Arctic from a weak polar vortex.

The failing of this argument is that Arctic Amplification occurred during periods when the Arctic polar vortex was strong. During the 1990s the Arctic polar vortex was strong and had the longest absence of SSW events during the 150-year record.[37] Yet the Arctic warmed twice as fast as the global average during the 1990s. A weak Arctic polar vortex would certainly provide some warming of the Arctic, but a weak polar vortex does not explain Arctic Amplification since such warming occurred in the 1990s when the Arctic polar vortex was strong.

Release of Methane from Permafrost: In the Arctic, average methane concentrations have been observed to range around 1950 ppb.[38] At the South Pole, methane concentrations typically range about 1825 ppb.[39] The difference of methane concentrations between the poles of 125 parts per billion of methane is only 7% more in the Arctic than Antarctica. Using the accepted radiative forcing equations from Myhre et al., adjusted for ERF.[40] the increase in heating from 1825 to 1950 ppb of methane is 0.08 W/m^2, which equates to a temperature increase of 0.024°C. A temperature increase of 0.024°C does not explain virtually any temperature difference between the North Pole and the South Pole and certainly does not explain even a small fraction of the 1.1°C to 1.5°C greater warming that has occurred at the North Pole since 1978.

These three feedbacks cannot explain why the Arctic is warming 25 times faster than Antarctica. Nor can they explain why Arctic temperatures have oscillated between warm and cool periods in sync with the Atlantic Multidecadal Oscillation. As will be covered below, **an examination of the facts supports the theory that the primary driver of Arctic Amplification is from ocean heat transfer, not CO_2, loss of polar ice albedo, a weakened polar vortex, or the release of methane from melting permafrost**. Oceans are primarily heated by the Sun, so the rapid warming of the Arctic can be explained today, and in past climate cycles, from solar heating of the oceans, which is significantly impacted by cloud cover.

Some have speculated that the greater land area in the Northern Hemisphere is responsible for Arctic Amplification since land heats and cools faster than the oceans. This does not, however, seem like a plausible explanation. This may explain some of the warming of the Arctic, but such warming seems pale in comparison to the influence of the North Atlantic warm currents. London, Moscow, Krasnoyarsk, Siberia, and Winnipeg, Canada are on similar latitudes, yet the climate in London, in the North Atlantic, is considerably warmer than land-locked Moscow, Krasnoyarsk or Winnipeg. And although it is true that any location near the water will have more moderate temperatures, the contrast in average temperatures between London and these land-locked cities is extreme. The average annual temperature in London is around 50°F (10°C), while Moscow is 39°F (4°C), Winnipeg 28°F (-2°C), and Krasnoyarsk 21°F (-6°C). Only the warm North Atlantic current near London can explain these stark temperature contrasts.

Heat Transport to the Polar Regions

Since 1978, global warming, as measured by UAH Satellites, has been 0.25°C per decade in the Arctic, 0.12°C in the Tropics, and 0.1°C in Antarctica.[41] Atmospheric and oceanic heat transfer explains why the Arctic has warmed more than twice as fast as the Tropics and much faster than southern latitudes over the past 44 years.

The Sun shines nearly directly on the equator and the Tropics region from 23.5 degrees north to 23.5 degrees south latitude. Due to the curvature of the Earth, other regions of the world receive less direct sunlight. Therefore, the Tropics are the region of the most heating of the Earth (see Figure 70). Since nearly 77% of the Tropics is covered by the oceans, most of this solar energy is absorbed into the oceans. In contrast, the polar regions receive significantly less solar energy due to their higher latitudes, oblique angle of sunlight, and high reflective albedo of snow and ice cover. Because of the angle of the Sun in the polar regions, light is easily reflected out to space. **Yet the Arctic is warming faster than the Tropics. This can only be explained by the transport of heat from the Tropics to the Arctic.** Satellite observations confirm that the Tropics absorb more solar heat than they radiate to space at the top of the troposphere. Therefore, the Tropics would continue to heat if they did not move the heat out to other areas of the Earth. The Arctic has warmed twice as fast as the Tropics over the past 40 years. This is explained by the heat transferred from the Tropics to the polar regions, particularly the Arctic.

There are two primary modes of moving heat from the Tropics to the polar regions: 1) atmospheric circulations; and 2) ocean currents.

Atmospheric Circulation: The accepted model of meridional atmospheric heat transfer to the Polar regions from the Tropics is the *Hadley, Ferrel, and Polar Cells.* Both the Northern and Southern Hemispheres have these alternating (clockwise and counterclockwise) circulation loops (see Figure 70). Between the northern and southern Hadley cells is the *Intertropical Convergence Zone (ITCZ),* a location of intense rain. The movement of the ITCZ can impact monsoon rains.

As depicted in Figure 70, the Northern Hadley Cell is a counterclockwise circulation pattern which takes heat near the equator and at the ITCZ and lifts it high into the top of the troposphere (12 to 15 kilometers). This creates a strong wind that carries the heat northward high in the troposphere. As the wind cools, it drops back to Earth at about 30 degrees latitude north. The Southern Hadley Cell is a clockwise mirror of the Northern Hadley Cell extending from about the ITCZ to about 30 degrees south. The top of Hadley cells is cold as they are high in the troposphere. Consequently, water vapor condenses out of the air and the Hadley cells return dry air to the

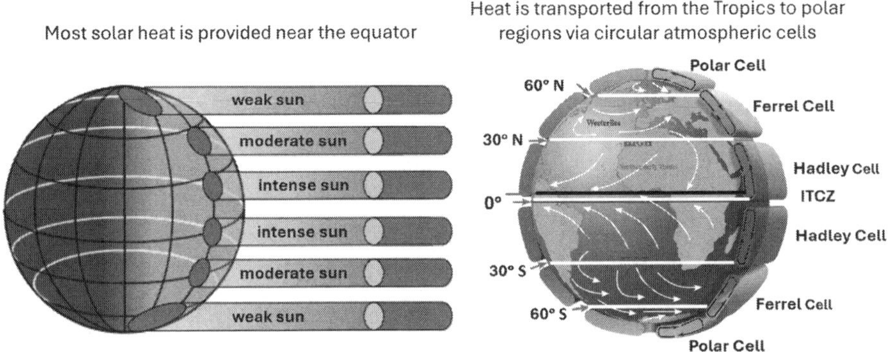

Figure 70 – Heat from the Sun and Movement of Heat in the Atmosphere. *Depicted on the left is the incoming sunlight to the Earth. Because of the curvature of the Earth, the Tropics from 23.5 degrees north to 23.5 degrees south receive the greatest amount of solar energy per area. The Tropics send heat to the poles through a series of alternating atmospheric circulation loops (clockwise and counterclockwise). On the right is depicted the Hadley Cells, Ferrel Cells, and Polar Cells which transport heat from the Tropics to the Poles.* Sources: (*Solar Angle of the Sun*) *Wikimedia Commons,* Solar angle of the Sun. https://upload.wikimedia.org/wikipedia/commons/a/aa/Solar_Angle_of_Incidence_on_Earth.png; (*Atmospheric Circulation Cells*) *Wikimedia Commons,* Atmospheric circulation cells: Modified by PW from globe image by Location_of_Cape_Verde_in_the_globe.svg: Eddo derivative work: Luan fala! [CC BY-SA 3.0]. https://commons.wikimedia.org/wiki/File:Earth_Global_Circulation_-_en.svg#Licensing.

surface. Deserts are often found at the termination of the Hadley Cells, around 30 degrees latitude north and south of the equator. Such deserts include the Saraha Desert in Africa, the Thar Desert in Asia, the Sonoran Desert in North America, the Great Victoria Desert in Australia, the Kalahari Desert in southern Africa, and the Atacama Desert in South America.

At about 30 degrees latitude north, the Northern Ferrel Cell begins as a clockwise circulation pattern which blows the heat north along the surface until it is taken up into the troposphere at about 60 degrees latitude north. The edge of the Arctic Polar region is at 60 degrees latitude north. The Southern Ferrel Cell is a counterclockwise mirror of the Northern Ferrel Cell spanning from about 30 degrees to 60 degrees latitude south. At the Polar regions, the counterclockwise Northern Polar Cell and the clockwise Southern Polar Cell take heat back up into the troposphere, where much of the heat is radiated out to space. **Like the radiator of a car, the radiation of heat out to space from the Polar regions is how the Earth Energy Budget is maintained. Excess heat from the Tropics is transferred to Polar regions where it is radiated out to space.**

The Earth's circumference is largest at the equator; therefore, the Earth spins faster at the equator than in regions farther north or south of the equator. The Earth spins at 1674 kilometers per hour at the equator, 1451 kilometers per hour at 30 degrees latitude, and 839 kilometers per hour at 60 degrees latitude. Because of this differential in speed, winds of the Hadley, Ferrel, and Polar Cells are not true north and true south. This differential in speed creates the Coriolis Effect, which moves winds of the Northern Hadley Cell and Southern Hadley Cell from east to west (the easterlies or trade winds), winds of the North Ferrel Cell and South Ferrel Cell from west to east (the westerlies), and winds of the Northern and Southern Polar Cell from east to west (the polar easterlies). Because the Ferrel Cell westerlies blow from southwest to northeast in the Northern Hemisphere and from northwest to southeast in the Southern Hemisphere, they carry warm winds to Alaska, Europe, New Zealand, and the west coast of South America. Accordingly, the highest rate of glacier melt in recent years has been in Alaska, the Alps, the Southern Alps in New Zealand, and the Andes. A study of glaciers in the Southern Alps of New Zealand and the European Alps has confirmed that the westerlies are enhancing glacier melt in these regions.[42]

The existence of the Hadley Cells, Ferrel Cells, and Polar Cells has been confirmed in observations. However, the atmosphere is a dynamic system, and random eddies, seasonal influences, and other impacts constantly move the locations and intensity of these cells and the ITCZ. Paleoclimate reconstructions of prior eras have shown the movement of deserts and wetland areas, which can be explained by the movement of the ITCZ. Javier Vinós, in his book *Climate of the Past, Present, and Future*, illustrates the movement of the ITCZ, accounting for different climates in the Holocene Climatic Optimum and Neoglacial times.[43]

Cooling of air in the high troposphere and the movement of the Hadley Cell winds on the surface back to the equator limit the amount of heat transferred to the poles, but observations have confirmed about 10% of the heat in Hadley Cells is transported toward the poles.[44] Latent heat transferred to the troposphere through evaporation and returning the latent heat to the Earth through condensation and rain is a major component of the Hadley Cells. Therefore, storms become an important mechanism in the transfer of heat in the atmosphere. Since Ferrel Cells have surface winds that blow northeast in the Northern Hemisphere and southeast in the Southern Hemisphere, they move warmer air to the poles.

According to esteemed Professor Dennis Hartmann of the Department of Atmospheric Science at the University of Washington, the Southern Hadley Cell is stronger on average than its northern counterpart because it extends slightly beyond the equator into the Northern Hemisphere.[45] According to *Wikipedia*: "This results in a strong Southern Hemisphere Hadley Cell relative to its northern counterpart which leads to a small net energy transport from the northern to the southern hemisphere; as a result, the transport of energy at the equator is directed southward on average, with an annual net transport of around 0.1 PW."[46] The Hadley Cells move more atmospheric heat south than north. Not only is there more initial heat transported to the Southern Hemisphere, but each year the Southern Hemisphere receives more solar energy than the Northern Hemiphere.[47] This is because the orbit of the Earth around the Sun is not a perfect circle but is elliptical. The Earth approaches closest to the Sun during the Southern Hemisphere summer when the tilt axis of the Earth angles the Southern Hemisphere closest to the Sun.

Despite the fact that the Southern Hemisphere receives more energy from the Sun and more initial atmospheric heat transport than the Northern Hemisphere, warmer temperatures are observed in the Northern Hemisphere. This discrepancy can be explained by additional northward flow of heat in ocean currents.

Ocean Currents: The oceanic meridional heat transfer from the Tropics to the polar regions is from ocean currents that move heated ocean water. The primary heat-moving currents include the so-called

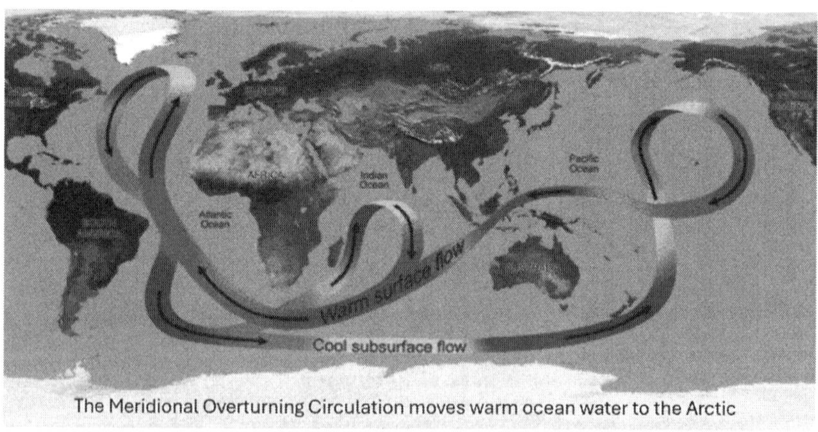

The Meridional Overturning Circulation moves warm ocean water to the Arctic

Figure 71 – The Meridional Overturning Circulation or MOC (aka, Thermohaline Ocean Circulation) Moves Heat to Northern Latitudes. *The currents of the Meridional Overturning Circulation transport sea water, heated by the Sun, to northern latitudes. The transport of heat to the oceans off Greenland and Europe is most pronounced, but the Northern Pacific Ocean past Japan, and along the west coast, is also heated by the MOC. The heating of the North Atlantic is enhanced by the Gulf Stream Circulation, and heating in the North Pacific is enhanced by the Kuroshio Circulation.* Source of the Meridional Overturning Source: Courtesy of NASA. https://www.jpl.nasa.gov/news/nasa-study-finds-atlantic-conveyor-belt-not-slowing/.

Thermohaline circulation, the *Gulf Stream* circulation, and the *Kuroshio* circulation. According to *Encyclopedia Britannica*'s description of the Thermohaline circulation: "A significant characteristic of the large-scale North Atlantic circulation is the poleward transport of heat. Heat is transferred in a northward direction through the North Atlantic. This heat is absorbed by the tropical waters of the Pacific and Indian oceans as well as of the Atlantic and is then transferred to the high latitudes, where it is finally given up to the atmosphere."[48]

The term "Thermohaline" is a misnomer. It was originally thought that this current was pushed by the sinking of heavier colder water (*thermo*) and heavier saltier water caused by evaporation (*haline*). Emeritus Professor of Physical Oceanography at MIT, Carl Wunsch, in 2002 published a paper which revealed the thermohaline mechanism would not work, and he then proposed wind and tidal forces move the currents.[49] Thermohaline Circulation is sometimes referred to as *the ocean conveyor belt* as it moves ocean water around the globe. A recent, more popular, and more appropriate name is the *Meridional Overturning Circulation (MOC)* since it does not limit the ocean current to temperature and salinity mechanisms that may not be a significant part of the current. Heat from the MOC moves northward to northern latitudes of the Atlantic and Pacific. The heat transfer is most pronounced in the oceans off Greenland and Europe, but the MOC current also warms the North Pacific Ocean past Japan and along the coast of North America. The warming in the North Atlantic is further enhanced by the Gulf Stream Ocean Circulation, which sends warm waters from the Caribbean northeasterly along the North American coast to the seas near Greenland and Europe. The North Pacific warming is enhanced by the Kuroshio Ocean Circulation, which sends warm waters from the Philippines northwest past the coast of Japan and over to Alaska.

The potential change in the Earth's Energy Balance from the meridional heat transport is worth noting. Because more heat is absorbed in the Tropics than is radiated out to space, the energy balance can only be maintained if the meridional heat transport moves enough heat to be radiated out to space in the polar regions. If this heat transport is disrupted, the Earth's Energy Balance could change, and this would result in a change in temperature. Andy May has pointed out that "since moving energy around does not alter the total energy within the system . . . this fact has caused many climate scientists to believe that changes in meridional transport cannot cause climate change, probably the most fundamental mistake of modern consensus climatology."[50]

Spanish scientist Javier Vinós has presented evidence that a major contributor to climate change is the movement of heat from the Tropics to the Arctic. Clouds and water vapor retains vastly more heat than any greenhouse gas. Because the Arctic is cold and has fewer clouds and limited water vapor, heat sent to the Arctic more easily radiates out to space. When more heat is transferred by the atmosphere and oceans to the Arctic, more heat will radiate out to space, which changes the Earth's Energy Balance. Vinós has shown that this loss of heat is especially strong during the winter when it is colder and there is lower water vapor and clouds in the Arctic skies. Known as the "Winter Gatekeeper Hypothesis," Vinós has set forth evidence for this phenomenon in his book, *Solving the Climate Puzzle*.[51] This is yet another natural cause of climate change (see Chapter 7).

An examination of ocean currents provides a clear explanation for Arctic warming with little or no warming in Antarctica (see Figure 72). The Gulf Stream Ocean Current carries warm water from the Caribbean to the North Atlantic. This is a powerful current and could explain why global warming in the Minoan Warm Period, Roman Warm Period, Medieval Warm Period, and Modern Warming was, and has been, most pervasive in Greenland and Europe. The Kuroshio Ocean Current carries warm water from the Philippines to the North Pacific. This could explain why Japan and China experienced warmer climates during the Minoan Warm Period, Roman Warm Period, Medieval Warm Period,

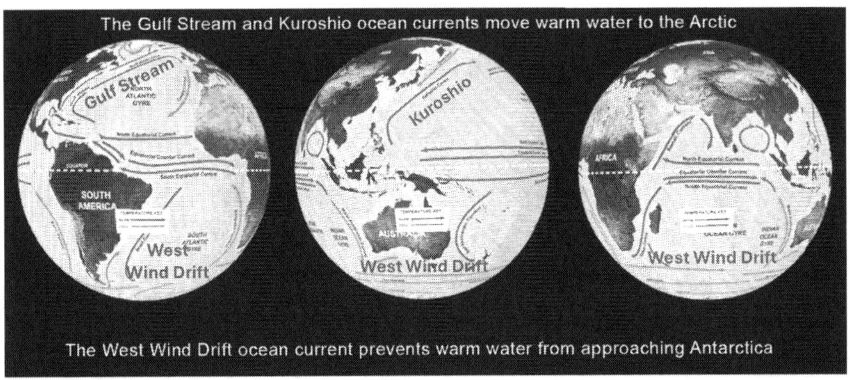

Figure 72 – Warm and Cold Ocean Currents. *Warming in the Arctic and not in Antarctica can be explained by ocean currents moving warm and cold water to different regions of the globe. The Gulf Stream Ocean Current moves warm water from the Caribbean into the North Atlantic past Greenland and Europe into the polar region. The Kuroshio Ocean Current moves warm water from the Philippines past Japan into the North Pacific. These are the regions experiencing the most warming today and in the Minoan, Roman, and Medieval Warm periods. The Antarctic Circumpolar Current (aka West Wind Drift) surrounds the continent of Antarctica and prevents warm waters from approaching the continent. Ice core records reveal that warming in Antarctica has been less pronounced than in Greenland in past warm climate cycles, just as we are experiencing today.* Source: Courtesy of NOAA. NOAA, "Science on a Sphere, Ocean Circulation (labeled currents). https://sos.noaa.gov/catalog/datasets/ocean-circulation-labeled-currents/.

and today in the Modern Warm Period. Because of the Gulf Stream Current, the Kuroshio Ocean Current, and the fact that winds in the Northern Hemisphere from the Ferrel Cell, known as the westerlies, blow in a northeast direction, it is not surprising that the greatest glacier melt in recent years has been in Greenland, Europe, and Alaska. These regions are north or northeast of the North Atlantic and North Pacific Ocean currents.

Generally, the records from Greenland ice cores and other Northern Hemisphere temperature proxies tend to show more pronounced and rapid temperature variations compared to Antarctica ice cores due to differences in regional climate dynamics (see Figures 68, 69, and 50 in Chapter 8). However, both Greenland and Antarctica ice cores indicate the presence of warmer periods corresponding to the Holocene Climatic Optimum, Minoan, Roman, and Medieval Warm Periods. University of Lisbon professor and influential atmospheric scientist José Peixoto and NOAA climatologist Abraham Oort, in their book *Physics of Climate*, propose that the Southern Ocean surrounds Antarctica which forms a substantial barrier to meridional transport of heat.[52] The Antarctic Circumpolar Current, also known as the West Wind Drift, is a cold ocean current that surrounds the continent of

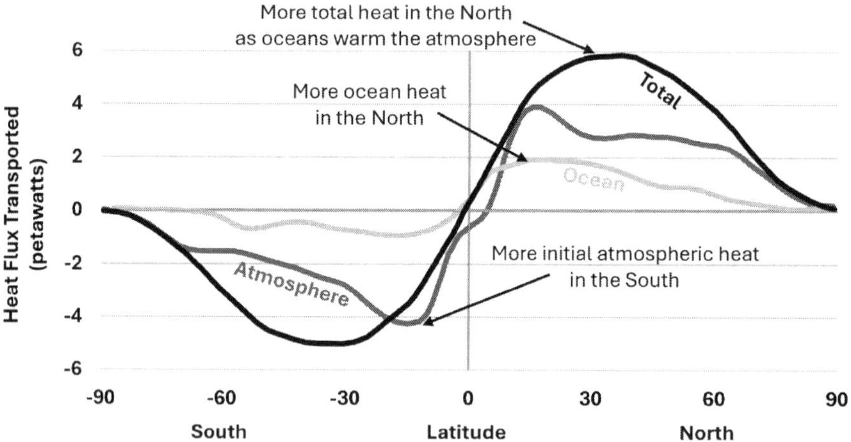

Figure 73 – Meridional Heat Transport from the Tropics to the Polar Regions. *A study of the meridional heat transport from the Tropics to the polar regions reveals 1) more initial atmospheric heat is in the Southern Hemisphere; 2) the oceanic heat flux is higher in the Northern Hemisphere; and 3) the Northern Hemisphere becomes warmer than the Southern Hemisphere as ocean heat flux is transferred to the atmosphere. As heat is released from the oceans, the Northern Hemisphere becomes warmer, which explains why the Arctic has warmed faster than the rest of the globe.* Source: Schmitt, Raymond W., "The Ocean's Role in Climate," *Oceanography,* 15 November 2018. https://doi.org/10.5670/oceanog.2018.225; and Yang, Haijun et al., "Decomposing the meridional heat transport in the climate system," *Climate Dynamics*, Volume 44, pp. 2751-2768, 21 October 2014. https://doi.org/10.1007/s00382-014-2380-5.

Antarctica and keeps warmer waters from approaching the continent (see Figure 72). This would explain why the warming has been less pronounced in Antarctica today and during past climate warm cycles.

Tectonic changes of the Earth altered ocean currents and initiated a poleward transport of heat to the Arctic.[53] According to Javier Vinós: 1) The Arctic Gateway opened about 55 million years ago initiating the transport of heat to the Arctic; 2) The Drake Passage opened between 30 and 20 million years ago, isolating Antarctica; 3) The Bering Strait opened about 5.3 million years ago, which provided another gateway to the Arctic; and 4) The Panama Gateway closed about 3 million years ago.[54] The closing of the Panama Gateway cut off the flow of water between the Pacific Ocean and Atlantic Ocean, forcing warm water of the tropical Atlantic to flow northward. This strengthened the Gulf Stream, sending more heat to the Arctic through the North Atlantic.

Fudan University (China) atmospheric and oceanic scientist Haijun Yang and coauthors have conducted a detailed study of the modern atmospheric and oceanic transfer of heat from the Tropics to the polar regions by latitude.[55] Figure 73 summarizes their findings which include 1) there is more initial atmospheric heat in the Southern Hemisphere, 2) there is more ocean heat in the Northern Hemisphere, and 3) the Northern Hemisphere ends up with more heat than the Southern Hemisphere as ocean heat is transferred from the oceans to the atmosphere. Due to the ITCZ being slightly in the Northern Hemisphere, the Hadley cells at the equator initially move more heat south than north, but as heat from the oceans is released into the atmosphere, the greater ocean heat in the Northern Hemisphere allows the atmosphere in the Northern Hemisphere to overtake the heat flux of the Southern Hemisphere. Although the heat flux from the oceans is less than from the atmosphere, ocean heat makes a real difference.

Theoretically, atmospheric heat flux, without oceanic flux, would see similar warming of the Northern and Southern Hemispheres because the Hadley and Ferrel Cells are similar in both regions. In fact, the Southern Hemisphere might be slightly warmer, since the Southern Hemisphere receives more solar energy, and the Southern Hadley Cell is stronger than its northern counterpart due to the ITCZ being slightly north of the equator. However, observational measurements show the Northern Hemisphere is warming much faster than the Southern Hemisphere, which is clear evidence of heat flux in ocean currents impacting the climate.

Extra heat flux in the oceans of the Northern Hemisphere explains why the Arctic has warmed faster than the rest of the world. Heat flux in the oceans also explains why the Northern Extra Tropics are warming faster than the Southern Extra Tropics (see Figure 68). The combination of the Meridional Overturning Circulation and other ocean currents transfer heat to the Northern Hemisphere, where it is released into the atmosphere (see Figure 73). The more pronounced ocean heat transfer to the north explains the discrepancy in warming between the North and South Poles as well as the Northern and Southern Extra Tropics.

Cyclical Ocean Oscillations

The climate is not linear but cyclical. There are several important ocean cycles that must be considered in assessing climate change. In the short term, there are changes in ocean temperatures which drive climate. El Niño and La Niña are the warm and cool phases of a recurring climate pattern across the tropical Pacific known as the El Niño-Southern Oscillation (ENSO). El Niño and La Niña ocean patterns occur in short bursts, usually lasting less than one year. Temperatures spike upward during an El Niño and down during a La Niña. It is interesting to note that warm El Niños form during periods of low winds and cool La Niñas form during periods of high winds. High

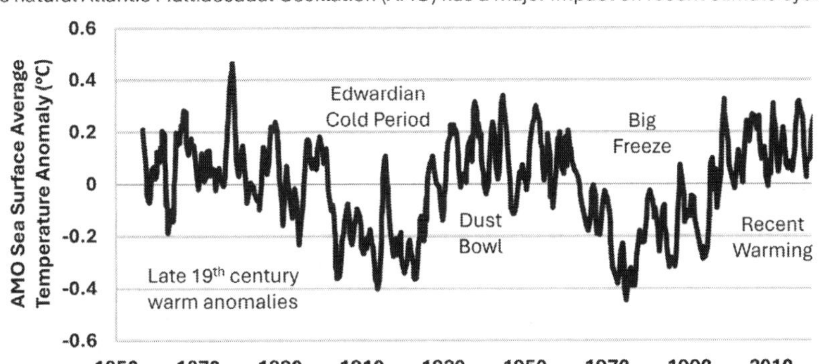

Figure 74 – Atlantic Multidecadal Oscillation (AMO) *– For hundreds of years there has been a natural ocean temperature cycle that has impacted climate in the North Atlantic known as the Atlantic Multidecadal Oscillation or AMO. The cycle is about 60 to 80 years gyrating between warm ocean temperatures to cold ocean temperatures about every 30 to 40 years and back again to warm temperatures about 30 to 40 years later.* Source: *Wikipedia,* "Atlantic Multidecadal Oscillation." https://en.wikipedia.org/wiki/Atlantic_multidecadal_oscillation.

winds increase evaporation which cools the ocean surface, while low winds result in less evaporative cooling. This may be an area warranting additional research.

Several papers have shown that most of the atmospheric warming in the last few decades was due to ENSO spikes in ocean temperature, which drove atmospheric temperatures. According to physicist Max Derakhshani, since 1979, warming from ocean "ENSO events" accounts for 72% or more of the global temperature variation.[56] According to Derakhshani, the combination of ENSO warming, along with modest warming from CO_2 radiative forcing, accounts for all the Earth's warming measured since 1979.

Longer and more significant cycles include the Atlantic Multidecadal Oscillation (AMO), which occurs every 60 to 80 years and the Pacific Multidecadal Oscillation (PDO), which occurs every 40 to 60 years. The AMO is a natural variation in Atlantic Ocean temperatures. The AMO is closely linked to the Meridional Overturning Circulation (MOC).[57] Although it is not certain why these cycles occur, some believe that when the MOC is strong, it brings warm water northward in the Atlantic contributing to a warm AMO phase. When the MOC is weak, less heat is transported and the AMO is in a cold phase. The AMO alternates between warm for 30 to 40 years, then cool for 30 to 40 years, then completing the cycle back to warm 60 to 80 years after the start of the previous warm period.

Atmospheric temperature records follow this pattern. **We had the heat waves of the end of the 19th century, then cooling up to 1910, then it was hot again during the 1930s and 1940s "Dust Bowl" period, then cooling during "The Big Freeze" scare of the 1970s, then shifting to warm again in recent years, which has become**

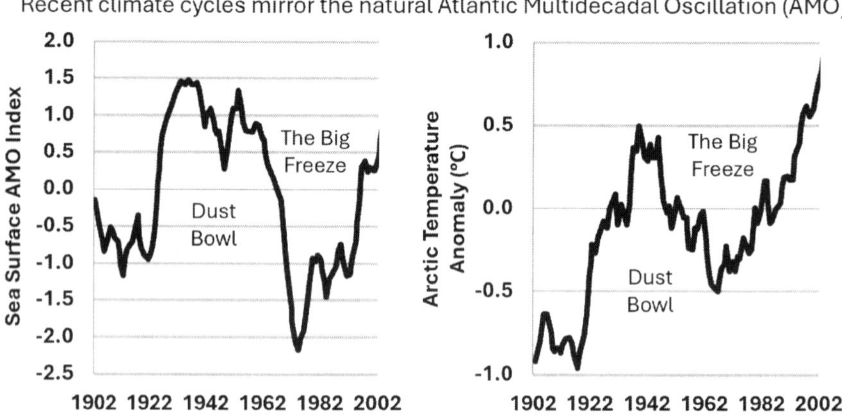

Figure 75 – Atlantic Multidecadal Oscillations (above left) and Arctic Temperatures (above right) Show Similar Temperature Cycles. *Global warming has been most extreme in the Arctic, and the temperature swings of the Arctic since 1902 mirror the natural temperature cycles of the AMO. The charts above display ocean temperature variations during phases of the AMO and atmospheric temperatures in the Arctic.* Source: *Chylék*, Petr et al., "Isolating the anthropogenic component of Arctic warming," *Geophysical Research Letters*, Volume 41, Issue 10, pp. 3569 to 3576, 28 May 2014. https://doi.org/10.1002/2014GL060184.

known as "Climate Change." The AMO warm cycle peaked in about 2010 and is expected to move into a cold phase in 2030.[58]

Because global warming is most extreme in the Arctic (see Figures 68 and 69), comparing temperature changes in the Arctic to the AMO is instructive. A study of Arctic warming by Los Alamos National Laboratory Earth physicist and Remote Sensing Team Leader Petr Chylék and coauthors reveals that more than half of Arctic warming is from natural variation, with less than half from anthropogenic causes.[59] The Chylék study also reports, "Arctic temperature variability cannot be reproduced without the AMO."[60] Observed temperatures in the Arctic since 1920 follow the same cyclical pattern as the AMO. Figure 75 displays temperature observations in the Arctic and ocean temperature variations in the AMO, which clearly shows the

Figure 76 – Climate Cycles – The Dust Bowl of the 1930s and 1940s and The Big Freeze Scare of the 1970s. *The heatwaves of the 1930 "Dust Bowl" and the cold "Big Freeze" of the 1970s are well documented in books, magazine articles, newspapers, and scientific publications of and from these times.*

influence of the AMO on Arctic warming. Chylék et al. also conclude that the AMO is a natural climate cycle. They conclude that studies "suggest the AMO has been present for many hundreds of years, indicating a high probability that it is a natural mode rather than a recent anthropogenic effect."[61] Confirming the cyclicality of these warm periods, University of Graz (Austria) polar climate researcher Florina R. Schalamon and coauthors two distinct warming periods in Greenland. "The first warming period (1922-1932) shows an average air temperature anomaly increase of 2.9°C across all stations considered in this study. The second period (1993-2007) exhibits a comparable warming of 3.1°C."[62]

During the warm AMO cycle of ocean temperatures in the 1930s and 1940s, atmospheric temperatures throughout North America and Europe were high. In a paper by Swedish Meteorological and Hydrological Institute atmospheric scientist and oceanographer Joakim Kjellsson and coauthors, investigated European heat waves in the 20th century. They found more heat waves in Europe between 1920 to 1950 and conclude this "may be related to the positive phase of the Atlantic Multidecadal Variation."[63] Many of the hottest temperatures on record in the United States, Canada, and Europe remain those experienced in the 1930s.

Known as the "Dust Bowl," the United States experienced dry and dusty conditions in the Great Plains States during the 1930s and early 1940s (see Figure 76). **Published in 1939, the book *The Grapes of Wrath* by John Steinbeck tells the story of countless numbers of farmers from the Midwest who saw their farms destroyed by heat, drought, and dust, moving to California in the 1930s and early 1940s to find a better life**. Many cite over-plowing, which led to soil erosion, as the primary cause of the massive dust storms. Dry conditions were also experienced, and this seems counter to the historical record where warmer climates are generally more humid and result in more precipitation. However, the Dust Bowl dry conditions were an extreme local event.

Warm conditions were seen throughout North America, but the dry conditions were only experienced in the Great Plains States. Conditions in the Northeast, Southeast, and West Coast were unusually hot, but not dryer than usual. For example, on July 19, 1934, Cincinnati, Ohio experienced a temperature of 99°F (37°C) combined with high humidity that made it feel like 110°F (43°C). On July 20, the temperature increased to 105°F (41°C). On July 21, the temperature reached 108°F (42°C), the highest temperature ever recorded in Cincinnati. Temperatures of 100°F (38°C) or higher were experienced for seven consecutive days. The high humidity made the heat more dangerous, and as many as 150 people died from heat-related causes in Cincinnati that week.

The year 1936 was another unusually hot summer in the United States and Canada. Many temperatures soared to all-time highs, with 12 states measuring 120°F (49°C) or higher. North Dakota measured 121°F (49°C) on July 6, 1936, and Wisconsin recorded 114°F (46°C) on July 6, 1936. In New York City, the temperature reached 106°F (41°C). By the end of the summer of 1936, 5,000 Americans and 1,100 Canadians died from heat-related causes. **A total of 23 of the 50 United States set high temperature records in the 1930s that remain records to this day. Many of the highest temperatures in Canada remain those recorded in the 1930s**. Such temperatures remain record highs for six of the thirteen providences and territories of Canada, including 113.0°F (45.0°C) in Saskatchewan on July 5, 1937; 111.9°F (44.4°C) in Manitoba on July 11 and 12, 1935; 108.0°F (42.2°C) in Ontario on July 11-13, 1936; 102.9°F (39.4°C) in New Brunswick on August 18, 1935; 100.9°F (38.3°C) in Nova Scotia on August 19, 1935; 98.1°F (36.7°C) in Prince Edward Island on August 19, 1935. High temperatures were also recorded in other areas of

Canada including 110°F (42°C) in Alberta on July 21, 1931 and 108°F (42°C) in Ontario on July 11, 12, and 13, 1936.

Europe was also extremely warm. A temperature of 119.3°F (48.5°C) was recorded in Catania, Italy on August 11, 1935. Seville, Spain recorded 116.6°F (47.0°C) on August 4, 1931; and 118.4°F (48.0°C) was recorded in Athens, Greece. Sweden recorded 101.8°F (38.8°C) in Ultuna on July 9, 1933 and Iceland reported a temperature of 86.9°C (30.5°C) in 1939, highs that have not been exceeded in these countries. Conqueyrac, France recorded 114.8°F (46.0°C) on August 12, 1930; Garmerdorf, Germany recorded 104.5°F (40.3°C) on July 27, 1935; and Uccles, Belgium recorded 101.8°F (38.8°C) on August 27, 1930.

There is a similar warming and cooling pattern in the Pacific Ocean known as the Pacific Decadal Oscillation, or PDO, which swings between cool and warm waters every 20 to 30 years. The cycles of the PDO are known to be natural as they are documented to go back about 1,000 years.[64] During the late 1930s and early 1940s, Australia experienced infamous heat waves. Two of the eight states and territories of Australia reported record temperatures in 1939 that have not been exceeded including 122.2°F (50.1°C) in New South Wales on January 11, 1939; and 109.0 (42.8°C) in the Australian Capital Territory on January 11, 1939. These heatwaves also impacted Victoria and South Australia.

There are no comprehensive temperature records from Asia in the 1930s. However, a study of available temperature records in China by Peking University professor and atmospheric scientist Weihong Qian and coauthors states, "In Northwest China, the highest temperature appeared over the period 1930s–1940s. Along the Yangtze River Valley in central eastern China and southwest China, interdecadal high temperature occurred from 1920s to 1940s and in [the] 1990s, but the drought climate mainly appeared from [the] 1920s to early 1940s. In South China, temperature remained at a high value over the period 1910s–1940s."[65] Warm phases of the PDO were from 1924 to 1946 and after 1977.[66] The PDO cycle peaked in a warm period in about 2005 and moved into the cold phase in 2024.[67] Cold phases of the PDO persisted from 1890 to 1924 and 1947 to 1976.[68] If past correlations of the AMO and PDO with atmospheric temperatures continue as expected, we should start to see the beginning of a cooling trend as the cold periods of the AMO and PDO converge in the 2030s.

In a study by climate scientist Markus G. Donat and coauthors titled, "Warm oceans caused hottest Dust Bowl years in 1934/1936," Dr. Donat and colleagues from the AERC Centre of Excellence for Climate System Science (Australia) explored the causes of the severe heat in North America in 1934 and 1936. They revealed that unusually warm sea surface temperatures occurring at exactly the same time in two very specific locations were likely responsible for creating the record-breaking heat in 1934 and 1936.[69] They site warm oceans in both the Atlantic and Pacific. It is worth noting these warm ocean events occurred during the warm phase of the Atlantic Multidecadal Oscillation and the warm phase of the Pacific Decadal Oscillation.

From the mid-1940s through the 1970s, the Earth experienced a cooling period in sync with the cool phases of the Atlantic Multidecadal Oscillation and Pacific Decadal Oscillation. More than 100 articles, publications, and a TV documentary highlighted the cooling period.[70] No less than three *Time* magazine issues (December 3, 1973; January 31, 1977; and June 24, 1979) and one in *Science News*, dated March 1, 1975, featured cover articles about the "Big Freeze" and a coming ice age (see Figure 76). *Newsweek* also published an article on the topic titled, "The Cooling World" on April 28, 1978. *US News and World Report* published an article titled, "CIA Report: Even U.S. Farms May Be Hit by Cooling Trend" (May 31, 1976). This article cites an August 1974 CIA report of cooling temperatures and the potential impact on agriculture, famine, and resulting political unrest.[71] Dozens of newspaper articles covered the cooling trend including reports in the *Washington Post*, *New York Times*, *L.A. Times*, *Boston Globe*, and *Chicago Tribune*. Leonard Nimoy, of *Star Trek* fame, hosted a TV documentary in May 1975 titled "The Coming Ice Age."

Because of the 35-year cooling period that existed from the mid-1940s into the 1970s, some scientists were predicting a drop into another ice age. In 1972, climatologist George Kukla from Columbia University and geologist Robley Matthews of Brown University organized a climate conference to discuss global cooling. They summarized the dangers of global cooling a paper in *Science* titled, "When Will the Present Ingerglacial End?"[72] They also wrote to President Richard Nixon advising him of the real threat and danger of global cooling, in a letter dated December 3, 1972.[73] In the letter, Kukla and Matthews wrote, "The present rate of cooling seems fast enough to bring glacial temperatures in about a century, if continuing at the present pace." This led Nixon to establish the Panel on the Present Interglacial in the State Department in February 1973 that was formed to deal with the cooling trend.

Just as climate alarmists of today have extrapolated off the top of the current temperature curve to forecast ever-continuing warming, climate alarmists during the cooling period extrapolated off the bottom of the cool trend. ***Any competent scientist knows you never extrapolate off the top or bottom of a curve in a cyclical system***. The temperature record, the Atlantic Multidecadal Oscillation, and the Pacific Decadal Oscillation are all cyclical.

The IPCC does not account for the impact of the AMO and PDO in their climate models, despite the overwhelming evidence of their impact on temperature cycles. According to Anthony Lupo, Professor of Atmospheric Science at the University of Missouri, "These cycles are missing from the IPCC climate models, which don't know how to represent them alone or in natural combinations, leading to inaccurate temperature attribution as well as poor estimation of climate sensitivity to greenhouse gases."[74] The next time both the AMO and PDO will converge in cold phases is in the 2030s. We should expect temperatures will decline as they did in the 1970s.

Oceans are not the source of the heat, but the carriers of heat. Ocean heating is primarily from the Sun, so the dominance of the oceans in warming the Earth points to the leading source of global warming as the Sun, not anthropogenic greenhouse gases. **Understanding the role the oceans and the Sun play in the climate provides clarity as to why we have had historical climate cycles in the past, including the Minoan Warm Period, cold Greek Dark Ages, Roman Warm Period, cold Dark Ages, Medieval Warm Period, Little Ice Age, and Modern Warming.** Each of these warm periods match the millennial Eddy Solar Cycles (see Chapters 5, 6, and 7). Ocean temperature cycles also explain current and recent temperature cycles including the hot 1930s, cold 1970s, and warmth of today.

Chapter 14:

Summary and Implications

There Is No Climate Crisis

John Adams said, "Facts are stubborn things; and whatever may be our wishes, our inclinations, or the dictates of our passions, they cannot alter the state of facts and evidence."[1] Many believe in the climate crisis narrative and will not listen to facts. They want to believe this because they wish to save the Earth, and the crisis narrative provides a roadmap, however misled, to accomplish their goal. They ignore facts and common sense about climate change because, as Mark Twain said, "It is easier to fool people than convince them that they have been fooled."[2] However, the scientific and historical facts are clear: there is no climate crisis, and this reality will, at some time, become clear. **Applying the definition of the scientific method provided by the great physicist Richard Feynman, it doesn't matter who tells you there is a climate crisis, it disagrees with experiment and observation and is therefore wrong.**[3] Any sincere and rational person who studies the evidence presented in this book must conclude there is global warming, but no climate crisis.

Perhaps the greatest strategic error made by climate alarmists is perpetuating the falsehood that warming is leading to more severe weather and increasing droughts. Meteorological theory suggests that warmer temperatures moderate the severity of storms and colder temperatures induce more severe weather (see Chapter 3). This is due to the faster warming of the Arctic, which moderates the contrast in temperature between the Arctic and the Tropics[4] and the moderation of the lapse rate as the top of the troposphere warms faster than the surface.[5] Eddy kinetic energy that drives storms is driven by the severity of the temperature contrast between the Arctic and the Tropics[6] (known as the Meridional Temperature Gradient) and convective energy that impacts storm severity is from the temperature contrast between the top and bottom of the troposphere[7] (known as the lapse rate). The decline in convective energy from this reduction in the lapse rate has led to less severe storms.[8] Furthermore, droughts are more frequent in colder weather because warm air holds more humidity and cold air is drier.

History has shown us convincingly that there were less severe storms in warm periods such as the Medieval Warm Period and Modern Warming and more severe storms during the Little Ice Age[9] (see Chapter 3). The historical record also confirms aridity and droughts are more common in colder climates.[10] The Medieval and Modern Warm Periods are times with more precipitation while the cold Dark Ages and the Little Ice Age both exhibited a decline in rain.[11] The modern temperature record shows storms and droughts are declining as temperatures warm (See Figures 4, 5, 6, and 8 in Chapter 3). Even the scientific reports from the IPCC state "evidence is lacking or the signal is not present, leading to low confidence of an emerging signal" with tropical cyclones, severe windstorms, river floods and coastal floods, aridity, hydrological drought, and agricultural and ecological drought.[12] **As the truth comes out that severe weather and droughts are not getting worse, the climate alarm narrative loses credibility, and the public will wonder what other lies have been told.**

The climate crisis narrative loses further trustworthiness as one predicted climate catastrophe after another fails to materialize (see

Table 5 in Chapter 3). Such failed apocalyptic climate prophecies include: islands of the Maldives will be underwater by 2028;[13] within a few years children in the UK "aren't going to know what snow is;"[14] end of New England maple syrup by 2008;[15] and multiple predictions of an ice-free Arctic by 2012,[16] 2014,[17] 2015,[18] 2016,[19] 2018,[20] and 2022.[21] Greta Thunberg even claimed, "Top scientist says humans will go extinct if we don't fix climate change by 2023."[22] None of these predictions have come true and the Maldives islands have gained rather than lost land since this prediction.[23] **It is only a matter of time before the climate cult loses followers as happened to millennial religious groups after proclaimed dates for the end of the world repeatedly passed.**

We can conclude from scientific evidence that anthropogenic increases in greenhouse gases produce modest warming that is not dangerous heating by any measure. Using the accepted Myhre radiative forcing equation,[24] adjusted for ERF, doubling CO_2 to 800 ppm from the 2013 level of about 400 ppm will result in an increase of 3 W/m^2 of incremental radiative forcing energy (see Table 11 in Chapter 9). Watts per square meter can be converted to temperature using the Stefan-Boltzmann equation.[25] In this case, 3 W/m^2 of radiation that is trapped by CO_2 from radiating out to space warms the Earth by 0.9°C (see Table 12 in Chapter 9). A separate determination can be performed with MODerate resolution atmospheric TRANsmission (MODTRAN) software. MODTRAN software is a program written to analyze electromagnetic radiation. Using the MODTRAN radiation transfer code yields a temperature increase of 0.9°C for an increase in CO_2 from 400 ppm to 800 ppm, without feedbacks.[26] This independent method validates the ERF adjusted Myhre radiative forcing equation and MODTRAN software, as both methods provide a temperature increase of 0.9° C.

The IPCC RPCP6.0 scenario (moderate-to-high emissions) forecasts a CO_2 level of 700 ppm by the end of this century.[27] Therefore, in this moderate-to-high scenario, anthropogenic greenhouse gas emissions from CO_2 will increase temperature by 0.6°C by the end of the 21st century—hardly a crisis. This calculation is made directly from the well-accepted Myhre radiative forcing equations, adjusted for ERF. Add in other anthropogenic greenhouse gases such as methane and nitrous oxide and the figure is 0.94°C. There is no climate catastrophe with a temperature increase of 0.94°C by the end of the century.

The entire climate crisis scare is based upon IPCC climate models that predict warming of 2.1°C between today and the year 2100 (3.2°C since pre-industrialization[28]). *This is 2.2-fold higher than 0.94°C.* They justify this 2.2× warming multiplier from the greenhouse effect of increases in water vapor and clouds as temperatures warm, and from lower albedo from melting ice caps as the Earth heats up. However, the IPCC, in its *AR6* report, states the effective water vapor feedback is 1.3 W/m^2 for each 1°C of warming,[29] and the estimated warming from melting ice caps is 0.3 W/m^2 for each 1°C in warming.[30] Since 1°C in warming is 3.4 W/m^2 and (1.3 W/m2 + 0.3 W/m^2) ÷ 3.4 W/m^2 = <1/2×, *the warming multiplier effect is less than 1/2×, not 2.2×!* As W.C. Fields once said, "If you can't dazzle them with brilliance, baffle them with bullshit."[31] Given the IPCC estimate of the positive feedback from water vapor and melting ice caps is less than 1/2×, using an amplification of 2.2× in climate models is certainly not brilliance.

But this is not all. For each 1°C in warming, there is an increase of 7% in water vapor and clouds.[32] In its *AR5* Report, the *IPCC estimated the feedback of clouds to be one of net cooling of -20 W/m^2.*[33] This is due to the fact that clouds act like a parasol in shading the Earth and reflect incoming solar energy out to space off their white reflective tops. Even dark clouds as viewed from below are brilliantly white on top as they reflect more sunlight out to space and block sunlight from reaching the Earth. The more current paper by L'Ecuyer et al. estimate a net cooling from clouds of -24.8 W/m^2 and is likely the most accurate estimate since it uses the most modern data sets from CloudSat and CALIPSO satellites.[34] Cooling of -24.8 W/m^2 x 7% = -1.7 W/m^2. This number of cooling by -1.7 W/m^2 for each +1°C in warming is verified from ISCCP satellite measurements (see Chapter 11, Figure 63).[35] Cooling of -1.7 W/m^2 more than offsets the +1.6 W/m^2 of warming from increased water vapor and melting ice caps. Therefore, clouds more than offset the 1/2× warming multiplier effect of water vapor and melting ice. Despite these facts, according to the IPCC, climate models use an average number of +0.6 W/m^2 of net warming from clouds.[36] Once again, "If you can't dazzle them with brilliance, baffle them with bullshit."[37] Having net warming feedback of clouds in IPCC climate models is not brilliance. The 2022 Nobel laureate in Physics, **Dr. John Clauser, identified low clouds as the primary thermostatic control of the climate.[38] Because of the cooling impact of clouds, Dr. Clauser has said, "There is no climate crisis."[39]**

When compared with actual temperature measurements from highly calibrated weather balloons and satellites, IPCC climate models forecast too much warming. Professor of Atmospheric Sciences, Dr. John Christy, has published a paper with Ross McKitrick that shows *climate model forecasts run on average 2.4× warmer than reality*[40] (see Figure 60 in Chapter 10). This overestimation of warming is likely because climate models assume the disproven hypothesis that water vapor, melting ice caps, and clouds will amplify warming by 2.2× and they ignore the cooling feedback of clouds and sulfate aerosols..

Between 1850 and 2021, the Earth's average temperature increased by 1.1°C.[41] CO_2 has increased in the atmosphere from 285 ppm to 417 ppm during that time.[42] Using the ERF adjusted Myhre radiative forcing equation[43] yields a temperature increase of 0.5°C. This suggests greenhouse gas radiative forcing from CO_2 represents less than half of the warming since 1850. Natural variation is evidently a significant contributor to recent and historical warming.

Our review of paleoclimate records and climate history in Chapters 5 and 6 provides the unmistakable imprint of natural variation and the reality of climate cycles. These past climate cycles do not correlate with CO_2 levels in the atmosphere. Over the past 500 million years, 70% of the time there was *no* correlation or a *negative correlation* between CO_2 concentrations in the atmosphere, and global temperature and CO_2 concentrations moved in the opposite direction of temperature 42% of the time during the last 50 million years.[44] For the past million years, there was no correlation or a negative correlation 87% of the time.[45] Between 1944 and 1976 the temperature dropped despite a five-fold increase in CO_2 emissions[46] (see Figure 49 in Chapter 8). Furthermore, CERES satellite data shows us that warming between 2001 to 2023 is not primarily from anthropogenic greenhouse gases. Satellite observations over this period have shown the Earth has lost 0.6 W/m^2 of longwave radiation,[47] which goes counter to the greenhouse effect of warming. If CO_2 greenhouse gas warming were the major source of climate change, then the amount of outgoing longwave radiation at the top of the atmosphere would be decreasing as it is trapped by the greenhouse effect, yet outgoing longwave radiation has been *increasing*.[48] **CO_2 is clearly not the primary driver of climate.**

On the other hand, Milankovitch cycles (see Figure 38 in Chapter 10) and solar cycles (see Figures 27, 28, 29, and 31 in Chapter 5, Figure 40 in Chapter 7, and Figure 58 in Chapter 10) show a close correlation with historical climate cycles. Temperatures peaked during the Holocene Climatic Optimum, 5,000 to 9,000 years ago, when the Milankovitch cycle of obliquity peaked at 24.5 degrees. Temperatures have been declining since as the Milankovitch cycle of obliquity has declined to 23.5 degrees today, and it continues to go down.[49] The Holocene Climatic Optimum, the Minoan Warm Period, cold Greek Dark Ages, warm Roman Optimum, cold Dark Ages, warm Medieval Optimum, cold Little Ice Age, and Modern Warming align perfectly with grand solar maximums and minimums and the corresponding sunspots and cosmic ray flux, as well as temperature and cosmic ray paleoclimate proxies of $\delta^{18}O$, $\Delta^{14}C$, and ^{10}Be levels found in ice core, ocean sediments, stalactites, and tree ring samples over the ages.[50] Such data supports the thesis that these centennial climate cycles are primarily driven by solar cycles. We learned in Chapter 7 that solar cycles modulate solar irradiance;[51] heating of the stratosphere;[52] the strength of the polar vortex, which regulates the flow of heat to the poles and out to space;[53] and cosmic ray flux that impacts the formation of cloud cover.[54] All of these forces impact the temperature of the Earth.

As expected with the millennial period of Eddy Solar Cycles, it was warm 1,000 years ago, cold 500 years ago, and warm today.[55] As Yogi Berra said, "It's déjà vu all over again."[56] Modern Warming has occurred during the largest solar maximum in 10,000 years.[57] The current warming is compounded by the fact that we have recently been in warm periods of the Atlantic Multidecadal Oscillation (AMO) and Pacific Decadal Oscillation (PDO). On a decadal scale, you can see these 20- to 40-year temperature swings of the AMO and PDO in the cold climate of 1910, followed by the hot "Dust Bowl" period of the 1930s and early 1940s, then the cold temperatures of "The Big Freeze" scare of the mid-1940s to 1970s, and the warm period since the 1980s[58] (see Chapter 13). How quickly we forget. High temperature records were set in the 1930s that have not been exceeded in 23 of the continental United States, six of 13 Canadian providences and territories, two of eight states and territories of Australia, and in Sweden and Iceland. Northern China also saw its highest temperatures in the 1930s and 1940s.[59]

Climate has always been cyclical, is now cyclical, and will continue to be cyclical.

The study of history in Chapter 6 teaches us the tragic consequences brought on from climate change. The misery caused by crop failures, famines, starvation, malnutrition, disease, and rebellion caused by hunger are repeated themes throughout history. Those forced to resort to cannibalism in China[60] and Jamestown, Virginia[61] as well as those selling family members into servitude in exchange for food in Mexico[62] dramatically paint the horrors of the Little Ice Age. But wait: all climate crisis episodes in the past occurred during cold times. Warm has always resulted in prosperity and optimal times. As George Santayana said, "Those who cannot remember the past are condemned to repeat it." Lessons of history teach us that we should celebrate the warm temperatures of our times. Warming is not producing a climate crisis. It is quite the opposite: warming has contributed greatly to modern prosperity. Yet many climate alarmists use the end of the Little Ice Age in 1850 as a starting point to claim that climate change began to lead to "disastrous warming." The Paris Climate Agreement has a goal to limit warming to 1.5°C from the temperature of 1850. However, anyone who studies history would know this is a counterproductive goal since warmth has always been a blessing to mankind and civilizations.

The media is constantly warning of coming catastrophes due to climate change. To discern the truth, it is wise to follow the advice attributed to the famous statistician Edwards Deming: "In God we trust, all others bring data."[63] The data shows that not only does increased CO_2 and warmer temperatures lead to a greener world[64] and abundant harvests,[65] they also lead to less severe storms. Chapter 3 documents that despite recent warming, there is a slight declining trend in hurricanes,[66] severe tornados are down significantly,[67] floods have not gotten worse,[68] heat waves have declined to a fraction of what they were in the 1930s,[69] droughts are less severe than in the past,[70] and acres burned in wildfires today are five times less than the 1920s and 1930s.

The benefits of CO_2 and a warming climate are clearly seen in the data, yet you never hear the press reporting on this *good* news. A saying associated with the sensational and emotionally charged style of journalism pioneered by William Randolph Hearst and Joseph Pulitzer says, "Bad news is good news, and good news is no news at all." **Nine times more people die from cold weather than hot weather, so climate change has saved many lives.**[71] Weather-related deaths are dramatically down from prior years. **In the 1920s, almost 250 people per million died of climate-related causes. By 2020, that number was about 5 deaths per million.**[72] Known as CO_2 fertilization, increased CO_2 allows plants to grow better and faster. Satellite measurements of total leaf area of the Earth have shown the Earth has greened by more than 20% since 1982,[73] and greening accelerated between 2001 to 2020.[74] A global figure of 20% is more than twice the size of the United States. Scientific papers report that 70% of this greening is from CO_2, followed by 9% due to nitrogen deposition, and 8% to global warming.[75]

Global warming can sound frightening as climate alarmists cite rising average global temperatures to paint a picture of gloom. In reality, recent climate change has been one of moderating temperatures. Temperature records reveal that we have not generally had hotter summers or more heat waves in recent decades, but rather more moderate winters due to the recent warming.[76] Average temperatures have gone up, but in a positive way, mostly from a reduction in cold rather than an increase of heat. And although the average temperature has gone up, **a close examination of the data shows heat waves have declined significantly since the 1930s and winters have consistently been warming for over a century.**[77] **This means the climate is moderating, which is good for agriculture and humanity.**

MIT Professor of Atmospheric Science, Richard Lindzen, one of the world's greatest experts on climate, said, **"What historians will definitely wonder about in future centuries is how deeply flawed logic, obscured by shrewd and unrelenting propaganda, actually enabled a coalition of powerful special interests to convince nearly everyone in the world that CO_2 from human industry was a dan-**

gerous planet-destroying toxin. It will be remembered as the greatest mass delusion in the history of the world—that CO_2, the life of plants, was considered for a time to be deadly poison."[78]

After closely examining climate change evidence using the scientific method of verifying or disregarding hypotheses based upon observational data, and witnessing the abhorrent and unfair treatment of skeptical scientists, I must agree with Professor Lindzen.

Many respected scientists, including Nobel Laureates Ivar Giaever and John Clauser, have stated there is no climate crisis. Brilliant and renowned physicists such as Freeman Dyson, Giaever, Clauser, and William Happer have all denounced the climate crisis. Since greenhouse gas warming is from quantum physics, the judgment of these experts should be taken seriously. The often-cited paper by Cook that "97% of scientists agree" only tells us that most scientists accept that anthropogenic greenhouse gas emissions have contributed to warming in recent years. As we have seen, CO_2 has contributed to less than one half of the warming since 1850, and it is not a crisis. The net impact of CO_2 warming has been and will continue to be beneficial throughout the twenty-first century.[79]

Although there is no climate crisis, the cost of climate hysteria places an unnecessary toll on humanity. A friend of mine advised his children not to have children of their own due to climate change. This is tragic as it deprives his children of life's greatest fulfillment and denies him and his wife the joy of grandchildren. It also has future consequences as depopulation leads to economic challenges, with an increasing number of retirees compared to fewer workers funding pensions. We have already seen the economic disaster of low birth rates unfold in Japan with nearly 30% of Japan's population age 65 or older in 2024.[80]

Air and water pollution, as well as poverty, are problems which need investment to find solutions, but many of the available resources to fight these challenges are being diverted to mitigating climate change. The U.S. Inflation Reduction Act dedicated $391 million to climate and related energy solutions, and Goldman Sachs estimates the total spending for green initiatives in this one U.S. Congressional bill is $1.2 trillion.[81] Janice Yellen, Secretary of the Treasury under President Biden, said that the global transition to a low-carbon economy requires $3 trillion in new capital each year through 2030.[82]

This massive spending has spurred a climate crisis industry. The EPA announced on February 21, 2025, that eight Non-Governmental Organizations (NGOs) had received $20 billion from the EPA's Greenhouse Gas Reduction Fund.[83] **As W. C. Fields once said, "You can fool some of the people some of the time—and that's enough to make a decent living."**[84] The Power Forward Communities was a recipient of $2 billion of the $20 billion from the EPA's Greenhouse Gas Reduction Fund. According to Lee Zeldon, Secretary of the EPA, the CEO of this NGO received $800,000 in compensation.[85] The Environmental Defense Fund NGO states on its website: "We are a global nonprofit organization tackling climate change—the greatest challenge of our time."[86] Fred Krupp, the president of the Environmental Defense Fund, is reported by *E&E News* to have received $922,022 in total compensation in 2021.[87] A decent living indeed!

Executive compensation is even greater in the renewable energy sector, which is sustained by taxpayer subsidies that are justified by the climate crisis narrative. Rebecca Kujawa, President and CEO of the wind energy company, NexEra Energy Resources (NEER), received $12,176,476 in compensation in fiscal year 2023.[88] Badri Kothandaraman, CEO of the solar energy company Enphase, received $19,527,534 in compensation in fiscal year 2024.[89]

Snapshot Summary of Climate Change

An overall, snapshot summary of the underlying primary causes of climate change are as follows:

1. The climate is complex and driven by many factors including Milankovitch cycles, cloud cover, solar cycles, stratospheric warming, polar vortex strength, greenhouse gases, aerosols, volcanoes, and Arctic ice. CO_2 is only one of these many drivers.
2. Anthropogenic (human caused) greenhouse gases, primarily CO_2, warm the Earth by restricting the amount of longwave infrared radiation from escaping out to space, which changes the Earth's Energy Budget. CO_2 radiative forcing can only account for less than 50% of total temperature increases since 1850. Natural causes are a major driver of the balance of warming.
3. The power of CO_2 to heat declines logarithmically with increases in concentration. Therefore, CO_2 becomes less of a driver of climate change in the future. Increases in anthropogenic greenhouse gases will continue to provide modest warming, but exhausting all fossil fuel reserves will only add 1.0°C in warming over the next 175 years. Such warming may potentially prevent another Little Ice Age by counteracting natural variation that is likely to drive colder temperatures in coming centuries.
4. 99.9% of the energy input to the Earth is from the Sun.
5. Ocean temperatures are a dominant driver of climate since oceans represent 70% of the Earth's surface area and absorb 90% of the Earth's solar energy due to the low albedo (reflectivity) of seawater. Oceans are a massive solar energy collector.
6. Oceans store and transport heat and transfer this heat to the atmosphere, primarily in the Arctic and Northern Hemisphere.
7. Over tens-of-thousands to one-hundred-thousand-year periods, the climate is driven by Milankovitch cycles that vary the amount of solar energy reaching the Earth due to the Earth's orbit, axial tilt, and axial rotational progression.

8. Other than long-term Milankovitch cycles, solar cycles and cloud cover are the largest modulators of solar heating of the oceans.
9. Low clouds are formed from water vapor, aerosols, and ions.
10. Cosmic rays strip electrons off molecules in the atmosphere producing charged ions which lead to a cascade of events that enhances the formation of sulfate cloud condensation nuclei. These ions help form cloud condensation nuclei by reducing the nucleation energy barrier to forming aerosol clusters of sufficient size to seed low clouds, primarily over oceans.
11. Solar cycles vary the magnetic field strength of the Sun, and this impacts cosmic ray flux by up to 20%. Variations in cosmic ray flux impact cloud cover over the oceans. Solar cycles also impact solar irradiance, the temperature of the stratosphere, and the strength of the polar vortex, which modulates the radiation of heat out to space from polar regions. All these mechanisms warm or cool the Earth.
12. The Eddy Solar Cycle is experienced on a millennial time scale. About every 1,000 years we have grand solar maximums followed by solar minimums 500 years later. We are currently in a solar maximum, as we were 1,000 years ago (Medieval Warm Period) and 2,000 years ago (Roman Warm Period). The Earth experienced solar minimums 500 years ago (Little Ice Age), 1,500 years ago (Dark Ages), and 2,500 years ago (Greek Dark Ages).
13. Low clouds are the most powerful thermostat of the Earth. As the Earth warms, more water vapor and sulfate aerosols from phytoplankton are produced, and these form more low clouds to cool the Earth. When the Earth is cold, there is less water vapor and less phytoplankton sulfate aerosols, producing fewer low clouds, which warms the Earth.

14. Natural periodic ocean temperature vacillations such as the Atlantic Multidecadal Oscillation (AMO) and Pacific Decadal Oscillation (PDO) produce climate cycles of between 20 to 40 years between hot and cold cycles. Recently, we have been in the convergence of the warm periods of the AMO and PDO and are experiencing a warm climate. The hot 1930s and 1940s "Dust Bowl" era was experienced the last time the AMO and PDO warm periods converged, and the 1970s "Big Freeze" occurred the last time the cold periods of the AMO and PDO converged.
15. Natural ocean temperature oscillation and solar cycles which drive cosmic ray flux, solar irradiance, stratospheric warming, and polar vortex strength are cyclical. We are just past the peak heating of the grand solar maximum and AMO ocean oscillation and the PDO ocean oscillation moved into its cold cycle in 2024. There should be cooling in the 2030s since all three of these cycles should be in cold or declining phases after 2031, and warming from increased levels of greenhouse gases will be too weak to offset the decline.
16. Other climate drivers such as volcanoes, ENSO ocean cycles, and melting ice caps have either small impacts on climate or produce only short-lived warming or cooling spikes in temperature of no more than a few years.

We Have Time to Build a Better World

CO_2 is not a pollutant; it is odorless, transparent, non-toxic, and the staff of life for plants. A crowded auditorium generally has about 2,000 ppm of CO_2. Astronauts breathe 5,000 ppm of CO_2 for months without side effects. CO_2 concentrations are as high as 8,000 ppm on submarines. An ideal level for plants, agriculture, and thus, humanity, is likely about 1,200 parts per million, which is about the level typically found in a moderately ventilated classroom and a CO_2-enhanced greenhouse. However, we could never get to 1200 ppm as burning all known fossil fuel reserves can only get us to 930ppm. The Effective Radiative Forcing (ERF) of CO_2 at 930 ppm is: (ERF = 4.33 x ln(930ppm/427 ppm) = 3.4 W/m^2 = 1.0°C in warming.

The warmth of 1.0°C is expected to have only a small negative impact on the world's economy[90] (see Figure 24 in Chapter 4). At the current rate of fossil fuel use, it would take about 175 years to burn all currently known fossil fuel reserves, thereby giving us ample time to transition away from these fuels. Much of the money spent on this imaginary climate crisis has been in so-called "green energy." It is absurd that these technologies are called "green energy" since they are being installed to reduce CO_2, the food source of plants. If they were effective at curtailing CO_2 emissions, they would lead to less greening of the Earth.[91] Solar and wind energy should be included in our energy power mix, but large-scale solar and wind energy deployments as urgent fixes to a non-existent climate crisis are not needed, damage the environment,[92] destabilize our electrical grid,[93] increase energy prices,[94] hurt the poor, [95] and destroy energy-intensive industries and their associated jobs.[96]

The beauty of nature is also impacted by the installation of green energy. There is spiritual refreshment to the soul in taking in the beauties of nature. Many beautiful landscapes are being destroyed by the blight of solar panel farms and windmills. Even seascapes are now being destroyed by offshore windmill farms. **The harm and blight on the landscape and seascape from wind and solar should be avoided, and the massive amounts of money spent on such technology should be diverted to natural gas, safe nuclear energy, and anti-pollution measures.** Yes, we should transition to EVs and alternative forms of energy, including hydrogen, synthetic fuels, and nuclear. But the reason is not to reduce CO_2—it is to reduce real pollutants and because we will one day run out of fossil fuels.

Natural gas is exceptionally clean, but the burning of oil and coal produces particulates which are pollutants. The role of wind and solar power as fossil fuel replacements are limited due to their poor energy densities and intermittent outages. Moreover, **the production of wind turbines and solar panels requires substantial quantities of mined materials, such as copper and rare Earth metals, which can have adverse environmental impacts due to the mining processes involved**. Copper and rare Earth metals are in limited supply and are also essential for electronics, EV deployment, and electrical grid expansion. As ore grades decline, more rock must be excavated, crushed, and processed to produce each ton of refined metal. We should focus on nuclear energy, natural gas, new energy discoveries, anti-pollution technologies, and EVs. Ultimately, small-scale, affordable nuclear fission reactors could meet our long-term energy needs.

The standard of living has been greatly increased due to the industrial revolution of the 19th century, electrification in the first half of the 20th century, and the computer revolution of the second half of the 20th century. Potentially, the greatest impact of all may well be the coming AI revolution of the 21st century. Artificial Intelligence (AI) deep learning and deep reasoning algorithms should provide significant advancements in economic productivity, accelerating scientific breakthroughs, and improved medical care. Investments in this important technology are paramount to a prosperous future. However, AI requires affordable uninterrupted electrical power, which is a complete mismatch with wind and solar energy. Servers in AI data centers use NVIDIA GPUs which draw 300-700 watts per unit, which is two to four times more than traditional computers. Cooling adds another 20% to 40% power requirement. For service providers such as OpenAI, Google, X, Meta, and Microsoft, electricity accounts for about 60% of their total operating budget.[97] Since power is the single largest operating cost of running an AI data center, affordable electricity prices are imperative.

AI also requires uninterrupted power 24 hours per day, 7 days per week. AI training clusters run continuously for weeks or months. An interruption of even minutes can crash expensive computations. Wind and solar are intermittent, they produce power only when the wind blows or the Sun shines. To rely on such renewable energy is only feasible with redundant backup or massive storage capabilities (e.g., batteries), which are costly solutions. Battery backup can be used today to prevent surges and dropouts of power for a few minutes to hours, but large-scale, overnight, multi-day, or seasonal battery backup remains limited and extremely expensive. Large data centers consume hundreds of megawatts of power, equivalent to a medium-sized city. Wind and solar output fluctuates sharply and cannot reliably sustain such massive continuous loads without backup power such as diesel or natural gas generators. Such redundancy is expensive. In addition, data centers are often located in urban centers with fiber optic backbones and a skilled high-tech workforce. Wind and solar farms require massive amounts of acreage and transmission lines to bring renewable power from remote areas is expensive, slow to permit, and often face community resistance. Despite the many virtue signaling statements of using wind and solar to power AI data centers, "renewables have seemingly dropped out of data center power conversations."[98]

Wind and solar power fluctuates without warning from random changes to wind and cloud cover. Electric supply must always match demand exactly. Sudden changes create imbalances that can cause frequency or voltage instability resulting in brownouts and rolling blackouts from voltage swings, flickering and equipment trips. Gas, coal, nuclear, and hydro powered spinning turbines turn with reliable rotational inertia that stabilized frequency (e.g. 60 Hz and 50 Hz) and can be ramped up and down quickly to allow supply to match demand. Solar and wind cannot easily be controlled in the same way. Adding wind and solar to an otherwise stable electric grid can lead to instability, a situation not conducive to the operation of AI data centers. Brownouts in California in August 2020 were caused in large part to the decline of solar power into the evening during a period of peak demand of air conditioning during a heat wave.[99] Silicon Valley in California, the center of the computer revolution, may well be forfeiting its opportunity to be a major center of AI due to the state's investment in wind and solar that has resulted in unstable and expensive electricity.

Not only is winning the AI race paramount to future prosperity, but its application in military applications leaves no doubt as to the criticality of winning the AI race. The military use of AI includes the guidance of missiles and munitions, the intelligence behind autonomous drones and battlefield robots, reconnaissance, cyber warfare, rapid scenario analysis, logistics, simulation and training, and rapid decision support. If the U.S. and its allies lose the AI race to adversaries, they could face higher casualties and weakened deterrence thus enhancing the chances of conflict or war. Winning the AI race is imperative and part of this will involve the deployment of reliable fossil-fuel electricity generation, especially clean burning gas-powered turbines. Hydro and nuclear power are good source of reliable power for AI data centers, but their permitting and implementation is too slow to be of any value in the near-term. Blocking reliable fossil fuel-powered electricity in favor of unreliable wind and solar installations to potentially solve an imaginary climate crisis would foolishly give an advantage to our adversaries in the AI race.

We should move forward with an energy transition, but we have time to do it properly and not rush into poor and unrealistic solutions which threaten our energy security, waste money, disproportionally hurt poor and third world countries, and cause additional threats to the environment, humanity, and the natural beauty of this Earth. We should focus our investments on new energy sources that are affordable, reliable, and reduce real pollution (not CO_2). Focusing spending on adapting to climate is far more prudent than wasting money in a futile attempt to control the climate. Money spent on reducing CO_2 would be better utilized on adaptation to climate, reducing pollution, saving endangered species, lifting the poor of the world out of poverty, and protecting the beauty of our natural world.

We need to invest in innovation and avoid wasting money on boondoggles such as carbon dioxide capture technologies. All these solutions will be created by people. Accordingly, we need to encourage, not discourage, our children to have children of their own, as it is these future generations who will innovate new technologies and adapt our world to the changing climate—regardless of whether it is warmer or colder.

I leave you with the words of Winston Churchill: "Truth is incontrovertible. Malice may attack it, ignorance may deride it, but in the end, there it is."[100] The evidence and truth are clear: there it is, there is no "climate crisis." Let us act accordingly and focus our attention on solving legitimate environmental and social problems while protecting the beauty of our precious Earth for our children and grandchildren.

Endnotes

Introduction

1 Roston, Eric and Rathi, Akshat, "Climate scientist reach 'unequivocal' consensus on human-made warming in landmark report," *PHYS.ORG,* 9 August 2021. https://phys.org/news/2021-08-climate-scientists-unequivocal-consensus-human-made.html.

2 *Our World in Data,* "Countries with more than 25 species at risk of losing more than 25% of their habitat by 2050." https://ourworldindata.org/grapher/habitat-loss-25-species.

3 WSJ Editorial Board, "The Real Cost of the Inflation Act Subsidies, $1.2 trillion," *The Wall Street Journal,* Opinion, 24 March 2023. https://www.wsj.com/articles/inflation-reduction-act-subsidies-cost-goldman-sachs-report-5623cd29.

4 Lomborg, Bjorn, "Welfare in the 21st century: Increasing development, reducing inequality, the impact of climate change, and the cost of climate policies," *Technological Forecasting and Social Change,* Volume 156, July 2020. https://doi.org/10.1016/j.techfore.2020.119981.

5 Lawder, David, "Yellen says $3 trillion needed annually for climate financing, far more than current level," *Reuters,* 27 July 2024. https://www.reuters.com/sustainability/sustainable-finance-reporting/yellen-says-3-trillion-needed-annually-climate-financing-far-more-than-current-2024-07-27/.

6 Audubon, "Birds and Clean Energy." https://www.audubon.org/news/wind-power-and-birds.

7 Botkin, Daniel B., "Forecasting the Effects of Global Warming on Biodiversity," *BioScience,* Volume 57, Issue 3, pp. 227-236, 1 March 2007, p. 231. https://doi.org/10.1641/B570306.

8 Wrightstone, Gregory, "Mass extinction lie exposed: life is thriving," *Inconvenient Blog,* 13 May, 2019. https://inconvenientfacts.xyz/blog/f/mass-extinction-lie-exposed-life-is-thriving.

9 Saban, Kristen E. and Wiens, John J., "Unpacking the extinction crisis: rates, patterns and causes of recent extinctions in plants and animals," *Biological Sciences,* Proceedings of the Royal Society B, 15 October 2025. https://doi.org/10.1098/rspb.2025.1717.

10 Stolte, Daniel, "Extinction rates have slowed across many plant and animal groups, study shows," *The University of Arizona News,* 22 October 2025.

https://news.arizona.edu/news/extinction-rates-have-slowed-across-many-plant-and-animal-groups-study-shows.

11 Richard, Kenneth, "During the last few hundred years, species extinctions primarily occurred due to habitat loss and predator introduction on islands. Extinctions have not been linked to a warming climate or higher CO_2 levels. In fact, since the 1870s, species extinction rates have been plummeting," *No Trick Zone,* 16 May 2019. https://notrickszone.com/2019/05/16/recent-studies-indicate-species-extinctions-decline-with-warming-mass-extinction-events-due-to-cooling/.

12 Katzner, Todd E. et al., "Impacts of onshore wind energy production on biodiversity," *Nature Reviews Biodiversity,* Volume 1, pp. 567-580, 8 September 2025. https://doi.org/10.1038/s44358-025-00078-1.

13 Sierra Club, "2022 Sierra Club Annual Report," https://vault.sierraclub.org/annual-report-2022/?homepage=.

14 Happer, William, "How to Think About Climate Change," CLINTEL-lezing, Amsterdam, 15 November 2021. https://www.youtube.com/watch?v=P-blYr-KjOVY.

15 US EPA, "EPA Awards $27B in Greenhouse Gas Reduction to Accelerate Clean Energy Solutions, Combat the Climate Crisis, and Save Families Money," *EPA News Releases,* 16 August 2024. https://www.epa.gov/newsreleases/epa-awards-27b-greenhouse-gas-reduction-fund-grants-accelerate-clean-energy-solutions.

16 US EPA, "ICYMI: Administrator Lee Zelden Finds Gold Bars from EPA at Stacey Abrams' Connected Group, Biden-Harris Ethical Red Flags," *EPA News Releases,* 21 February 2025. https://www.epa.gov/newsreleases/icymi-administrator-lee-zeldin-finds-gold-bars-epa-stacey-abrams-connected-group-biden.

17 Follett, Chelsea, "Sri Lanka Crisis Reveals the Dangers of Green Utopianism," Foundation for Economic Education (FEE), 20 July 2022. https://fee.org/articles/sri-lanka-crisis-reveals-the-dangers-of-green-utopianism/.

18 Hansen, Stein, "Debt for Nature Swaps," The World Bank, February 1988. https://documents1.worldbank.org/curated/en/823691493257754828/pdf/Debt-for-nature-swaps-overview-and-discussion-of-key-issues.pdf.

19 Kutukwa, Kudzai, "How the Attacks on Farming and Bitcoin Are Connected," *Bitcoin Magazine,* 30 July 2022. https://bitcoinmagazine.com/business/attacks-on-farmers-and-bitcoin-are-connected.

20 *Reliefweb*, "Sri Lanka Food Security Crisis—Humanitarian Needs and Priorities 2022," 9 June 2020. https://reliefweb.int/report/sri-lanka/sri-lanka-food-security-crisis-humanitarian-needs-and-priorities-2022-june-sept-2022-ensita.

21 World Health Organization, "Household Air Pollution," 16 October 2024. https://www.who.int/news-room/fact-sheets/detail/household-air-pollution-and-health.

22 Gritt, Brian, *In the Dark: Fixing Energy Policies That Hurt People & the Planet,* Altamira Studio, 1 January 2023, 132 pp. https://altamira.studio/itd.

23 Kekesi, Alex, "Haitian Deforestation," *NASA*, 25 October 2002. https://svs.gsfc.nasa.gov/2640?utm_source=chatgpt.com.

24 Min, Brian et al., "Beyond Access: 1.18 Billion in Energy Poverty Despite Rising Electricity Access," *United Nations,* 12 June 2024. https://www.iea.org/reports/sdg7-data-and-projections/access-to-electricity.

25 IEA, "Access to Electricity." https://www.iea.org/reports/sdg7-data-and-projections/access-to-electricity.

26 Epstein, Alex, *The Moral Case for Fossil Fuels,* Penguin Group, New York, NY, 2014, p. 38: Kathryn Hall, "Kathryn's Story: Power Up, Gambia," ISBN 978-1-59184-744-1.

27 Abnett, Kate, "Rich Countries May Have Met $100 Billion Climate Goal Last Year – OECD," *Reuters*, 16 November 2023. https://www.reuters.com/sustainability/sustainable-finance-reporting/rich-countries-may-have-met-100-bln-climate-goal-last-year-oecd-2023-11-16/; See also Lawder, David, "Yellen Says $3 Trillion Needed Annually for Climate Financing, Far More Than Current Level," *Reuters*, 27 July 2024. https://www.reuters.com/sustainability/sustainable-finance-reporting/yellen-says-3-trillion-needed-annually-climate-financing-far-more-than-current-2024-07-27/.

28 Tol, Richard, "The Economic Impacts of Climate Change," *Review of Environmental Economics and Policy,* Volume 12, Number 1, pp. 4-25, Winter 2018, p. 11. https://doi.org/10.1093/reep/rex027.

29 World Health Organization, "Air Pollution in China." https://www.who.int/china/health-topics/air-pollution?utm_source=chatgpt.com.

30 Landau, Erica K., "India Is Challenging China for Most Toxic Air in the World," *VICE Newsletters,* 15 February 2017. https://www.vice.com/en/article/india-is-challenging-china-for-most-toxic-air-in-the-world/.

31 Bhattacharji, Chetan, "Only Seven Countries Meet WHO Air Quality Standard; Most Polluted are Chad and Bangladesh–Northern India Also Dominates," *Health Policy Watch,* 3 November 2025. https://healthpolicy-watch.news/northern-india-dominates-global-air-pollution-rankings/.

32 Epstein, Alex, *The Moral Case for Fossil Fuels,* Penguin Group, New York, NY, 2014, p. 52, ISBN 978-1-59184-744-1.

33 U.S. Department of Energy, "Fuel Properties Comparison." Office of Energy Efficiency and Renewable Energy, October 2024. https://afdc.energy.gov/fuels/

properties.

34 Usman, Muhammad et al., "SI engine performance, lubricant oil deterioration, and emissions: A comparison of liquid and gaseous fuel," *Advances in Mechanical Engineering*, 1 June 2014. https://doi.org/10.1177/1687814020930451.

35 Epstein, Alex, *The Moral Case for Fossil Fuels,* Penguin Group, New York, NY, 2014, p. 153, ISBN 978-1-59184-744-1.

36 Mills, Mark, "Mark Mills–The energy transition delusion: inescapable mined realities," *SKAGEN Funds New Years Conference 2023,* 16 January 2023. https://www.youtube.com/watch?v=sgOEGKDVvsg.

37 Epstein, Alex, *The Moral Case for Fossil Fuels,* Penguin Group, New York, NY, 2014, p. 49, ISBN 978-1-59184-744-1.

38 Decker, Kris De, "How much energy does it take (on average) to produce 1 kilogram of the following materials?", *Low Tech Magazine,* 14 June 2009. https://solar.lowtechmagazine.com/2009/06/how-much-energy-does-it-take-on-average-to-produce-1-kilogram-of-the-following-materials/.

39 Rooij, Dricus De, "Silicon (SI) for solar cells: how is it produced," *Sino Voltaics.* https://sinovoltaics.com/learning-center/solar-cells/silicon-si-solar-cells-produced/.

40 Epstein, Alex, *The Moral Case for Fossil Fuels,* Penguin Group, New York, NY, 2014, pp, 155-160, ISBN 978-1-59184-744-1.

41 Manheimer, Wallace, "Wallace Manheimer: Mass Delusions," *Tom Nelson Podcast,* # 143, 31 August 2023. https://www.youtube.com/watch?v=hgo-9cH9aUkM.

42 Heymann, Eric, "German industrial production: The decline is not over yet," *Deutsch Bank Research,* 13 Feb 2024. www.dbresearch.de/PROD/RPS_DE-PROD/PROD0000000000531908/German_industrial_production%3A_The__decline_is_not_.PDF?undefined&realload=oaSmAyQUxIjYyK8o7xesJBiNfEysYVJTpW4dexd2fdI168vJfisv9nhqb04CijZA.

43 Goreham, Steve, "Europe's Impossible Choice: AI Development or Net Zero," *Watts Up With That?,* 18 September 2025. https://wattsupwiththat.com/2025/09/18/europe-ai-development-or-net-zero/.

44 Lomborg, Bjorn, "The enormous cost of green policies," post on social media site X, 4 January 2025. https://x.com/BjornLomborg/status/1875529420049424746.

45 Goreham, Steve, "Exploding Energy Prices in California," *Master Resource,* March 12, 2024. https://www.masterresource.org/goreham-steve/exploding-energy-prices-california/.

46 Bettarelli, Luca et al., "Energy Inflation and Inequality," *Energy Economics,* Volume 124, August 2023. https://doi.org/10.1016/j.eneco.2023.106823.

47 Marshall, Claire, "Kenya's Ogiek people being evicted for carbon credits: lawyers," *BBC News,* 9 November 2023. https://www.bbc.co.uk/news/world-africa-67352067.amp.

48 Clifford, Catherine, "Biden says global warming topping 1.5 degrees in the next 10 to 20 years is scarier than nuclear war." *CNBC,* 11 September 2023. https://www.cnbc.com/2023/09/11/biden-global-warming-even-more-frightening-than-nuclear-war.html?msockid=1b73056173826cd3033f11f7729e6d87

49 Benardete, Doris, *Mark Twain Wit and Wisecracks,* The Peter Pauper Press, New York, 1961, p. 54.

50 Feynman, Richard, "If it disagrees with experiment, it's wrong." Cornell University, 1964. https://youtu.be/b240PGCMwV0.

51 Twain, Mark, "What gets us into trouble…" *Goodreads.* https://www.goodreads.com/quotes/738123-what-gets-us-into-trouble-is-not-what-we-don-t.

52 Crok, Marcel and May, Andy, *The Frozen Climate Views of the IPCC, an Analysis of AR6,* Clintel Foundation, Amsterdam, The Netherlands, 2023, p. 83, ISBN: 979-8-89074-861-4.

53 IPCC, "Climate change widespread, rapid, and intensifying," 9 August 2021. https://www.ipcc.ch/2021/08/09/ar6-wg1-20210809-pr/.

54 Roston, Eric and Rathi, Akshat, "The IPCC states that humanity is unequivocally to blame for rising temperatures," *Fortune,* 9 August 2021. https://fortune.com/2021/08/09/ipcc-report-climate-change-humans-responsible-warming-emissions/; See also Roston, Eric and Rathi, Akshat, "Climate scientist reach 'unequivocal' consensus on human-made warming in landmark report," *PHYS.ORG,* 9 August 2021. https://phys.org/news/2021-08-climate-scientists-unequivocal-consensus-human-made.html.

55 Jones, Nicola, "How the World Passed a Carbon Threshold and Why It Matters," *Yale Environment 360,* 26 January 2017. https://e360.yale.edu/features/how-the-world-passed-a-carbon-threshold-400ppm-and-why-it-matters?_bhlid=38186d09213e68ec77ed9db9b24aa48cba37d1c1&utm_campaign=when-the-glitter-of-an-economic-boom-clashes-with-the-environment&utm_medium=newsletter&utm_source=thedeepview.

56 Rosane, Olivia, "A File of Shame: World on Track for 3.2 Degrees Celsius of Warming, Latest IPCC Report Warns," *EcoWatch,* 4 April 2022. https://www.ecowatch.com/ipcc-report-climate-change.html; See also United Nations, "UN emissions report: World on course for more than 3-degree spike, even if climate commitments are met," *UN News,* 26 November 2019. https://news.un.org/en/story/2019/11/1052171.

57 IPCC, "Climate change widespread, rapid, and intensifying," 9 August 2021. https://www.ipcc.ch/2021/08/09/ar6-wg1-20210809-pr/.

58 Myhre, Gunnar et al., "New estimates of radiative forcing due to well mixed greenhouse gases," *Geophysical Research Letters,* Volume 25, Issue 14, pp. 2715-2718, 15 July1998. https://doi.org/10.1029/98GL01908. See the "Logarithmic Decline of CO_2 Warming" section of Chapter 9 for the proper use of the Myhre equations.

59 McKitrick, Ross and Christy, John, "Pervasive Warming Bias in CMIP6 Tropospheric Layers," *Earth and Space Science,* September 2020, Volume 7, Issue 9, 15 July 2020. https://doi.org/10.1029/2020EA001281.

60 Vinós, Javier, *Solving the Climate Puzzle: The Sun's Surprising Role,* Critical Science Press, Madrid, 2023, p. 34, ISBN: 978-84-12867-7-0.

61 Dewitte, Steven and Clerbaux, Nicholas, "Decadal Changes of Earth's Outgoing Longwave Radiation," *Remote Sensing,* Volume 10, Issue 10, 25 September 2018, p. 1543. https://doi.org/10.3390/rs10101539; See also, Qualglia, Ilaria and Visioni, Daniele, "Modeling 2020 regulatory changes in international shipping changes in international shipping emissions helps explain anomalous 2023 warming," *EDS Letters*, Volume 15, Issue 6, pp. 1527-1441, 28 November 2024. https://doi.org/10.5194/esd-15-1527-2024; and Nikolov, Ned and Zeller, Karl F., "Role of Earth's Albedo Variations and Top-of-the Atmosphere Energy Imbalance in Recent Warming: New Insights from Satellite and Surface Observations," *MDPI Geomatics 2024,* 4(3), pp. 311-341, 20 August 2024. https://doi.org/10.3390/geomatics4030017.

62 Vinós, Javier, *Solving the Climate Puzzle: The Sun's Surprising Role,* Critical Science Press, Madrid, 2023, p. 34, ISBN: 978-84-12867-7-0.

63 Tselioudis, George et al., "Contraction of the World's Storm-Cloud Zones the Primary Contributor to the 21st Century Increase in the Earth's Sunlight Absorption," *Geophysical Research Letters,* Volume 52, Issue 11, 8 June 2025. https://doi.org/10.1029/2025GL114882.

64 Sagan, Carl, *Cosmos*, Ballantine Books: New York, NY, USA, 1985. ISBN: 978-0394502946.

65 Brown, Patrick, "As a scientist, I am not allowed to tell the full truth about climate change," *New York Post,* Opinion, 5 September 2023. https://nypost.com/2023/09/05/as-a-scientist-im-not-allowed-to-tell-the-full-truth-about-climate-change/.

66 Climategate emails were more than 1,000 emails that were illegally obtained from the Climatic Research Unit at the University of East Anglia in the UK in November 2009. The batch contained emails from international scientists who are contributors to the United Nations International Panel on Climate Change. To support the climate crisis narrative, some of these emails suggested manipulating data, hiding data, undermining the peer review process, and pressuring journals to fire editors who allowed papers that were counter to the climate

crisis narrative to be published. For specific Climategate emails see Bradley, Robert Jr., "Climategate Turns 15!" *Watts Up With That,* 27 November 2024. https://wattsupwiththat.com/2024/11/27/climategate-turns-15/.

67 Stossel, John, "Scientist admits the 'overwhelming consensus' on the climate change crisis is 'manufactured,'" *New York Post*, 9 August 2023. https://nypost.com/2023/08/09/climate-scientist-admits-the-overwhelming-consensus-is-manufactured/.

68 Curry, Judith, "Judith Curry, Opening Statement to the Space, Science, and Competitiveness Senate Subcommittee on Climate Change," *American Rhetoric*, 8 December 2015. https://www.americanrhetoric.com/speeches/judithcurryclimatechangecongress.htm.

69 Svensmark, Henrik, "It is impossible for me to get funding for this kind of work," *Tom Nelson Podcast,* #44, 21 November 2022. https://www.youtube.com/watch?v=sZBWEKCW2Fc.

70 Heller, Tony, "Dr. William Gray funding cut," @HellerClimate, X post, 13 March 2024, https://x.com/TonyClimate/status/1768106051650699491.

71 Handy, Rayan Maye, "Colorado State hurricane forecasts may end due to lack of funds," *USA Today,* 27 November 2013. https://www.usatoday.com/story/weather/2013/11/27/hurricane-forecast-funding-william-gray/3766095/.

72 Overland, Indra, and Sovacool, Benjamin K., "The misallocation of climate research funding," *Energy & Social Science,* Volume 62, 101349, April 2020. https://doi.org/10.1016/j.erss.2019.101349.

73 Sinclair, Upton, "It is difficult to get anybody to understand something…" Rowan Simpson Follow the Money. https://rowansimpson.com/quotes/salary/.

74 Biezen, Michel van, "Astronomy Chapter 9.1–Earth's Atmosphere," Online *YouTube* Course, https://www.youtube.com/watch?v=dw3vQ6hguWg.

75 Dyson, Freeman, "Is Carbon Dioxide Making The World Greener?" *Conversations That Matter,* Institute for Advanced Studies, 9 June 2015. https://www.youtube.com/watch?v=BQHhDxRuTkI.

76 Lindzen, Richard, "2018 Annual GWPF Lecture Prof Richard Lindzen Global Warming for the Two Cultures," The Global Warming Policy Foundation, Institution of Mechanical Engineers, London, 8 October 2018. https://www.youtube.com/watch?v=IOKElp_jGLQ.

77 Clauser, John, "Nobel Laureate John Clauser: Climate Models Miss Key Variable," American Thought Leaders, *The Epoch Times,* 6 September 2023. https://youtu.be/CvqIqy8dUvA?si=2lkV0c6ehw3FqcRs.

78 Happer, William, "How to Think About Climate Change," CLINTEL-lezing, Amsterdam, 15 November 2021. https://www.youtube.com/watch?v=P-blYr-KjOVY.

79 Curry, Judith, "2024 GWPF Lecture – Judith Curry – Climate Uncertainty and Risk," GWPF Annual Lecture, 2 May 2024. https://www.youtube.com/watch?v=iqsZV8i3O1E.

80 Koonin, Steven, "Hot or Not: Steven Koonin Questions Conventional Climate Science and Methodology," *Uncommon Knowledge,* Hoover Institution, 15 August 2023. https://www.youtube.com/watch?v=l90FpjPGLBE.

81 Koonin, Steven, "Hot or Not: Steven Koonin Questions Conventional Climate Science and Methodology," *Uncommon Knowledge,* Hoover Institution, 15 August 2023. https://www.youtube.com/watch?v=l90FpjPGLBE.

82 Soon, Willie, "Willie Soon: The 'art' of calculating the total solar irradiation (TSI) since 1700," European Institute for Climate and Energy (EIKE), 16th International EIKE Climate and Energy Conference, IKEK-16, 22 July 2024. https://youtube.com/watch?v=qwurHDivqaQ.

83 Shaviv, Nir, "What role has the Sun played in climate change? What does this mean for us?" European Climate and Energy Institute (EIKE), 8 December 2022. https://youtu.be/hRFIzVB4Qss?si=NBxs5eu1qxHwl4Li.

84 Svensmark, Henrik, "The Influence of Cosmic Rays on Climate," European Climate and Energy Institute (EIKE), 13 April 2023. https://youtu.be/_hRH-gz55-zA?si=6Frd-QKq3kjHFRus.

85 Zharkova, Valentina, "How the Sun affects temperatures on Earth," *Conversations That Matter,* 10 October 2019. https://www.youtube.com/watch?v=-JyyuouPSNEA.

86 Vahrenholt, Fritz and Lüning, Sebastion, *The Neglected Sun: Why the Sun Precludes Climate Catastrophe,* The Heartland Institute, Arlington Heights, IL, 2015, ISBN 13-978-1-934791-54-7.

87 *Wikipedia*, "Atmospheric Optics," Cloud Coloration. https://en.wikipedia.org/wiki/Atmospheric_optics.

88 Viterito, Arthur, "Viterito/Kamis/Yim/Catt: Impacts of Geothermal Energy on Climate," *Tom Nelson Podcast,* #181, 23 December 2023. https://www.youtube.com/watch?v=lbDbA32fNek.

89 Nelson, Michael and Nelson, David B., "Decoupling CO_2 from Climate Change," *International Journal of Geosciences,* Volume 15, pp. 246-269, 26 March 2024. https://doi.org/10.4236/ijg.2024.153015.

Chapter 1 - The Scientific Consensus

1 Sheppard, Kate, "Obama: We Don't Have Time for a Meeting of the Flat Earth Society," *The Atlantic,* June 2013. https://www.theatlantic.com/technology/archive/2013/06/obama-we-dont-have-time-for-a-meeting-of-the-flat-earth-society/277222/.

2 America Reports, "Biden calls climate change deniers 'Neanderthals' during border speech," *Fox News,* 29 February 2024. https://www.foxnews.com/politics/biden-calls-climate-change-deniers-neanderthals-during-border-speech-texas.

3 CLINTEL, "World Climate Declaration," CLINTEL Foundation. https://clintel.org/world-climate-declaration/.

4 Giaever, Ivar, *Brainy Quote,* Ivar Giaever Quotes. https://www.brainyquote.com/quotes/ivar_giaever_736288.

5 CO2 Coalition, "Nobel Laureate John Clauser Elected to CO2 Coalition Board of Directors", CO2 Coalition, 5 May 2023, https://co2coalition.org/publications/nobel-laureate-john-clauser-elected-to-co2-coalition-board-of-directors/.

6 Dumitro, Eli, "The Enhancement and Standardization of Climate-Related Disclosures for Investors," U.S. Securities and Exchange Commission Filing No. S7-10-22, 21 April 2022. Includes quote from William Happer. https://www.sec.gov/comments/s7-10-22/s71022-284613.htm.

7 Christy, John R., "Climate Models for Policy? ... a Bridge Too Far," 14th International EIKE Climate and Energy Conference, European Institute for Climate and Energy (EIKE), 17 February 2022. https://youtu.be/XltFOh7Cg2U?si=G-BuFwI3SV9vpjYyk.

8 Global Warming Petition Project, http://www.petitionproject.org/.

9 Koonin, Steven E., *Unsettled*, BenBella Books, Dallas, Texas, 2021, ISBN 978-1-95066-579-2.

10 Koonin, Steven E., "Joe Rogan Experience #1776 – Steven E. Koonin," *The Joe Rogan Experience,* 27 June 2024. https://www.youtube.com/watch?v=OBjX0O7gOmw Minutes 14:00 to 15:45.

11 Koonin, Steven E., "Climate Science Is Not Settled," *Wall Street Journal,* 19 September 2014. https://www.wsj.com/articles/climate-science-is-not-settled-1411143565?msockid=1b73056173826cd3033f11f7729e6d87.

12 Koonin, Steven, E., "2021 Annual GWPF Lecture, Steven Koonin, Unsettled," The Global Warming Policy Foundation, 18 November 2021. https://www.youtube.com/watch?v=6Tz1MiX1p5I&t=11s; See also, Koonin, Steven E., "Joe Rogan Experience #1776 – Steven E. Koonin," *The Joe Rogan Experience,* 27 June 2024. https://www.youtube.com/watch?v=OBjX0O7gOmw; and Koonin, Steven E., "Hot or Not: Steven Koonin Questions Conventional Climate Science and Methodology," *Uncommon Knowledge*, Hoover Institution, 15 August 2023. https://www.youtube.com/watch?v=l90FpjPGLBE.

13 Tol, Richard, "Climate Change: Mr. Obama, 97 percent of experts is a bogus number," Opinion, *Fox News*, 28 May 2015. https://www.foxnews.com/opin-

ion/climate-change-mr-obama-97-percent-of-experts-is-a-bogus-number?msockid=1b73056173826cd3033f11f7729e6d87.

14 Cook, John et al., "Quantifying the consensus on anthropogenic global warming in the scientific literature," *Environmental Research Letters*, Volume 8, Number 2, 15 May 2013. https://doi.org/10.1088/1748-9326/8/2/024024.

15 Wrightstone, Gregory, "97% Consensus – What Consensus?" CO2 Coalition, 31 October 2021. https://co2coalition.org/2021/10/31/97-consensus-what-consensus/.

16 Verheggen, Bart et al., "Scientific Views About Attribution of Global Warming," ACS, *Environmental Science & Technology*, Volume 48, Issue 16, pp. 8963-8971, 27 January 2014. https://doi.org/10.1021/es501998e.

17 Sarv, Hannes, "INTERVIEW. Dr. Judith Curry on Global Warming: Where is the Danger?" *Freedom Research,* 20 August 2025. https://www.freedom-research.org/p/interview-dr-judith-curry-on-global.

18 Lindzen, Richard S., "Falsification of climate models used in experimentation and scenario construction," *WIREs Climate Change.* https://www.parliament.uk/globalassets/documents/commons-committees/energy-and-climate-change/Professor-Richard-Lindzen-IPC0047.pdf?utm_source=chatgpt.com.

19 Happer, William, "The Staff of the Best Schools Interviews William Happer," *Independent Institute,* 24 March 2020. https://www.independent.org/article/2020/03/24/the-staff-of-the-best-schools-interviews-william-happer/.

20 Shaviv, Nir, "Up to two-thirds of the warming comes from the Sun," *Science Matters*, 30 April 2025. https://rclutz.com/2025/04/?utm_source=chatgpt.com.

21 Roston, Eric and Rathi, Akshat, "The IPCC states that humanity is unequivocally to blame for rising temperatures," *Fortune*, 9 August 2021. https://fortune.com/2021/08/09/ipcc-report-climate-change-humans-responsible-warming-emissions/; See also, Roston, Eric and Rathi, Akshat, "Climate scientist reach 'unequivocal' consensus on human-made warming in landmark report," *PHYS.ORG,* 9 August 2021. https://phys.org/news/2021-08-climate-scientists-unequivocal-consensus-human-made.html.

22 Farnsworth, Stephen J. and Lichter, Samuel Robert, "The Structure of Evolving US Scientific Opinion on Climate Change and Its Potential Consequences," *APSA 2009 Toronto Meeting Paper,* 26 pp., 18 September 2009. https://papers.ssrn.com/sol3/papers.cfm?abstract_id=1450804.

23 Doran, Peter T. and Zimmerman, Maggie Kendall, "Examining the Scientific Consensus on Climate Change," *Transactions American Geophysical Union,* Volume 90, Issue 3, pp. 22-23, 3 June 2011. https://doi.org/10.1029/2009EO030002.

24 Stenhouse, Neil et al., "Meteorologists' Views About Global Warming," *Bulletin of the American Meteroological Society,* pp. 1029-1040, 1 July 2014. https://doi.org/10.1175/BAMS-D-13-00091.1

25 Maibach, Edward et al., "A 2016 National Survey of Broadcast Meteorologists: Initial Findings: Member Views on Climate Change: Initial Findings," George Mason University, Center for Climate Change Communication, Fairfax, VA, March 2016. http://dx.doi.org/10.13140/RG.2.2.11394.89281.

26 Lynas, Mark et al., "Greater than 99% consensus on human caused climate change in the peer-reviewed scientific literature," *Environmental Research Letters,* Volume 6, Number 11, 19 October 2021. https://doi.org/10.1088/1748-9326/ac2966.

27 Dentelski, David et al., "Re-examining the Consensus on the Anthropogenic Contribution to Climate Change," *Climate*, Volume 11, Issue 11, 30 September 2023. https://doi.org/10.3390/cli11110215.

28 Robson, John, "The 97 Percent Consensus Myth Revisited," *Climate Nexus Discussion*, 1 January 2023. https://climatediscussionnexus.com/videos/the-97-percent-consensus-myth-revisited/.

Chapter 2 - IPCC

1 IPCC, "The role of the IPCC and key elements of the IPCC assessment process, Geneva," 4 February 2010. https://www.ipcc.ch/site/assets/uploads/2018/04/role_ipcc_key_elements_assessment_process_04022010.pdf.

2 IPCC, "About the IPCC." https://www.ipcc.ch/about/.

3 Koonin, Steven E., "Joe Rogan Experience #1776 – Steven E. Koonin," *The Joe Rogan Experience*, 27 June 2024. https://www.youtube.com/watch?v=OBjX-0O7gOmw.

4 United Nations, "Secretary-General's Opening Remarks at Press Conference on Climate," United Nations, 27 July 2023. https://www.un.org/sg/en/content/sg/speeches/2023-07-27/secretary-generals-opening-remarks-press-conference-climate.

5 Hays, Gabriel, "Al Gore goes on 'unhinged' rant about 'rain bombs,' boiled oceans, and other climate threats at Davos," *Fox News,* 18 January 2023. https://www.foxnews.com/media/al-gore-goes-unhinged-rant-about-rain-bombs-boiled-oceans-other-climate-threats-davos?msockid=1b73056173826cd3033f11f7729e6d87.

6 United Nations, "Secretary-General Calls Latest IPCC Climate Report 'Code Red for Humanity,' Stressing 'Irrefutable' Evidence of Human Influence," United Nations Press Release, 9 August 2021. https://press.un.org/en/2021/sgsm20847.doc.htm.

7 *Discover*, "Introduction: Global Climate Explained," Stephen H. Schneider interview, October 1989. p. 47. https://www.justfacts.com/document/global_warming_stephen_schneider_discover_magazine_october_1989.pdf.

8 Ebbs, Sephanie, "Unequivocal that human influence has warmed the planet, UN climate panel finds," *ABC News*, 9 August 2021. https:// https://abcnews.go.com/Politics/unequivocal-human-influence-warmed-planet-climate-panel-finds/story?id=79314089.

9 "Kimapolitik verteilt das Weltvermögen neu," Ottmar Edenhofer interview, *Neue Zürcher Zeitung*, 14 November 2010. https://www.nzz.ch/klimapolitik_verteilt_das_weltvermoegen_neu-ld.1003523.

10 IIPCC, Climate *Change 2021: The Physical Science Basis. Contribution of Working Group I to the Sixth Assessment Report of the Intergovernmental Panel on Climate Change,* "Summary for Policymakers," Cambridge University Press, Cambridge, United Kingdom and New York, NY, USA, Section A.1.8, Figure SPM.1, p. 6, 9 August 2021. https://www.ipcc.ch/report/ar6/wg1/downloads/report/IPCC_AR6_WGI_SPM.pdf.

11 IPCC, *Climate Change 2023: Synthesis Report, Contribution of Working Groups I, II, and III to the Sixth Assessment Report of the Intergovernmental Panel on Climate Change,* "2023: Summary for Policymakers," Geneva, Switzerland, Section B.2.1, p. 15, 2003. https://www.ipcc.ch/report/ar6/syr/downloads/report/IPCC_AR6_SYR_SPM.pdf.

12 Zhao, Qi et al., "Global regional and national burden of mortality associated with non-optimal ambient temperatures from 2000 to 2019: a three-stage modeling study," *The Lancet*, Volume 5, Issue 7, E415-E425, July 2021. https://doi.org/10.1016/S2542-5196(21)00081-4.

13 World Meteorological Organization, "Rate and impact of climate change surges dramatically in 2011-2020," World Meteorological Organization Press Release, 5 December 2023. https://wmo.int/news/media-centre/rate-and-impact-of-climate-change-surges-dramatically-2011-2020.

14 IPCC, *The Physical Science Basis, Contribution of Working Group I to the Sixth Assessment Report on the Intergovernmental Panel on Climate Change,* "Chapter 11 - Weather and Climate Extreme Events in a Changing Climate," Cambridge University Press, Cambridge, United Kingdom and New York, NY, USA, Section 11.4.2, p. 1569, 9 August 2021. https://www.ipcc.ch/report/ar6/wg1/downloads/report/IPCC_AR6_WGI_Chapter11.pdf.

15 IPCC, *The Physical Science Basis, Contribution of Working Group I to the Sixth Assessment Report on the Intergovernmental Panel on Climate Change,* "Chapter 11 - Weather and Climate Extreme Events in a Changing Climate," Cambridge University Press, Cambridge, United Kingdom and New York, NY, USA, Section 11.7.1.2., p. 1585, August 2021. https://www.ipcc.ch/report/ar6/

wg1/downloads/report/IPCC_AR6_WGI_Chapter11.pdf.

16 IPCC, *The Physical Science Basis, Contribution of Working Group I to the Sixth Assessment Report on the Intergovernmental Panel on Climate Change,* "Summary for Policymakers," Cambridge University Press, Cambridge, United Kingdom and New York, NY, USA, Section B.2.4., p. 1585, 9 August 2021. https://www.ipcc.ch/report/ar6/wg1/downloads/report/IPCC_AR6_WGI_SPM.pdf.

17 Ma, Ning et al., "Recent Decline in Global Ocean Evaporation Due to Wind Stilling," *Geophysical Research Letters,* Volume 52, Issue 4, 19 February 2025. https://doi.org/10.1029/2024GL114256.

18 Ma, Ning et al., "Recent Decline in Global Ocean Evaporation Due to Wind Stilling," *Geophysical Research Letters,* Volume 52, Issue 4, 19 February 2025. https://doi.org/10.1029/2024GL114256.

19 Tuleya, Robert E., "Impact of Upper-Tropospheric Temperature Anomalies and Vertical Wind Shear on Tropical Cyclone Evolution Using an Idealized Version of the Operational GFDL Hurricane Model," *Journal of the Atmospheric Sciences,* Volume 73, Issue 10, pp. 3803-3820, 1 October 2016, p. 3803. https://doi.org/10.1175/jas-d-16-0045.1.

20 Tuleya, Robert E., "Impact of Upper-Tropospheric Temperature Anomalies and Vertical Wind Shear on Tropical Cyclone Evolution Using an Idealized Version of the Operational GFDL Hurricane Model," *Journal of the Atmospheric Sciences*, Volume 73, Issue 10, pp. 3803-3820, 1 October 2016 p. 3816. https://doi.org/10.1175/jas-d-16-0045.1.

Chapter 3 - Climate Catastrophes

1 Koonin, Steven E., *Unsettled*, BenBella Books, Dallas, TX, 2021, ISBN 978-1-95066-579-2.

2 Alimonti, G. et al., "A critical assessment of extreme events trends in times of global warming," *European Physical Journal Plus,* 13 January 2022, Volume 137, Article Number 112. https://doi.org/10.1140/epjp/s13360-021-02243-9.

3 Klotzbach, Philip J. et al., "Trends in Tropical Cyclone Activity 1990–2021," *Geophysical Research Letters,* Volume 49, Issue 6, e2021GL095774, 28 March 2022. https://doi.org/10.1029/2021GL095774.

4 Anderson, Kerby, "Hurricanes and Climate," *Point of View*, 2 October 2018. https://pointofview.net/viewpoints/hurricanes-and-climate/.

5 Tu, Shifei et al., "Decreasing trend in destructive potential of tropical cyclones in the South India Ocean since the mid-1990s," *Nature Communications Earth & Environment,* Volume 5, p. 543, 30 December 2024. https://doi.org/10.1038/s43247-024-01683-2.

6 Tu, Shifei et al., "Decreasing trend in destructive potential of tropical cyclones in the South India Ocean since the mid-1990s," *Nature Communications Earth & Environment,* Volume 5, p. 543, 30 December 2024, Table 1. https://doi.org/10.1038/s43247-024-01683-2.

7 IPCC, *Climate Change 2021: The Physical Science Basis. Contributions of Working Group I to the Sixth Assessment Report of the Intergovernmental Panel on Climate Change,* "Chapter 12 - Climate Change Information for Regional Impact and for Risk Assessment," Cambridge University Press, Cambridge, UK and New York, NY, USA, Table 12.12, p. 1856, 9 August 2021. https://doi.org/10.1017/9781009157896.014.

8 Zuohao, Cao and Cai, Huaqing, "Trend Analysis of U.S. Tornado Activity Frequency," *Atmosphere*, Volume 13, p. 498, 2022. https://doi.org/10.3390/atmos13030498.

9 Nouri, Niloufar et al., "Explaining the trend and variability in the United States tornado records using climate teleconnections and shifts in observational practices," *Nature Scientific Reports,* Volume 11, Article Number: 1741, 18 January 2021. https://doi.org/10.1038/s41598-021-81143-5.

10 Zhang, Jinhui et al., "Time trends in losses from major tornadoes in the United States," *Weather and Climate Extremes,* Elsevier, 41, 5 June 2023. https://doi.org/10.1016/j.wace.2023.100579.

11 IPCC, *Climate Change 2021: The Physical Science Basis. Contributions of Working Group I to the Sixth Assessment Report of the Intergovernmental Panel on Climate Change,* "Chapter 12 - Climate Change Information for Regional Impact and for Risk Assessment," Cambridge University Press, Cambridge, UK and New York, NY, USA, Table 12.12, p. 1856, 9 August 2021. https://doi.org/10.1017/9781009157896.014.

12 IPCC, "Climate change widespread, rapid, and intensifying," 9 August 2021. https://www.ipcc.ch/2021/08/09/ar6-wg1-20210809-pr/.

13 The United States Environmental Protection Agency (EPA) Climate Change Indicators Heat Waves. EPA, "Climate Change Indicators: High and Low Temperatures," 19 January 2021. https://19january2021snapshot.epa.gov/climate-indicators/climate-change-indicators-high-and-low-temperatures_.html.

14 Qian, Weihong, and Zhu, Yafen, "Climate Change in China from 1880 to 1998 and Its Impact on the Environmental Condition," *Nature Climate Change,* Volume 50, pp. 419-444, September 2001. https://doi.org/10.1023/A:1010673212131.

15 IPCC, *The Physical Science Basis, Contribution of Working Group I to the Sixth Assessment Report on the Intergovernmental Panel on Climate Change,* "Chapter 11 - Weather and Climate Events in a Changing Climate," Cambridge University Press, Cambridge, UK and New York, NY, USA, 11.3.2, pp.

1548-1550, 9 August 2021. https://www.ipcc.ch/report/ar6/wg1/chapter/chapter-11/?os=wtmbLooZOwcJ&ref=app&utm_source=chatgpt.com.

16 Harmer, Andrew, "What Does The IPCC's Sixth Assessment WG1 Report Warn Us About the Health Impacts of Global Warming?" *ANDREWHARMER.ORG, Climate Change,* 25 October 2021. https://andrewharmer.org/2021/10/25/what-does-the-ipccs-sixth-assessment-report-wg-1-warn-us-about-the-health-impacts-of-global-warming/.

17 Vicente, Servio M. et al., "Global drought trends and future projections," *Philosophical Transactions,* Royal Society A, 380:20201085, 18, August 2022. https://doi.org/10.1098/rsta.2021.0285.

18 Zhu, Zaichun, et al., "Greening of the Earth and its drivers," *Nature Climate Change,* Volume 6, pp. 791-755, 25 April 2016. https://doi.org/10.1038/nclimate3004; See also, Chen, Xin et al., "The global greening continues despite increased drought stress since 2020," *Global Ecology and Conservation,* Volume 49, January 2024. https://doi.org/10.1016/j.gecco.2023.e02791.

19 IPCC, *Climate Change 2021: The Physical Science Basis. Contributions of Working Group I to the Sixth Assessment Report of the Intergovernmental Panel on Climate Change,* "Chapter 12 - Climate Change Information for Regional Impact and for Risk Assessment," Cambridge University Press, Cambridge, UK and New York, NY, USA, Table 12.12, p. 1856, 9 August 2021. https://doi.org/10.1017/9781009157896.014.

20 Zhu, Zaichun, et, al, "Greening of the Earth and its drivers," *Nature Climate Change,* Volume 6, pp. 791–755, 25 April 2016. https://doi.org/10.1038/nclimate3004; See also, Piao, Shilong, et. al., "Characteristics, drivers and feedbacks of global greening" *Nature Reviews,* Volume 1, pp. 14–27, 9 December 2019. https://doi.org/10.1038/s43017-019-0001-x.

21 Chen, Xin, et al., "The global greening continues despite increased drought stress since 2020," *Global Ecology and Conservation,* Volume 49, January 2024. https://doi.org/10.1016/j.gecco.2023.e02791.

22 Heller, Tony, "Burn Acreage Fraud in the 2023 National Climate Assessment," *Real Climate Science,* 30 November 2023. https://realclimatescience.com/2023/11/burn-acreage-fraud-in-the-2023-national-climate-assessment/#gsc.tab=0.

23 CO_2 Science, "Fires (Relationship to Global Warming)," www.co2science.org/subject/f/firegw.php.

24 Wrightstone, Gregory, *A Very Convenient Warming,* Silver Crown Productions, LLC, 2023, p. 97, ISBN: 9781545614105.

25 IPCC, *Climate Change 2021: The Physical Science Basis. Contributions of Working Group I to the Sixth Assessment Report of the Intergovernmental Panel on Climate Change,* "Chapter 12 - Climate Change Information for Regional

Impact and for Risk Assessment," Cambridge University Press, Cambridge, UK and New York, NY, Table 12.12, p. 1856, 9 August 2021. https://doi.org/10.1017/9781009157896.014.

26 IPCC, *The Physical Science Basis, Contribution of Working Group I to the Sixth Assessment Report on the Intergovernmental Panel on Climate Change,* "Chapter 11 - Weather and Climate Extreme Events in a Changing Climate," Cambridge University Press, Cambridge, UK and New York, NY, USA, Chapter 11, Section 11.5.2, p. 1568, 9 August 2021. https://www.ipcc.ch/report/ar6/wg1/downloads/report/IPCC_AR6_WGI_Chapter11.pdf.

27 Pielke, Roger, Jr., "21st Century Global Disasters: A fresh update on global disasters since 2000," *The Honest Broker.* https://rogerpielkejr.substack.com/p/21st-century-global-disasters?utm_campaign=post&utm_medium=web.

28 IPCC, *Climate Change 2021: The Physical Science Basis. Contributions of Working Group I to the Sixth Assessment Report of the Intergovernmental Panel on Climate Change,* "Chapter 12 - Climate Change Information for Regional Impact and for Risk Assessment," Cambridge University Press, UK and New York, NY, USA, Table 12.12, p. 1856, 9 August 2021. https://doi.org/10.1017/9781009157896.014.

29 Rantanen, Mika, "The Arctic has warmed nearly four times faster than the globe since 1979," *Nature Communications,* Volume 3, Article Number: 168, 11 August 2022. https://doi.org/10.1038/s43247-022-00498-3.

30 Viterito, Arthur, "Viterito/Kamis/Yim/Catt: Impacts of Geothermal Energy on Climate," *Tom Nelson Podcast,* #181, December 23, 2023. https://www.youtube.com/watch?v=lbDbA32fNek.

31 Lindzen, Richard, "MIT Professor Richard Lindzen: Climate Models vs. Measured Values," *EIKE, VII Internationale Klima – und Energiekonference,* Mannheim, Germany, 10 April 2014. https://youtu.be/rXw_HipM9Eg?si=ftTgLGZHDWNeDy9E.

32 Gertler, Charles G. and O'Gorman, Paul A., "Changing available energy for extratropical cyclones and associated convection in Northern Hemisphere summer," *PNAS*, Volume 116, Issue 10, pp. 4105-4110, 19 February 2019. https://doi.org/10.1073/pnas.1812312116.

33 Chang, Edmund K. et al., "Observed and projected decrease in Northern Hemiphere extratropical cyclone activity in summer and its impact on maximum temperature," *Geophysical Research Letters,* Volume 43, pp. 2200-2208, 12 March 2016. https://doi.org/10.1002/2016GL068172.

34 Shaw, T. A., "Storm track processes and the opposing influences of climate change," *Nature Geoscience,* 29 Review Article, 29 August 2016, p. 2. https://doi.org/10.1038/NGEO2783.

35 Emanuel, Kerry A., "The Theory of Hurricanes." *Annual Review of Fluid Mechanics,* Volume 23, pp. 179-196, January 1991, p. 183. https://doi.org/10.1146/annurev.fl.23.010191.001143.

36 Dai, Aiguo and Deng, Jiechun, "Arctic Amplification Weakens the Variability of Daily Temperatures over Northern Middle-High Latitude," *Journal of Climate,* Volume 34, 1 April 2021. https://doi.org/10.1175/JCLI-D-20-0514.1; See also, Martinez, A. and Iglesias, G., "Global Wind Energy Resources Decline Under Climate Change," *Energy*, Volume 288, 1 February 2024. https://doi.org/10.1016/j.energy.2023.129765.

37 Robins, Jim, "Global 'Stilling': Is Climate Change Slowing Down the Wind?" *Yale Environment 360,* 13 September 2022. https://e360.yale.edu/features/global-stilling-is-climate-change-slowing-the-worlds-wind.

38 Ma, Ning et al., "Recent Decline in Global Ocean Evaporation Due to Wind Stilling," *Geophysical Research Letters,* Volume 52, Issue 4, 19 February 2025. https://doi.org/10.1029/2024GL114256.

39 Ma, Ning et al., "Recent Decline in Global Ocean Evaporation Due to Wind Stilling," *Geophysical Research Letters,* Volume 52, Issue 4, 19 February 2025. https://doi.org/10.1029/2024GL114256.

40 Black, Peter G. et al., "Air-Sea Exchange in Hurricanes: Synthesis of Observations from the Coupled Boundary Layer Air-Sea Transfer Experiment." *Bulletin of the American Meteorological Society,* Volume 88, Issue 3, pp. 357-374, 1 March 2007, p. 362. https://doi.org/10.1175/BAMS-88-3-357.

41 Trenberth, K. E., "Atmospheric Moisture Recycling: Role of Advection and Local Evaporation," *Journal of Climate,* Volume 12, Issue 5, pp. 1368-1381, 1 May 1999, p. 1370. https://doi.org/10.1175/1520-0442(1999)012%3C1368:AMRROA%3E2.0.CO;2.

42 Christy, J. R. and McNider, R.T., "Satellite bulk tropospheric temperatures as a metric for climate sensitivity," *Asia-Pacific Journal of Atmospheric Sciences,* Volume 53, Issue 4, pp. 511-518, 29 November 2017. https://doi.org/10.1007/s13143-017-0070-z.

43 Steiner, A. K. et al., "Observed Temperature Changes in the Troposphere and Stratosphere from 1979 to 2018," *Journal of Climate,* Research Article, 1 October 2020, p. 1. https://doi.org./10.1175/JCLI-D-19-0998.1.

44 Bony, S. et al., "How well do we understand and evaluate climate change feedback processes?" *Journal of Climate Change,* Volume 19, Issue 15, pp. 3445-3482, 2006. https://doi.org/10.1175/JCLI3819.1.

45 The Penn State Meteorology Department, "Lapse Rates," Learning Weather, https://learningweather.psu.edu/node/92.

46 Tu, Shifei et al., "Decreasing trend in destructive potential of tropical

cyclones in the South India Ocean since the mid-1990s," *Nature Communications Earth & Environment,* Volume 5, p. 543, 30 December 2024. https://doi.org/10.1038/s43247-024-01683-2.

47 Tuleya, Robert E., "Impact of Upper-Tropospheric Temperature Anomalies and Vertical Wind Shear on Tropical Cyclone Evolution Using an Idealized Version of the Operational GFDL Hurricane Model," *Journal of the Atmospheric Sciences,* Volume 73, Issue 10, pp. 3803-3820, 1 October 2016, p. 3803. https://doi.org/10.1175/jas-d-16-0045.1.

48 Tuleya, Robert E., "Impact of Upper-Tropospheric Temperature Anomalies and Vertical Wind Shear on Tropical Cyclone Evolution Using an Idealized Version of the Operational GFDL Hurricane Model," *Journal of the Atmospheric Sciences,* Volume 73, Issue 10, pp. 3803-3820, 1 October 2016 p. 3816. https://doi.org/10.1175/jas-d-16-0045.1.

49 Shewchuk, John, "Global Warming Inhibits Hurricane Activity as Indicated by Decreasing Tropical CAPE," *John@EJplace.com.* https://www.climatecraze.com/doc/GlobalWarmingInhibitsHurricaneActivity.pdf.

50 Chang, Edmund K. et al., "Observed and projected decrease in Northern Hemiphere extratropical cyclone activity in summer and its impact on maximum temperature," *Geophysical Research Letters, Vol*ume 43, pp. 2200-2208, 12 March 2016. https://doi.org/10.1002/2016GL068172.

51 McCabe, Kristy, "How Does Climate Change Effect Thunderstorms?" Met Matters, *Royal Meteorological Society,* 20 July 2023. https://www.rmets.org/metmatters/how-does-climate-change-affect-thunderstorms.

52 White, Sam, *A Cold Welcome,* Harvard Press, Cambridge, Massachusetts, 2017, p. 236. ISBN 9780674244900.

53 Wheeler, D. et al., "Atmospheric circulation and storminess derived from Royal Navy logbooks: 1685 to 1750," *Climatic Change,* Volume 101, pp. 257-280, July 2010. https://doi.org/10.1007/s10584-00t.9-9732-x; See also, Trouet, V. et al., "North Atlantic storminess and Atlantic Meridional Overturning Circulation during the last Millennium: Reconciling contradictory proxy records of NAO variability," *Global and Planetary Change,* Volumes 84-85, pp. 48-55, p. 51, March 2012. https://doi.org/10.1016/j.gloplacha.2011.10.003.

54 Trouet, V. et al., "North Atlantic storminess and Atlantic Meridional Overturning Circulation during the last Millennium: Reconciling contradictory proxy records of NAO variability," *Global and Planetary Change,* Volumes 84-85, pp. 48-55, p. 51-53, March 2012. https://doi.org/10.1016/j.gloplacha.2011.10.003.

55 Meeker, Loren D. and Mayewski, Paul A., "A 1400-year high-resolution record of atmospheric circulation over the North Atlantic and Asia," *The Holocene,* Volume 12, Issue 3, April 2002. https://doi.

org/10.1191/0959683602hl542ft; See also, Dawson, A. G. et al., "Greenland (GISP2) ice core and historical indicators of complex North Atlantic climate changes during the fourteenth century," *The Holocene,* Volume 17, Issue 4, May 2007. https://doi.org/10.1177/0959683607077010.

56 Szkornik, Katie et al., "Aeolian sand movement and relative sea-level rise in Ho Bugt, western Denmark, during the Little Ice Age," *The Holocene,* Volume 18, Issue 6, 1 September 2008. https://doi.org/10.1177/0959683608091800; See also, Clemmensen, Lars, B. et al., "Sedimentology, stratigraphy and landscape evolution of a Holocene coastal dune system, Lodbjerg, NW, Jutland, Denmark," *Sedimentology*, Volume 48, Issue 1, pp. 3-27, 18 July 2008. https://doi.org/10.1111/j.1365-3091.2001.00345.x; See also, Aagaard, Troels, et al., "Environmental controls on coastal dune formation: Skallingen Spit, Denmark," *Geomorphology*, Volume 83, Issues 1-2, pp. 29-47, January 2007. https://doi.org/10.1016/j.geomorph.2006.06.007.

57 Hansom, J.D. and Hall, A.M., "Magnitude and frequency of extra-tropical North Atlantic cyclones: A chronology from cliff-top storm deposits," *Quaternary International*, Volume 195, Issues 1-2, pp. 42-52, 15 February 2009. https://doi.org/10.1016/j.quaint.2007.11.010.

58 Dejong, Rixt et al., "Storminess variation during the last 6500 years as reconstructed from an ombrotrophic peat bog in Halland, southwest Sweden," *Journal of Quaternary Science,* Volume 21, Issue 8, pp. 905-919, 12 May 2006. https://doi.org/10.1002/jqs.1011.

59 Jelgersma, S. et al., "Holocene storm surge signatures in the coastal dunes of the western Netherlands," *Marine Geology,* Volume 125, Issues 1-2, pp. 95-110, June 1995. https://doi.org/10.1016/0025-3227(95)00061-3.

60 Gilberson, D. D., et al., "Sand-drift and Soil Formation Along an Exposed North Atlantic Coastline: 14,000 Years of Diverse Geomorphological, Climatic and Human Impacts," *Journal of Archaeological Science*, Volume 26, Issue 4, pp. 439-469, April 1999. https://doi.org/10.1006/jasc.1998.0360; See also, Wilson, Peter, "Holocene coastal dune development on the South Erradale peninsula, Wester Ross, Scottland," *Scottish Journal of Geology*, Volume 38, Issue 1, pp. 5-13, 1 May 2002. https://doi.org/10.1144/sjg38010005.

61 Sorrel, P. et al., "Evidence for millennial-scale climate events in the sedimentary infilling of a macrotidal estuarine system, the Seine estuary (NW France)," *Quaternary Science Reviews,* Volume 28, Issues 5-6, pp. 499-516, March 2009. https://doi.org/10.1016/j.quascirev.2008.11.009.

62 Clarke, Michele et al., "Late-Holocene sand invasion and North Atlantic storminess along the Aquitaine coast, southwest France," *The Holocene,* Volume 12, Issue 2, February 2002. https://doi.org/10.1191/0959683602hl539rr.

63 Borja, F. et al., "Holocene Aeolian phases and human settlements along

the Atlantic coast of southern Spain," *The Holocene*, Volume 9, Issue 3, April 1999. https://doi.org/10.1191/095968399668924476; See also, Clarke, Michele L. and Rendell, Helen, M., "Effects of storminess and sand supply and the North Atlantic Oscillation on sand invasion and coastal dune accretion in western Portugal," *The Holocene,* Volume 16, Issue 3, April 2006. https://doi.org/10.1191/0959683606hl932rp.

64 Dezileau, L. et al., "Intense storm activity during the Little Ice Age on the French Mediterranean coast," *Palaeogeography, Palaeoclimatology, Palaeoecology*, Volume 299, Issues 1-2, pp. 289-297, January 2011. https://doi.org/10.1016/j.palaeo.2010.11.009.

65 Bramante, James F. et al., "Increased typhoon activity in the Pacific deep tropics driven by Little Ice Age circulation changes," *Nature Geoscience,* Volume 13, pp. 806-811, 16 November 2020. https://doi.org/10.1038/s41561-020-00656-2.

66 Chen, Zhi et al., "Evolution of Storm Surges over the Little Ice Age Indicated by Aeolian Sand Records on the Coast of the Beibu Gulf, China," *Water,* Volume 13, Issue 14, 14 July 2021. https://doi.org/10.3390/w13141941.

67 Winkler, T.S. et al., "More Frequent Hurricane Passage Across the Bahamian Archipelago During the Little Ice Age," *Paleoceanography and Paleoclimatology,* Volume 38, Issue 11, 21 November 2023. https://doi.org/10.1029/2023PA004623.

68 Jelgersma, S. et al., "Holocene storm surge signatures in the coastal dunes of the western Netherlands," *Marine Geology,* Volume 125, Issues 1-2, pp. 95-110, June 1995. https://doi.org/10.1016/0025-3227(95)00061-3.

69 Winkler, T.S. et al., "More Frequent Hurricane Passage Across the Bahamian Archipelago During the Little Ice Age," *Paleoceanography and Paleoclimatology,* Volume 38, Issue 11, 21 November 2023. https://doi.org/10.1029/2023PA004623.

70 Bramante, James F. et al., "Increased typhoon activity in the Pacific deep tropics driven by Little Ice Age circulation changes," *Nature Geoscience,* Volume 13, pp. 806-811, 16 November 2020. https://doi.org/10.1038/s41561-020-00656-2.

71 Dezileau, L. et al., "Intense storm activity during the Little Ice Age on the French Mediterranean coast," *Palaeogeography, Palaeoclimatology, Palaeoecology*, Volume 299, Issues 1-2, pp. 289-297, January 2011. https://doi.org/10.1016/j.palaeo.2010.11.009.

72 Degroot, Dagomar, *The Frigid Golden Age: Climate Change, the Little Ice Age, and the Dutch Republic, 1560-1720,* Cambridge University Press, Cambridge, UK, 2018, p. 44-45, ISBN: 987 110 829 7639.

73 Raible, C. C. et al., "Extreme midlatitude cyclones and their implications

for precipitation and wind speed extremes in simulations of the Maunder Minimum versus present day conditions," *Climate Dynamics,* Volume 28, pp. 409-423, 27 October 2006. https://doi.org/10.1007/s00382-006-0188-7.

74 Trouet, V. et al., "North Atlantic storminess and Atlantic Meridional Overturning Circulation during the last Millennium: Reconciling contradictory proxy records of NAO variability," *Global and Planetary Change,* Volumes 84-85, pp. 48-55, p. 53, March 2012. https://doi.org/10.1016/j.gloplacha.2011.10.003.

75 Colorado State University, "Understanding Plant Water Use: Evapotranspiration (ET)." https://coagmet.colostate.edu/extended_etr_about.php?utm_source=chatgpt.com.

76 Indiana University, "Rising carbon dioxide is causing plants to have fewer pores, releasing less water to the atmosphere." *ScienceDaily,* 4 March 2011. www.sciencedaily.com/releases/2011/03/110303111624.htm.

77 Lammersma, Emmy I. et al., "Global CO_2 rise leads to reduced maximum stomatal conductance in Florida vegetation," *PNAS*, Volume 108, Issue 10, pp. 4035-4040, 17 February 2011. https://doi.org/10.1073/pnas.1100371108.

78 Colorado State University, "Understanding Plant Water Use: Evapotranspiration (ET)." https://coagmet.colostate.edu/extended_etr_about.php?utm_source=chatgpt.com.

79 Sazib, Naxmus, et al., "Leveraging NASA Soil Moisture Active Passive for Assessing Fire Susceptibility and Potential Impacts Over Australia and California," *IEEE Journal of Selected Topics in Applied Earth Observations and Remote Sensing*, Volume 15, 20 December 2021. https://dx.doi.org/10.1109/JSTARS.2021.3136756.

80 NIDIS (National Integrated Drought Information System), "Topofire," Drought.gov. www.drought.gov/data-maps-tools/topofir.e

81 Behringer, Wolfgang, *A Cultural History of Climate,* Polity Press, Cambridge, UK, 2010, p. 88, ISBN 13: 978-0-7456-4528-5.

82 Tan, L. et al., "Climate patterns in North Central China during the last 1800 years and their possible driving force," *Climate of the Past*, Volume 7, Issue 3, pp. 685-692, 4 July 2011, p. 688. https://doi.org/10.5194/cp-7-685-2011.

83 Thunberg, Greta, "Transcript: Greta Thunberg's Speech At The U.N. Climate Action Summit," *NPR*, 23, September 2019. https://www.npr.org/2019/09/23/763452863/transcript-greta-thunbergs-speech-at-the-u-n-climate-action-summit.

84 Thunberg, Greta, "Greta Thunberg Ted Talk Transcript: School Strike For Climate," *rev*,12 December 2018. https://www.rev.com/transcripts/greta-thunberg-ted-talk-transcript-school-strike-for-climate.

85 Wrightstone, Gregory, "Mass extinction lie exposed: life is thriving," *Inconvenient Blog,* 13 May 2019. https://inconvenientfacts.xyz/blog/f/mass-extinction-lie-exposed-life-is-thriving.

86 Endangered Species International, "Extinct Species Chart." https://www.endangeredspeciesinternational.org/overview5.html.

87 Saban, Kristen E. and Wiens, John J., "Unpacking the extinction crisis: rates, patterns and causes of recent extinctions in plants and animals, *Biological Sciences,* Proceedings of the Royal Society B, 15 October 2025. https://doi.org/10.1098/rspb.2025.1717.

88 Stolte, Daniel, "Extinction rates have slowed across many plant and animal groups, study shows," *The University of Arizona News,* 22 October 2025. https://news.arizona.edu/news/extinction-rates-have-slowed-across-many-plant-and-animal-groups-study-shows.

89 Richards, Kenneth, "Recent Studies Indicate Species Extinctions Decline With Warming—Mass Extinction Events Due to COOLING," *NoTrickZone,* 16 May 2019. https://notrickszone.com/2019/05/16/recent-studies-indicate-species-extinctions-decline-with-warming-mass-extinction-events-due-to-cooling/.

90 IPCC, Working Group II Contribution to the Intergovernmental Panel on Climate Change, Fourth Assessment Report, Climate Change 2007: Climate Change Impacts, Adaptation and Vulnerability, 6 April 2007, "Executive Summary," p. 213, 6 April 2007. https://www.ipcc.ch/site/assets/uploads/2018/02/ar4-wg2-chapter4-1.pdf.

91 Thomas, C. D. et al., "Extinction Risk from Climate Change," *Nature,* Volume 145, pp. 145-148, 8 January 2004. https://doi.org/10.1038/nature02121.

92 Johnston, Jason Scott, "Global Warming Advocacy Science: A Cross Examination," Institute for Law and Economic Research Paper, Number 10-08, Program on Law, Environment and Economy, University of Pennsylvania Law School, pp. 82, 22 May 2010, pp. 62-65. https://papers.ssrn.com/sol3/papers.cfm?abstract_id=1612851.

93 Lewis, Owen T., "Climate Change, Species-Area Curves and the Extinction Crisis," *Philosophical Transactions: Biological Sciences, Royal Society,* Volume 361, Number 1465, pp. 163-171, 29 January 2006, pp. 163-167. https://www.jstor.org/stable/20209616.

94 Lewis, Owen T., "Climate Change, Species-Area Curves and the Extinction Crisis," *Philosophical Transactions: Biological Sciences, Royal Society,* Volume 361, Number 1465, pp. 163-171, 29 January 2006, p. 167. https://www.jstor.org/stable/20209616.

95 Lewis, Owen T., "Climate Change, Species-Area Curves and the Extinction Crisis," *Philosophical Transactions: Biological Sciences, Royal Society,* Volume

361, Number 1465, pp. 163-171, 29 January 2006, p. 168. https://www.jstor.org/stable/20209616.

96 Lewis, Owen T., "Climate Change, Species-Area Curves and the Extinction Crisis," *Philosophical Transactions: Biological Sciences, Royal Society,* Volume 361, Number 1465, pp. 163-171, 29 January 2006, p. 168. https://www.jstor.org/stable/20209616.

97 Botkin, Daniel B., "Forecasting the Effects of Global Warming on Biodiversity," *BioScience*, Volume 57, Issue 3, pp. 227-236, 1 March 2007. https://doi.org/10.1641/B570306.

98 Botkin, Daniel B., "Forecasting the Effects of Global Warming on Biodiversity," *BioScience*, Volume 57, Issue 3, pp. 227-236, 1 March 2007, p. 231. https://doi.org/10.1641/B570306.

99 Botkin, Daniel B., "Forecasting the Effects of Global Warming on Biodiversity," *BioScience*, Volume 57, Issue 3, pp. 227-236, 1 March 2007, p. 231. https://doi.org/10.1641/B570306.

100 Dorman, Carsten F., "Promising the future? Global change projections of species distribution," *Basic and Applied Ecology,* Volume 8, pp. 387-397, 13 June 2006. https://doi.org/10.1016/j.baae.2006.11.001.

101 Araújo, Miguel and Rahbek, Carsten, "How Does Climate Change Affect Biodiversity," *Science*, Volume 313, Issue 5792, pp. 1396-1397, 8 September 2006, p. 1396. https://doi.org/10.1126/science.1131758.

102 Johnston, Jason Scott, "Global Warming Advocacy Science: A Cross Examination," Institute for Law and Economic Research Paper, Number 10-08, Program on Law, Environment, and Economy, University of Pennsylvania Law School, pp. 82, 22 May 2010, pp. 65. https://papers.ssrn.com/sol3/papers.cfm?abstract_id=1612851.

103 Steinbauer, Manuel J., "Accelerated increase in plant species richness on mountain summits is linked to warming," *Nature*, Volume 556, pp. 231-234, 4 April 2018. https://doi.org/10.1038/s41586-018-0005-6.

104 Crockford, Susan J., "Testing the hypothesis that routine sea ice coverage of 3–5 mkm2 results in a greater than 30% decline in population size of polar bears (Ursus maritimus)" *PeerJ Preprints,* 19 January 2017. https://doi.org/10.7287/peerj.preprints.2737v3.

105 DeMaster, Douglas P. and Stirling, Ian, "Ursus Maritimus," *The American Society of Mammologists, Mammalian Species,* Issue, 145, pp. 1-7, 8 May 1981, p. 4. https://doi.org/10.2307/3503828.

106 Wiig, Ø. et al., "Ursus maritimus. The IUCN Red List of Threatened Species 2015," 27 August 2015. https://www.iucnredlist.org/species/22823/14871490; See also, *Polar Bear Science,* "How are polar bears doing

15 years after the IUCN declared them vulnerable to extinction?" 10 May 2021. https://polarbearscience.com/2021/05/10/how-are-polar-bears-doing-15-years-after-the-iucn-declared-them-vulnerable-to-extinction/.

107 Lough, J. M. and Barnes, D. J., "Environmental controls on growth of the massive coral Porites," *Journal of Experimental Marine Biology and Ecology,* Volume 245, Issue 2, pp. 225-243, 15 March 2000. https://doi.org/10.1016/S0022-0981(99)00168-9.

108 "Corals Like It Hot," Great Barrier Reef Science Commentary. https://platogbr.com/corals-like-it-hot/#_edn4.

109 Australian Institute of Marine Science, "Long-Term Monitoring Program Annual Summary Report of Coral Reef Condition 2023/2024," 7 August 2024. https://www.aims.gov.au/monitoring-great-barrier-reef/gbr-condition-summary-2023-24.

110 Australian Institute of Marine Science, "Long-Term Monitoring Program Annual Summary Report of Coral Reef Condition 2024/2025," 6 August 2025. https://www.aims.gov.au/monitoring-great-barrier-reef/gbr-condition-summary-2024-25.

111 Australian Institute of Marine Science, "Long-Term Monitoring Program Annual Summary Report of Coral Reef Condition 2023/2024," 7 August 2024. https://www.aims.gov.au/monitoring-great-barrier-reef/gbr-condition-summary-2023-24.

112 Ridd, Peter, "The Great Barrier Reef Swindle: A Note from Peter Ridd," *Jennifermarohasy.com,* 19 July 2007. https://jennifermarohasy.com/2007/07/the-great-great-barrier-reef-swindle-a-note-from-peter-ridd/.

113 McGrath, Matt, "Climate change: Global sea level rise could be bigger than expected," *BBC*, 20 May 2019. https://www.bbc.com/news/science-environment-48337629.

114 NOAA, "Global Regional Trends Comparison (4 Mian Regions, various subregions)," *TIDES & CURRENTS.* https://tidesandcurrents.noaa.gov/sltrends/globalregionalcomparison.html.

115 Dangendorf, Sönke, "Reassessment of 20th century global mean sea level rise," *PNAS*, Volume 114, Number 23, pp 5946-5951, 6 June 2017. https://doi.org/10.1073/pnas.1616007114.

116 Jeon Taehan, et al., "Global sea level change signatures observed by GRACE satellite gravimetry," *Nature Scientific Reports,* Volume 8, Number 13519, 10 September 2018. https://doi.org/10.1038/s41598-018-31972-8.

117 University of Colorado, "Most Recent GMSL Release," Sea Level Research Group, 2023_rel2. https://sealevel.colorado.edu.

118 University of Colorado, "Most Recent GMSL Release," Sea Level Research

Group, 2023_rel2. https://sealevel.colorado.edu; See also, Everything Climate, "Antarctic Ice Melt," 7 March 2025. https://everythingclimate.com/topics/antarctic-ice-melt/.

119 Pultarova, Tereza, "Sea level rise slowed down in 2022. NASA says it's just a blip," *SPACE.com,* 22 March 2023. https://www.space.com/sea-level-rise-slow-down-2022.

120 Frederikse, Thomas et al., "The cause of sea-level rise since 1900," *Nature,* Volume 584, pp. 393-397, 19 August 2020. https://doi.org/10.1038/s41586-020-2591-3.

121 Frederikse, Thomas et al., "The cause of sea-level rise since 1900," *Nature,* Volume 584, pp. 393–397, 19 August 2020. https://doi.org/10.1038/s41586-020-2591-3.

122 Voortman, Hessel G. and De Vos, Rob, "A Global Perspective on Local Sea Level Changes," *Journal of Marine Science and Engineering,* Volume 13, Issue 9, 27 August 2025. https://doi.org/10.3390/jmse13091641.

123 Voortman, Hessel G. and De Vos, Rob, "A Global Perspective on Local Sea Level Changes," *Journal of Marine Science and Engineering,* Volume 13, Issue 9, 27 August 2025. https://doi.org/10.3390/jmse13091641.

124 Hoge, Bob, "Wait, What? Climate Scientist Says Sea Level Rising at Nowhere Near the Level They Claim," *RedState*, 4 September 2025. https://redstate.com/bobhoge/2025/09/04/wait-what-climate-scientist-says-sea-level-rising-at-nowhere-near-the-level-they-claim-n2193605.

125 IPCC, *Climate Change 2021: The Physical Science Basis. Contributions of Working Group I to the Sixth Assessment Report of the Intergovernmental Panel on Climate Change,* "Chapter 9 - Ocean, Cryosphere and Sea Level Change," Cambridge University Press, Cambridge, UK and New York, NY, USA, Chapter 9, Section 9.6.1.2, Table 9.5, p. 1289, 9 August 2021. https://www.ipcc.ch/report/ar6/wg1/downloads/report/IPCC_AR6_WGI_Chapter09.pdf

126 Holzhauser, Hanspeter et al., "Glacier and lake level variations in west-central Europe over the last 3500 years," *The Holocene,* Vol 15, No.6, pp. 789–801, 2005. https://doi.org/10.1191/0959683605hl853ra.

127 Grinsted, A. et al., "Reconstructing sea level from paleo and projected temperatures 200 to 2100 AD," *Climate Dynamics,* Volume 34, pp. 461–472, 6 January 2009. https://doi.org/10.1007/s00382-008-0507-2.

128 Mayewski, Paul A. et al., "Holocene climate variability," *Quaternary Research,* Volume 62, Issue 3, pp. 243-255, November 2004. https://doi.org/10.1016/j.yqres.2004.07.001; See also, Marcott, Shaun A., "A Reconstruction of Regional and Global Temperature for the Past 11,300 Years," *Science*, Volume 339, Number 6124, pp. 1198-1201, 8 May 2013. https://doi.org/10.1126/science.1228026.

129 Creel, Roger C. et al., "Global mean sea level likely higher than present during the Holocene," *Nature Communications,* Volume 15, Article Number 10731, 30 December 2024. https://doi.org/10.1038/s41467-024-54535-0; See also, Vacchi, Matteo et al., "Sea level since the Last Glacial Maximum from the Atlantic coast of Africa," *Nature Communications,* Volume 7, Article number 1486, 10 February 2025. https://doi.org/10.1038/s41467-025-56721-0.

130 Zhang, Yuxin et al., "Evolutionary dynamics of Island shoreline in the context of climate change: Insights from extensive empirical evidence," *International Journal of Digital Earth,* Volume 17, 18-March-2024. https://doi.org/10.1080/17538947.2024.2329816.

131 Zhang, Yuxin et al., "Evolutionary dynamics of Island shoreline in the context of climate change: Insights from extensive empirical evidence," *International Journal of Digital Earth,* Volume 17, 18-March-2024, p. 1. https://doi.org/10.1080/17538947.2024.2329816.

132 McVeigh, Karen, "It's absolutely guaranteed: the best and worst case scenarios for sea level rise," *The Guardian,* 26 June 2023. https://www.theguardian.com/environment/2023/jun/26/its-absolutely-guaranteed-the-best-and-worst-case-scenarios-for-sea-level-rise.

133 Morrison, Chris, "Islands That Climate Alarmists Said Would Soon 'Disappear' Due to Rising Sea Found to Have Grown in Size," *Climate-Science.press*, 8 April 2024. https://climate-science.press/2024/04/08/islands-that-climate-alarmists-said-would-soon-disappear-due-to-rising-sea-found-to-have-grown-in-size/.

134 Shihab, Hussein, "Threat to Islands," *Canberra Times,* September 26, 1988.

135 Morrison, Chris, "Islands That Climate Alarmists Said Would Soon 'Disappear' Due to Rising Sea Found to Have Grown in Size," *Climate-Science.press,* 8 April 2024. https://climate-science.press/2024/04/08/islands-that-climate-alarmists-said-would-soon-disappear-due-to-rising-sea-found-to-have-grown-in-size/.

136 Narwani, Depa, "DAMAC awards AED544 million contract for 120-villa Maldives resort," *Hotelier*, 20 March 2023. https://www.hoteliermiddleeast.com/business/damac-awards-aed544-million-contract-for-120-villa-maldives-resort#:~:text=The%20contract%2C%20valued%20at%20approximately%20AED544%20million%2C%20was,also%20come%20with%20private%20pools%20and%20ocean%20views.

137 Heller, Tony, "Children Just Won't Know What an Honest Scientist Is," *Real Climate Science*, 28 February 2018. https://realclimatescience.com/2018/02/children-just-wont-know-what-an-honest-scientist-is/#gsc.tab=0.

138 Delingpole, James, "The telling tale of Glacier National Park's 'gone by 2020' signs," *New York Post,* 10 January 2020. https://nypost.com/2020/01/10/

the-telling-tale-of-glacier-national-parks-gone-by-2020-signs/.

139 *Albuquerque Journal*, "Climate Outlook: No Maple Syrup," December 2001. https://www.newspapers.com/search/results/?keyword=Albuquerque+-Journal%2C+"Climate+Outlook%3A+No+Maple+Syrup%2C"+December+2001.

140 Graham, Gary, "Maple Syrup Production Statistics," *The Ohio State University,* October 2016. https://mapleresearch.org/pub/grahamstats2016-2/.

141 Gore, Al, "Al Gore Nobel Prize Lecture," The Nobel Prize. https://www.nobelprize.org/prizes/peace/2007/gore/lecture/.

142 McMahon, Jeff, "We Have Five Years to Save Ourselves From Climate Change, Harvard Scientist Says," *Forbes*, January 15, 2018. https://www.forbes.com/sites/jeffmcmahon/2018/01/15/carbon-pollution-has-shoved-the-climate-backward-at-least-12-million-years-harvard-scientist-says/.

143 Heller, Tony, "NASA Ice-Free Prophesy Update," *Real Climate Science*, 11 June 2018. https://realclimatescience.com/2018/06/nasa-ice-free-prophesy-update/#gsc.tab=0.

144 Markley, Edward J., "The Arctic Is Screaming: Ominous Arctic Melt Worries Experts," The Select Committee on Energy Independence and Global Warming, Representative Edward J. Markey, Chairman. https://www.markey.senate.gov/imo/media/globalwarming/resources/articles_id=0018.html.

145 Vidal, John, "Arctic expert predicts final collapse of sea ice within four years," *The Guardian*, 17 September 2012. https://www.theguardian.com/environment/2012/sep/17/arctic-collapse-sea-ice.

146 Amos, Jonathan, "Arctic summers ice-free by 2013," *BBC News*, 12 December 2007. http://news.bbc.co.uk/2/hi/7139797.stm.

147 NSIDC, "Arctic sea ice extend levels off; 2023 minimum set," National Snow and Ice Data Center, 24 September 2024.https://nsidc.org/sea-ice-today/analyses/arctic-sea-ice-extent-levels-2024-minimum-set.

148 Stern, Harry L., "Regime Shift in Arctic Ocean Sea-Ice Extent," *Physical Research Letters,* Volume 52, 8 March 2025. https://doi.org/10.1029/2024GL114546.

149 Balmakov, Roman, "Inconvenient Truth: 32 Climate Predictions Proven False—Facts Matter," *The Epoch Times,* 2024. https://www.youtube.com/watch?v=E1e5HAZo4iw&t=763s.

150 Bowden, John, "Ocasio-Cortez: 'World will end in 12 years' if climate change not addressed," *The Hill*, 22 January 2019. https://thehill.com/policy/energy-environment/426353-ocasio-cortez-the-world-will-end-in-12-years-if-we-dont-address/.

151 Balmakov, Roman, "Inconvenient Truth: 32 Climate Predictions Poven

False–Facts Matter," *The Epoch Times,* 2024. https://www.youtube.com/watch?v=E1e5HAZo4iw&t=763s.

Chapter 4 - Benefits of CO_2 and Warming

1 Hille, Karl B., "Carbon Dioxide Fertilization Greening the Earth, Study Finds," *NASA*," 26 April 2016. https://www.nasa.gov/feature/goddard/2016/carbon-dioxide-fertilization-greening-Earth.

2 Zhu, Zaichun et al., "Greening of the Earth and Its Drivers," *Nature Climate Change,* Volume 6, pp. 791-755, 25 April 2016. https://doi.org/10.1038/nclimate3004; See also, Piao, Shilong et al., "Characteristics, Drivers and Feedbacks of Global Greening," *Nature Reviews,* Volume 1, pp. 14-27, 9 December 2019. https://doi.org/10.1038/s43017-019-0001-x.

3 Lammertsma, Emmy I. et al., "Global CO_2 rise leads to reduced maximum stomatal conductance in Florida vegetation," *PNAS*, Volume 108, Issue 10, pp. 4035-4040, 17 February 2011. https://doi.org/10.1073/pnas.1100371108.

4 Chen, Xin et al., "The global greening continues despite increased drought stress since 2020," *Global Ecology and Conservation,* Volume 49, January 2024. https://doi.org/10.1016/j.gecco.2023.e02791.

5 Schmitt, Kellie, "Less Nutritious Crops: Another Result of Rising CO_2," *Hopkins Bloomberg Public Health,* 27 September 2024, https://magazine.publichealth.jhu.edu/2024/less-nutritious-crops-another-result-rising-co2.

6 Dunn, Bruce, "Greenhouse Carbon Dioxide Supplementation," OSU Extension, Id: HLS-6723, September 2023. https://extension.okstate.edu/fact-sheets/greenhouse-carbon-dioxide-supplementation.html.

7 Taylor, Charles A. and Schlenker, Wolfram, "Environmental Drivers of Agricultural Productivity Growth: CO_2 Fertilization of US Field Crops," National Bureau of Economic Research, Cambridge, MA, January 2023. https://www.nber.org/system/files/working_papers/w29320/w29320.pdf.

8 Taylor, Charles A. and Schlenker, Wolfram, "Environmental Drivers of Agricultural Productivity Growth: CO_2 Fertilization of US Field Crops," National Bureau of Economic Research, Cambridge, MA, January 2023. https://www.nber.org/system/files/working_papers/w29320/w29320.pdf.

9 Steinbauer, Manuel J., "Accelerated increase in plant species richness on mountain summits is linked to warming," *Nature*, Volume 556, pp. 231-234, 4 April 2018. https://doi.org/10.1038/s41586-018-0005-6.

10 Zhao, Qi et al, "Global regional and national burden of mortality associated with non-optimal ambient temperatures from 2000 to 2019: a three-stage modeling study," *The Lancet,* Volume 5, Issue 7, E415-E425, July 2021. https://doi.org/10.1016/S2542-5196(21)00081-4.

11 Lomborg, Bjorn, "Climate Change Is Not an Apocalyptic Threat—Let's Address It Smartly," Hoover Institution, Stanford University, September 2024. https://www.hoover.org/sites/default/files/research/docs/Lomborg_ClimateChange_web_240903.pdf.

12 US Global Change Research Program (USGCRP), "Climate Science Special Report: Fourth National Climate Assessment, Volume 1, Chapter 6" ("Temperature Changes in the United States"), USGCRP, Washington, D.C. https://science2017.globalchange.gov/chapter/6/.

13 Tol, Richard S., "The Economic Impacts of Climate Change," *Review of Environmental Economics and Policy,* Volume 12, Number 1, Winter 2018, pp. 4-25. https://doi.org/10.1093/reep/rex027.

14 IPCC, "Climate change widespread, rapid, and intensifying," 9 August 2021. https://www.ipcc.ch/2021/08/09/ar6-wg1-20210809-pr/.

15 Tol, Richard S., "The Economic Impacts of Climate Change," *Review of Environmental Economics and Policy,* Volume 12, Number 1, Winter 2018, pp. 4-25, p. 8. https://doi.org/10.1093/reep/rex027.

16 Tol, Richard S., "The Economic Impacts of Climate Change," *Review of Environmental Economics and Policy,* Volume 12, Number 1, Winter 2018, pp. 4-25, p. 16. https://doi.org/10.1093/reep/rex027.

17 Barrage, Lint and Nordhaus, William, "Policies, projections, and the social cost of carbon: Results from the DICE-2023 model," *PNAS*, Volume 121, No. 13, 15 January 2024, p. 3. https://doi.org/10.1073/pnas.2312030121.

18 Nordhaus, William D., "Economic aspects of global warming in a post-Copenhagen environment," *PNAS*, Volume 107, No. 26, pp. 11721-11726, 14, June 2010, Figure 2. https://doi.org/10.1073/pnas.1005985107; See also, Barrage, Lint and Nordhaus, William, "Policies, projections, and the social cost of carbon: Results from the DICE-2023 model," *PNAS*, Volume 121, No. 13, 15 January 2024, Figure 2. https://doi.org/10.1073/pnas.2312030121.

19 The Conference Board, "Global Economic Outlook, 2027-2036 (% change)," Economy, Strategy & Finance Center. https://www.conference-board.org/topics/global-economic-outlook.

20 Stern, Nicholas, *The Economics of Climate Change, The Stern Review, Grantham Research Institute on Climate Change and the Environment,* The London School of Economics and Political Science, 30 October 2006. https://www.lse.ac.uk/granthaminstitute/publication/the-economics-of-climate-change-the-stern-review/.

21 Keen, Steve, "None So Blind As Those Who Will Not See," *Building a New Economics,* 25 May 2022. https://profstevekeen.substack.com/p/none-so-blind-as-those-who-will-not.

22 Keen, Steve, Collum, Dave, and Fleckenstein, Bill, "The Ultimate Climate Change Debate, Dave Collum, Steve Keen, and Bill Fleckenstein," *Zero Hedge,* Minutes 14:39 – 14:48, 16 November 2024. https:// HYPERLINK "http://www.youtube.com/watch?v=SdBXPC6EjjQ"www.youtube.com/watch?v=SdBXP-C6EjjQ.

23 Myhre, Gunnar, et. al., "New estimates of radiative forcing due to well mixed greenhouse gases," *Geophysical Research Letters,* Volume 25, Issue 14, pp. 2715-2718, 15 July 1998, p. 2718. https://doi.org/10.1029/98GL01908.

24 Cummins, Eoin P. et al., "Carbon dioxide-sensing in organisms and its implications for human disease," *Cellular and Molecular Life Sciences,* Volume 71, Issue 5, pp. 831-845, 18 Sep 2013. https://doi.org/10.1007/s00018-013-1470-6.

25 Viterito, Arthur, "Viterito/Kamis/Yim/Catt: Impacts of Geothermal Energy on Climate," *Tom Nelson Podcas*t, #181, December 23, 2023. https://www.youtube.com/watch?v=lbDbA32fNek.

26 NOAA Global Monitoring Laboratory reports ~1800 to ~1850 at the South Pole for 2023 to 2024. https://gml.noaa.gov.

27 Myhre, Gunnar et al., "New estimates of radiative forcing due to well mixed greenhouse gases," *Geophysical Research Letters*, Volume 25, Issue 14, pp. 2715-2718, 15 July1998, p. 2718. https://doi.org/10.1029/98GL01908

28 Archer, David et al., "Ocean methane hydrates as a slow tipping point in the global carbon cycle," *PNAS*, Volume 106, Issue 49, pp. 20596-20601, 8 December 2009. https://doi.org/10.1073/pnas.0800885105.

29 Archer, David et al., "Ocean methane hydrates as a slow tipping point in the global carbon cycle," *PNAS*, Volume 106, Issue 49, pp. 20596-20601, 8 December 2009. https://doi.org/10.1073/pnas.0800885105.

30 Stern, Harry L., "Regime Shift in Arctic Ocean Sea-Ice Extent," *Physical Research Letters,* Volume 52, 8 March 2025. https://doi.org/10.1029/2024GL114546.

31 Keen, Steve, "None So Blind As Those Who Will Not See," *Building a New Economics*, 25 May 2022. https://profstevekeen.substack.com/p/none-so-blind-as-those-who-will-not.

32 Boyer, Tim et al., "Changes in freshwater content in the North Atlantic Ocean 1955-2006," *Geophysical Research Letters,* Volume 34, Issue 16, 21 August 2007. https://doi.org/10.1029/2007GL030126.

33 Wunsch, Carl, "What is the Thermohaline Circulation?" *Science*, Vol 298, 8 November 2002, pp. 1179-1181. https://www.science.org/doi/10.1126/science.1079329.

34 Deidov, Dan et al., "Resilience of the Gulf Stream path on decadal and longer timescales," *Nature Scientific Reports,* Volume 9, Article Number 11549, 9 August 2019. https://doi.org/10.1038/s41598-019-48011-9.

35 NSIDC, "Sea Ice Index, Version 4," Data set ID: GO2135. https://nsidc.org/data/g02135/versions/4.

36 Stern, Harry L., "Regime Shift in Arctic Ocean Sea-Ice Extent," *Physical Research Letters,* Volume 52, 8 March 2025. https://doi.org/10.1029/2024GL114546.

37 Dilley, David, "Global Warming will be dead by 2030," *Tom Nelson Podcast,* #216, May 14, 2024. https://www.youtube.com/watch?v=NeFePI1nW1Y.

38 Hille, Karl B., "Carbon Dioxide Fertilization Greening the Earth, Study Finds," *NASA,*" 26 April 2016. https://www.nasa.gov/feature/goddard/2016/carbon-dioxide-fertilization-greening-Earth.

39 Gertler, Charles G. and O'Gorman, Paul A., "Changing available energy for extratropical cyclones and associated convection in Northern Hemisphere summer," *PNAS*, Volume 116, Issue 10, pp. 4105-4110, 19 February 2019. https://doi.org/10.1073/pnas.1812312116.

40 Dai, Aiguo and Deng, Jiechun, "Arctic Amplification Weakens the Variability of Daily Temperatures Over Northern Middle-High Latitude," *Journal of Climate*, Volume 34, 1 April 2021. https://doi.org/10.1175/JCLI-D-20-0514.1

41 Robins, Jim, "Global 'Stilling': Is Climate Change Slowing Down the Wind?" *YaleEnvironment 360,* 13 September 2022. https://e360.yale.edu/features/global-stilling-is-climate-change-slowing-the-worlds-wind.

42 Bony, S. et al., "How well do we understand and evaluate climate change feedback processes?" *Journal of Climate Change,* Volume 19, Issue 15, pp. 3445-3482, 2006. https://doi.org/10.1175/JCLI3819.1.

43 Pielke, Roger Jr., "Global Weather Disaster Losses as a Percent of Global GDP: 1990-2023," Cited by Dr. Matthew M. Wielicki @MatthewWielicki X post. https://x.com/matthewwielicki/status/1889678185136038250?s=51&t=qWx0UbKJuQfOx7mMEjSxsw.

Chapter 5 - Past Climate Change Cycles

1 Roston, Eric and Rathi, Akshat, "Climate scientists reach 'unequivocal' consensus on human-made warming in landmark report," *PHYS.ORG*, 9 August 2021. https://phys.org/news/2021-08-climate-scientists-unequivocal-consensus-human-made.html.

2 Mayewski, Paul A., et al., "Holocene climate variability," *Quaternary Research,* Volume 62, Issue 3, pp. 243-255, November 2004. https://doi.org/10.1016/j.yqres.2004.07.001; See also, Marcott, Shaun A., "A Reconstruction of Regional and Global Temperature for the Past 11,300 Years," *Science*, Volume 339, Number 6124, pp. 1198-1201, 8 May 2013. https://doi.org/10.1126/science.1228026.

3 Fagan, Brian, *The Little Ice Age, How Climate Made History, 1300-1850,* Basic Books, New York, 2000, p. 17, ISBN: -10-465-02272-3.

4 Keigwin, Lloyd D., "The Little Ice Age and Medieval Warm Period in the Sargasso Sea," *Science*, Volume 274, Issue 5292, pp. 1504-1508, 29 November 1996. https://doi.org/10.1126/science.274.5292.1504.

5 Easterbrook, Don J., "Temperature Fluctuations in Greenland and the Arctic," *Evidence Based Climate Science* (Second Edition, 2016).

6 Grootes, P. M., M. Stuvier, "Oxygen-18/16 variability in Greenland snow and ice with 10-3 to 10-5-year time resolution, *Journal of Geophysical Research,* Volume 102, Issue C12, pp. 26455-26470, 30 November 1997. https://agupubs.onlinelibrary.wiley.com/doi/pdf/10.1029/97JC00880.

7 Sicre, Marie Alexandrine et al., "Decadal variability of sea surface temperatures off North Iceland over the last 2000 years," *Earth and Planetary Science Letters*, Volume 268, Issue 1-2, pp. 137-142, April 2008. https://doi.org/10.1016/j.epsl.2008.01.011.

8 Patterson, William P., "Two millennia of North Atlantic seasonality and implications for Norse colonies," *PNAS*, Volume 107, Issue 12, pp. 5306-5310, 8 March 2010. https://doi.org/10.1073/pnas.0902522107.

9 Ljungqvist, Fredrik, "A new reconstruction of temperature variability in the extra-tropical Northern Hemisphere during the last two millennia," *Geografiska Annaler: Series A, Physical Geography*, Volume 92, Issue 3, pp. 339-351, 6 September 2010, https://doi.org/10.1111/j.1468-0459.2010.00399.x; See also, *The Hockey Schtick,* "Paper: Roman & Medieval Warming Period Temps Reached or Exceeded the 20th Century," 25 September 2010. https://hockeyschtick.blogspot.com/2010/09/paper-roman-medieval-warming-period.html.

10 Moberg, A. et al., "Highly variable Northern Hemisphere temperature reconstructions from low- and high-resolution proxy data," *Nature*, Volume 433, pp. 613-617, 10 February 2005. https://doi.org/10.1038/nature03265.

11 Moberg, A. et al., "Highly variable Northern Hemisphere temperature reconstructions from low- and high-resolution proxy data," *Nature*, Volume 433, pp. 613-617, 10 February 2005, p. 613. https://doi.org/10.1038/nature03265.

12 Moberg, A. et al., "Highly variable Northern Hemisphere temperature reconstructions from low- and high-resolution proxy data," *Nature*, Volume 433, pp. 613-617, 10 February 2005. https://doi.org/10.1038/nature03265.

13 Loehle, Craig, "A 2,000-Year Global Temperature Reconstruction Based on Non-Tree Ring Proxies," *Energy and Environment,* Volume 18, Issue 7, pp. 1049-1058. https://doi.org/10.1260/095830507782616797.

14 Margaritelli, G. et al., "Persistent warm Mediterranean surface waters during the Roman period," *Nature Scientific Reports*, 26 June 2020, p. 1. https://

doi.org/10.1038/s41598-020-67281-2.

15 Keigwin, Lloyd D., "The Little Ice Age and Medieval Warm Period in the Sargasso Sea," *Science New Series,* Vol. 274 No. 5292, pp. 1504-1508, 29 November, 1996. https://doi.org/10.1126/science.274.5292.1504.

16 He, Yuxin et al., "Solar influenced late Holocene temperature changes on the northern Tibetan Plateau," *Chinese Science Bulletin*, Volume 58, Issue 9, pp.1053-1059, 13 January 2013, p. 1. https://doi.org/10.1007/s11434-012-5619-8.

17 Worbes, Martin, "Mensuration – Tree-Ring Analysis," *Encyclopedia of Forest Sciences,* pp. 586-599, 2004. https://doi.org/10.1016/B0-12-145160-7%2F00059-4.

18 Liu,Yu et al., "Amplitudes, rates periodicities and causes of temperature variations in the past 2485 years and future trends over the central-eastern Tibetan Plateau," *Chinese Science Bulletin*, Volume 56, pp. 2986-2994, 14 July 2011. https://doi.org/10.1007/S11434-011-4713-7.

19 Liu,Yu et al., "Amplitudes, rates periodicities and causes of temperature variations in the past 2485 years and future trends over the central-eastern Tibetan Plateau," *Chinese Science Bulletin*, Volume 56, pp. 2986-2994, 14 July 2011. https://doi.org/10.1007/S11434-011-4713-7.

20 Behringer, Wolfgang, *A Cultural History of Climate,* Polity Press, Cambridge, UK, 2010, ISBN 13: 978-0-7456-4528-5.

21 Behringer, Wolfgang, *A Cultural History of Climate*, Polity Press, Cambridge, UK, 2010, p. 86, ISBN 13: 978-0-7456-4528-5.

22 Humlum, Ole et al., "Identifying natural contributions to late Holocene climate change," *Global and Planetary Change*, Volume 79, Issue 1, pp. 145-156, 9 September 2011. https://doi.org/10.1016/j.gloplacha.2011.09.005.

23 Nesje, Atle et al., "Norwegian mountain glaciers in the past, present, and future," *Global and Planetary Change*, Volume 60, Issue 102, pp. 10-27, January 2008. https://doi.org/10.1016/j.gloplacha.2006.08.004.

24 Thompson, L. G., "Tropical Climate Instability: The Last Glacial Cycle from a Qinghai-Tibetan Ice Core," *Science*, Volume 276, Number 5320, pp. 1821-1825, 20 January 1997. https://doi.org/10.1126/science.276.5320.1821; See also, Yang, Bao et al., "Temperature changes on the Tibetan Plateau during the past 600 years inferred from ice cores and tree rings," *Global and Planetary Change,* Volume 69, Issues 1-2, pp. 71-78, October 2009. https://doi.org/10.1016/j.gloplacha.2009.07.008.

25 Chepstow-Lusty, A. J. et al., "Putting the rise of the Inca Empire within a climatic and land management context," *European Geosciences Union,* Volume 5, Issue 3, pp. 375-388, 22 July 2009. https://doi.org/10.5194/cp-5-375-2009.

26 Thompson, L. G., "Annually Resolved Ice Core Records of Tropical Climate Variability over the Past ~1800 Years," *Science*, Volume 340, Number 6135, pp. 945-950, 4 April 2013. https://doi.org/10.1126/science.1234210.

27 Lorrey, Andrew et al., "The Little Ice Age climate of New Zealand reconstructed from Southern Alps cirque glaciers: a synoptic type approach," *Climate Dynamics,* Volume 42, pp. 3039-3060, 24 July 2013. https://doi.org/10.1007/s00382-013-1876-8.

28 Luckman, B. H., "The Little Ice Age in the Canadian Rockies," *Geomorphology*, Volume 32, Issues 3-4, pp. 357-384, March 2000. https://doi.org/10.1016/S0169-555X(99)00104-X.

29 Hafner, Albert, "Archaeological Discoveries on Schnidejoch and at Other Ice Sites in the European Alps," *Arctic*, Volume 65, Issue 5, May 2012. http://dx.doi.org/10.14430/arctic4193; See also, Grosjean, Martin, "Ice-borne prehistoric finds in the Swiss Alps reflect Holocene glacier fluctuations," *Journal of Quaternary Science*, Volume 22, Issue 3, pp. 203-207, 21 February 2007. https://doi.org/10.1002/jqs.1111.

30 Nicolussi, Kurt and Patzelt, Gemot, "Discovery of early Holocene wood and peat on the forefield of the Pasterze Glacier, Eastern Alps, Austria," *The Holocene*, Volume 10, Issue 2, February 2000. https://doi.org/10.1191/095968300666855842.

31 Hare, Gregory P., "The Archaeology of Yukon Ice Patches: New Artifacts, Observations, and Insights," *Arctic*, Volume 65, Supplement 1, pp. 118-135, 20 May 2011. https://www.academia.edu/24595228/The_Archaeology_of_Yukon_Ice_Patches_New_Artifacts_Observations_and_Insights.

32 Holzhauser, Hanspeter et al., "Glacier and lake-level variations in west-central Europe over the last 3500 years," *The Holocene*, Volume 15, Number 6, pp. 789-801, 30 November 2004. https://doi.org/10.1191/0959683605hl853ra.

33 Holzhauser, Hanspeter et al., "Glacier and lake-level variations in wet-central Europe over the last 3500 years," *The Holocene,* Volume 15, Number 6, pp. 789-801, 30 November 2004. https://doi.org/10.1191/0959683605hl853ra.

34 Holzhauser, Hanspeter et al., "Glacier and lake-level variations in wet-central Europe over the last 3500 years," *The Holocene,* Volume 15, Number 6, pp. 789-801, 30 November 2004. https://doi.org/10.1191/0959683605hl853ra.

35 Behringer, Wolfgang, *A Cultural History of Climate,* Polity Press, Cambridge, UK, 2010, p. 66, ISBN 13: 978-0-7456-4528-5.

36 Holzhauser, Hanspeter et al., "Glacier and lake-level variations in wet-central Europe over the last 3500 years," *The Holocene*, Volume 15, Number 6, pp. 789-801, 30 November 2004. https://doi.org/10.1191/0959683605hl853ra.

37 Grove, Jean M. and Switsur, Roy, "Glacial Geological Evidence for the Me-

dieval Warm Period," *Climatic Change,* Volume 26, pp 143-169, March 1994. https://doi.org/10.1007/BF01092411 . Also see Behringer, Wolfgang, *A Cultural History of Climate,* Polity Press, Cambridge, UK, 2010, p. 75, ISBN 13: 978-0-7456-4528-5.

38 Holzhauser, Hanspeter et al., "Glacier and lake-level variations in west-central Europe over the last 3500 years," *The Holocene,* Volume 15, Number 6, pp. 789-801, 30 November 2004. https://doi.org/10.1191/0959683605hl853ra.

39 Ladurie, Emmanuel Le Roy, "Writing the History of Climate," *The Territory of the Historian,* Chicago, pp. 287-291, 1979. Also see Behringer, A Cultural History of Climate, Polity Press, Cambridge, UK, 2010, p. 90, ISBN 13: 978-0-7456-4528-5.

40 Fagan, Brian, *The Little Ice Age, How Climate Made History, 1300-1850,* Basic Books, New York, 2000, p. 123, ISBN: 10-465-02272-3.

41 Holzhauser, Hanspeter et al., "Glacier and lake-level variations in west-central Europe over the last 3500 years," *The Holocene*, Volume 15, Number 6, pp. 789-801, 30 November 2004. https://doi.org/10.1191/0959683605hl853ra.

42 Behringer, Wolfgang, *A Cultural History of Climate*, Polity Press, Cambridge, UK, 2010, p. 75, ISBN 13: 978-0-7456-4528-5.

43 Grinsted, A. et al., "Reconstructing sea level from paleo and projected temperatures 200 to 2100 AD," *Climate Dynamics*, Volume 34, pp. 461-472, 6 January 2009. https://doi.org/10.1007/s00382-008-0507-2.

44 Grinsted, A. et al., "Reconstructing sea level from paleo and projected temperatures 200 to 2100 AD," *Climate Dynamics,* Volume 34, pp. 461-472, 6 January 2009, p. 468. https://doi.org/10.1007/s00382-008-0507-2.

45 Kemp, Andrew C., "Climate related sea-level variations over the past two millennia," *PNAS*, Volume 108, Issue 27, pp. 11017-11022, 20 June 2011. https://doi.org/10.1073/pnas.1015619108.

46 Mayewski, Paul A. et al., "Holocene climate variability," *Quaternary Research,* Volume 62, Issue 3, pp. 243-255, November 2004. https://doi.org/10.1016/j.yqres.2004.07.001; See also, Marcott, Shaun A., "A Reconstruction of Regional and Global Temperature for the Past 11,300 Years," *Science*, Volume 339, Number 6124, pp. 1198-1201, 8 May 2013. https://doi.org/10.1126/science.1228026.

47 Vacchi, Matteo et al., "Sea level since the Last Glacial Maximum from the Atlantic coast of Africa," *Nature Communications,* Volume 7, Article number 1486, 10 February 2025. https://doi.org/10.1038/s41467-025-56721-0.

48 Creel, Roger C. et al., "Global mean sea level likely higher than present during the Holocene," *Nature Communications,* Volume 15, Article Number 10731, 30 December 2024. https://doi.org/10.1038/s41467-024-54535-0.

49 Easterbrook, Don J., "Temperature Fluctuations in Greenland and the Arctic," *Evidence-Based Climate Science* (Second Edition), 2016, ISBN: 9780128045893.

50 Behringer, Wolfgang, *A Cultural History of Climate*, Polity Press, Cambridge, UK, 2010, p. 76, ISBN 13: 978-0-7456-4528-5.

51 Weber, Wilfried, *Die Entwicklung der nördlichen Weinbaugrenze in Europa,* Trier 1980. See also, Behringer, Wolfgang, *A Cultural History of Climate,* Polity Press, Cambridge, UK, 2010, p. 94, ISBN 13: 978-0-7456-4528-5.

52 Fagan, Brian, *The Little Ice Age, How Climate Made History, 1300-1850,* Basic Books, New York, 2000, p. 17, ISBN: 10-465-02272-3.

53 Brady, Howard Thomas, *Mirrors and Mazes: A guide through the climate change debate,* 2nd Edition, Creative Commons, Canberra, Australia, 2017, p. 35, ISBN 10: 1-5466-2911-4.

54 Holmsen, Andreas, *Norske Historie,* Oslo 1961. Also see Behringer, Wolfgang, *A Cultural History of Climate,* Polity Press, Cambridge, UK, 2010, p. 78, ISBN 13: 978-0-7456-4528-5.

55 Holmsen, Andreas, *Norske Historie,* Oslo 1961. Also see Behringer, Wolfgang, *A Cultural History of Climate,* Polity Press, Cambridge, UK, 2010, p. 78, ISBN 13: 978-0-7456-4528-5.

56 Lamb, H. H., "Our Changing Climate, Past and Present," *Weather,* Volume 14, Issue 10, pp. 299-318, October 1959. https://doi.org/10.1002/j.1477-8696.1959.tb00533.x.

57 Easterbrook, Don J., "Temperature Fluctuations in Greenland and the Arctic," *Evidence-Based Climate Science* (Second Edition), 2016, ISBN: 9780128045893.

58 Highweald Wine of England, "The History of English Winemaking: The origins of English Vineyards." https://www.highwealdwine.com/news/9080/.

59 Fagan, Brian, *The Little Ice Age, How Climate Made History, 1300-1850,* Basic Books, New York, 2000, p. 2, ISBN: 10-465-02272-3.

60 Patterson, William, P. et al., "Two millennia of North Atlantic seasonality and implications for Norse colonies," *PNAS*, Volume 107, Issue 12, pp. 5306-5310, 3 April 2009. https://doi.org/10.1073/pnas.0902522107.

61 Behringer, Wolfgang, *A Cultural History of Climate,* Polity Press, Cambridge, UK, 2010, p. 83, ISBN 13: 978-0-7456-4528-5.

62 Behringer, Wolfgang, *A Cultural History of Climate,* Polity Press, Cambridge, UK, 2010, p. 83, ISBN 13: 978-0-7456-4528-5.

63 Easterbrook, Don J., "Temperature Fluctuations in Greenland and the Arctic," *Evidence-Based Climate Science* (Second Edition), 2016, ISBN:

9780128045893.

64 Behringer, Wolfgang, *A Cultural History of Climate*, Polity Press, Cambridge, UK, 2010, p. 98, ISBN 13: 978-0-7456-4528-5.

65 Dansgaard, Willi et al., "Climate Changes, Norsemen and Modern Man," *Nature*, Volume 255, pp. 24-28, 1975. https://doi.org/10.1038/255024a0. See also, Behringer, Wolfgang, *A Cultural History of Climate*, Polity Press, Cambridge, UK, 2010, p. 98, ISBN 13: 978-0-7456-4528-5.

66 Zhang, De'er, "Evidence for the Existence of the Medieval Warm Period in China," *Climatic Change*, Volume 26, pp. 289-297, March 1994. https://doi.org/10.1007/BF01092419; See also, Behringer, Wolfgang, *A Cultural History of Climate*, Polity Press, Cambridge, UK, 2010, p. 78, ISBN 13: 978-0-7456-4528-5.

67 Zhang, De'er, "Evidence for the Existence of the Medieval Warm Period in China," *Climatic Change*, Volume 26, pp. 289-297, March 1994. https://doi.org/10.1007/BF01092419.

68 Aono, Yasuyuki and Kazui, Keiko, "Phenological data series of cherry tree flowering in Kyoto, Japan, and its application to reconstruction of springtime temperatures since the 9th century," *International Journal of Climatology*, Volume 28, Issue 7, pp. 905-914, 1 August 2007. https://doi.org/10.1002/joc.1594; See also, White, Sam, "The Real Little Ice Age," *Journal of Interdisciplinary History*, Volume 44, pp. 327-352, 1 August 2007. 1 November 2013, p. 338. https://doi.org/10.1162/JINH_A_00574; and Brady, Howard Thomas, *Mirrors and Mazes: A guide through the climate change debate*, 2nd Edition, Creative Commons, Canberra, Australia, 2017, p. 36, ISBN 10: 1-5466-2911-4.

69 Brooks, Stephen J., "Fossil midges (Diptera: Chironomidae) as paleoclimatic indicators for the Eurasian region," *Quaternary Science Reviews*, Volume 25, Issues 15-16, pp. 1894-1910, August 2006. https://doi.org/10.1016/j.quascirev.2005.03.021.

70 Davis, B. A. S. et al., "The temperature of Europe during the Holocene reconstructed from pollen data," *Quaternary Science Reviews*, Volume 22, Issues 15-17, pp. 1701-1716, July-August 2003. https://doi.org/10.1016/S0277-3791(03)00173-2.

71 Buckland, Paul C. and Wagner, Pat E., "Is There an Insect Signal for the 'Little Ice Age,'" *Climatic Change*, Volume 48, pp. 127-149, January 2001, p. 139. https://doi.org/10.1023/A:1005690117593.

72 Lamb, Hubert H., *Climate, History and the Modern World*, Routledge, London, 19 July 1995, p. 173, ISBN 978-0234-3365-2. See also, Behringer, Wolfgang, *A Cultural History of Climate*, Polity Press, Cambridge, UK, 2010, p. 78, ISBN 13: 978-0-7456-4528-5.

73 Heiri, Oliver et al., "A chironomid-based Holocene summer air temperature

reconstruction from the Swiss Alps," *The Holocene,* Volume 13, Issue 4, May 2003. https://doi.org/10.1191/0959683603hl640ft.

74 Buckland, Paul C. and Wagner, Pat E., "Is There an Insect Signal for the 'Little Ice Age,'" *Climatic Change,* Volume 48, pp. 127-149, January 2001, p. 140. https://doi.org/10.1023/A:1005690117593.

75 Buckland, Paul C. and Wagner, Pat E., "Is There an Insect Signal for the 'Little Ice Age,'" *Climatic Change,* Volume 48, pp. 127-149, January 2001, p. 143. https://doi.org/10.1023/A:1005690117593.

76 Behringer, Wolfgang, *A Cultural History of Climate,* Polity Press, Cambridge, UK, 2010, p. 78, ISBN 13: 978-0-7456-4528-5.

77 Brooks, Stephen J. and Birks, H.J.B., "Chironomid-inferred air temperatures from Lateglacial and Holocene sites in Northwest Europe: progress and problems," *Quaternary Science Reviews,* Volume 20, Issues 16-17, pp. 1723-1741, October-November 2001. https://doi.org/10.1016/S0277-3791(01)00038-5.

78 Buckland, Paul C. and Wagner, Pat E., "Is There an Insect Signal for the 'Little Ice Age,'" *Climatic Change,* Volume 48, pp. 127-149, January 2001, p. 139. https://doi.org/10.1023/A:1005690117593.

79 Brown, Patrick, "As a scientist, I am not allowed to tell the full truth about climate change," *New York Post,* Opinion, 5 September 2023. https://nypost.com/2023/09/05/as-a-scientist-im-not-allowed-to-tell-the-full-truth-about-climate-change/.

80 Buckland, Paul C. and Wagner, Pat E., "Is There an Insect Signal for the 'Little Ice Age,'" *Climatic Change,* Volume 48, pp. 127-149, January 2001, p. 145. https://doi.org/10.1023/A:1005690117593.

81 Buckland, Paul C. and Wagner, Pat E., "Is There an Insect Signal for the 'Little Ice Age,'" *Climatic Change,* Volume 48, pp. 127-149, January 2001, p. 139. https://doi.org/10.1023/A:1005690117593.

82 Buckland, Paul C. and Wagner, Pat E., "Is There an Insect Signal for the 'Little Ice Age,'" *Climatic Change,* Volume 48, pp. 127-149, January 2001, p. 141. https://doi.org/10.1023/A:1005690117593.

83 Heiri, Oliver et al., "A chironomid-based Holocene summer air temperature reconstruction from the Swiss Alps," *The Holocene,* Volume 13, Issue 4, May 2003. https://doi.org/10.1191/0959683603hl640ft.

84 Kullman, Leif, "Tree line (Pinus sylvestris) landscape evolution in the Swedish Scande—a 40-year demographic effort viewed in a broader temporal context," *Norsk Geografisk Tidsskrift,* Volume 68, Issue 3, 14 April 2014. https://doi.org/10.1080/00291951.2014.904402.

85 Jäger, Helmut, Einführung in die Umweltgeschichte, Darmstadt, 1994, p. 26. See also, Behringer, Wolfgang, *A Cultural History of Climate,* Polity Press,

Cambridge, UK, 2010, p. 65, ISBN 13: 978-0-7456-4528-5.

86 Lamb, Hubert H., *Climate, History, and the Modern World,* Routledge, London, 19 July 1995, p. 170, ISBN 978-0234-3365-2. See also, Behringer, Wolfgang, *A Cultural History of Climate,* Polity Press, Cambridge, UK, 2010, p. 77, ISBN 13: 978-0-7456-4528-5.

87 Hiller, Achim et al., "Medieval climate warming recorded by radiocarbon dated alpine tree line shift on the Kola Peninsula/Russia," *The Holocene*, Volume 11, Issue 4, pp. 491-497, May 2001. http://dx.doi.org/10.1191/095968301678302931.

88 Helama, Samuli et al., "Subaerially preserved remains of pine stem wood as indicators of late Holocene timberline fluctuations in Fennoscandia, with comparisons of tree-ring and 14C dated depositional histories," *Review of Paleobotany and Palynology,* Volume 278, July 2020. https://doi.org/10.1016/j.revpalbo.2020.104223.

89 Arseneult, Dominique and Payette, Serge, "Reconstruction of Millennial Forest Dynamics from Tree Remains in a Subarctic Tree Line Peatland," *Ecology, Ecological Society of America,* Volume 78, Issue 6, 1 September 1997. https://doi.org/10.1890/0012-9658(1997)078[1873:ROMFDF]2.0.CO;2.

90 Büntgen, Ulf et al., "Common Era tree line fluctuations and their implications for climate reconstructions," *Global & Planetary Change,* Volume 219, 125757, December 2022. https://doi.org/10.1016/j.gloplacha.2022.103979.

91 White, Sam, *A Cold Welcome,* Harvard Press, Cambridge, Massachusetts, 2017, p. 192. ISBN 9780674244900.

92 White, Sam, *A Cold Welcome,* Harvard Press, Cambridge, Massachusetts, 2017, p. 161. ISBN 9780674244900. See also, White, Sam, "A Cold Welcome: The Little Ice Age and America's Colonial Beginnings," The National Socio-Environmental Synthesis Center, 25 October 2016. https://www.youtube.com/watch?v=OXrZxYs9gdM&t=1601s Minute 31:45 to 32:11.

93 White, Sam, *A Cold Welcome*, Harvard Press, Cambridge, Massachusetts, 2017, p. 168. ISBN 9780674244900.

94 White, Sam, *A Cold Welcome,* Harvard Press, Cambridge, Massachusetts, 2017, p. 182. ISBN 9780674244900. See also, White, Sam, "A Cold Welcome: The Little Ice Age and America's Colonial Beginnings," The National Socio-Environmental Synthesis Center, 25 October 2016. https://www.youtube.com/watch?v=OXrZxYs9gdM&t=1601s Minute 31:45 to 32:11.

95 Patowary, Kaushik, "The Frost Fairs of River Thames," *AMUSING PLANET,* 7 April 2018. https://www.amusingplanet.com/2018/04/the-frost-fairs-of-river-thames.html.

96 Hessayon, Ariel, "The Little Ice Age and the River Thames frost fairs of the

seventeenth century," *Historical Essays,* 18 February 2022. https://arielhessayon.substack.com/p/the-little-ice-age-and-the-river.

97 *Wikipedia*, "River Thames frost fairs," https://n.wikipedia.org/wiki/River_Thames_frost_fairs.

98 De Castella, Tom, "Frost fair: When an elephant walked on the frozen River Thames," *BBC News,* 28 January 2014. https://www.bbc.com/news/magazine-25862141.

99 Johnson, Ben, "The Thames Frost Fairs," *Historic UK.* https://www.historic-uk.com/HistoryUK/HistoryofEngland/The-Thames-Frost-Fairs/; See also, Wikipedia, "River Thames frost fairs," https://n.wikipedia.org/wiki/River_Thames_frost_fairs.

100 *Wikipedia*, "River Thames frost fairs," https://n.wikipedia.org/wiki/River_Thames_frost_fairs.

101 History Posts, "When the Thames Froze Over: The Little Ice Age Thames Frost Fairs," 30 November 2017. https://justhistoryposts.com/2017/11/30/when-*the*-thames-froze-over-the-little-ice-age-thames-frost-fairs/.

102 Wikipedia, "River Thames frost fairs," https://n.wikipedia.org/wiki/River_Thames_frost_fairs.

103 Hessayon, Ariel, "The Little Ice Age and the River Thames frost fairs of the seventeenth century," *Historical Essays,* 18 February 2022. https://arielhessayon.substack.com/p/the-little-ice-age-and-the-river.

104 Eugippius, *The Life of Saint Severinus,* Cambridge, MA, 1914, p. 23. See also, Behringer, Wolfgang, *A Cultural History of Climate,* Polity Press, Cambridge, UK, 2010, p. 64, ISBN 13: 978-0-7456-4528-5.

105 White, Sam, *A Cold Welcome*, Harvard Press, Cambridge, Massachusetts, 2017, p. 132, ISBN 9780674244900.

106 White, Sam, *A Cold Welcome,* Harvard Press, Cambridge, Massachusetts, 2017, p. 133, ISBN 9780674244900.

107 Brunet, Gustave et al., *Mémoires-Journaux de Pierre de l'Estoile, Paris: Librairie des Bibliophiles, 1881*, 9:42. See also, White, Sam, *A Cold Welcome*, Harvard Press, Cambridge, Massachusetts, 2017, p. 133, ISBN 9780674244900.

108 White, Sam, *A Cold Welcome,* Harvard Press, Cambridge, Massachusetts, 2017, p. 78, ISBN 9780674244900.

109 Camuffo, Dario, "Freezing of the Venetian Lagoon Since the 9th Century A.D. in Comparison to the Climate of Western Europe and England," *Climate Change,* Volume 10, pp. 43-66, 4 August 1987, p. 45. https://doi.org/10.1007/BF00140556.

110 Rijks Museum, "Hendrick Avercamp," from the series, *Dutch Masters.*

https://www.rijksmuseum.nl/en/stories/dutch-masters/story/hendrick-avercamp-10.

111 Franke, Heinrich and Trauzettel, Rolf, *Das Chinesische Kaiser-reich,* Frankfurt/Main 1968, pp. 117-187. See also, Behringer, Wolfgang, *A Cultural History of Climate,* Polity Press, Cambridge, UK, 2010, p. 65, ISBN 13: 978-0-7456-4528-5.

112 Chang, Chia-chang, *The Reconstruction of Climate in China for Historical Times,* Science Press, Beijing, 174pp., 1988. See also, Behringer, Wolfgang, *A Cultural History of Climate,* Polity Press, Cambridge, UK, 2010, p. 90, ISBN 13: 978-0-7456-4528-5.

113 Parker, Geoffrey, *Global Crisis: War, Climate Change, and Catastrophe in the Seventeenth Century,* Yale University Press, April 30, 2013, pp. 904, p. xxv, ISBN: 978-0300-153231.

114 Catherine, "Tag: Boston Harbor Frozen Over," *Good Morning Gloucester,* 27 January 2022. https://goodmorninggloucester.com/tag/boston-harbor-frozen-over/.

115 Gaffin, Adam, "Listen, sonny, lemme tell ya: In the old days, it got so cold, people could walk on the harbor," *Universal HUB,* 21 February 2015. https://www.universalhub.com/2015/listen-sonny-lemme-tell-ya-old-days-it-got-so-cold.

116 *Travelinlightoparad,* "Why did Boston Harbor use to freeze in from at least 1614–1857, and now it doesn't?" https://www.reddit.com/r/boston/comments/12lcj17/why_did_boston_harbor_used_to_freeze_in_from_at/.

117 Berry, Melissa, "Genealogy Tip: Diaries (Part 3)," *Genealogy Bank,* 15 August 2022. https://www.genealogybank.com/blog/genealogy-tip-diaries-part-3.html.

118 Snow, Edward Rowe, "When Boston Harbor Froze Over," *Boston Herald,* 25 October 1970, p. 41. https://www.genealogybank.com/blog/genealogy-tip-diaries-part-3.html.

119 Snow, Edward Rowe, "When Boston Harbor Froze Over," *Boston Herald,* 25 October 1970, p. 41. https://www.genealogybank.com/blog/genealogy-tip-diaries-part-3.html.

120 Snow, Edward Rowe, "When Boston Harbor Froze Over," *Boston Herald,* 25 October 1970, p. 41. https://www.genealogybank.com/blog/genealogy-tip-diaries-part-3.html.

121 White, Sam, *A Cold Welcome,* Harvard Press, Cambridge, Massachusetts, 2017, p. 119, ISBN 9780674244900.

122 White, Sam, *A Cold Welcome,* Harvard Press, Cambridge, Massachusetts, 2017, p. 119, ISBN 9780674244900.

123 Parker, Geoffrey, *Global Crisis: War, Climate Change and Catastrophe in the Seventeenth Century,* Yale University Press, April 30, 2013, pp. 904, p. xxv, ISBN: 978-0300-153231.

124 DePold, Hans, "So Cold, the Rich Had to Beg," *Bolton Historical Society,* 6 December 1999. https://boltoncthistory.org/so-cold-the-rich-had-to-beg/.

125 DePold, Hans, "So Cold, the Rich Had to Beg," *Bolton Historical Society,* 6 December 1999. https://boltoncthistory.org/so-cold-the-rich-had-to-beg/.

126 DePold, Hans, "So Cold, the Rich Had to Beg," *Bolton Historical Society,* 6 December 1999. https://boltoncthistory.org/so-cold-the-rich-had-to-beg/.

127 George Washington's Mount Vernon, "Ice, A Most Luxurious Crop," https://www.mountvernon.org/the-estate-gardens/ice.

128 Vinós, Javier, *Solving the Climate Puzzle: The Sun's Surprising Role,* Critical Science Press, Madrid, 2023, p. 174, ISBN: 978-84-125867-7-0.

129 Indermühle, A. et al., "Holocene carbon-cycle dynamics based on CO_2 trapped in ice at Taylor Dome, Antarctica," *Nature*, Volume 398, pp. 121–126, 11 March 1999. https://doi.org/10.1038/18158.

130 Zhong, Y. et al., "Centennial-scale climate change from decadally-paced explosive volcanism: a coupled se ice-ocean mechanism," *Climate Dynamics,* Volume 37, pp. 2373-2387, 31 December 2010. https://doi.org/10.1007/s00382-010-0967-z.

131 Miller, Gifford H. et al., "Abrupt onset of the Little Ice Age triggered by volcanism and sustained by sea-ice/ocean feedbacks," *Geophysical Research Letters,* Volume 39, Issue 2, 31 January 2012. https://doi.org/10.1029/2011GL050168.

132 Robock, Alan, "Volcanic eruptions and climate change," *Reviews of Geophysics*, Volume 38, Issue 2, pp. 191-219, 1 May 2000. https://doi.org/10.1029/1998RG000054.

133 Hegerl, Gabriele C. et al., "Detection of volcanic, solar and greenhouse gas signals in paleo-reconstructions of Northern Hemispheric temperature," *Geophysical Research Letters,* Volume 30, Issue 5, 12 March 2003, Figure 3. https://doi.org/10.1029/2002GL016635; See also, Vinós, Javier, *Solving the Climate Puzzle, The Sun's Surprising Role*, Critical Science Press, Madrid, 2023, p. 164, ISBN: 978-84-125867-7-0.

134 Crowley, T. J. et al., "Technical details concerning development of a 1200 yr proxy index for global volcanism," *Earth System Science Data,* Volume 5, pp. 187-197, 23 May 2013. Figure B., p. 194. https://doi.org/10.5194/essd-5-187-2013; See also, Vinós, Javier, *Solving the Climate Puzzle, The Sun's Surprising Role,* Critical Science Press, Madrid, 2023, p. 177, ISBN: 978-84-125867-7-0.

135 White, Sam, *A Cold Welcome,* Harvard Press, Cambridge, Massachusetts,

2017, p. 161. ISBN 9780674244900.

136 Crowley, T. J. et al., "Technical details concerning development of a 1200 yr proxy index for global volcanism," *Earth System Science Data*, Volume 5, pp. 187-197, 23 May 2013. Figure B., p. 194. https://doi.org/10.5194/essd-5-187-2013; See also, Vinós, Javier, *Solving the Climate Puzzle, The Sun's Surprising Role, Critical Science Press*, Madrid, 2023, p. 177, ISBN: 978-84-125867-7-0.

137 Crowley, T. J., et al., "Technical details concerning development of a 1200 yr proxy index for global volcanism," *Earth System Science Data,* Volume 5, pp. 187-197, 23 May 2013. Figure B., p. 194. https://doi.org/10.5194/essd-5-187-2013; See also, Vinós, Javier, *Solving the Climate Puzzle, The Sun's Surprising Role, Critical Science Press,* Madrid, 2023, p. 177, ISBN: 978-84-125867-7-0.

138 Crok, Marcel and May, Andy, *The Frozen Climate Views of the IPCC, an analysis of AR6,* CLINTEL Foundation, Amsterdam, The Netherlands, 2023. p. 83, ISBN: 979-8-89074-861-4.

139 NASA, "Chilly Temperatures During the Maunder Minimum, Earth Observatory." https://www.earthobservatory.nasa.gov/images/7122/chilly-temperatures-during-the-maunder-minimum.

140 Crok, Marcel and May, Andy, *The Frozen Climate Views of the IPCC, an analysis of AR6,* CLINTEL Foundation, Amsterdam, The Netherlands, 2023, p. 81, ISBN: 979-8-89074-861-4. Citing: Kerr, R., "A variable Sun paces millennial climate," *Science*, Volume 294, Issue 5546, pp. 1431-1443, 16 November 2001. https://doi.org/10.1126/science.294.5546.1431b; Neff, U. et al., "Strong coherence between solar variability and monsoon in Oman between 9 and 6 kyr ago," *Nature*, Volume 411, pp. 290-293, 1 May 2001. https://doi.org/10.1038/35077048; Ogurtsov, M.G., "Long-period cycles of the Sun's activity recorded in direct solar data and proxies," *Solar Physics*, Volume 211, pp. 371-392, December 2002. https://doi.org/10.1023/A:1022411209257.

141 Bond, Gerald et al., "Persistent solar influence on North Atlantic climate during the Holocene," *Science*, 294 (5549), pp. 2130-2136, 15 November 2001. https://doi.org/10.1126/science.1065680 Also see Crok, Marcel and May, Andy, *The Frozen Climate Views of the IPCC, an analysis of AR6,* CLINTEL Foundation, Amsterdam, The Netherlands, 2023, p. 81, ISBN: 979-8-89074-861-4.

142 Crok, Marcel and May, Andy, *The Frozen Climate Views of the IPCC, an analysis of AR6,* CLINTEL Foundation, Amsterdam, The Netherlands, 2023, p. 81, ISBN: 979-8-89074-861-4. Citing: Willard, D.A. et al., "Impact of millennial-scale Holocene climate variability on eastern North American terrestrial ecosystems: Pollen-based climate reconstruction," *Global and Planetary Change,* Volume 47, Issue 1, pp. 17-35, May 2005. https://doi.org/10.1016/j.gloplacha.2004.11.017; Springer, G.S. et al., "Solar forcing of Holocene droughts in a stalagmite record from West Virginia in east-central North America," *Geophysical Research Letters,* Volume 35, pp 1-5, 9 June 2008.

https://doi.org/10.1029/2008GL034971; Bernal, J.P. et al., "High-resolution Holocene South American monsoon history recorded by a speleothem from Bouverá Cave, Brazil," *Earth and Planetary Science Letters,* Volume 450, pp. 186-196, 15 September 2016. https://doi.org/10.1016/j.epsl.2016.06.008; Killian, R. et al., "A review of Glacial and Holocene paleoclimate records from southernmost Patagonia (49-55°S)," *Quaternary Science Reviews,* Volume 53, pp. 1-23, 15 October 2012. https://doi.org/10.1016/j.quascirev.2012.07.017; Bush, M.B. et al., "A 17,000-year history of Andean climate and vegetation change from Laguna de Chochos, Peru," *Quaternary Science,* Volume 20, Issue 7-8, pp. 703-714, 15 December 2005. https://doi.org/10.1002/jqs.983; Crosta, X, "Holocene long- and short-term climate changes off Adélie Land, East Antarctica," *Geochemistry, Geophysics, Geosystems,* Volume 8, Issue 11, pp. 1-15, 30 November 2007. https://doi.org/10.1029/2007GC001718; Homgren, Karin, et al., "Persistent millennial-scale climatic variability over the past 25,000 years in Southern Africa," *Quaternary Science Reviews,* Volume 22, pp. 2311-2326, November/December 2003 https://doi.org/10.1016/S0277-3791(03)00204-X; Xielhofer, C. et al., "Western Mediterranean hydro-climatic consequences of Holocene ice-rafted debris (Bond) events," Climate of the Past, Volume 15, pp. 463-475, 20 March 2019. https://doi.org/10.5194/cp-15-463-2019; Fleitmann, Dominik et al., "Holocene forcing of the Indian monsoon recorded in a stalagmite from southern Omen," *Science,* Volume 300, Issue 5626, pp. 1737-1739, 13 Jun 2003. https://doi.org/10.1126/science.1083130; Thamban, M. et al, *Journal of Oceanography,* Volume 63, pp. 1009-1020, December 2007. https://doi.org/10.1007/s10872-007-0084-8; Wang, Y. et al., "The Holocene Asian Monsoon: Links to Solar Changes and North Atlantic Climate," *Science,* Volume 308, Volume 5723, pp. 854-857, 6 May 2005. https://doi.org/10.1126/science.1106296; McGowan, H.A. et al., "Evidence of solar and tropical-ocean forcing of hydroclimate cycles in southeastern Australia for the past 6500 years," *Geophysical Research Letters,* Volume 37, Issue 10, 27 May 2010. https://doi.org/10.1029/2010GL042918; Fletcher, W.J. et al., "Mid-Holocene emergence of a low-frequency millennial oscillation in western Mediterranean climate: Implications for past dynamics of the North Atlantic atmospheric westerlies," *The Holocene,* Volume 23, Issue 2, pp. 153-166, February 2013. https://doi.org/10.1177/0959683612460783; Mangini, A. et al., "Persistent influence of the North Atlantic hydrography on central European winter temperature during the last 9000 years," *Geophysical Research Letters,* Volume 34, Issue 2, 18 January 2007. https://doi.org/10.1029/2006GL028600; and Ojala, A.E.K. et al., "Effects of solar forcing and North Atlantic oscillation on the climate of continental Scandinavia during the Holocene," *Quaternary Science Reviews,* Volume 112, pp. 153-171, 15 March 2015. https://doi.org/10.1016/j.quascirev.2015.01.021.

143 Holzhauser, Hanspeter et al., "Glacier and lake-level variations in west-central Europe over the last 3500 years," *The Holocene,* Volume 15, Num-

ber 6, pp. 789-801, 30 November 2004. https://doi.org/10.1191/0959683605hl-853ra.

144 Svensmark, Henrik et al., "Atmospheric Ionization and Cloud Radiative Forcing," *Nature Scientific Reports,* Volume 11, Article Number 19668, 11 October 2021. https://doi.org/10.1038/s41598-021-99033-1.

145 Burton, Robert, "The Anatomy of Melancholy," Oxford, 1621. Also see Behringer, Wolfgang, *A Cultural History of Climate,* Polity Press, Cambridge, UK, 2010, p. 115, ISBN 13: 978-0-7456-4528-5.

146 Fagan, Brian, *The Little Ice Age: How Climate Made History, 1300-1850,* Basic Books, New York, 2000, p. 90, ISBN: 10-465-02272-3.

147 Deming, David, "We have to get rid of the Medieval Warm Period!" *ScootleRoyale,* 29 October 2010. www.youtube.com/watch?v=u1rj00BoItw; See also, Heller, Tony, "Erasing the Medieval Warm Period," *Real Climate Science,* 34 December 2018. https://realclimatescience.com/2018/12/erasing-the-medieval-warm-period/.

148 Bradley, Robert Jr., "Climategate Turns 15!," *Watts Up With That?* 27 November 2024. https://wattsupwiththat.com/2024/11/27/climategate-turns-15/.

149 IPCC, *Climate Change: 2001: The Scientific Basis, Contribution of Working Group I to the Third Assessment Report of the Intergovernmental Panel on Climate Change,* "Summary for Policymakers," Cambridge University Press, Cambridge, UK, p. 3, Figure 1(b), January 2001. https://www.ipcc.ch/site/assets/uploads/2018/02/WG1_TAR-FRONT.pdf.

150 Mann, Michael E. et al., "Northern hemisphere temperatures during the past millennium: Inferences, uncertainties, and limitations," *Geophysical Research Letters,* Volume 26, Issue 6, pp. 759-762, 15 Marth 1999. https://doi.org/10.1029/1999GL900070.

151 McIntyre, Stephen and McKitrick, Ross, "Hockey sticks, principal components and spurious significance," *Geophysical Research Letters,* Volume 32, Issue 3, 12 February 2005. https://doi.org/10.1029/2004GL021750.

152 Crok, Marcel and May, Andy, *The Frozen Climate Views of the IPCC, an analysis of AR6,* CLINTEL Foundation, Amsterdam, The Netherlands, 2023, p. 5, ISBN: 979-8-89074-861-4. Cited: Soon, Willie and Baliunas, Sallie, "Proxy Climatic and Environmental Changes of the Past 1000 Years," *Climate Research*, Volume 23, Number 2, pp. 89-110, 2003. https://doi.org/10.3354/cr023089 (the doi is disabled); See also, https://www.researchgate.net/publication/253767761_Climatic_and_Environmental_Changes_of_the_Past_1000_Years); McIntyre, S. and McKitrick, R., "Corrections to the Mann et al. (1998) proxy data base and Northern Hemispheric average temperature series," *Energy and Environment,* Volume 14, Number 6, pp. 751-771, November 2003. https://doi.org/10.1260/095830503322793632; McIntyre, Steven and McK-

intrick, Ross, "The M&M Critique of MGH98 Norther Hemisphere Climate Index: Update and Implications," *Energy and Environment*, Volume 16, pp. 69-100, January 2005. https://doi.org/10.1260/0958305053516226; McIntyre, Steven and McKitrick, Ross, "Hockey sticks, Principal Components and Spurious Significance," *Geophysical Research Letters,* Volume 32, Number 3, LO3710, 12 February 2005. https://doi.org/10.1029/2004GL021750; McShane, B. B. and Wyner, A. J., "Rejoinder," *The Annals of Applied Statistics,* Volume 5, Number 1, pp. 99-123, March 2011. https://www.jstor.org/stable/23024822; McShane, B. B. and Wyner, A. J., "A statistical analysis of multiple temperature proxies: Are reconstructions of surface temperature over the last 1000 years reliable?" *The Annals of Applied Statistics*, Volume 5, Number 1, pp. 5-44, 20 April 2011. https://doi.org/10.48550/arXiv.1104.4002; Montford, A.W. *The Hockey Stick Illusion*, Stacey International, London, 482 pp., 2010. ISBN-13: 978-1-906768-35-5.

153 Spencer, Roy, *The Great Global Warming Blunder*, Encounter Books, New York, 2010, p. 10, ISBN-13: 978-1-59403-373-5.

154 Bell, Larry, "Michael Mann And The ClimateGate Whitewash: Part One," *Forbes*, 28 July, 2011. https://www.forbes.com/sites/larrybell/2011/06/28/michael-mann-and-the-climategate-whitewash-part-one/.

155 Jones, P. D., and Mann, M.E., "Climate over past millennia," *Reviews of Geophysics*, Volume 42, Issue 2, 6 May 2004, Section 7. Conclusions and Summary. https://doi.org/10.1029/2003RG000143.

156 Grinsted, A. et al., "Reconstructing sea level from paleo and projected temperatures 200 to 2100 AD," *Climate Dynamics,* Volume 34, pp. 461-472, 6 January 2009. https://doi.org/10.1007/s00382-008-0507-2.

157 Grinsted, A. et al., "Reconstructing sea level from paleo and projected temperatures 200 to 2100 AD," *Climate Dynamics,* Volume 34, pp. 461-472, 6 January 2009, p. 470. https://doi.org/10.1007/s00382-008-0507-2.

158 Brady, Howard Thomas, *Mirrors and Mazes: A guide through the climate change debate,* 2nd Edition, Creative Commons, Canberra, Australia, 2017, p. 41, ISBN-10: 1-5466-2911-4.

159 *The Higher Education*, "Heated discussions," 25 March 2010. https://www.timeshighereducation.com/features/heated-discussions/410938.article.

160 Durkin, Martin, *Climate the Movie*, produced by Tom Nelson, 2024, 53:00 to 54:00 and 1:00:00 to 1:00:01, https://www.youtube.com/watch?v=A24fWmNA6lM.

161 IPCC, *Climate Change 2021: The Physical Science Basis. Contribution of Working Group I to the Sixth Assessment Report of the Intergovernmental Panel on Climate Change*, "Summary for Policymakers," Cambridge University Press, Cambridge, United Kingdom and New York, NY, USA, Section A.1.8, Figure

SPM.1, p. 6, 9 August 2021.. https://www.ipcc.ch/report/ar6/wg1/ downloads/report/IPCC_AR6_WGI_SPM.pdf.

162 White, Sam, "The Real Little Ice Age," *Journal of Interdisciplinary History,* Volume 44, pp. 327-352, 1 November 2013, p. 328. https://doi.org/10.1162/JINH_A_00574.

163 Crok, Marcel and May, Andy, *The Frozen Climate Views of the IPCC, an analysis of AR6,* CLINTEL Foundation, Amsterdam, The Netherlands, 2023, p. 33, ISBN: 979-8-89074-861-4.

164 Crok, Marcel and May, Andy, *The Frozen Climate Views of the IPCC, an analysis of AR6,* CLINTEL Foundation, Amsterdam, The Netherlands, 2023, p. 35, ISBN: 979-8-89074-861-4.

165 Büntgen, Ulf. et al., "Prominent role of volcanism in Common Era climate variability and human history," *Dendrochronologia*, v. 64, p. 125757, 2020. https://doi.org/10.1016/j.dendro.2020.125757.

166 Crok, Marcel and May, Andy, *The Frozen Climate Views of the IPCC, an analysis of AR6,* CLINTEL Foundation, Amsterdam, The Netherlands, 2023, p. 37, ISBN: 979-8-89074-861-4.

167 Crok, Marcel and May, Andy, *The Frozen Climate Views of the IPCC, an analysis of AR6,* CLINTEL Foundation, Amsterdam, The Netherlands, 2023, p. 37, ISBN: 979-8-89074-861-4.

168 Crok, Marcel and May, Andy, *The Frozen Climate Views of the IPCC, an analysis of AR6,* CLINTEL Foundation, Amsterdam, The Netherlands, 2023, p. 37, ISBN: 979-8-89074-861-4.

169 Crok, Marcel and May, Andy, *The Frozen Climate Views of the IPCC, an analysis of AR6,* CLINTEL Foundation, Amsterdam, The Netherlands, 2023, p. 38, ISBN: 979-8-89074-861-4.

170 Crok, Marcel and May, Andy, *The Frozen Climate Views of the IPCC, an analysis of AR6,* CLINTEL Foundation, Amsterdam, The Netherlands, 2023, p. 38, ISBN: 979-8-89074-861-4.

171 White, Sam, "The Real Little Ice Age," *Journal of Interdisciplinary History,* Volume 44, pp. 327-352, 1 November 2013, p. 353. https://doi.org/10.1162/JINH_A_00574.

172 Marcott, Shaun A. et al., "A Reconstruction of the Regional and Global Temperature for the Past 11,300 years," *Science*, Volume 339, Issue 6124, 8 March 2013, pp. 1198-1201. https://doi.org/10.1126/science.1228026; Rosenthal, Yair et al., "Pacific Ocean Heat Content During the Past 10,000 Years" *Science*, Volume 342, Number 617, 1 November 2013. https://doi.org/10.1126/science.1240837 and; Humlum et al., "Identifying natural contributions to late Holocene climate change." *Global and Planetary Change*, Volume 79,

Issue 1, pp. 145-156, 09 September 2011. https://doi.org/10.1016/j.gloplacha.2011.09.005; Kaufman, D. McKay, N. Routson, C. et al., "A global database of Holocene paleotemperature records," *Science Data,* Volume 7, Issue 115, 12 August 2020. https://doi.org/10.1038/s41597-020-0445-3.

173 Vinós, Javier, "Periodicities in solar variability and climate change: A simple model," *Energy Matters, Energy, Environment and Policy,* posted by Evan Mearns, 11 May 2016. https://euanmearns.com/periodicities-in-solar-variability-and-climate-change-a-simple-model/.

174 Marcott, Shaun A. et al., "A Reconstruction of the Regional and Global Temperature for the Past 11,300 years," *Science,* Volume 339, Issue 6124, 8 March 2013, pp. 1198-1201. https://doi.org/10.1126/science.1228026; Rosenthal, Yair et al., "Pacific Ocean Heat Content During the Past 10,000 Years" *Science,* Volume 342, Number 617, 1 November 2013. https://doi.org/10.1126/science.1240837; Humlum et al., "Identifying natural contributions to late Holocene climate change." *Global and Planetary Change,* Volume 79, Issue 1, pp. 145-156, 09 September 2011. https://doi.org/10.1016/j.gloplacha.2011.09.005.

175 Vinós, Javier, *Solving the Climate Puzzle, The Sun's Surprising Role,* Critical Science Press, Madrid, 2023, p. 151, ISBN: 978-84-125867-7-0.

176 Walcott-George, Caleb K. et al., "Deglaciation of the Prudhoe Dome in northwestern Greenland in response to Holocene warming," *Nature Geoscience,* 5 January 2026. https://doi.org/10.1038/s41561-025-01889-9.

177 Viterito, Arthur, "Viterito/Kamis/Yim/Catt: Impacts of Geothermal Energy on Climate," *Tom Nelson Podcast,* #181, December 23, 2023. https://www.youtube.com/watch?v=lbDbA32fNek.

178 White, Sam, "The Real Little Ice Age," *Journal of Interdisciplinary History,* Volume 44, pp. 327-352, 1 November 2013, p. 336. https://doi.org/10.1162/JINH_A_00574.

179 Lutz, Ashley, "Map of Day: Pretty Much Everyone Lives in the Northern Hemisphere," *Business Insider,* 4 May 2012. https://www.businessinsider.com/90-of-people-live-in-the-northern-hemisphere-2012-5.

180 Yang, Bao et al., "General characteristics of temperature variation in China during the last two millennia," *Geophysical Research Letters,* Volume 29, Issue 9, pp. 38-1 – 28-4, 11 May 2002. https://doi.org/10.1029/2001GL014485.

181 Tan, Ming et al., "Cyclic rapid warming on a centennial-scale revealed by a 2,650-year stalagmite record of warm season temperature," *Geophysical Research Letters,* Volume 30, Issue 12, 19 June 2003. https://doi.org/10.1029/2003GL017352.

182 Ge, Q.-S. et al., "Temperature variation through 2000 years in China: An uncertainty analysis of reconstruction and regional difference," *Geophysical Research Letters,* Volume 37, Issue 3, 9 February 2010. https://doi.

org/10.1029/2009GL041281.

183 Zhang, Pingzhong et al., "A Test of Climate, Sun, and Culture Relationships from an 1,810-Year Chinese Cave Record," *Science*, Volume 322, Issue 5903, pp. 940-942, 7 November 2008. https://doi.org/10.1126/science.1163965.

184 Park, J., "A modern pollen-temperature calibration data set from Korea and quantitative temperature reconstructions for the Holocene," *The Holocene,* Volume 21, Issue 7, pp. 1125-1135, 16 May 2011. https://doi.org/10.1177/0959683611400462.

185 Vahrenholt, Fritz and Lüning, Sebastion, *The Neglected Sun: Why the Sun Precludes Climate Catastrophe,* The Heartland Institute, Arlington Heights, IL, 2015, p. 145, ISBN-13: 978-1-934791-54-7.

186 Vahrenholt, Fritz and Lüning, Sebastion, *The Neglected Sun: Why the Sun Precludes Climate Catastrophe,* The Heartland Institute, Arlington Heights, IL, 2015, p. 145, ISBN-13: 978-1-934791-54-7.

187 Crok, Marcel and May, Andy, *The Frozen Climate Views of the IPCC, an analysis of AR6*, CLINTEL Foundation, Amsterdam, The Netherlands, 2023, pp. 25-26, ISBN: 979-8-89074-861-4.

188 Rosenthal, Yair et al., "Pacific Ocean Heat Content During the Past 10,000 Years," *Science*, Volume 342, Number 617, 1 November 2013. https://doi.org/10.1126/science.1240837.

189 Esterbrook, Don J., "Geologic Evidence of Recurring Climate Cycles and Their Implications for Cause of Global Climate Change—the Past is the Key to the Future," *Evidence-Based Climate Science*, 2011, pp. 3-51. https://est.ufba.br/sites/est.ufba.br/files/kim/medievalwarmperiod.pdf.

190 Watts, Anthony, "More evidence that the Medieval Warm Period was global," *Watts Up With That,* 12 February 2018. See https://wattsupwiththat.com/2018/02/12/more-evidence-that-the-medieval-warming-period-was-global-not-regional/.

Chapter 6 - Civilizations and Climate Change

1 Zhang, Yuxin et al., "Evolutionary dynamics of Island shoreline in the context of climate change: Insights from extensive empirical evidence," *International Journal of Digital Earth*, Volume 17, 18 March 2024, p. 1. https://doi.org/10.1080/17538947.2024.2329816.

2 Morrison, Chris, "Islands That Climate Alarmists Said Would Soon 'Disappear' Due to Rising Sea Found to Have Grown in Size," *Climate-Science.press*, 8 April 2024. https://climate-science.press/2024/04/08/islands-that-climate-alarmists-said-would-soon-disappear-due-to-rising-sea-found-to-have-grown-in-size/; See also, Senjupta, Meghna, "Drivers of shoreline change on Pacific coral reef islands: linking island change to processes," *Regional Environ-*

mental Change, Volume 23, Article Number 110, 19 August 2023. https://doi.org/10.1007/s10113-023-02103-5.

3 Behringer, Wolfgang, *A Cultural History of Climate,* Polity Press, Cambridge, UK, 2010, ISBN 13: 978-0-7456-4528-5.

4 Bevan, Andrew et al., "Holocene fluctuations in human population demonstrate repeated links to food production and climate," *PNAS*, 20 November 2017. https://doi.org/10.1073/pnas.1709190114.

5 Kennett, Douglas and Marwan, Norbert, "Climatic volatility, agricultural uncertainty, and the formation, and consolidation and breakdown, of preindustrial agrarian states," *Philosophical Transactions of the Royal Society A*, 28 November 2015. https://doi.org/10.1098/rsta.2014.0458.

6 Steinhilber, F. et al., "9,400 years of cosmic radiation and solar activity from ice cores and tree rings," *Proceedings of the National Academy of Sciences,* Volume 109, Issue 16, pp. 5967-5971, 2 April 2012. https://doi.org/10.1073/pnas.1118965109; See also, Fleitmann, Dominick et al., "Holocene Forcing of the Indian Monsoon Recorded in a Stalagmite from Southern Oman," *Science*, Volume 300, Issue 5626, pp. 1737-1739, 13 June 2003. https://doi.org/10.1126/science.1083130.

7 Berger, A. and Loutre, M.F., "Isolation values for the climate of the last 10 million years," *Quaternary Science Reviews,* Volume 10, Issue 4, pp. 297-317, 1991. https://doi.org/10.1016/0277-3791(91)90033-Q.

8 Mayewski, Paul A. et al., "Holocene climate variability," *Quaternary Research*, Volume 62, Issue 3, pp. 243-255, November 2004. https://doi.org/10.1016/j.yqres.2004.07.001; See also, Marcott, Shaun A., "A Reconstruction of Regional and Global Temperature for the Past 11,300 Years," *Science*, Volume 339, Number 6124, pp. 1198-1201, 8 May 2013. https://doi.org/10.1126/science.1228026.

9 Kutzback, J. E. and Liu, Z., "Response of the African Monsoon to Orbital Forcing and Ocean Feedbacks in the Middle Holocene," *Science*, Volume 278, Issue 5337, pp. 440-443, 17 October 1997. https://doi.org/10.1126/science.278.5337.440; See also, Fleitmann, Dominick et al., "Holocene Forcing of the Indian Monsoon Recorded in a Stalagmite from Southern Oman," *Science*, Volume 300, Issue 5626, pp. 1737-1739, 13 June 2003. https://doi.org/10.1126/science.1083130; and Wang, Yongjin et al., "The Holocene Asian Monsoon: Links to Solar Changes and North Atlantic Climate," *Science*, Volume 308, Number 5723, pp. 854-857, 6 May 2005. https://doi.org/10.1126/science.1106296.

10 Kaufman et al., "Holocene thermal maximum in the Western Arctic (0-180°W)," *Quaternary Science Reviews,* Volume 23, Issues 5-6, pp. 529-560, March 2004. https://doi.org/10.1016/j.quascirev.2003.09.007.

11 Yu, Fengling, "Reconstruction of the East Asian monsoon variability since

the mid-Holocene from the Pearl River estuary," Durham E-Thesis, Durham University, Department of Geology, Durham, UK, June 2009, p. 147. https://etheses.dur.ac.uk/69/1/Thesis_FYU.pdf?DDD14+.

12 Kutzback, J. E. and Liu, Z., "Response of the African Monsoon to Orbital Forcing and Ocean Feedbacks in the Middle Holocene," *Science*, Volume 278, Issue 5337, pp. 440-443, 17 October 1997.

13 Lawder, David, "Yellen says $3 trillion needed annually for climate financing, far more than current level," *Reuters News Agency*, 27 July 2024. https://www.reuters.com/sustainability/sustainable-finance-reporting/yellen-says-3-trillion-needed-annually-climate-financing-far-more-than-current-2024-07-27/.

14 Lomborg, Bjorn, "Welfare in the 21st century: Increasing development, reducing inequality, the impact of climate change, and the cost of climate policies," *Technological Forecasting and Social Change*, Volume 156, July 2020. https://doi.org/10.1016/j.techfore.2020.119981.

15 Bar-Matthews, Miryam and Ayalon, Avner, "Mid-Holocene climate variations revealed by high-resolution speleothem records from Soreq Cave, Israel and their correlation with cultural changes," *The Holocene,* Volume 21, Issue 1, pp. 163-171, January 2011. http://dx.doi.org/10.1177/0959683610384165.

16 Langgut, Dafna et al., "Vegetation and Climate Changes During the Bronze and Iron Ages (~3600-600 BCE) in the Southern Levant Based on Palynological Records," Cambridge University Press, 9 February 2016. https://doi.org/10.2458/azu_rc.57.18555.

17 Weiss, H. et al., "The Genesis and Collapse of Third Millennium North Mesopotamian Civilization," *Science*, Volume 261, Issue 5124, pp. 995-1004, 20 August 1993. https://doi.org/10.1126/science.261.5124.995.

18 Bar-Matthews, Miryam and Ayalon, Avner, "Mid-Holocene climate variations revealed by high-resolution speleothem records from Soreq Cave, Israel and their correlation with cultural changes," *The Holocene,* Volume 21, Issue 1, pp. 163-171, January 2011. http://dx.doi.org/10.1177/0959683610384165.

19 Steinhilber, F. et al., "9,400 years of cosmic radiation and solar activity from ice cores and tree rings," *Proceedings of the National Academy of Sciences,* Volume 109, Issue 16, pp. 5967-5971, 2 April 2012. https://doi.org/10.1073/pnas.1118965109.

20 An, Zhisheng et al., "Asynchronous Holocene optimum of the East Asian monsoon," *Quaternary Science Reviews*, Volume 19, Issue 8, pp. 743-762, April 2000. https://doi.org/10.1016/S0277-3791(99)00031-1.

21 Kaniewski David et al., "Environmental Roots of the Late Bronze Age Crisis," *PLOS ONE*, August 14, 2013. https://doi.org/10.1371/journal.pone.0071004.

22 Klengel, Horst, "Hengerjahre in Hatti," *Altorientalische Forschungen Volume 1*, pp. 165-174, 1974. See also, Behringer, Wolfgang, *A Cultural History of Climate*, Polity Press, Cambridge, UK, 2010, p. 56, ISBN 13: 978-0-7456-4528-5.

23 Kaniewski David, et al., "Environmental Roots of the Late Bronze Age Crisis," *PLOS ONE*, August 14, 2013. https://doi.org/10.1371/journal.pone.0071004.

24 Cline, Eric H., *1177 B.C.: The Year Civilization Collapsed,* Princeton University Press, Princeton, 2014, p. 11, ISBN: 978-0-691-14089-6.

25 Behringer, Wolfgang, *A Cultural History of Climate*, Polity Press, Cambridge, UK, 2010, p. 58, ISBN 13: 978-0-7456-4528-5.

26 Drake, Brandon L., "The Influence of Climatic Change on the Late Bronze Age Collapse and the Greek Dark Ages," *Journal of Archaeological Science,* Volume 39, Issue 6, June 2012, pp. 1862 to 1870. https://doi.org/10.1016/j.jas.2012.01.029.

27 Drake, Brandon L., "The Influence of Climatic Change on the Late Bronze Age Collapse and the Greek Dark Ages," *Journal of Archaeological Science,* Volume 39, Issue 6, June 2012, pp. 1; 1862 to 1870. https://doi.org/10.1016/j.jas.2012.01.029.

28 Steinhilber, F. et al., "Total Solar Irradiance During the Holocene," *Geophysical Research Letters,* Volume 36, Issue 19, 2 October 2009. https://doi.org/10.1029/2009GL04014.

29 Liu, Xiaokang et al., "New insights on Chinese cave $\delta^{18}O$ records and their paleoclimatic significance," *Earth-Science Reviews,* Volume 207, August 2020. https://doi.org/10.1016/j.earscirev.2020.103216.

30 Fang, Jin-qi, and Xie, Zhiren, "Deforestation in preindustrial China: The Loess Plateau region as an example," *Chemosphere*, Volume 29, Issue 5, pp. 983-999, September 1994. https://doi.org/10.1016/0045-6535(94)90164-3.

31 Yu, Fengling, "Reconstruction of the East Asian monsoon variability since the mid-Holocene from the Pearl River estuary," Durham E-Thesis, Durham University, Department of Geology, Durham, UK., June 2009, p. 147. https://etheses.dur.ac.uk/69/1/Thesis_FYU.pdf?DDD14+.

32 Margaritelli, G. et al., "Persistent Warm Mediterranean surface waters during the Roman period," *Nature Scientific Reports,* 26 June 2020. https://doi.org/10.1038/s41598-020-67281-2.

33 Oesterdiekhoff, Georg W., "Geographische Bedingungen der Weltgeschichte. Die Wechselwirkung von Klima, Bevölkerungswachstum and Landnutzung in der Evolution der Hochkultren," Zeithschrift für agrargeschichte und Agararsoziologie, 47 (1999), p. 126. See also, Behringer, Wolfgang, *A Cultural History of Climate,* Polity Press, Cambridge, UK, 2010, p. 61, ISBN 13:

978-0-7456-4528-5.

34 Harper, Kyle, *The Fate of Rome: Climate, Disease, and the End of an Empire*, Princeton University Press, April 2018, p. 30, ISBN: 978-0-691-16683-4.

35 Franke, Heinrich and Trauzettel, Rolf, *Das Chinesische Kaiser-reich*, Frankfurt/Main 1968, pp. 117-187. See also, Behringer, Wolfgang, *A Cultural History of Climate,* Polity Press, Cambridge, UK, 2010, p. 62, ISBN 13: 978-0-7456-4528-5.

36 Steinhilber, F. et al., "Total Solar Irradiance During the Holocene," *Geophysical Research Letters*, Volume 36, Issue 19, 2 October 2009. https://doi.org/10.1029/2009GL040142. (see Figure 41 in Chapter 7 for a chart of the Steinhilber data). See also, Harper, Kyle, *The Fate of Rome: Climate, Disease, and the End of an Empire* Princeton University Press, April 2018, p. 44, ISBN: 978-0-691-16683-4.

37 Harper, Kyle, *The Fate of Rome: Climate, Disease, and the End of an Empire*, Princeton University Press, April 2018, p. 45, ISBN: 978-0-691-16683-4.

38 Harper, Kyle, *The Fate of Rome: Climate, Disease, and the End of an Empire*, Princeton University Press, April 2018, p. 45, ISBN: 978-0-691-16683-4.

39 Harper, Kyle, *The Fate of Rome: Climate, Disease, and the End of an Empire*, Princeton University Press, April 2018, p. 53, ISBN: 978-0-691-16683-4.

40 Harper, Kyle, *The Fate of Rome: Climate, Disease, and the End of an Empire*, Princeton University Press, April 2018, p. 53, ISBN: 978-0-691-16683-4.

41 Harper, Kyle, *The Fate of Rome: Climate, Disease, and the End of an Empire*, Princeton University Press, April 2018, p. 52, ISBN: 978-0-691-16683-4.

42 Harper, Kyle, *The Fate of Rome: Climate, Disease, and the End of an Empire*, Princeton University Press, April 2018, p. 53, ISBN: 978-0-691-16683-4.

43 Harper, Kyle, *The Fate of Rome: Climate, Disease, and the End of an Empire*, Princeton University Press, April 2018, p. 131, ISBN: 978-0-691-16683-4.

44 Yang, Bao et al., "General characteristics of temperature variation in China during the last two millennia," *Geophysical Research Letters,* Volume 29, Issue 9, pp. 38-1-28-4, 11 May 2002. https://doi.org/10.1029/2001GL014485.

45 Tan, Ming et al., "Cyclic rapid warming on a centennial scale revealed by a 2,650-year stalagmite record of warm season temperature," *Geophysical Research Letters,* Volume 30, Issue 12, 19 June 2003. https://doi.org/10.1029/2003GL017352.

46 Ge, Q.-S. et al., "Temperature variation through 2000 years in China: An uncertainty analysis of reconstruction and regional difference," *Geophysical Research Letters,* Volume 37, Issue 3, 9 February 2010. https://doi.org/10.1029/2009GL041281.

47 Zhang, Pingzhong et al., "A Test of Climate, Sun, and Culture Relationships from an 1810-Year Chinese Cave Record," *Science*, Volume 322, Issue 5903, pp. 940-942, 7 November 2008. https://doi.org/10.1126/science.1163965.

48 Behringer, Wolfgang, *A Cultural History of Climate*, Polity Press, Cambridge, UK, 2010, p. 64, ISBN 13: 978-0-7456-4528-5.

49 "Severinus von Noricum," Lexikon des Mittelalters 7, 1999, pp. 1805-1806. See also, Behringer, Wolfgang, *A Cultural History of Climate,* Polity Press, Cambridge, UK, 2010, p. 64, ISBN 13: 978-0-7456-4528-5.

50 Lamb, Hubert H., Climate, History and the Modern World, Routledge, London, 19 July 1995, p. 160, ISBN 978-0234-3365-2. See also, Behringer, Wolfgang, *A Cultural History of Climate,* Polity Press, Cambridge, UK, 2010, pp. 64-65, ISBN 13: 978-0-7456-4528-5.

51 Behringer, Wolfgang, *A Cultural History of Climate,* Polity Press, Cambridge, UK, 2010, p. 66, ISBN 13: 978-0-7456-4528-5.

52 Behringer, Wolfgang, *A Cultural History of Climate,* Polity Press, Cambridge, UK, 2010, p. 67, ISBN 13: 978-0-7456-4528-5.

53 Harper, Kyle, *The Fate of Rome: Climate, Disease, and the End of an Empire,* Princeton University Press, April 2018, ISBN: 978-0-691-16683-4. See also, Temin, Peter, "Review, The Fate of Rome: Climate, Disease, and the End of an Empire," *Economic History Association*, April 2018. https://eh.net/book_reviews/the-fate-of-rome-climate-disease-and-the-end-of-an-empire/.

54 Marx, Werner et al., "Climate and the Decline and Fall of the Western Roman Empire: A Bibliometric View on an Interdisciplinary Approach to Answer a Most Classic Historical Question," *Climate*, Volume 6, Issue 4, 15 November 2018. https://doi.org/10.3390/cli6040090.

55 Harper, Kyle, *The Fate of Rome: Climate, Disease, and the End of an Empire,* Princeton University Press, April 2018, p. 131, ISBN: 978-0-691-16683-4.

56 Harper, Kyle, *The Fate of Rome: Climate, Disease, and the End of an Empire,* Princeton University Press, April 2018, p. 131, ISBN: 978-0-691-16683-4.

57 Behringer, Wolfgang, *A Cultural History of Climate,* Polity Press, Cambridge, UK, 2010, p. 65, ISBN 13: 978-0-7456-4528-5.

58 Harper, Kyle, *The Fate of Rome: Climate, Disease, and the End of an Empire,* Princeton University Press, April 2018, p. 131, ISBN: 978-0-691-16683-4.

59 Harper, Kyle, *The Fate of Rome: Climate, Disease, and the End of an Empire,* Princeton University Press, April 2018, p. 192, ISBN: 978-0-691-16683-4.

60 Harper, Kyle, *The Fate of Rome: Climate, Disease, and the End of an Empire,* Princeton University Press, April 2018, pp. 192-194, ISBN: 978-0-691-16683-4.

61 Harper, Kyle, *The Fate of Rome: Climate, Disease, and the End of an Empire,*

Princeton University Press, April 2018, p. 174, ISBN: 978-0-691-16683-4.

62 Harper, Kyle, *The Fate of Rome: Climate, Disease, and the End of an Empire,* Princeton University Press, April 2018, p. 220, ISBN: 978-0-691-16683-4.

63 Harper, Kyle, *The Fate of Rome: Climate, Disease, and the End of an Empire,* Princeton University Press, April 2018, p. 220, ISBN: 978-0-691-16683-4.

64 Harper, Kyle, *The Fate of Rome: Climate, Disease, and the End of an Empire,* Princeton University Press, April 2018, p. 219, ISBN: 978-0-691-16683-4.

65 Gregory of Tours, *The History of the Franks*, Penguin, London, 1974. See also, Behringer, Wolfgang, *A Cultural History of Climate*, Polity Press, Cambridge, UK, 2010, p. 68, ISBN 13: 978-0-7456-4528-5.

66 Steinhilber, F. et al., "Total Solar Irradiance during the Holocene," *Geophysical Research Letters*, Volume 36, Issue 19, 2 October 2009. https://doi.org/10.1029/2009GL040142 (see Figure 41 in Chapter 7 for a chart of the Steinhilber data). See also, Harper, Kyle, *The Fate of Rome: Climate, Disease, and the End of an Empire,* Princeton University Press, April 2018, p. 254, ISBN: 978-0-691-16683-4.

67 Le Roy, M. IS, "Calendar-Dated Glacier Variations in Western European Alps During the Neoglacial: The Mer de Glace Record, Mont Blanc Massif," *Quaternary Science Reviews,* Volume 108, pp. 1-22, 15 January 2015.

68 Wikipedia, "Volcanic winter of 536." https://en.wikipedia.org/wiki/Volcanic_winter_of_536; See also, *Procopius, Volume 2: History of the [Vandalic] Wars, Books III and IV*, Translated by Dewing, Henry Bronson, London, England, William Heinemann, 1916, p. 329. ISBN 978-0-674-99054-8.

69 Wikipedia, "Volcanic winter of 536." https://en.wikipedia.org/wiki/Volcanic_winter_of_536; See also, Larsen, L. B. et al., "New ice core evidence for a volcanic cause of the A.D. 536 dust veil," *Geophysical Research Letters*, Volume 35, Issue 4, 29 February 2008. https://doi.org/10.1029/2007GL032450

70 Harper, Kyle, *The Fate of Rome: Climate, Disease, and the End of an Empire,* Princeton University Press, April 2018, p. 256, ISBN: 978-0-691-16683-4.

71 Harper, Kyle, *The Fate of Rome: Climate, Disease, and the End of an Empire,* Princeton University Press, April 2018, p. 256, ISBN: 978-0-691-16683-4.

72 Harper, Kyle, *The Fate of Rome: Climate, Disease, and the End of an Empire,* Princeton University Press, April 2018, p. 257, ISBN: 978-0-691-16683-4.

73 Harper, Kyle, *The Fate of Rome: Climate, Disease, and the End of an Empire,* Princeton University Press, April 2018, p. 257, ISBN: 978-0-691-16683-4.

74 Gunn, Joel D., "The Impact of the 536 CE Little Ice Age on Mesoamerica with Dr. Joel D. Gunn," https://youtu.be/_GZH3Cqc2MM?si=4yvyDT-VYP3h_VQkd Minute 8:00 to 9:00 and 32:00 to 32:30.

75 Behringer, Wolfgang, *A Cultural History of Climate,* Polity Press, Cambridge, UK, 2010, p. 69, ISBN 13: 978-0-7456-4528-5.

76 Harper, Kyle, *The Fate of Rome: Climate, Disease, and the End of an Empire,* Princeton University Press, April 2018, p. 248, ISBN: 978-0-691-16683-4.

77 Behringer, Wolfgang, *A Cultural History of Climate,* Polity Press, Cambridge, UK, 2010, p. 69, ISBN 13: 978-0-7456-4528-5.

78 Yang, Bao et al., "General characteristics of temperature variation in China during the last two millennia," *Geophysical Research Letters,* Volume 29, Issue 9, pp. 38-1-28-4, 11 May 2002. https://doi.org/10.1029/2001GL014485.

79 Tan, Ming et al., "Cyclic rapid warming on a centennial scale revealed by a 2,650-year stalagmite record of warm season temperature," *Geophysical Research Letters*, Volume 30, Issue 12, 19 June 2003. https://doi.org/10.1029/2003GL017352.

80 Zhang, Pingzhong et al., "A Test of Climate, Sun, and Culture Relationships from an 1,810-Year Chinese Cave Record," *Science*, Volume 322, Issue 5903, pp. 940-942, 7 November 2008. https://doi.org/10.1126/science.1163965.

81 Zhang, Pingzhong et al., "A Test of Climate, Sun, and Culture Relationships from an 1,810-Year Chinese Cave Record," *Science*, Volume 322, Issue 5903, pp. 940-942, 7 November 2008. https://doi.org/10.1126/science.1163965.

82 Steinhilber, F. et al., "Total Solar Irradiance during the Holocene," *Geophysical Research Letters,* Volume 36, Issue 19, 2 October 2009. https://doi.org/10.1029/2009GL040142 (see Figure 41 in Chapter 7 for a chart of the Steinhilber data).

83 Zhang, Pingzhong et al., "A Test of Climate, Sun, and Culture Relationships from an 1,810-Year Chinese Cave Record," *Science*, Volume 322, Issue 5903, pp. 940-942, 7 November 2008. https://doi.org/10.1126/science.1163965.

84 Yancheva, Gergana et al., "Influence of the intertropical convergence zone on East Asian monsoon," *Nature*, Volume 445, pp. 74-77, 4 January 2007. https://doi.org/10.1038/nature05431.

85 Hodell, David A. et al., "Possible Role of Climate in the Collapse of Classic Mayan Civilization," *Nature*, Volume 375, pp. 391-394, 1 June 1995. https://doi.org/10.1038/375391a0; See also, Behringer, Wolfgang, *A Cultural History of Climate,* Polity Press, Cambridge, UK, 2010, pp. 70-71, ISBN 13: 978-0-7456-4528-5.

86 Yancheva, Gergana et al., "Influence of the intertropical convergence zone on East Asian monsoon," *Nature*, Volume 445, pp. 74-77, 4 January 2007. https://doi.org/10.1038/nature05431.

87 Webster, James W. et al., "Stalagmite evidence from Belize indicating significant droughts at the time of Preclassic Abandonment, the Maya Hiatus,

and the Classic Maya collapse," *Paleogeography, Palaeoclimatology, Paleoecology,* Volume 250, Issues 1-4, pp. 1-17, 25 June 2007. https://doi.org/10.1016/j.palaeo.2007.02.022.

88 Hodell, David A., "Solar Forcing of Drought Frequency in the Maya Lowlands," *Science,* Volume 292, Number 5530, pp. 1367-1370, 18 May 2001. https://doi.org/10.1126/science.1057759.

89 Steinhilber, F. et al., "Total Solar Irradiance during the Holocene," *Geophysical Research Letters,* Volume 36, Issue 19, 2 October 2009. https://doi.org/10.1029/2009GL040142; See also, Figure 41 in Chapter 7 for a chart of the Steinhilber data.

90 Fagan, Brian, *The Little Ice Age: How Climate Made History, 1300-1850,* Basic Books, New York, 2000, p. 16, ISBN: 10-465-02272-3.

91 Fagan, Brian, *The Little Ice Age: How Climate Made History, 1300-1850,* Basic Books, New York, 2000, p. 12, ISBN: 10-465-02272-3.

92 Behringer, Wolfgang, *A Cultural History of Climate,* Polity Press, Cambridge, UK, 2010, p. 81, ISBN 13: 978-0-7456-4528-5.

93 Fagan, Brian, *The Little Ice Age: How Climate Made History, 1300-1850,* Basic Books, New York, 2000, p. 33, ISBN: 10-465-02272-3.

94 Fagan, Brian, *The Little Ice Age: How Climate Made History, 1300-1850,* Basic Books, New York, 2000, p. 33, ISBN: 10-465-02272-3.

95 Stoob, Heinz, Forschungen zum Städtewesen in Europa, Cologne, 1970. See also, Behringer, Wolfgang, A *Cultural History of Climate*, Polity Press, Cambridge, UK, 2010, p. 81, ISBN 13: 978-0-7456-4528-5.

96 Behringer, Wolfgang, *A Cultural History of Climate,* Polity Press, Cambridge, UK, 2010, p. 80, ISBN 13: 978-0-7456-4528-5.

97 De'er, Zhang, "Evidence for the Existence of the Medieval Warm Period in China," *Climatic Change,* Volume 26, pp. 289-297, March1994. https://doi.org/10.1007/BF01092419; See also, Behringer, Wolfgang, *A Cultural History of Climate,* Polity Press, Cambridge, UK, 2010, p. 78, ISBN 13: 978-0-7456-4528-5.

98 Fagan, Brian, *The Little Ice Age: How Climate Made History, 1300-1850,* Basic Books, New York, 2000, pp.19-20, ISBN: 10-465-02272-3.

99 Fagan, Brian, *The Little Ice Age: How Climate Made History, 1300-1850,* Basic Books, New York, 2000, p. 17, ISBN: 10-465-02272-3.

100 Behringer, Wolfgang, *A Cultural History of Climate*, Polity Press, Cambridge, UK, 2010, p. 82, ISBN 13: 978-0-7456-4528-5.

101 Behringer, Wolfgang, *A Cultural History of Climate,* Polity Press, Cambridge, UK, 2010, p. 82, ISBN 13: 978-0-7456-4528-5.

102 Behringer, Wolfgang, *A Cultural History of Climate*, Polity Press, Cambridge, UK, 2010, p. 83, ISBN 13: 978-0-7456-4528-5.

103 Yang, Bao et al., "General characteristics of temperature variation in China during the last two millennia," *Geophysical Research Letters,* Volume 29, Issue 9, pp. 38-1-28-4, 11 May 2002. https://doi.org/10.1029/2001GL014485.

104 Ge, Q.-S. et al., "Temperature variation through 2000 years in China: An uncertainty analysis of reconstruction and regional difference," *Geophysical Research Letters,* Volume 37, Issue 3, 9 February 2010. https://doi.org/10.1029/2009GL041281.

105 Zhang, Pingzhong et al., "A Test of Climate, Sun, and Culture Relationships from an 1810-Year Chinese Cave Record," *Science*, Volume 322, Issue 5903, pp. 940-942, 7 November 2008. https://doi.org/10.1126/science.1163965.

106 Zhang, Pingzhong et al., "A Test of Climate, Sun, and Culture Relationships from an 1810-Year Chinese Cave Record," *Science*, Volume 322, Issue 5903, pp. 940-942, 7 November 2008. https://doi.org/10.1126/science.1163965.

107 Zhang, Pingzhong et al., "A Test of Climate, Sun, and Culture Relationships from an 1,810-Year Chinese Cave Record," *Science*, Volume 322, Issue 5903, pp. 940-942, 7 November 2008. https://doi.org/10.1126/science.1163965.

108 Tan, Ming et al., "Cyclic rapid warming on a centennial scale revealed by a 2,650-year stalagmite record of warm season temperature," *Geophysical Research Letters,* Volume 30, Issue 12, 19 June 2003. https://doi.org/10.1029/2003GL017352.

109 Ledru, M.-P. et al., "The medieval climate anomaly and the Little Ice Age in the eastern Ecuadorian Andes," *Climate of the Past Discussion,* Volume 9, Issue 1, pp. 307-321, 5 February 2013. https://doi.org/10.5194/cp-9-307-2013.

110 Chepstow-Lusty, A. J. et al., "Putting the rise of the Inca Empire within a climatic and land management context," *Climate of the Pas*t, Volume 5, Issue 3, pp. 375-388, 22 July 2009. https://doi.org/10.5194/cp-5-375-2009.

111 Steinhilber, F. et al., "Total Solar Irradiance during the Holocene," *Geophysical Research Letters,* Volume 36, Issue 19, 2 October 2009. https://doi.org/10.1029/2009GL040142; See also, Figure 41 in Chapter 7 for a chart of the Steinhilber data.

112 Kaniewski, D. E. et al., "The medieval climate anomaly and the Little Ice Age in coastal Syria inferred from pollen-derived paleoclimatic patterns," *Global and Planetary Change,* Volume 78, Issues 3-4, pp. 178-187, August-September 2011. https://doi.org/10.1016/j.gloplacha.2011.06.010.

113 Vinós, Javier, *Solving the Climate Puzzle: The Sun's Surprising Role*, Critical Science Press, Madrid, 2023, p. 173, ISBN: 978-84-125867-7-0.

114 Fagan, Brian, *The Little Ice Age: How Climate Made History, 1300-1850,*

Basic Books, New York, 2000, p. 83, ISBN: 10-465-02272-3.

115 Fagan, Brian, *The Little Ice Age: How Climate Made History, 1300-1850,* Basic Books, New York, 2000, p. 83, ISBN: 10-465-02272-3.

116 Rensberger, Moyce, "Aztec Sacrifices Laid to Hunger, Not Just Religion," *The New York Times*, 19 February 1977. https://www.nytimes.com/1977/02/19/archives/aztec-sacrifices-laid-to-hunger-not-just-religion.html; See also, Mexicolore, "The human cost of famine," AZTECS. https://www.mexicolore.co.uk/aztecs/aztec-life/human-cost-of-famine.

117 White, Sam, *A Cold Welcome,* Harvard Press, Cambridge, Massachusetts, 2017, p. 192. ISBN 9780674244900.

118 White, Sam, *A Cold Welcome,* Harvard Press, Cambridge, Massachusetts, 2017, p. 207. ISBN 9780674244900.

119 White, Sam, *A Cold Welcome,* Harvard Press, Cambridge, Massachusetts, 2017, pp. 160-161. ISBN 9780674244900.

120 Yang, Bao et al., "General characteristics of temperature variation in China during the last two millennia," *Geophysical Research Letters,* Volume 29, Issue 9, pp. 38-1-28-4, 11 May 2002. https://doi.org/10.1029/2001GL014485.

121 Chang, Chia-chang, "The Reconstruction of Climate in China for Historical Times," Beijing 1988. See also, Behringer, Wolfgang, *A Cultural History of Climate,* Polity Press, Cambridge, UK, 2010, p. 90, ISBN 13: 978-0-7456-4528-5.

122 Behringer, Wolfgang, *A Cultural History of Climate,* Polity Press, Cambridge, UK, 2010, p. 114, ISBN 13: 978-0-7456-4528-5.

123 Wagner, B.B., "The Little Ice Age and Its Giant Impact on Human History," *Ancient Origins*, 25 September, 2020. https://www.ancient-origins.net/human-origins-science/little-ice-age-0014312.

124 Wakeman, Frederic, "China and the Seventeenth-Century Crisis," Late Imperial China 7, 1986, pp. 1-26. See also, Behringer, Wolfgang, *A Cultural History of Climate,* Polity Press, Cambridge, UK, 2010, p. 113, ISBN 13: 978-0-7456-4528-5.

125 Zhang, Pingzhong et al., "A Test of Climate, Sun, and Culture Relationships from an 1,810-Year Chinese Cave Record," *Science*, Volume 322, Issue 5903, pp. 940-942, 7 November 2008. https://doi.org/10.1126/science.1163965.

126 Parker, Geoffrey, *Global Crisis: War, Climate Change and Catastrophe in the Seventeenth Century,* Yale University Press, 30 April, 2013, 904 pp., p. xxvii, ISBN: 978-0300-153231.

127 Burton, Robert, *The Anatomy of Melancholy,* Oxford, 1621. See also, Behringer, Wolfgang, *A Cultural History of Climate,* Polity Press, Cambridge, UK, 2010, p. 115, ISBN 13: 978-0-7456-4528-5.

128 Wagner, B.B., "The Little Ice Age and Its Giant Impact on Human History," *Ancient Origins*, 25 September 2020. https://www.ancient-origins.net/human-origins-science/little-ice-age-0014312.

129 Parker, Geoffrey, *Global Crisis: War, Climate Change and Catastrophe in the Seventeenth Century*, Yale University Press, 30 April 30, 2013, 904 pp., ISBN: 978-0300-153231.

130 Parker, Geoffrey, *Global Crisis: War, Climate Change and Catastrophe in the Seventeenth Century,* Yale University Press, April 30, 2013, 904 pp., pp. xxiv-xxv, ISBN: 978-0300-153231.

131 Parker, Geoffrey, *Global Crisis: War, Climate Change and Catastrophe in the Seventeenth Century,* Yale University Press, April 30, 2013, 904 pp., p. xxv, ISBN: 978-0300-153231.

132 Parker, Geoffrey, *Global Crisis: War, Climate Change and Catastrophe in the Seventeenth Century,* Yale University Press, April 30, 2013, 904 pp., p. 3, ISBN: 978-0300-153231.

133 Parker, Geoffrey, *Global Crisis: War, Climate Change and Catastrophe in the Seventeenth Century,* Yale University Press, April 30, 2013, 904 pp., p. 4, ISBN: 978-0300-153231.

134 Parker, Geoffrey, *Global Crisis: War, Climate Change and Catastrophe in the Seventeenth Century*, Yale University Press, April 30, 2013, 904 pp., p. xxviii, ISBN: 978-0300-153231.

135 Gertler, Charles G. and O'Gorman, Paul A., "Changing available energy for extratropical cyclones and associated convection in Northern Hemisphere summer," *PNAS*, Volume 116, Issue 10, pp. 4105-4110, 19 February 2019. https://doi.org/10.1073/pnas.1812312116.

136 Martinez, A. and Iglesias, G., "Global wind energy resources decline under climate change," *Energy*, Volume 288, 1 February 2024. https://doi.org/10.1016/j.energy.2023.129765; See also, Dai, Aiguo and Deng, Jiechun, "Arctic Amplification Weakens the Variability of Daily Temperatures over Northern Middle-High Latitude," *Journal of Climate,* Volume 34, 1 April 2021. https://doi.org/10.1175/JCLI-D-20-0514.1.

137 Bony, S. et al., "How well do we understand and evaluate climate change feedback processes?" *Journal of Climate Change*, Volume 19, Issue 15, pp. 3445-3482, 2006. https://doi.org/10.1175/JCLI3819.1.

138 White, Sam, *A Cold Welcome,* Harvard Press, Cambridge, Massachusetts, 2017, p. 236. ISBN 9780674244900.

139 Degroot, Dagomar, *The Frigid Golden Age: Climate Change, the Little Ice Age, and the Dutch Republic, 1560-1720*, Cambridge University Press, Cambridge, UK, 2018, pp. 44, 120-121, ISBN: 987 110 829 7639.

140 Wheeler, D. et al., "Atmospheric circulation and storminess derived from Royal Navy logbooks: 1685 to 1750," *Climatic Change,* Volume 101, pp. 257-280, July 2010. https://doi.org/10.1007/s10584-00t.9-9732-x; See also, Trouet, V. et al., "North Atlantic storminess and Atlantic Meridional Overturning Circulation during the last Millennium: Reconciling contradictory proxy records of NAO variability," *Global and Planetary Change,* Volumes 84-85, pp. 48-55, p. 53, March 2012. https://doi.org/10.1016/j.gloplacha.2011.10.003.

141 Meeker, Loren D. and Mayewski, Paul A., Pa 1400-year high-resolution record of atmospheric circulation over the North Atlantic and Asia, *The Holocene,* Volume 12, Issue 3, April 2002. https://doi.org/10.1191/0959683602hl542ft; See also, Dawson, A. G. et al., "Greenland (GISP2) ice core and historical indicators of complex North Atlantic climate changes during the fourteenth century," *The Holocene,* Volume 17, Issue 4, May 2007. https://doi.org/10.1177/0959683607077010.

142 Szkornik Katie et al., "Aeolian sand movement and relative sea-leel rise in Ho Bugt, western Denmark, during the Little Ice Age," *The Holocene,* Volume 18, Issue 6, 1 September 2008. https://doi.org/10.1177/0959683608091800; See also, Clemmensen, Lars, B. et al., "Sedimentology, stratigraphy and landscape evolution of a Holocene coastal dune system, Lodbjerg, NW, Jutland, Denmark," *Sedimentology,* Volume 48, Issue 1, pp. 3-27, 18 July 2008. https://doi.org/10.1111/j.1365-3091.2001.00345.x; See also, 2001, and Aagaard, Troels, et al., "Environmental controls on coastal dune formation: Skallingen Spit, Denmark, *Geomorphology,* Volume 83, Issues 1-2, pp. 29-47, January 2007. https://doi.org/10.1016/j.geomorph.2006.06.007.

143 Hansom, J.D. and Hall, A. M., "Magnitude and frequency of extra-tropical North Atlantic cyclones: A chronology from cliff-top storm deposits," *Quaternary International,* Volume 195, Issues 1-2, pp. 42-52, 15 February 2009. https://doi.org/10.1016/j.quaint.2007.11.010.

144 Dejong, Rixt et al., "Storminess variation during the last 6500 years as reconstructed from an ombrotrophic peat bog in Halland, southwest Sweden," *Journal of Quaternary Science,* Volume 21, Issue 8, pp. 905-919, 12 May 2006. https://doi.org/10.1002/jqs.1011.

145 Jelgersma, S. et al., "Holocene storm surge signatures in the coastal dunes of the western Netherlands," *Marine Geology,* Volume 125, Issues 1-2, pp. 95-110, June 1995. https://doi.org/10.1016/0025-3227(95)00061-3.

146 Gilberson, D. D. et al., "Sand-drift and Soil Formation Along an Exposed North Atlantic Coastline: 14,000 Years of Diverse Geomorphological, Climatic and Human Impacts," *Journal of Archaeological Science,* Volume 26, Issue 4, pp. 439-469, April 1999. https://doi.org/10.1006/jasc.1998.0360; See also, Wilson, Peter, "Holocene coastal dune development on the South Erradale peninsula, Wester Ross, Scottland," *Scottish Journal of Geology,* Volume 38, Issue 1, pp.

5-13, 1 May 2002. https://doi.org/10.1144/sjg38010005.

147 Sorrel, P. et al., "Evidence for millennial-scale climate events in the sedimentary infilling of a macrotidal estuarine system, the Seine estuary (NW France)," *Quaternary Science Reviews,* Volume 28, Issues 5-6, pp. 499-516, March 2009. https://doi.org/10.1016/j.quascirev.2008.11.009.

148 Clarke, Michele. et al., "Late-Holocene sand invasion and North Atlantic storminess along the Aquitaine coast, southwest France," *The Holocene*, Volume 12, Issue 2, February 2002. https://doi.org/10.1191/0959683602hl539rr.

149 Borja, F. et al., "Holocene Aeolian phases and human settlements along the Atlantic coast of southern Spain," *The Holocene,* Volume 9, Issue 3, April 1999. https://doi.org/10.1191/095968399668924476; See also, Clarke, and Michele L. and Rendell, Helen, M., "Effects of storminess and sand supply and the North Atlantic Oscillation on sand invasion and coastal dune accretion in western Portugal," *The Holocene,* Volume 16, Issue 3, April 2006. https://doi.org/10.1191/0959683606hl932rp.

150 Dezileau, L. et al., "Intense storm activity during the Little Ice Age on the French Mediterranean coast," *Paleogeography, Palaeoclimatology, Paleoecology,* Volume 299, Issues 1-2, pp. 289-297, January 2011. https://doi.org/10.1016/j.palaeo.2010.11.009.

151 Bramante, James F. et al., "Increased typhoon activity in the Pacific deep tropics driven by Little Ice Age circulation changes," *Nature Geoscience,* Volume 13, pp. 806-811, 16 November 2020. https://doi.org/10.1038/s41561-020-00656-2.

152 Chen, Zhi et al., "Evolution of Storm Surges over the Little Ice Age Indicated by Aeolian Sand Records on the Coast of the Beibu Gulf, China," *Water,* Volume 13, Issue 14, 14 July 2021. https://doi.org/10.3390/w13141941.

153 Winkler, T.S. et al., "More Frequent Hurricane Passage Across the Bahamian Archipelago During the Little Ice Age," *Paleoceanography and Paleoclimatology,* Volume 38, Issue 11, 21 November 2023. https://doi.org/10.1029/2023PA004623.

154 Parker, Geoffrey, *Global Crisis: War, Climate Change and Catastrophe in the Seventeenth Century*, Yale University Press, April 30, 2013, 904, pp., p. 4, ISBN: 978-0300-153231.

155 Parker, Geoffrey, *Global Crisis: War, Climate Change and Catastrophe in the Seventeenth Century*, Yale University Press, April 30, 2013, 904 pp., p. 4, ISBN: 978-0300-153231.

156 Parker, Geoffrey, *Global Crisis: War, Climate Change and Catastrophe in the Seventeenth Century,* Yale University Press, April 30, 2013, 904 pp., p. 4, ISBN: 978-0300-153231.

157 Reid, Anthony, "The Seventeenth Century Crisis in South East Asia," *Modern Asian Studies*, 24, 1990, pp. 639-659. Behringer, Wolfgang, *A Cultural History of Climate,* Polity Press, Cambridge, UK, 2010, p.113, ISBN 13: 978-0-7456-4528-5.

158 Parker, Geoffrey, *Global Crisis: War, Climate Change and Catastrophe in the Seventeenth Century,* Yale University Press, April 30, 2013, 904 pp., p. 4, ISBN: 978-0300-153231.

159 Parker, Geoffrey, *Global Crisis: War, Climate Change and Catastrophe in the Seventeenth Century,* Yale University Press, April 30, 2013, 904 pp., p. 4, ISBN: 978-0300-153231.

160 Parker, Geoffrey, *Global Crisis: War, Climate Change and Catastrophe in the Seventeenth Century,* Yale University Press, April 30, 2013, 904 pp., p. 4, ISBN: 978-0300-153231.

161 Parker, Geoffrey, *Global Crisis: War, Climate Change and Catastrophe in the Seventeenth Century,* Yale University Press, April 30, 2013, 904 pp., p. 4, ISBN: 978-0300-153231.

162 Parker, Geoffrey, *Global Crisis: War, Climate Change and Catastrophe in the Seventeenth Century,* Yale University Press, April 30, 2013, 904 pp., p. 5, ISBN: 978-0300-153231.

163 Parker, Geoffrey, *Global Crisis: War, Climate Change and Catastrophe in the Seventeenth Century,* Yale University Press, April 30, 2013, 904 pp., p. 5, ISBN: 978-0300-153231.

164 Parker, Geoffrey, *Global Crisis: War, Climate Change and Catastrophe in the Seventeenth Century,* Yale University Press, April 30, 2013, 904 pp., p. 6, ISBN: 978-0300-153231.

165 Parker, Geoffrey, *Global Crisis: War, Climate Change and Catastrophe in the Seventeenth Century,* Yale University Press, April 30, 2013, 904 pp., p. 7, ISBN: 978-0300-153231.

166 Parker, Geoffrey, *Global Crisis: War, Climate Change and Catastrophe in the Seventeenth Century,* Yale University Press, April 30, 2013, 904 pp., p. 7, ISBN: 978-0300-153231.

167 Parker, Geoffrey, *Global Crisis: War, Climate Change and Catastrophe in the Seventeenth Century,* Yale University Press, April 30, 2013, 904 pp, p. 4, ISBN: 978-0300-153231.

168 Parker, Geoffrey, *Global Crisis: War, Climate Change and Catastrophe in the Seventeenth Century,* Yale University Press, April 30, 2013, 904 pp., p. 5, ISBN: 978-0300-153231.

169 Parker, Geoffrey, *Global Crisis: War, Climate Change and Catastrophe in the Seventeenth Century,* Yale University Press, April 30, 2013, 904 pp., p. 5, ISBN:

978-0300-153231.

170 Parker, Geoffrey, *Global Crisis: War, Climate Change and Catastrophe in the Seventeenth Century,* Yale University Press, April 30, 2013, 904 pp., p. 5, ISBN: 978-0300-153231.

171 Parker, Geoffrey, *Global Crisis: War, Climate Change and Catastrophe in the Seventeenth Century,* Yale University Press, April 30, 2013, 904 pp., p. 6, ISBN: 978-0300-153231.

172 Parker, Geoffrey, *Global Crisis: War, Climate Change and Catastrophe in the Seventeenth Century,* Yale University Press, April 30, 2013, 904 pp., p. 6, ISBN: 978-0300-153231.

173 Parker, Geoffrey, *Global Crisis: War, Climate Change and Catastrophe in the Seventeenth Century,* Yale University Press, April 30, 2013, 904 pp., p. 6, ISBN: 978-0300-153231.

174 Parker, Geoffrey, *Global Crisis: War, Climate Change and Catastrophe in the Seventeenth Century,* Yale University Press, April 30, 2013, 904 pp., p. 6, ISBN: 978-0300-153231.

175 Parker, Geoffrey, *Global Crisis: War, Climate Change and Catastrophe in the Seventeenth Century,* Yale University Press, April 30, 2013, 904 pp., p. 6, ISBN: 978-0300-153231.

176 Parker, Geoffrey, *Global Crisis: War, Climate Change and Catastrophe in the Seventeenth Century,* Yale University Press, April 30, 2013, 904 pp., p. 7, ISBN: 978-0300-153231.

177 Parker, Geoffrey, *Global Crisis: War, Climate Change and Catastrophe in the Seventeenth Century,* Yale University Press, April 30, 2013, 904 pp., p. 7, ISBN: 978-0300-153231.

178 Parker, Geoffrey, *Global Crisis: War, Climate Change and Catastrophe in the Seventeenth Century,* Yale University Press, April 30, 2013, 904 pp., p. 7, ISBN: 978-0300-153231.

179 Parker, Geoffrey, *Global Crisis: War, Climate Change and Catastrophe in the Seventeenth Century*, Yale University Press, April 30, 2013, 904 pp., pp. 7-8, ISBN: 978-0300-153231.

180 Parker, Geoffrey, *Global Crisis: War, Climate Change and Catastrophe in the Seventeenth Century,* Yale University Press, April 30, 2013, 904 pp., p. 8, ISBN: 978-0300-153231.

181 Parker, Geoffrey, *Global Crisis: War, Climate Change and Catastrophe in the Seventeenth Century,* Yale University Press, April 30, 2013, 904 pp., p. 67, ISBN: 978-0300-153231.

182 Parker, Geoffrey, *Global Crisis: War, Climate Change and Catastrophe in the Seventeenth Century,* Yale University Press, April 30, 2013, 904 pp., p. 17, ISBN:

978-0300-153231.

183 Parker, Geoffrey, *Global Crisis: War, Climate Change and Catastrophe in the Seventeenth Century,* Yale University Press, April 30, 2013, 904 pp., p. 57, ISBN: 978-0300-153231.

184 Parker, Geoffrey, *Global Crisis: War, Climate Change and Catastrophe in the Seventeenth Century,* Yale University Press, April 30, 2013, 904 pp., p. 57, ISBN: 978-0300-153231.

185 Parker, Geoffrey, *Global Crisis: War, Climate Change and Catastrophe in the Seventeenth Century,* Yale University Press, April 30, 2013, 904 pp., p. 65, ISBN: 978-0300-153231.

186 Parker, Geoffrey, *Global Crisis: War, Climate Change and Catastrophe in the Seventeenth Century,* Yale University Press, April 30, 2013, 904 pp., p. 19, ISBN: 978-0300-153231.

187 Parker, Geoffrey, *Global Crisis: War, Climate Change and Catastrophe in the Seventeenth Century,* Yale University Press, April 30, 2013, 904 pp., p. 20, ISBN: 978-0300-153231.

188 Parker, Geoffrey, *Global Crisis: War, Climate Change and Catastrophe in the Seventeenth Century,* Yale University Press, April 30, 2013, 904 pp., p. 22, ISBN: 978-0300-153231.

189 Parker, Geoffrey, *Global Crisis: War, Climate Change and Catastrophe in the Seventeenth Century,* Yale University Press, April 30, 2013, 904 pp., p. 22, ISBN: 978-0300-153231.

190 Parker, Geoffrey, *Global Crisis: War, Climate Change and Catastrophe in the Seventeenth Century,* Yale University Press, April 30, 2013, 904 pp., p. 22, ISBN: 978-0300-153231.

191 Parker, Geoffrey, *Global Crisis: War, Climate Change and Catastrophe in the Seventeenth Century,* Yale University Press, April 30, 2013, 904 pp., p. 55, ISBN: 978-0300-153231.

192 Parker, Geoffrey, *Global Crisis: War, Climate Change and Catastrophe in the Seventeenth Century,* Yale University Press, April 30, 2013, 904 pp., p. 26, ISBN: 978-0300-153231.

193 Parker, Geoffrey, *Global Crisis: War, Climate Change and Catastrophe in the Seventeenth Century*, Yale University Press, April 30, 2013, 904 pp., p. 135, ISBN: 978-0300-153231.

194 Parker, Geoffrey, *Global Crisis: War, Climate Change and Catastrophe in the Seventeenth Century*, Yale University Press, April 30, 2013, 904 pp., p. 146, ISBN: 978-0300-153231.

195 Tan, Ming et al., "Cyclic rapid warming on a centennial scale revealed by a 2,650-year stalagmite record of warm season temperature," *Geo-*

physical Research Letters, Volume 30, Issue 12, 19 June 2003. https://doi.org/10.1029/2003GL017352.

196 Ge, Q.-S. et al., "Temperature variation through 2000 years in China: An uncertainty analysis of reconstruction and regional difference," *Geophysical Research Letters,* Volume 37, Issue 3, 9 February 2010. https://doi.org/10.1029/2009GL041281.

197 Zhang, Pingzhong et al., "A Test of Climate, Sun, and Culture Relationships from an 1,810-Year Chinese Cave Record," *Science*, Volume 322, Issue 5903, pp. 940-942, 7 November 2008. https://doi.org/10.1126/science.1163965.

198 Zhang, Pingzhong et al., "A Test of Climate, Sun, and Culture Relationships from an 1,810-Year Chinese Cave Record," *Science*, Volume 322, Issue 5903, pp. 940-942, 7 November 2008. https://doi.org/10.1126/science.1163965.

199 Parker, Geoffrey, *Global Crisis: War, Climate Change and Catastrophe in the Seventeenth Century*, Yale University Press, April 30, 2013, 904 pp., p. 118, ISBN: 978-0300-153231.

200 Parker, Geoffrey, *Global Crisis: War, Climate Change and Catastrophe in the Seventeenth Century*, Yale University Press, April 30, 2013, 904 pp., p. 122, ISBN: 978-0300-153231.

201 Parker, Geoffrey, *Global Crisis: War, Climate Change and Catastrophe in the Seventeenth Century*, Yale University Press, April 30, 2013, 904 pp., p. 123, ISBN: 978-0300-153231.

202 Parker, Geoffrey, *Global Crisis: War, Climate Change and Catastrophe in the Seventeenth Century,* Yale University Press, April 30, 2013, 904 pp., p. 125, ISBN: 978-0300-153231.

203 Parker, Geoffrey, *Global Crisis: War, Climate Change and Catastrophe in the Seventeenth Century*, Yale University Press, April 30, 2013, 904 pp., p. 125, ISBN: 978-0300-153231.

204 Parker, Geoffrey, *Global Crisis: War, Climate Change and Catastrophe in the Seventeenth Century*, Yale University Press, April 30, 2013, 904 pp., p. 126, ISBN: 978-0300-153231.

205 Parker, Geoffrey, *Global Crisis: War, Climate Change and Catastrophe in the Seventeenth Century,* Yale University Press, April 30, 2013, 904 pp., p. 126, ISBN: 978-0300-153231.

206 Parker, Geoffrey, *Global Crisis: War, Climate Change and Catastrophe in the Seventeenth Century,* Yale University Press, April 30, 2013, 904 pp., p. 126, ISBN: 978-0300-153231.

207 Quotations from Tong, Disorder, 83, 84, and Kessler, K'ang-his, 15, Memorial by Wei Yijie in 1660. See also, Parker, Geoffrey, *Global Crisis: War, Climate Change and Catastrophe in the Seventeenth Century*, Yale University Press,

April 30, 2013, 904 pp., p. 129, ISBN: 978-0300-153231.

208 Parker, Geoffrey, *Global Crisis: War, Climate Change and Catastrophe in the Seventeenth Century*, Yale University Press, April 30, 2013, 904 pp., p. 132, ISBN: 978-0300-153231.

209 Qin Huitian, Wuli Tongkao, 1761. See also, Parker, Geoffrey, *Global Crisis: War, Climate Change and Catastrophe in the Seventeenth Century,* Yale University Press, April 30, 2013, 904 pp., p. 130, ISBN: 978-0300-153231.

210 Parker, Geoffrey, *Global Crisis: War, Climate Change and Catastrophe in the Seventeenth Century*, Yale University Press, April 30, 2013, 904 pp., p. 134, ISBN: 978-0300-153231.

211 Wagner, B.B., "The Little Ice Age and Its Giant Impact on Human History," *Ancient Origins,* 25 September 2020. https://www.ancient-origins.net/human-origins-science/little-ice-age-0014312.

212 Parker, Geoffrey, *Global Crisis: War, Climate Change and Catastrophe in the Seventeenth Century,* Yale University Press, April 30, 2013, 904 pp., p. 485, ISBN: 978-0300-153231.

213 Parker, Geoffrey, *Global Crisis: War, Climate Change and Catastrophe in the Seventeenth Century,* Yale University Press, April 30, 2013, 904 pp., p. 485, ISBN: 978-0300-153231.

214 Parker, Geoffrey, *Global Crisis: War, Climate Change and Catastrophe in the Seventeenth Century,* Yale University Press, April 30, 2013, 904 pp., p. 486, ISBN: 978-0300-153231.

215 Parker, Geoffrey, *Global Crisis: War, Climate Change and Catastrophe in the Seventeenth Century,* Yale University Press, April 30, 2013, 904 pp., p. 494, ISBN: 978-0300-153231.

216 Parker, Geoffrey, *Global Crisis: War, Climate Change and Catastrophe in the Seventeenth Century,* Yale University Press, April 30, 2013, 904 pp., p. 494, ISBN: 978-0300-153231.

217 Parker, Geoffrey, *Global Crisis: War, Climate Change and Catastrophe in the Seventeenth Century,* Yale University Press, April 30, 2013, 904 pp., p. 193, ISBN: 978-0300-153231.

218 Parker, Geoffrey, *Global Crisis: War, Climate Change and Catastrophe in the Seventeenth Century,* Yale University Press, April 30, 2013, 904 pp., p. 193, ISBN: 978-0300-153231.

219 Parker, Geoffrey, *Global Crisis: War, Climate Change and Catastrophe in the Seventeenth Century*, Yale University Press, April 30, 2013, 904 pp., p. 188, ISBN: 978-0300-153231.

220 Parker, Geoffrey, *Global Crisis: War, Climate Change and Catastrophe in the Seventeenth Century,* Yale University Press, April 30, 2013, 904 pp., p. 197,

ISBN: 978-0300-153231.

221 Parker, Geoffrey, *Global Crisis: War, Climate Change and Catastrophe in the Seventeenth Century,* Yale University Press, April 30, 2013, 904 pp., p. 204, ISBN: 978-0300-153231.

222 Parker, Geoffrey, *Global Crisis: War, Climate Change and Catastrophe in the Seventeenth Century*, Yale University Press, April 30, 2013, 904 pp., p. 204, ISBN: 978-0300-153231.

223 Parker, Geoffrey, *Global Crisis: War, Climate Change and Catastrophe in the Seventeenth Century,* Yale University Press, April 30, 2013, 904 pp., p. 209, ISBN: 978-0300-153231.

224 White, Sam, *The Climate of Rebellion in the Early Modern Ottoman Empire,*" Cambridge University Press, Cambridge, UK, 2012, ISBN: 9780511844058.

225 Parker, Geoffrey, *Global Crisis: War, Climate Change and Catastrophe in the Seventeenth Century*, Yale University Press, April 30, 2013, 904 pp., p. 327, ISBN: 978-0300-153231.

226 Baker, J. N. L., "The climate of England in the seventeenth century," *Quarterly Journal of the Royal Meteorological Society*, LVIII, pp. 421-439, 1932. https://ui.adsabs.harvard.edu/link_gateway/1932QJRMS..58..421B/doi:10.1002/qj.49705824705; See also, Parker, Geoffrey, *Global Crisis: War, Climate Change and Catastrophe in the Seventeenth Century*, Yale University Press, April 30, 2013, 904 pp., p. 330, ISBN: 978-0300-153231.

227 Parker, Geoffrey, *Global Crisis: War, Climate Change and Catastrophe in the Seventeenth Century,* Yale University Press, April 30, 2013, 904 pp., p. 333, ISBN: 978-0300-153231.

228 Parker, Geoffrey, *Global Crisis: War, Climate Change and Catastrophe in the Seventeenth Century*, Yale University Press, April 30, 2013, 904 pp., p. 360, ISBN: 978-0300-153231.

229 Parker, Geoffrey, *Global Crisis: War, Climate Change and Catastrophe in the Seventeenth Century*, Yale University Press, April 30, 2013, 904 pp., p. 374, ISBN: 978-0300-153231.

230 Macfarlane, Diary of Ralph Josseline, pp. 125, 129, 9 May 1648 and 28 June 1648. See also, Parker, Geoffrey, *Global Crisis: War, Climate Change and Catastrophe in the Seventeenth Century,* Yale University Press, April 30, 2013, 904 pp., p. 374, ISBN: 978-0300-153231.

231 Hessayon, Ariel, "The Little Ice Age and the River Thames frost fairs of the seventeenth century," *Historical Essays*, 18 February 2022. https://arielhessayon.substack.com/p/the-little-ice-age-and-the-river.

232 Hessayon, Ariel, "The Little Ice Age and the River Thames frost fairs of the

seventeenth century," *Historical Essays,* 18 February 2022. https://arielhessay-on.substack.com/p/the-little-ice-age-and-the-river.

233 Parker, Geoffrey, *Global Crisis: War, Climate Change and Catastrophe in the Seventeenth Century*, Yale University Press, April 30, 2013, 904 pp., p. 213, ISBN: 978-0300-153231.

234 Zillhardt, G. and Heberle, Hans, Der dreissigjährige Krieg in zietgenössischer Darstellung, Hans Heberle's "Zeytregister" (168-1672), The Thirty Years' War in contemporary representation: Hans Heberle's "Zeytregister" (1618-1672), 319 pp., 1975, p. 117. See also, Parker, Geoffrey, *Global Crisis: War, Climate Change and Catastrophe in the Seventeenth Century*, Yale University Press, April 30, 2013, 904 pp., p. 220, ISBN: 978-0300-153231.

235 Parker, Geoffrey, *Global Crisis: War, Climate Change and Catastrophe in the Seventeenth Century*, Yale University Press, April 30, 2013, 904 pp., p. 267, ISBN: 978-0300-153231.

236 Parker, Geoffrey, *Global Crisis: War, Climate Change and Catastrophe in the Seventeenth Century,* Yale University Press, April 30, 2013, 904 pp., p. 280, ISBN: 978-0300-153231.

237 Parker, Geoffrey, *Global Crisis: War, Climate Change and Catastrophe in the Seventeenth Century,* Yale University Press, April 30, 2013, 904 pp., p. 283, ISBN: 978-0300-153231.

238 Parker, Geoffrey, *Global Crisis: War, Climate Change and Catastrophe in the Seventeenth Century,* Yale University Press, April 30, 2013, 904 pp., p. 286, ISBN: 978-0300-153231.

239 Parker, Geoffrey, *Global Crisis: War, Climate Change and Catastrophe in the Seventeenth Century,* Yale University Press, April 30, 2013, 904 pp., p. 435, ISBN: 978-0300-153231.

240 Parker, Geoffrey, *Global Crisis: War, Climate Change and Catastrophe in the Seventeenth Century*, Yale University Press, April 30, 2013, 904 pp., p. 287, ISBN: 978-0300-153231.

241 Moote, *The Revolt*, p. 368. See also, Parker, Geoffrey, *Global Crisis: War, Climate Change and Catastrophe in the Seventeenth Century*, Yale University Press, April 30, 2013, 904 pp., p. 291, ISBN: 978-0300-153231.

242 Parker, Geoffrey, *Global Crisis: War, Climate Change and Catastrophe in the Seventeenth Century,* Yale University Press, April 30, 2013, 904 pp., p. 291, ISBN: 978-0300-153231.

243 Le Roy Ladurie, *Historire humaine et compare du climat,* 3 Volumes, Paris, 2004-2009, Volume I, pp. 337 to 339. See also, Parker, Geoffrey, *Global Crisis: War, Climate Change and Catastrophe in the Seventeenth Century*, Yale University Press, April 30, 2013, 904 pp., p. 294, ISBN: 978-0300-153231.

244 Parker, Geoffrey, *Global Crisis: War, Climate Change and Catastrophe in the Seventeenth Century,* Yale University Press, April 30, 2013, 904 pp., p. 297, ISBN: 978-0300-153231.

245 Parker, Geoffrey, *Global Crisis: War, Climate Change and Catastrophe in the Seventeenth Century,* Yale University Press, April 30, 2013, 904 pp., p. 310, ISBN: 978-0300-153231.

246 Parker, Geoffrey, *Global Crisis: War, Climate Change and Catastrophe in the Seventeenth Century,* Yale University Press, April 30, 2013, 904 pp., p. 317, ISBN: 978-0300-153231.

247 Parker, Geoffrey, *Global Crisis: War, Climate Change and Catastrophe in the Seventeenth Century,* Yale University Press, April 30, 2013, 904 pp., p. 314, ISBN: 978-0300-153231.

248 Parker, Geoffrey, *Global Crisis: War, Climate Change and Catastrophe in the Seventeenth Century,* Yale University Press, April 30, 2013, 904 pp., p. 318, ISBN: 978-0300-153231.

249 Behringer, Wolfgang, A *Cultural History of Climate,* Polity Press, Cambridge, UK, 2010, p. 162, ISBN 13: 978-0-7456-4528-5.

250 Fagan, Brian, *The Little Ice Age, How Climate Made History, 1300-1850,* Basic Books, New York, 2000, p. 163, ISBN: 10-465-02272-3.

251 Weis, Eberhard, "Frankreich von 1661 bis 1789," Theodor Schieder, *Handbuch der Europäischen Geschichte*, vol. 4, Stuttgart, 1968, p. 270. See also, Behringer, Wolfgang, *A Cultural History of Climate,* Polity Press, Cambridge, UK, 2010, p. 163, ISBN 13: 978-0-7456-4528-5.

252 Stahle, D.W. et al., "The Lost Colony and Jamestown Droughts," *Science,* Volume 280, Number 5363, pp. 564-567, 1998, p. 567. https://doi.org/10.1126/science.280.5363.564 . See also, Parker, Geoffrey, *Global Crisis: War, Climate Change and Catastrophe in the Seventeenth Century,* Yale University Press, April 30, 2013, 904 pp., p. 447, ISBN: 978-0300-153231.

253 White, Sam, *A Cold Welcome*, Harvard Press, Cambridge, Massachusetts, 2017, p. 119, ISBN 9780674244900.

254 White, Sam, *A Cold Welcome,* Harvard Press, Cambridge, Massachusetts, 2017, pp. 109-131 and 239-249, ISBN 9780674244900.

255 White, Sam, *A Cold Welcome*, Harvard Press, Cambridge, Massachusetts, 2017, p. 245, ISBN 9780674244900.

256 White, Sam, *A Cold Welcome,* Harvard Press, Cambridge, Massachusetts, 2017, pp. 246-247, ISBN 9780674244900.

257 White, Sam, *A Cold Welcome*, Harvard Press, Cambridge, Massachusetts, 2017, pp. 246-247, ISBN 9780674244900.

258 White, Sam, *A Cold Welcome,* Harvard Press, Cambridge, Massachusetts, 2017, p. 145. ISBN 9780674244900.

259 *Winthtrop Papers,* 3 Volumes, Boston, 1929-1943, Volume III, p. 240. See also, Parker, Geoffrey, *Global Crisis: War, Climate Change and Catastrophe in the Seventeenth Century,* Yale University Press, April 30, 2013, 904 pp., p. 449, ISBN: 978-0300-153231.

260 Parker, Geoffrey, *Global Crisis: War, Climate Change and Catastrophe in the Seventeenth Century,* Yale University Press, April 30, 2013, 904 pp., p. 446, ISBN: 978-0300-153231.

261 Morton, Thomas, *New English Canaan, or New Canaan containing an abstract of New England,* London, 1637, pp. 94-95. See also, Parker, Geoffrey, *Global Crisis: War, Climate Change and Catastrophe in the Seventeenth Century,* Yale University Press, April 30, 2013, 904 pp., p. 448, ISBN: 978-0300-153231.

262 Villalba, R., "Climatic fluctuations in northern Patagonia during the last 1,000 years as inferred from tree-ring records," *Quaternary Research,* Volume 34, pp. 349-360, 24 April 1990, pp. 355-356. https://doi.org/10.1016/0033-5894(90)90046-N; See also, Parker, Geoffrey, *Global Crisis: War, Climate Change and Catastrophe in the Seventeenth Century,* Yale University Press, April 30, 2013, 904 pp., p. 446, ISBN: 978-0300-153231.

263 Parker, Geoffrey, *Global Crisis: War, Climate Change and Catastrophe in the Seventeenth Century,* Yale University Press, April 30, 2013, 904 pp., p. 83, ISBN: 978-0300-153231.

264 Parker, Geoffrey, *Global Crisis: War, Climate Change and Catastrophe in the Seventeenth Century,* Yale University Press, April 30, 2013, 904 pp., p. 83, ISBN: 978-0300-153231.

265 Parker, Geoffrey, *Global Crisis: War, Climate Change and Catastrophe in the Seventeenth Century,* Yale University Press, April 30, 2013, 904 pp., p. 77, ISBN: 978-0300-153231.

266 Parker, Geoffrey, *Global Crisis: War, Climate Change and Catastrophe in the Seventeenth Century,* Yale University Press, April 30, 2013, 904 pp., p. 77, ISBN: 978-0300-153231.

267 Parker, Geoffrey, *Global Crisis: War, Climate Change and Catastrophe in the Seventeenth Century,* Yale University Press, April 30, 2013, 904 pp., p. 77, ISBN: 978-0300-153231.

268 Behringer, Wolfgang, *A Cultural History of Climate,* Polity Press, Cambridge, UK, 2010, p. 113, ISBN 13: 978-0-7456-4528-5.

269 Behringer, Wolfgang, *A Cultural History of Climate,* Polity Press, Cambridge, UK, 2010, pp. 128-129, ISBN 13: 978-0-7456-4528-5.

270 Behringer, Wolfgang, *A Cultural History of Climate,* Polity Press, Cam-

bridge, UK, 2010, p. 130, ISBN 13: 978-0-7456-4528-5.

271 Parker, Geoffrey, *Global Crisis: War, Climate Change and Catastrophe in the Seventeenth Century,* Yale University Press, April 30, 2013, 904 pp., p. 9, ISBN: 978-0300-153231.

272 White, Sam, *A Cold Welcome*, Harvard Press, Cambridge, Massachusetts, 2017, p. 128, ISBN 9780674244900.

273 Reagan, Romany, "Ice & Fire: How a Folk Demonology in the 'Little Ice Age' Led to the Witch Hunts of the 16th & 17th Centuries," *Blackthorn & Stone,* 9 July 2020. https://blackthornandstone.com/2020/07/09/ice-fire-how-a-folk-demonology-in-the-little-ice-age-led-to-the-witch-hunts-of-the-16th-17th-centuries/.

274 Reagan, Romany, "Ice & Fire: How a Folk Demonology in the 'Little Ice Age' Led to the Witch Hunts of the 16th & 17th Centuries," *Blackthorn & Stone*, 9 July 2020. https://blackthornandstone.com/2020/07/09/ice-fire-how-a-folk-demonology-in-the-little-ice-age-led-to-the-witch-hunts-of-the-16th-17th-centuries/.

275 The State Energy & Environmental Impact Center, "Local Suits Against Oil Companies," NYU School of Law, New York, NY. https://stateimpactcenter.org/issues/climate-action/local-suits-against-oil-companies?utm_source=chatgpt.com.

276 Parker, Geoffrey, *Global Crisis: War, Climate Change and Catastrophe in the Seventeenth Century*, Yale University Press, April 30, 2013, 904 pp., p. 512, ISBN: 978-0300-153231.

277 Parker, Geoffrey, *Global Crisis: War, Climate Change and Catastrophe in the Seventeenth Century,* Yale University Press, April 30, 2013, 904 pp., p. 509, ISBN: 978-0300-153231.

278 Parker, Geoffrey, *Global Crisis: War, Climate Change and Catastrophe in the Seventeenth Century,* Yale University Press, April 30, 2013, 904 pp., p. 509, ISBN: 978-0300-153231.

279 Parker, Geoffrey, *Global Crisis: War, Climate Change and Catastrophe in the Seventeenth Century,* Yale University Press, April 30, 2013, 904 pp., p. 509, ISBN: 978-0300-153231.

280 Piao, Shilong et al., "Characteristics, drivers and feedbacks of global greening," *Nature Reviews,* Volume 1, pp. 14-27, 9 December 2019. https://doi.org/10.1038/s43017-019-0001-x.

281 *Wikipedia*, "Medieval demography." https://en.wikipedia.org/wiki/Medieval_demography#cite_note-7.

282 *Worldometer*, "World Population by Year." https://www.worldometers.info/world-population/world-population-by-year/?utm_source=chatgpt.com.

283 *Worldometer*, "World Population by Year." https://www.worldometers.info/world-population/world-population-by-year/?utm_source=chatgpt.com.

284 Curry, Judith, "Judith Curry: Their weakest argument is that global warming is dangerous," *Michael Shellenberger Podcast,* 30 October 2025. https://www.public.news/p/judith-curry-their-weakest-argument.

Chapter 7 - Significance of the Sun

1 Marcott, Shaun A. et al., "A Reconstruction of the Regional and Global Temperature for the Past 11,300 years," *Science*, Volume 339, Issue 6124, pp. 1198-1201, 8 March 2013. https://doi.org/10.1126/science.1228026.

2 Humlum, Ole et al., "Identifying natural contributions to late Holocene climate change," *Global and Planetary Change*, Volume 79, Issue 1, pp. 145-156, 09 September 2011. https://doi.org/10.1016/j.gloplacha.2011.09.005.

3 Rosenthal, Yair et al., "Pacific Ocean Heat Content During the Past 10,000 Years," *Science*, Volume 342, Number 617, 1 November 2013. https://doi.org/10.1126/science.1240837.

4 Kaufman, D. et al., "A global database of Holocene paleotemperature records," *Scientific Data,* Volume 7, Issue 115, 12 August 2020. https://doi.org/10.1038/s41597-020-0445-3.

5 Bohleber, Pascal et al., "New glacier evidence for ice-free summits during the life of the Tyrolean Iceman," *Scientific Reports,* Volume 10, Article Number: 20513, 17 December 2020. https://doi.org/10.1038/s41598-020-77518-9.

6 Creel, Roger C. et al., "Global mean sea level likely higher than present during the Holocene," *Nature Communications*, Volume 15, Article Number 10731, 30 December 2024. https://doi.org/10.1038/s41467-024-54535-0; See also, Vacchi, Matteo, et. al., "Sea level since the Last Glacial Maximum from the Atlantic coast of Africa," *Nature Communications,* Volume 7, Article number 1486, 10 February 2025. https://doi.org/10.1038/s41467-025-56721-0.

7 Vinós, Javier, "Periodicities in solar variability and climate change: A simple model," *Energy Matters, Energy, Environment and Policy*, posted by Evan Mearns, 11 May 2016. https://euanmearns.com/periodicities-in-solar-variability-and-climate-change-a-simple-model/.

8 Crok, Marcel and May, Andy, *The Frozen Climate Views of the IPCC, an analysis of AR6*, CLINTEL Foundation, Amsterdam, The Netherlands, 2023, p. 83, ISBN: 979-8-89074-861-4.

9 Steinhilber, F. et al., "9,400 years of cosmic radiation and solar activity from ice cores and tree rings," *Proceedings of the National Academy of Sciences,* Volume 109, Issue 16, pp. 5967-5971, 2 April 2012. https://doi.org/10.1073/pnas.1118965109.

10 Vinós, Javier, "Periodicities in solar variability and climate change: A simple model," *Energy Matters, Energy, Environment and Policy,* posted by Evan Mearns, 11 May 2016. https://euanmearns.com/periodicities-in-solar-variability-and-climate-change-a-simple-model/.

11 Fagan, Brian, *The Little Ice Age: How Climate Made History, 1300-1850,* Basic Books, New York, 2000, p. 121, ISBN: 10-465-02272-3.

12 Eddy, J.A., "The maunder minimum," *Science*, Volume 192, Issue 4245, pp. 1189-1202, 18 June 1976. https://doi.org/10.1126/science.192.4245.1189; See also, Fleitmann, Dominik et al., "Holocene forcing of the Indian monsoon recorded in a stalagmite from southern Omen," *Science*, Volume 300, Issue 5626, pp. 1737-1739, 13 June 2003. https://doi.org/10.1126/science.1083130.

13 Esper, Jan et al., "Low-Frequency Signals in Long Tree-Ring Chronologies for Reconstructing Past Temperature Variability," *Science*, Volume 295, No. 5563, 22 March 2002. https://doi.org/10.1126/science.1066208; See also, Moberg, A. et al., "Highly variable Northern Hemisphere temperature reconstructions from low-and high-resolution proxy data," *Nature*, Volume 433, pp. 613-617, 10 February 2005. https://doi.org/10.1038/nature03265; and Ljungqvist, Fredrik, "A new reconstruction of temperature variability in the extra-tropical northern hemisphere during the last two millennia," *Geografiska Annaler: Series A, Physical Geography,* Volume 92, Issue 3, pp. 339-351, 6 September 2010, https://doi.org/10.1111/j.1468-0459.2010.00399.x.

14 Ammann, Caspar M. et al., "Solar influence on climate during the past millennium: Results from transient simulations with the NCAR Climate System Model," *PNAS*, Volume 104, Issue 10, pp. 3713-3718, 6 March 2007. https://doi.org/10.1073/pnas.0605064103.

15 IPCC, *The Physical Science Basis, Contribution of Working Group I to the Sixth Assessment Report on the Intergovernmental Panel on Climate Change,* "Chapter 7 - The Earth's Energy Budget, Climate Feedbacks and Climate Sensitivity," Cambridge University Press, Cambridge, UK and New York, NY, USA, Section 7.3, p. 962, 9 August 2021. https://www.ipcc.ch/report/ar6/wg1/downloads/report/IPCC_AR6_WGI_Chapter07.pdf.

16 Scafetta, Nicola, "Empirical assessment of the role of the Sun in climate change using balanced multi-proxy solar records," *Geoscience Frontiers,* Volume 14, Issue 6, November 2023. https://doi.org/10.1016/j.gsf.2023.101650.

17 Scafetta, Nicola, "Empirical assessment of the role of the Sun in climate change using balanced multi-proxy solar records," *Geoscience Frontiers,* Volume 14, Issue 6, November 2023. https://doi.org/10.1016/j.gsf.2023.101650.

18 Schmutz, Werner K., "Changes in the Total Solar Irradiance and climatic effect," *Journal of Space Weather and Space Climate,* Volume 11, Article Number 40, 22 July 2021. https://doi.org/10.1051/swsc/2021016.

19 Vieira, L. E. A. et al., "Evolution of the solar irradiance during the Holocene," *Astronomy and Astrophysics*, Volume 531, July 2011. https://doi.org/10.1051/0004-6361/201015843.

20 Krivova, N.A. et al., "Towards a long-term record of solar total and spectral irradiance," *Journal of Atmospheric and Solar-Terrestrial Physics,* Volume 73, Issue 2-3, pp. 223-234, February 2011. https://doi.org/10.1016/j.jastp.2009.11.013.

21 Indermühle, A. et al., "Holocene carbon-cycle dynamics based on CO_2 trapped in ice at Taylor Dome, Antarctica," *Nature*, Volume 398, pp. 121–126, 11 March 1999. https://doi.org/10.1038/18158.

22 Ljungqvist, Fredrik et al., "Centennial-Scale Temperature Change in Last Millennium Simulations and Proxy-Based Reconstructions," *Journal of Climate*, Volume 32, pp. 2441-2482, May 2019. https://doi.org/10.1175/JCLI-D-18-0525.1.

23 Scafetta, Nicola, "Empirical assessment of the role of the Sun in climate change using balanced multi-proxy solar records," *Geoscience Frontiers,* Volume 14, Issue 6, November 2023. https://doi.org/10.1016/j.gsf.2023.101650.

24 Shaviv, Nir, "Using the oceans as a calorimeter to quantify the solar radiative forcing," *Journal of Geophysical Research: Space Physics,* Volume 113, Issue A11, 4 November 2008. https://doi.org/10.1029/2007JA012989.

25 Scafetta, Nicola, "Empirical assessment of the role of the Sun in climate change using balanced multi-proxy solar records," *Geoscience Frontiers*, Volume 14, Issue 6, November 2023. https://doi.org/10.1016/j.gsf.2023.101650.

26 Steinhilber, F. et al., "Total Solar Irradiance during the Holocene," *Geophysical Research Letters*, Volume 36, Issue 19, 2 October 2009. https://doi.org/10.1029/2009GL040142; See also, Bond, Gerald et al., "Persistent solar influence on North Atlantic climate during the Holocene," *Science*, 294 (5549), pp 2130-2136, 15 Nov 2001. https://doi.org/10.1126/science.1065680 ; and Neff, U. et al., "Strong coherence between solar variability and monsoon in Oman between 9 and 6 kyr ago," *Nature*, Volume 411, pp. 290-293, 1 May 2001. https://doi.org/10.1038/35077048.

27 Gray, L.J. et al., "Solar Influences on Climate," *Reviews of Geophysics*, Volume 48, Issue 4, 30 October 2010. https://doi.org/10.1029/2009RG000282.

28 Soukharev, B. E. and Hood, L. L., "Solar cycle variation of stratospheric ozone: Multiple regression analysis of long-term satellite data sets and comparisons with models," *Journal of Geophysical Research: Atmospheres,* Volume 111, Issue D20, 31 October 2006. https://doi.org/10.1029/2006JD007107.

29 Shindell, Drew et al., "Solar Cycle Variability, Ozone, and Climate," *Science*, Volume 284, Number 5412, pp. 305-308, 9 April 1999. https://doi.org/10.1126/science.284.5412.305.

30 Georgieva, Katya and Vereteneko, Svetlana, "Solar influences on the Earth's atmosphere: solved and unsolved questions," *Frontiers in Astronomy and Space Sciences*, Volume 10, 21 December 2023, pp. 11-12. https://doi.org/10.3389/fspas.2023.1244402.

31 Georgieva, Katya and Vereteneko, Svetlana, "Solar influences on the Earth's atmosphere: solved and unsolved questions," *Frontiers in Astronomy and Space Sciences,* Volume 10, 21 December 2023, p. 12. https://doi.org/10.3389/fspas.2023.1244402.

32 Vinós, Javier, *Solving the Climate Puzzle: The Sun's Surprising Role*, Critical Science Press, Madrid, 2023, ISBN: 978-84-125867-7-0.

33 Vinós, Javier, *Solving the Climate Puzzle: The Sun's Surprising Role*, Critical Science Press, Madrid, 2023, p. 193, ISBN: 978-84-125867-7-0.

34 Vinós, Javier, *Solving the Climate Puzzle: The Sun's Surprising Role,* Critical Science Press, Madrid, 2023, p. 264, ISBN: 978-84-125867-7-0.

35 Vinós, Javier, *Solving the Climate Puzzle: The Sun's Surprising Role*, Critical Science Press, Madrid, 2023, p. 269, ISBN: 978-84-125867-7-0.

36 NASA, "Chilly Temperatures During the Maunder Minimum," *Earth Observatory*. https://www.earthobservatory.nasa.gov/images/7122/chilly-temperatures-during-the-maunder-minimum.

37 Tselioudis, George et al., "Contraction of the World's Storm-Cloud Zones the Primary Contributor to the 21st Century Increase in the Earth's Sunlight Absorption," *Geophysical Research Letters*, Volume 52, Issue 11, 8 June 2025. https://doi.org/10.1029/2025GL114882.

38 Vinós, Javier, *Solving the Climate Puzzle: The Sun's Surprising Role*, Critical Science Press, Madrid, 2023, ISBN: 978-84-12867-7-0, p. 72.

39 Burton, Robert, *The Anatomy of Melancholy,* Oxford, 1621. Also see Behringer, Wolfgang, *A Cultural History of Climate*, Polity Press, Cambridge, UK, 2010, p. 115, ISBN 13: 978-0-7456-4528-5.

40 Fagan, Brian, *The Little Ice Age: How Climate Made History, 1300-1850,* Basic Books, New York, 2000, p. 90, ISBN: 10-465-02272-3.

41 Bradzdil, Rudolf, "Historical Climatology in Europe—the State of the Art," *Climatic Change,* Volume 70, Issue 3, pp. 363-430, June 2005. https://doi.org/10.1007/s10584-005-5924-1.

42 White, Sam, *A Cold Welcome,* Harvard Press, Cambridge, Massachusetts, 2017, p. 218. ISBN 9780674244900.

43 Neuberger, Hans, "Climate in Art," *Weather*, Volume 25, Issue 2, pp. 46-56, February 1970. https://doi.org/10.1002/j.1477-8696.1970.tb03232.x

44 Binkhof, Tim, "The Climate Canvasses of the Little Ice Age," *JSTOR Daily,*

Art & Art History, 6 December 2023. https://daily.jstor.org/climate-canvasses-of-the-little-ice-age/; See also, White, Sam, "The Real Little Ice Age," *Journal of Interdisciplinary History,* Volume 44, pp. 327-352, 1 November 2013, p. 338. https://doi.org/10.1162/JINH_A_00574.

45 McCormick, Michael et al., "Climate change during and after the Roman Empire: Reconstructing the Past from Scientific and Historical Evidence," *Journal of Interdisciplinary History,* Volume XLIII, Issue 1, pp. 169-220, Autumn 2012, p. 175. https://doi.org/10.1162/JINH_a_00379.

46 McCormick, Michael et al., "Climate Change during and after the Roman Empire: Resconstructing the Past from Scientific and Historical Evidence," *Journal of Interdisciplinary History,* Volume XLIII, Issue 1, pp. 169-220, Autumn 2012, p. 185. https://doi.org/10.1162/JINH_a_00379.

47 McCormick, Michael et al., "Climate Change during and after the Roman Empire: Resconstructing the Past from Scientific and Historical Evidence," *Journal of Interdisciplinary History,* Volume XLIII, Issue 1, pp. 169-220, Autumn 2012, p. 191. https://doi.org/10.1162/JINH_a_00379.

48 McCormick, Michael et al., "Climate Change during and after the Roman Empire: Resconstructing the Past from Scientific and Historical Evidence," *Journal of Interdisciplinary History,* Volume XLIII, Issue 1, pp. 169-220, Autumn 2012, p. 195. https://doi.org/10.1162/JINH_a_00379.

49 Duckworth. P. D., *The Environment of Britain in the First Millennium A.D.,* Cambridge University Press, London, 2000, ISBN: 0-7156-2903-4.

50 Ogilvie, Astrid and Farmer, Graham, *Climates of the British Isles, Documenting the Medieval Climate,* Routledge, 1997, pp. 112-133. ISBN: 978-1315-87-0793.

51 Hilderbrandt, Sybille, "Vikings grew barley in Greenland," *ScienceNordic,* 3 February 2012. https://www.sciencenordic.com/agriculture-archaeology-denmark/vikings-grew-barley-in-greenland/1447746.

52 Campbell, Bruce, *The Great Transition: Climate, Disease and Society in the Late-Medieval World,* Cambridge University Press, Cambridge, UK, pp. 488, July 2016. ISBN: 978-0-521-19588-1.

53 Campbell, Bruce, *The Great Transition: Climate, Disease and Society in the Late-Medieval World,* Cambridge University Press, Cambridge, UK, pp. 488, July 2016. ISBN: 978-0-521-19588-1.

54 IPCC, *The Physical Science Basis, Contribution of Working Group 1 to the Fifth Assessment Report of the Intergovernmental Panel on Climate Change,* "Chapter 7 - Clouds and Aerosols," Cambridge University Press, Cambridge, UK, Section 7.2.1.2, pp. 580-581, 30 September 2013. https://www.ipcc.ch/site/assets/uploads/2018/02/WG1AR5_Chapter07_FINAL-1.pdf.

55 IPCC, *The Physical Science Bais, Contribution of Working Group 1 to the Fifth Assessment Report of the Intergovernmental Panel on Climate Change,* "Chapter 7 - Clouds and Aerosols," Cambridge University Press, Cambridge, UK, Executive Summary, p. 574, 30 September 2013. https://www.ipcc.ch/site/assets/uploads/2018/02/WG1AR5_Chapter07_FINAL-1.pdf.

56 Shindell, D. T. et al., "Solar forcing of regional climate change during the Maunder Minimum," *Science,* Volume 294, Issue 5549, pp. 2149-2152, 7 December 2001. https://doi.org/10.1126/science.1064363.

57 Potgieter, Marius S., "Solar Modulation of Cosmic Rays," *Living Reviews in Solar Physics,* Volume 10, Article Number 3, 13 June 2013, Figure 6. https://doi.org/10.12942/lrsp-2013-3.

58 Svensmark, Henrik and Friis-Christensen, Eigil, "Variation of cosmic ray flux and global cloud coverage—a missing link in solar-climate relationships," *Journal of Atmospheric and Solar-Terrestrial Physics,* Volume 59, Issue 11, pp. 1225-1232, July 1997. https://doi.org/10.1016/S1364-6826(97)00001-1.

59 *Wikipedia,* "Global atmospheric electrical circuit," https://en.wikipedia.org/wiki/Global_atmospheric_electrical_circuit.

60 *Wikipedia,* "Global atmospheric electrical circuit," https://en.wikipedia.org/wiki/Global_atmospheric_electrical_circuit.

61 Grieder, Peter K. F., Cosmic Rays at Earth, Elsevier, 2001, "Chapter 2—Cosmic Rays in the Atmosphere," pp. 55-303. https://doi.org/10.1016/B978-044450710-5/50004-X.

62 Baxilevskaya, G. A. et al., "Cosmic Ray Induced Ion Production in the Atmosphere," *Space Science Reviews,* Volume 137, pp. 149-173, 22 March 2008. https://doi.org/10.1007/s11214-008-9339-y.

63 Baxilevskaya, G. A. et al., "Cosmic Ray Induced Ion Production in the Atmosphere," *Space Science Reviews,* Volume 137, pp. 149-173, 22, March 2008, p. 150. https://doi.org/10.1007/s11214-008-9339-y.

64 Enghoff, Martin B., "Aerosol nucleation induced by a high energy particle beam," *Geophysical Research Letters,* Volume 38, Issue 9, 12 May 2011. https://doi.org/10.1029/2011GL047036.

65 Kirkby, Jasper et al., "Role of sulfuric acid, ammonia and galactic cosmic rays in atmospheric aerosol nucleation," *Nature,* volume 476, pp. 429-433, 24 August 2011. https://doi.org/10.1038/nature10343.

66 Pierce, J. R. and Adams, P. J., "Can cosmic rays affect cloud condensation nuclei by altering new particle formation rates?" *Geophysical Research Letters,* Volume 36, Issue 9, 13 May 2009. https://doi.org/10.1029/2009GL037946; See also, *Physicsworld,* "Physicists claim further evidence of link between cosmic rays and cloud formation," 9 September 2013. https://physicsworld.com/a/

physicists-claim-further-evidence-of-link-between-cosmic-rays-and-cloud-formation/.

67 Svensmark, H. et al., "Increased Ionization Supports Growth of Aerosols into Cloud Condensation Nuclei," *Nature Communications,* Volume 8, Article Number 2199, 19 December 2017. https://doi.org/10.1038/s41467-017-02082-2.

68 Svensmark, H. et al., "Cosmic Ray Decreases Effect Atmosphere Aerosols and Clouds," *Geophysical Research Letters,* 1 August 2009. https://doi.org/10.1029/2009GL038429.

69 Svensmark, Henrik et al., "Atmospheric Ionization and Cloud Radiative Forcing," *Nature Scientific Reports,* Volume 11, Article Number:19668, 11 October 2021. https://www.nature.com/articles/s41598-021-99033-1.

70 Svensmark, Henrik et al., "Atmospheric Ionization and Cloud Radiative Forcing," *Nature Scientific Reports,* Volume 11, Article Number:19668, 11 October 2021. https://www.nature.com/articles/s41598-021-99033-1. See also, Svensmark, Hendrik and Shaviv, Nir, "Hendrik Svensmark & Nir Shaviv—How the Sun regulates our climate," EIKE, 2023. https://www.youtube.com/watch?v=PHiVlf5PXRA.

71 Usoskin, I.G. et al., "Latitudinal dependence of low cloud amount on cosmic ray induced ionization," *Geophysical Research Letters,* Volume 31, Issue 16, August 2004. https://doi.org/10.1029/2004GL019507.

72 Pallé, E. et al., "Changes in Earth's Reflectance over the Past Two Decades," *Science,* 304 (5675): pp. 1299-1301. http://doi.org/10.1126/science.1094070.

73 Shaviv, Nir J., "Using the oceans as a calorimeter to quantify the solar radiative forcing," *Journal of Geophysical Research,* Volume 113, Issue A11, 4 November 2008. https://doi.org/10.1029/2007JA012989.

74 Neff, U. et al., "Strong coherence between solar variability and monsoon in Oman between 9 and 6 kyr ago," *Nature,* Volume 411, pp. 290-293, 1 May 2001. https://doi.org/10.1038/35077048.

75 Shaviv, Nir, "The Milky Way Galaxy's Spiral Arms and Ice-Age Epochs and the Cosmic Ray Connection," *ScienceBits.* http://www.sciencebits.com/ice-ages.

76 Shaviv, Nir J. and Veizer, Ján, "Celestial driver of Phanerozoic climate," *GSA TODAY,* July 2003. https://www.atmos.washington.edu/academics/classes/2003Q4/211/articles_optional/CelestialDriver.pdf.

77 IPCC, *The Physical Science Basis, Contribution of Working Group I to the Sixth Assessment Report on the Intergovernmental Panel on Climate Change,* "Chapter 7 - The Earth's Energy Budget, Climate Feedbacks and Climate Sensitivity," Cambridge University Press, Cambridge, UK and New York, NY, USA, Section 7.3, p. 962, 9 August 2021. https://www.ipcc.ch/report/ar6/wg1/down-

loads/report/IPCC_AR6_WGI_Chapter07.pdf.

78 Svensmark, Henrik, "It is impossible for me to get funding for this kind of work," *Tom Nelson Podcast,* #44, 21 November 2022. https://www.youtube.com/watch?v=sZBWEKCW2Fc.

79 Shaviv, Nir, "The Critique of Knud Jahnke and a New Meteor Exposure Age Analysis," *ScienceBits.* http://www.sciencebits.com/JahnkeResponse.

80 Shaviv, Nir, "The Critique of Knud Jahnke and a New Meteor Exposure Age Analysis," *ScienceBits.* http://www.sciencebits.com/JahnkeResponse.

81 Shaviv, Nir, "The Critique of Knud Jahnke and a New Meteor Exposure Age Analysis," *ScienceBits.* http://www.sciencebits.com/JahnkeResponse.

82 Shaviv, Nir, "The Critique of Knud Jahnke and a New Meteor Exposure Age Analysis," *ScienceBits.* http://www.sciencebits.com/JahnkeResponse.

83 Shaviv, Nir, "The Critique of Knud Jahnke and a New Meteor Exposure Age Analysis," *ScienceBits.* http://www.sciencebits.com/JahnkeResponse.

84 SkepticalScience.com, "What's the link between cosmic rays and climate change?" https://skepticalscience.com/print.php?r=26.

85 Solanki, S. K. and Krivova, N. A., "Can solar variability explain global warming since 1970?" *Journal of Geophysical Research: Space Physics,* Volume 108, Issue A5, 21 May 2003. https://doi.org/10.1029/2002JA009753.

86 Agee, Ernest M. et al., "Relationship of Lower-Troposphere Cloud Cover and Cosmic Rays: An Updated Perspective," *Journal of Climate,* Volume 25: Issue 3, pp. 1057 to 1060, 1 February 2012. https://doi.org/10.1175/JCLI-D-11-00169.1.

87 Nikolov, Ned and Zeller, Karl F., "Role of Earth's Albedo Variations and Top-of-the-Atmosphere Energy Imbalance in Recent Warming: New Insights from Satellite and Surface Observations," *MDPI Geomatics,* 4(3), pp. 311-341, 20 August 2024. https://doi.org/10.3390/geomatics4030017.

88 Laken, B. A. et al., "Cosmic rays linked to rapid mid-latitude cloud changes," *Atmospheric Chemistry and Physics,* Volume 10, Issue 22, pp. 10941-10948, 24 November 2010. https://doi.org/10.5194/acp-10-10941-2010.

89 Gordon, Hamish et al., "Causes and importance of new particle formation in the present-day and preindustrial atmospheres," *Journal of Geophysical Research: Atmospheres,* Volume 122, Issue 16, pp. 8739-8760, 10 July 2017. https://doi.org/10.1002/2017JD026844.

90 *Our World in Data,* "Global sulfur dioxide (SO_2) emissions by world region"; "Sulfur dioxide emissions from all sectors." https://ourworldindata.org/grapher/so-emissions-by-world-region-in-million-tonnes.

91 Wang, Hai et al., "Atmosphere teleconnections from abatement of China

aerosol emissions and exacerbate Northeast Pacific warm blob events," *PNAS,* Volume 121, Issue 21, e2313797121, 6 May 2024. https://doi.org/10.1073/pnas.2313797121.

92 Yuan, Tianle et al., "Abrupt reduction in shipping emissions as inadvertent geoengineering termination shock produces substantial radiative warming," *Nature, Communications Earth & Environment,* Volume 5, Article 281, 30 May 2024. https://doi.org/10.1038/s43247-024-01442-3.

93 Pierce, J. R. and Adams, P. J., "Can cosmic rays affect cloud condensation nuclei by altering new particle formation rates?" *Geophysical Research Letters,* Volume 36, Issue 9, 13 May 2009. https://doi.org/10.1029/2009GL037946.

94 Svensmark, H. et al., "Cosmic Ray Decreases Effect Atmosphere Aerosols and Clouds," *Geophysical Research Letters,* 1 August 2009. https://doi.org/10.1029/2009GL038429; See also, Svensmark, H. et al., "Increased ionization supports growth of aerosols into cloud condensation nuclei," *Nature Communications,* Volume 8, Article Number 2199, 19 December 2017. https://doi.org/10.1038/s41467-017-02082-2; and Svensmark, Henrik et al., "Atmospheric Ionization and Cloud Radiative Forcing," *Nature Scientific Reports,* Volume 11, Article Number 19668, 11 October 2021. https://www.nature.com/articles/s41598-021-99033-1.

95 Kristjánsson, J. E. et al., "Cosmic rays, cloud condensation nuclei and clouds—a reassessment using MODIS data," *Atmospheric Chemistry and Physics,* Volume 8, Issue 24, 11 December 2008. https://doi.org/10.5194/acp-8-7373-2008.

96 Kniveton, D. R., "Precipitation, cloud cover and Forbush decreases in galactic cosmic rays," *Journal of Atmospheric and Solar-Terrestrial Physics,* Volume 66, Issues 13-14, pp. 1135-1142, September 2004. https://doi.org/10.1016/j.jastp.2004.05.010.

97 Todd, Martin C. and Kniveton, Dominic R., "Short-term variability in satellite-derived cloud cover and galactic cosmic rays: an update," *Journal of Atmospheric and Solar-Terrestrial Physics,* Volume 66, Issues 13-14, pp. 1205-1211, September 2004. https://doi.org/10.1016/j.jastp.2004.05.002.

98 Harrison, Giles R. and Ambaum, Maarten, H. P., "Observing Forbush Decreases in Cloud at Shetland," *Journal of Atmospheric and Solar-Terrestrial Physics,* Volume 72, Issue 18, pp. 1408-1414, December 2010. https://doi.org/10.1016/j.jastp.2010.09.025.

99 Alley, Richard, "Richard Alley—4.6 Billion Years of Earth's Climate History: The Role of CO_2," National Academy of Sciences, 152nd Annual Meeting, 2016. https://www.youtube.com/watch?v=ujkcTZZlikg.

100 Niels Bohr Institute, "Greenland ice cores reveal warm climate of the past," University of Copenhagen, 22 January 2013. https://nbi.ku.dk/english/news/

news13/greenland-ice-cores-reveal-warm-climate-of-the-past.

101 Cooper, Alan et al., "A global environmental crisis 42,000 years ago," *Science*, Volume 371, Number 6531, pp. 811-818, 19 February 2021. https://doi.org/10.1126/science.abb8677.

102 Courtillot, Vincent et al., "Are there connections between the Earth's magnetic field and climate?" *Earth and Planetary Science Letters*, Volume 253, Issues 3-4, pp. 328-339, 20 January 2007. https://doi.org/10.1016/j.epsl.2006.10.032.

103 Campuzano, S. A., "New perspectives in the study of the Earth's magnetic field and climate connection: The use of transfer entropy," *PLOS One,* Volume 15, Issue 11, 15 November 2018. https://doi.org/10.1371/journal.pone.0207270.

104 Gallet, Yves et al., "Does Earth's magnetic field secular variation control centennial climate change?" *Earth and Planetary Science Letters,* Volume 236, Issues 1-2, pp. 229-247, 30 July 2005. https://doi.org/10.1016/j.epsl.2005.04.045.

105 Zharkova, Valentina, "Modern Grand Solar Minimum will lead to terrestrial cooling," *Temperature*, Volume 7: Issue 3, pp. 217-222. https://doi.org/10.1080/23328940.2020.1796243.

106 Zhao, Junwei et al., "Detection of Equatorward Meridional Flow and Evidence of Double-cell Meridional Circulation inside the Sun," *The Astrophysical Journal Letters,* Volume 774, Issue 3, 27 August 2013. https://doi.org/10.1088/2041-8205/774/2/L29.

107 Usoskin, I. and Kovaltsov, G., "A two-wave dynamo model by Zharkova et. al. (2015) disagrees with data on long-term solar variability," Cornell University, 17 December 2015. https://doi.org/10.48550/arXiv.1512.05516.

108 Zharkova, V. et al., "Reply to Comment on paper 'On a role of quadruple component of magnetic field in defining solar activity in grand cycles' by Usokin (2017)," Northumbria University, 2 October 2017. https://researchportal.northumbria.ac.uk/ws/portalfiles/portal/14102501/Answer_zhark_jastp_rev_v2.pdf.

109 Zharkova, V. V. et al., "Probing the Double Dynamo Model with Solar-Terrestrial Activity in the Past Millennia," Proceedings of the Thirteenth Workshop "Solar Influences on the Magnetosphere, Ionosphere and Atmosphere, September 2021. https://doi.org/10.31401/WS.2021.proc.24.

110 Zharkova, Valentina, "#42 Valentina Zharkova: 'In next 30 years, global warming probably will be the last thing in our mind,'" *Tom Nelson Podcast,* #42, 14 November 2022. https://www.youtube.com/watch?v=LYOMKLD-beYE&t=3558s.

Chapter 8 - Carbon Dioxide

1 Nelson, Michael and Nelson, David B., "Decoupling CO_2 from Climate Change," *International Journal of Geosciences,* Volume 15, pp. 246-269, 26 March 2024. https://doi.org/10.4236/ijg.2024.153015.

2 IPCC, *Climate Change 2021: The Physical Science Basis: Contribution of Working Group I to the Sixth Assessment Report of the Intergovernmental Panel on Climate Change,* "Summary for Policymakers," Cambridge University Press, Cambridge UK and New York, NY, USA, Section A.2, p. 5, 9 August 2021. https://www.ipcc.ch/report/ar6/wg1/downloads/report/IPCC_AR6_WGI_SPM.pdf.

3 Dagsvik, John K. and Moen, Sigmund H., "To what extent are temperature levels changing due to greenhouse gas emissions?" Statistisk sentralbyra, Statistics Norway, Discussion Paper 1007, September 2023. https://www.ssb.no/en/natur-og-miljo/forurensning-og-klima/artikler/to-what-extent-are-temperature-levels-changing-due-to-greenhouse-gas-emissions/_/attachment/inline/5a3f4a9b-3bc3-4988-9579-9fea82944264:f63064594b9225f9d7dc458b-0b70a646baec3339/DP1007.pdf.

4 Vinós, Javier, *Solving the Climate Puzzle: The Sun's Surprising Role,* Critical Science Press, Madrid, 2023, p. 214, ISBN: 978-84-125867-7-0.

5 Marcott, Shaun A. et al., "A Reconstruction of the Regional and Global Temperature for the Past 11,300 Years," *Science,* Volume 339, Issue 6124, 8 March 2013, pp. 1198-1201. https://doi.org/10.1126/science.1228026.

6 Kaufman, D. et al., "A global database of Holocene paleotemperature records," *Scientific Data 7,* 115, 12 August 2020. https://doi.org/10.1038/s41597-020-0445-3.

7 Bohleber, Pascal, et al., "New glacier evidence for ice-free summits during the life of the Tyrolean Iceman," *Scientific Reports,* Volume 10, Article Number 20513, 17 December 2020. https://doi.org/10.1038/s41598-020-77518-9.

8 Vinós, Javier, *Solving the Climate Puzzle: The Sun's Surprising Role,* Critical Science Press, Madrid, 2023, p. 151, ISBN: 978-84-125867-7-0.

9 Vinós, Javier, *Solving the Climate Puzzle: The Sun's Surprising Role,* Critical Science Press, Madrid, 2023, p. 151, ISBN: 978-84-125867-7-0.

10 Hannon, Renee, "The Holocene CO_2 Dilemma," Andy May Petrophysicist, May 26, 2023. https://andymaypetrophysicist.com/2023/05/26/the-holocene-co2-dilemma/.

11 Viterito, Arthur et al., "Viterito/Kamis/Yim/Catt: Impacts of Geothermal Energy on Climate," *Tom Nelson Podcast,* #181, 23 December 2023. https://www.youtube.com/watch?v=lbDbA32fNek.

12 Myhre, Gunnar et al., "New estimates of radiative forcing due to well mixed

greenhouse gases," *Geophysical Research Letters,* Volume 25, Issue 14, pp. 2715-2718, 15 July1998, p. 2718. https://doi.org/10.1029/98GL01908.

13 Clark, Ian, "CO_2 lags temperature rises in ice core data"; Skeptical views of CO_2 driven global warming. http://hyperphysics.phy-astr.gsu.edu/hbase/thermo/icecore.html.

14 Vakulenko, N. V. et al., "Evidence for the Leading Role of Temperature Variations Relative to Greenhouse Gas Concentration Variations in the Vostok Ice Core Record," *Doklady Earth Sciences*, Volume 396, Issue 5, pp. 663-667, June 2004. https://www.researchgate.net/publication/295305841_Evidence_for_the_leading_role_of_temperature_variations_in_comparison_with_the_greenhouse_gases_concentration_variations_in_the_Vostok_ice_core_record.

15 Vakulenko, N. V. et al., "Evidence for the Leading Role of Temperature Variations Relative to Greenhouse Gas Concentration Variations in the Vostok Ice Core Record," *Doklady Earth Sciences,* Volume 396, Issue 5, pp. 663-667, June 2004. https://www.researchgate.net/publication/295305841_Evidence_for_the_leading_role_of_temperature_variations_in_comparison_with_the_greenhouse_gases_concentration_variations_in_the_Vostok_ice_core_record.

16 Caillon, Nicolas et al., "Timing of Atmospheric CO_2 and Antarctic Temperature Changes Across Termination III," *Science*, Volume 399, Issue 5613, pp. 1728-1731, 14 March 2003. https://doi.org/10.1126/science.1078758.

17 Fischer, Hubertus et al., "Ice Core Recorded of Atmospheric CO_2 Around the Last Three Glacial Terminations," *Science*, Volume 283, Issue 5408, pp. 1712-1714, 12 March 1999. https://doi.org/10.1126/science.283.5408.1712.

18 Monnin, Eric et al., "Atmospheric CO_2 Concentrations over the Last Glacial Termination," *Science*, Volume 291, Issue 5501, pp. 112-114, 5 January 2001. https://doi.org/10.1126/science.291.5501.112.

19 Pedro, J.B. et al., "Tightened constraints on the time-lag between Antarctic temperature and CO_2 during the last deglaciation," *Climate Past,* Volume 8, Issue 4, pp. 1213-1221, 23 July 2012. https://doi.org/10.5194/cp-8-1213-2012.

20 University of Southampton, "Carbon release from the ocean helped end the Ice Age," 12 February 2015. https://www.southampton.ac.uk/news/2015/02/ice-age-carbon.page.

21 Zeebe, Richard E. and Wolf-Gladrow, Dieter A., "CO_2 in Seawater: Equilibrium, Kinetics, Isotopes," *Elsevier Oceanography Series,* Volume 65, 2001. ISBN: 9780444509468.

22 *... and Then There's Physics,* "Henry's Law," 28 October 2013. https://andthentheresphysics.wordpress.com/2013/10/28/henrys-law/.

23 Sarmiento, Jorge L. and Gruber, Nicolas, *Ocean Biogeochemical Dynamics,* Princeton University Press, 2006. ISBN: 9780691017075.

24 University of Southampton, "Carbon release from the ocean helped end the Ice Age," 12 February 2015. https://www.southampton.ac.uk/news/2015/02/ice-age-carbon.page; See also, Bae, J. W. B. et al., "CO_2 storage and release in the deep Southern Ocean on millennial to centennial timescales," *Nature*, Volume 562, pp. 569-573, 24 October, 2018. https://doi.org/10.1038/s41586-018-0614-0.

25 Belt, Simon T. et al., "Identification of paleo Arctic winter sea ice limits and the marginal ice zone: Optimized biomarker-based reconstructions of late Quaternary Arctic sea ice," *Earth and Planetary Science Letters,* Volume 431, pp. 127-139, 1 December 2015. https://doi.org/10.1016/j.epsl.2015.09.020.

26 Yu, J. et al., "Sequestration of carbon in the deep Atlantic during the last glaciation," *Nature Geoscience,* Volume 8, pp. 319-324, 8 February 2016. https://doi.org/10.1038/ngeo2657.

27 Koutsoyiannis, D. et al., "On Hens, Eggs, Temperatures and CO_2: Causal Links in Earth's Atmosphere," *Science*, Volume 5, Issue 3, 13 September 2023. https://doi.org/10.3390/sci5030035.

Chapter 9 - Science of Greenhouse Gases

1 Vinós, Javier, *Solving the Climate Puzzle: The Sun's Surprising Role,* Critical Science Press, Madrid, 2023, p. 34, ISBN: 978-84-12867-7-0.

2 Qualglia, Ilaria and Visioni, Daniele, "Modeling 2020 regulatory changes in international shipping changes in international shipping emissions helps explain anomalous 2023 warming," *EDS Letters,* Volume 15, Issue 6, pp. 1527-1441, 28 November 2024. https://doi.org/10.5194/esd-15-1527-2024; See also Dewitte, Steven and Clerbaux, Nicholas, "Decadal Changes of Earth's Outgoing Longwave Radiation," *Remote Sensing,* Volume 10, Issue 10, pp. 153-1546, 25 September 2018. https://doi.org/10.3390/rs10101539.

3 Tselioudis, George et al., "Contraction of the World's Storm-Cloud Zones the Primary Contributor to the 21st Century Increase in the Earth's Sunlight Absorption," *Geophysical Research Letters,* Volume 52, Issue 11, 8 June 2025. https://doi.org/10.1029/2025GL114882.

4 L'Ecuyer, Tristan S. et al., "Reassessing the Effect of Cloud Type on Earth's Energy Balance in the Age of Active Spaceborne Observations; Part I: Top of Atmosphere and Surface," *Journal of Climate,* Volume 32, Issue 19, pp. 6197-6217, 1 October 2019. https://doi.org/10.1175/JCLI-D-18-0753.1.

5 Nikolov, Ned and Zeller, Karl F., "Role of Earth's Albedo Variations and Top-of-the-Atmosphere Energy Imbalance in Recent Warming: New Insights from Satellite and Surface Observations," *MDPI Geomatics*, Volume 4 Issue 3, pp. 311-341, 20 August 2024. https://doi.org/10.3390/geomatics4030017.

6 Tselioudis, George et al., "Contraction of the World's Storm-Cloud Zones

the Primary Contributor to the 21st Century Increase in the Earth's Sunlight Absorption," *Geophysical Research Letters,* Volume 52, Issue 11, 8 June 2025. https://doi.org/10.1029/2025GL114882.

7 Myhre, Gunnar et al., "New estimates of radiative forcing due to well mixed greenhouse gases," *Geophysical Research Letters,* Volume 25, Issue 14, pp. 2715-2718, 15 July1998. https://doi.org/10.1029/98GL01908 ; see also the "Logarithmic Decline of CO_2 Marginal Warming" section of Chapter 9 for the proper use of the Myhre equations.

8 Qualglia, Ilaria and Visioni, Daniele, "Modeling 2020 regulatory changes in international shipping changes in international shipping emissions helps explain anomalous 2023 warming," *EDS Letters,* Volume 15, Issue 6, pp. 1527-1441, 28 November 2024. https://doi.org/10.5194/esd-15-1527-2024.

9 L'Ecuyer, Tristan S.et al., "Reassessing the Effect of Cloud Type on Earth's Energy Balance in the Age of Active Spaceborne Observations; Part I: Top of Atmosphere and Surface," *Journal of Climate*, Volume 32, Issue 19, pp. 61217-6217, 1 October 2019. https://doi.org/10.1175/JCLI-D-18-0753.1.

10 Nikolov, Ned and Zeller, Karl F., "Role of Earth's Albedo Variations and Top-of-the-Atmosphere Energy Imbalance in Recent Warming: New Insights from Satellite and Surface Observations," *MDPI Geomatics 2024,* Volume 4 Issue 3, pp. 311-341, 20 August 2024. https://doi.org/10.3390/geomatics4030017.

11 Myhre, Gunnar et al., "New estimates of radiative forcing due to well mixed greenhouse gases," *Geophysical Research Letters,* Volume 25, Issue 14, pp. 2715-2718, 15 July 1998, p. 2718. https://doi.org/10.1029/98GL01908.

12 IPCC, *The Physical Science Basis, Contribution of Working Group 1 to the Fifth Assessment Report of the Intergovernmental Panel on Climate Change,* "Chapter 8 - Anthropogenic and Natural Radiation," Cambridge University Press, Cambridge UK, Sections 8.3.2.1, 8.3.2.2, 8.3.2.3, 8.3.2.4, and Table 8.2, pp. 676-678, 30 September 2013. http://www.ipcc.ch/site/assets/uploads/2018/02/WG1AR5_Chapter08_FINAL.pdf.

13 Myhre, Gunnar et al., "New estimates of radiative forcing due to well mixed greenhouse gases," *Geophysical Research Letters,* Volume 25, Issue 14, pp. 2715-2718, 15 July 1998. https://doi.org/10.1029/98GL01908.

14 Forster, Piers M., et al., "Recommendations for diagnosing effective radiative forcing from climate models for CMIP6," *Journal of Geophysical Research: Atmospheres,* Volume 121, Issue 20, pp. 12,460-12,475, 8 October 2016. https://doi.org/10.1002/2016JD025320E.

15 Feldman, D.R. et al., "Observational determination of surface radiative forcing by CO_2 from 2000 to 2010," *Nature Research Letter,* Volume 25, pp. 339-343, 25 February 2015. https://doi.org/10.1038/nature14240.

16 Wijngaarden, William van and Happer, William, "Infrared Forcing by

Greenhouse Gases," 18 June 2019. https://co2coalition.org/wp-content/uploads/2022/03/Infrared-Forcing-by-Greenhouse-Gases-2019-Revised-3-7-2022.pdf.

17 Wijngaarden, William van and Happer, William, "Infrared Forcing by Greenhouse Gases," 18 June 2019, p. 47. https://co2coalition.org/wp-content/uploads/2022/03/Infrared-Forcing-by-Greenhouse-Gases-2019-Revised-3-7-2022.pdf.

18 Wilson, Derrek J. and Gea-Banacloche, Julio, "Simple model to estimate the contribution of atmospheric CO_2 to the Earth's greenhouse effect," *American Journal of Physics,* Volume 80, Issue 4, pp. 306-315, 1 April 2012, pg. https://doi.org/10.1119/1.3681188.

19 Rasmussen, Carl Edward, "9 Atmospheric Carbon Dioxide Growth Rate," *Climate Cooperation,* 5 June 2025. https://mlg.eng.cam.ac.uk/carl/climate/CO2growth.html?utm_source=chatgpt.com.

20 Biezen, Michel van, "Astronomy Chapter 9.1—Earth's Atmosphere," Online YouTube Course. https://www.youtube.com/watch?v=dw3vQ6hguWg.

21 Happer, William, "How to Think About Climate Change," CLINTEL-lezing, Amsterdam, 15 November 2021, https://www.youtube.com/watch?v=P-blYr-KjOVY; See also Wijngaarden, William van, "William van Wijngaarden: Is Global Warming Hot Air?" *Tom Nelson Podcast,* #56, 23 December 2022. https://www.youtube.com/watch?v=WfwnKWIWPzk/.

22 Happer, William, "CO_2, The Gas of Life," Old Guard of Summit, New Jersey, Springfield, New Jersey, 3 October 2023. https://youtu.be/tXJ7UZjFDHU?si=-jnY-PTVlTLZZyFF9.

23 Wilson, Derrek J. and Gea-Banacloche, Julio, "Simple model to estimate the contribution of atmospheric CO_2 to the Earth's greenhouse effect," *American Journal of Physics,* Volume 80, Issue 4, pp. 306-315, 1 April 2012, pg. 314. https://doi.org/10.1119/1.3681188.

24 Wilson, Derrek J. and Gea-Banacloche, Julio, "Simple model to estimate the contribution of atmospheric CO_2 to the Earth's greenhouse effect," *American Journal of Physics,* Volume 80, Issue 4, pp. 306-315, 1 April 2012, pg. 3.14. https://doi.org/10.1119/1.3681188.

25 Stefan-Boltzmann Law Calculator, *BYJU'S.* https://byjus.com/stefan-boltzmann-law-calculator/.

26 Rasmussen, Carl Edward, "9 Atmospheric Carbon Dioxide Growth Rate," *Climate Cooperation,* 5 June 2025. https://mlg.eng.cam.ac.uk/carl/climate/CO2growth.html?utm_source=chatgpt.com.

27 Marston, Brad, "The Quantum Physics of Climate Change," *Aspen Physics,* 14 September 2017. https://youtu.be/LHJjDjEMihg?si=Iz31J0Knq6USrW1v.

28 Happer, William, "How to Think About Climate Change," CLINTEL-lezing, Amsterdam, 15 November 2021. https://www.youtube.com/watch?v=P-blYr-KjOVY.

29 *…and Then There's Physics,* "Effective emissions height," 5 March 2014. https://andthentheresphysics.wordpress.com/2014/03/05/effective-emission-height/; See also Wilson, Derrek J. and Gea-Banacloche, Julio, "Simple model to estimate the contribution of atmospheric CO_2 to the Earth's greenhouse effect," *American Journal of Physics,* Volume 80, Issue 4, pp. 306-315, 1 April 2012, pg. 314. https://doi.org/10.1119/1.3681188.

30 Britannica, "Normal Lapse Rate." https://www.britannica.com/science/normal-lapse-rate; See also, Nugent, Alison and DeCou, David, *Atmospheric Processes and Phenomena*, "Chapter 5: Atmospheric Stability," Pressbooks. https://pressbooks-dev.oer.hawaii.edu/atmo/chapter/chapter-5-atmospheric-stability/.

31 Held, Issac M. and Soden, Brian J., "Water Vapor Feedback and Global Warming," *Annual Review of Environment and Resources,* Volume 25, pp. 441-475, November 2000, pg. 447. https://doi.org/10.1146/annurev.energy.25.1.441.

32 IPCC, "Climate change widespread, rapid, and intensifying," 9 August 2021. https://www.ipcc.ch/2021/08/09/ar6-wg1-20210809-pr/.

33 "Historical Greenhouse Gas Levels in the Atmosphere," https://www.eea.europa.eu/data-and-maps/daviz/atmospheric-concentration-of-carbon-dioxide-5/download.table.

34 The IPCC estimated CO_2 increased radiative forcing by 1.82 W/m^2 between 1750 to 2013. This represents a change from 278 ppm to 400 ppm. This calculation is slightly overstated as they used a CO_2 radiative forcing coefficient of 5.00. The coefficient, adjusted for Effective Radiative Forcing and overlapping infrared radiation absorption bands, is 4.33. See IPCC, *The Physical Science Basis, Contribution of Working Group 1 to the Fifth Assessment Report of the Intergovernmental Panel on Climate Change*, "Chapter 8 - Anthropogenic and Natural Radiation," Cambridge University Press, Cambridge, UK, Section 8.3.2.1, pp. 676-677, 30 September 2012. http://www.ipcc.ch/site/assets/uploads/2018/02/WG1AR5_Chapter08_FINAL.pdf.

35 Happer, William, "The Staff of the Best Schools Interviews William Happer," Independent Institute, 24 March 2020. https://www.independent.org/article/2020/03/24/the-staff-of-the-best-schools-interviews-william-happer./.

36 Energy Institute, "Statistical Review of Energy," Download the data, 2024. https://www.energyinst.org/statistical-review.

37 Carbon Tracker, "Finally We Have a Global Registry of Fossil Fuels," 19 September 2022. https://carbontracker.org/finally-we-have-a-global-registry-of-fossil-fuels./.

38 Rasmussen, Carl Edward, "9 Atmospheric Carbon Dioxide Growth Rate,"

Climate Cooperation, 5 June 2025. https://mlg.eng.cam.ac.uk/carl/climate/CO2growth.html?utm_source=chatgpt.com.

39 Tol, Richard S., "The Economic Impacts of Climate Change," *Review of Environmental Economics and Policy*, Volume 12, Number 1, Winter 2018, pp. 4-25. https://doi.org/10.1093/reep/rex027.

40 Rasmussen, Carl Edward, "9 Atmospheric Carbon Dioxide Growth Rate," *Climate Cooperation*, 5 June 2025. https://mlg.eng.cam.ac.uk/carl/climate/CO2growth.html?utm_source=chatgpt.com.

41 Wijngaarden, William van, "William van Wijngaarden: Is Global Warming Hot Air?", *Tom Nelson Podcast,* #56, 23 December 2022. https://www.youtube.com/watch?v=WfwnKWIWPzk/.

42 EFCTC, "2023: HFCs contribute 1.3% to climate warming influence of greenhouse gases," 19, November 2024. https://www.fluorocarbons.org/news/2023-hfcs-contribute-1-3-to-climate-warming-influence-of-greenhouse-gases/.

43 EFCTC, "2023: HFCs contribute 1.3% to climate warming influence of greenhouse gases," 19, November 2024. https://www.fluorocarbons.org/news/2023-hfcs-contribute-1-3-to-climate-warming-influence-of-greenhouse-gases/.

44 United Nations UNDP, "Kigali Amendment," Global Chemicals and Waste Hub. https://www.undp.org/chemicals-waste/montreal-protocol/kigali-amendment.

45 Zhang, Xu, "Atmospheric observation and emission estimation of HFC-125 and HFC-32 in China from four representative cities," *Science of the Total Environment,* Volume 951, 15 November 2024. https://doi.org/10.1016/j.scitotenv.2024.175575.

46 Fernando, Anton M. et al., "Trends in atmospheric HFC-23 (CHF3) and HFC-134a abundances," Elsevier, 2019. https://www.sciencedirect.com/science/article/pii/S0022407318308963.

47 Zhang, Xu, "Atmospheric observation and emission estimation of HFC-125 and HFC-32 in China from four representative cities," *Science of the Total Environment*, Volume 951, 15 November 2024. https://doi.org/10.1016/j.scitotenv.2024.175575.

48 Myhre, Gunnar et al., "New estimates of radiative forcing due to well mixed greenhouse gases," *Geophysical Research Letters,* Volume 25, Issue 14, pp. 2715-2718, 15 July 199. https://doi.org/10.1029/98GL01908.

49 IPCC, *The Physical Science Basis, Contribution of Working Group 1 to the Fifth Assessment Report of the Intergovernmental Panel on Climate Change,* "Chapter 8 - Anthropogenic and Natural Radiation," Cambridge University

Press, Cambridge, UK, Sections 8.3.2.1, 8.3.2.2, 8.3.2.3, 8.3.2.4, and Table 8.2, pp. 676-678, 30 September 2013. http://www.ipcc.ch/site/assets/uploads/2018/02/WG1AR5_Chapter08_FINAL.pdf.

50 IPCC, *The Physical Science Basis, Contribution of Working Group 1 to the Fifth Assessment Report of the Intergovernmental Panel on Climate Change*, "Chapter 8 - Anthropogenic and Natural Radiation," Cambridge University Press, Cambridge, UK, Sections 8.3.2.1, 8.3.2.2, 8.3.2.3, 8.3.2.4, and Table 8.2, pp. 676-678, 30 September 2013. http://www.ipcc.ch/site/assets/uploads/2018/02/WG1AR5_Chapter08_FINAL.pdf.

51 Jones, Nicola, "How the World Passed a Carbon Threshold and Why it Matters," *YaleEnvironment360*, 26 January 2017. https://e360.yale.edu/features/how-the-world-passed-a-carbon-threshold-400ppm-and-why-it-matters?_bhlid=38186d09213e68ec77ed9db9b24aa48cba37d1c1&utm_campaign=when-the-glitter-of-an-economic-boom-clashes-with-the-environment&utm_medium=newsletter&utm_source=thedeepview.

Chapter 10 - Climate Feedbacks

1 United Nations, "UN emissions report: World on course for more than 3-degree spike, even if climate commitments are met," *UN News*, 26 November 2019. https://news.un.org/en/story/2019/11/1052171.

2 IPCC, "Climate change widespread, rapid, and intensifying," 9 August 2021. https://www.ipcc.ch/2021/08/09/ar6-wg1-20210809-pr/.

3 Feynman, Richard, "If it disagrees with experiment, it's wrong." Cornell University, 1964. *YouTube*. https://youtu.be/b240PGCMwV0.

4 Lindzen, Richard S., "The Absurdity of the Conventional Global Warming Narrative," Massachusetts Institute of Technology, 26 April 2022, slide 3. https://co2coalition.org/wp-content/uploads/2022/10/The-Absurdity-of-the-Conventional-Global-Warming-Narrative.pdf.

5 Cummins, Eoin P. et al., "Carbon dioxide-sensing in organisms and its implications for human disease," *Cellular and Molecular Life Sciences*, Volume 71, Issue 5, pp. 831-845, 18 Sep 2013. https://doi.org/10.1007/s00018-013-1470-6.

6 *Wikipedia*, "Atmospheric Optics," Cloud Coloration. https://en.wikipedia.org/wiki/Atmospheric_optics.

7 Biezen, Michel van, "Astronomy Chapter 9.1—Earth's Atmosphere," Online *YouTube* Course. https://www.youtube.com/watch?v=dw3vQ6hguWg.

8 IPCC, "1. The Climate System: An Overview," Lead Coordinating Author, Baede, A.P.M., p. 90. https://www.ipcc.ch/site/assets/uploads/2018/03/TAR-01.pdf#:~:text=On%20the%20other%20hand%2C%20most%20clouds%20are,-compensates%20for%20the%20greenhouse%20effect%20of%20clouds.

9 Stephens, Graeme L. et al., "An update on Earth's energy balance in light of the latest global observations," *Nature Geoscience,* Volume 5, pp. 691-696, 23 September 2012, p. 693. https://doi.org/10.1038/ngeo1580.

10 IPCC, *The Physical Science Basis, Contribution of Working Group 1 to the Fifth Assessment Report of the Intergovernmental Panel on Climate Change,* "Chapter 7 - Clouds and Aerosols," Cambridge University Press, Cambridge, UK, Section 7.2.1.2, pp. 580-58, 30 September 2013. https://www.ipcc.ch/site/assets/uploads/2018/02/WG1AR5_Chapter07_FINAL-1.pdf.

11 L'Ecuyer, Tristan S. et al., "Reassessing the Effect of Cloud Type on Earth's Energy Balance in the Age of Active Spaceborne Observations. Part I: Top of Atmosphere and Surface," *Journal of Climate,* Volume 32, Issue 19, pp. 61217-6217, 1 October 2019. https://doi.org/10.1175/JCLI-D-18-0753.1.

12 Archer, David et al., "Ocean methane hydrates as a slow tipping point in the global carbon cycle," *PNAS,* Volume 106, Issue 49, pp. 20596-20601, 8 December 2009. https://doi.org/10.1073/pnas.0800885105.

13 Dessler, Andrew E. and Sherwood, Steven, "A Matter of Humidity," *Science,* Volume 323, No. 5917, Issue 5917, pp. 1020-1021, 20 February 200. https://doi.org/10.1126/science.1171264.

14 Soden, Brian J., and Held, Isaac M., "An Assessment of Climate Feedbacks in Coupled Ocean-Atmosphere Models," *Journal of Climate*, pp. 3354-3360, 15 July 2006. https://doi.org/10.1175/JCLI3799.1.

15 Soden, Brian J., and Held, Isaac M., "An Assessment of Climate Feedbacks in Coupled Ocean-Atmosphere Models," *Journal of Climate,* pp. 3354-3360, 15 July 2006. https://doi.org/10.1175/JCLI3799.1.

16 IPCC, *The Physical Science Basis, Contribution of Working Group I to the Sixth Assessment Report on the Intergovernmental Panel on Climate Change,* "Chapter 7 - The Earth's Energy Budget, Climate Feedbacks and Climate Sensitivity," Cambridge University Press, Cambridge UK and New York, NY, USA, Section 7.4.2.2, 9, p. 957, 9 August 2021. https:// www.ipcc.ch/report/ar6/wg1/chapter/chapter-7/.

17 IPCC, *The Physical Science Basis, Contribution of Working Group I to the Sixth Assessment Report on the Intergovernmental Panel on Climate Change,* "Chapter 7 - The Earth's Energy Budget, Climate Feedbacks and Climate Sensitivity," Cambridge University Press, Cambridge UK and New York, NY, USA, Section 7.4.2.2, 9, p. 957, 9 August 2021. https:// www.ipcc.ch/report/ar6/wg1/chapter/chapter-7/.

18 IPCC, *The Physical Science Basis, Contribution of Working Group I to the Sixth Assessment Report on the Intergovernmental Panel on Climate Change,* "Chapter 7 - The Earth's Energy Budget, Climate Feedbacks and Climate Sensitivity," Cambridge University Press, Cambridge UK and New York, NY, USA,

Section 7.4.2.4, 9, p. 961, 9 August 2021. www.ipcc.ch/report/ar6/wg1/chapter/chapter-7/.

19 Parkinson, Claire L., "A 40-y record reveals gradual Antarctica sea ice increases followed by decreases at rates far exceeding the rates seen in the Arctic," *PNAS*, Volume 116, Issue 29, 1 July 2019, pp. 14414 to 14423. https://doi.org/10.1073/pnas.1906556116.

20 NSIDC, "Arctic sea ice extend levels off; 2023 minimum set," National Snow and Ice Data Center, 24 September 2024. https://nsidc.org/sea-ice-today/analyses/arctic-sea-ice-extent-levels-2024-minimum-set.

21 IPCC, *The Physical Science Basis, Contribution of Working Group 1 to the Fifth Assessment Report of the Intergovernmental Panel on Climate Change,* "Chapter 7 - Clouds and Aerosols," Cambridge University Press, Cambridge, UK, Section 7.2.4, p. 586, 30 September 2013. https://www.ipcc.ch/site/assets/uploads/2018/02/WG1AR5_Chapter07_FINAL-1.pdf.

22 IPCC, *The Physical Science Basis, Contribution of Working Group I to the Sixth Assessment Report on the Intergovernmental Panel on Climate Change,* "Chapter 8 – Water Cycle Changes," Cambridge University Press, Cambridge, UK and New York, NY, USA, Section 8.2.1, p. 1065, 9 August 2021. https://www.ipcc.ch/report/ar6/wg1/downloads/report/IPCC_AR6_WGI_Chapter08.pdf.

23 Dessler, Andrew E. and Sherwood, Steven, "A Matter of Humidity," *Science*, Volume 323, No. 5917, Issue 5917, pp. 1020-1021, 20 February 2009. https://doi.org/10.1126/science.1171264.

24 *Climate4you,* "Relative Atmospheric Humidity," *Greenhouse Gases, Atmospheric Water.* https://www.climate4you.com/GreenhouseGasses.htm#Atmospheric%20water%20vapor.

25 Milton, Joseph, "Why Winds Are Slowing," *Nature*, 1668 Access, 17 October 2010. https://doi.org/10.1038/news.2010.5 3.

26 *Climate4you,* "Relative Atmospheric Humidity," *Greenhouse Gases, Atmospheric Water.* https://www.climate4you.com/GreenhouseGasses.htm#Atmospheric%20water%20vapor.

27 May, Andy, "Atmospheric Water Vapor (TPW) and Climate Change," *Andy May Petrophysicist,* 21 March 2023. https://andymaypetrophysicist.com/2023/03/21/atmospheric-water-vapor-tpw-and-climate-change.

28 May, Andy, "Atmospheric Water Vapor (TPW) and Climate Change," *Andy May Petrophysicist,* 21 March 2023. https://andymaypetrophysicist.com/2023/03/21/atmospheric-water-vapor-tpw-and-climate-change.

29 Happer, William, "How to Think About Climate Change," Hillsdale College, Phoenix, Arizona, 19 February 2021. https://youtu.be/CA1zUW4u-

OSw?si=vn-GpQz_E9rz0T7L.

30 Wentz, Frank J. et al., "How Much More Rain Will Global Warming Bring?" *Science*, Volume 317, 13 July 2007. https://doi.org/10.1126/science.1140746.

31 L'Ecuyer, Tristan S. et al., "Reassessing the Effect of Cloud Type on Earth's Energy Balance in the Age of Active Spaceborne Observations. Part I: Top of Atmosphere and Surface," *Journal of Climate,* Volume 32, Issue 19, pp. 61217-6217, 1 October 2019. https://doi.org/10.1175/JCLI-D-18-0753.1.

32 Sciare, J. et al., "Interannual variability of atmospheric dimethylsulfide in the southern Indian Ocean," *Journal of Geophysical Research*, Volume 105, Issue D21, pp. 26,369 to 26,377, 16 November 2000. https://doi.org/10.1029/2000jD900236.

33 Thomas, M. A. et al., "Quantification of DMS aerosol-cloud-climate interactions using the ECHAM5-HAMMOZ model in a current climate scenario," *Atmospheric Chemistry and Physics,* Volume 10, pp. 7425-7438, 10 Aug 2010. https://doi.org/10.5194/acp-10-7425-2010.

34 United Nations, "UN emissions report: World on course for more than 3-degree spike, even if climate commitments are met," *UN News*, 26 November 2019. https://news.un.org/en/story/2019/11/1052171.

35 IPCC, "Climate change widespread, rapid, and intensifying," 9 August 2021. https://www.ipcc.ch/2021/08/09/ar6-wg1-20210809-pr/.

36 Vieira, L. E. A. et al., "Evolution of the solar irradiance during the Holocene," *Astronomy and Astrophysics,* Volume 531, July 2011. https://doi.org/10.1051/0004-6361/201015843.

37 Ljungqvist, Fredrik, "A new reconstruction of temperature variability in the extra-tropical Northern Hemisphere during the last two millennia," *Geografiska Annaler: Series A, Physical Geography,* Volume 92, Issue 3, pp. 339-351, 6 September 2010. https://doi.org/10.1111/j.1468-0459.2010.00399.x.

38 IPCC, "Climate change widespread, rapid, and intensifying," 9 August 2021. https://www.ipcc.ch/2021/08/09/ar6-wg1-20210809-pr/.

39 Crok, Marcel and May, Andy, *The Frozen Climate Views of the IPCC, an analysis of AR6,* CLINTEL Foundation, Amsterdam, The Netherlands, 2023, p. 83, ISBN: 979-8-89074-861-4.

40 Moberg, A. et al., "Highly variable Northern Hemisphere temperature reconstructions from low-and high-resolution proxy data," *Nature*, Volume 433, pp. 613-617, 10 February 2005. https://doi.org/10.1038/nature03265.

41 Esper, Jan et al., "Low-Frequency Signals in Long Tree-Ring Chronologies for Reconstructing Past Temperature Variability," *Science*, Volume 295, No. 5563, 22 March 2002. https://doi.org/10.1126/science.1066208.

42 Crok, Marcel and May, Andy, *The Frozen Climate Views of the IPCC, an*

analysis of AR6, CLINTEL Foundation, Amsterdam, The Netherlands, 2023, p. 83, ISBN: 979-8-89074-861-4.

43 Solanki, S. K. and Krivona, N.A., "Can solar variability explain global warming since 1970?," *Journal of Geophysical Research,* Volume 108, Issue A5, 21 May 2003. https://doi.org/10.1029/2002JA009753.

44 Vahrenholt, Fritz and Lüning, Sebastion, *The Neglected Sun, Why the Sun Precludes Climate Catastrophe*, The Heartland Institute, Arlington Heights, IL, 2015, p. 137, ISBN 13-978-1-934791-54-7. Citations: Scafetta, N., "Empirical evidence for a celestial origin of the climate oscillations and its implications," *Journal of Atmospheric and Solar-Terrestrial Physics*, Volume 72, pp. 951-970, 2010. https://doi.org/10.1016/j.gsf.2023.101650; Scafetta, Nicola and West, Bruce J., "Phenomenological reconstructions of the solar signature in the Northern Hemisphere surface temperature records since 1600," *Journal of Geophysical Research,* Volume 112, Issue D24, 03 November 2007. https://doi.org/10.1029/2007JD008437; Scafetta, Nicola. and West, Bruce J., "Is climate sensitive to solar variability?" *Physics Today,* Volume 61, Issue 3, pp. 50-51, March 2008. https://doi.org/10.1063/1.2897951; Beer, J., Mende, W. and Stellmacher R., "The role of the sun in climate forcing," *Quaternary Science Reviews,* Volume 19, Issues 1-5, pp. 403-415, 1 January 2000. https://doi.org/10.1016/S0277-3791(99)00072-4; Rao, U. R., "Contribution of changing galactic cosmic ray flux to global warming." *Current Science*, Volume 100, Issue 2, pp. 223-225, 25 January 2011. https://www.jstor.org/stable/24073049; Shaviv, Nir J., "On climate response to changes in the cosmic ray flux and radiative budget," *Journal of Geophysical Research,* Volume 110, Issue A8, pp. 1-15, 23 August 2005. https://doi.org/10.1029/2004JA010866; Shaviv, Nir J., "Using the oceans as a calorimeter to quantify the solar radiative forcing," *Journal of Geophysical Research,* Volume 113, Issue A11, pp. 1-13, 4 November 2008. https://doi.org/10.1029/2007JA012989; El-Borie, Ali et al., "Spectral analysis of solar variability and their possible role on global warming," *Journal of Environmental Protection*, Volume 1, Number 2, June 2010. http://dx.doi.org/10.4236/jep.2010.12014; De Jager, C. and Duhau, Silvia, "The variable solar dynamo and the forecast of solar activity; influence on terrestrial surface temperature." *Global Warming in the 21st Century*, In J. M. Cossia (ed), Nova Science, Hauppauge, NY, pp. 77-106, January 2011. https://www.semanticscholar.org/paper/THE-VARIABLE-SOLAR-DYNAMO-AND-THE-FORECAST-OF-SOLAR-Jager-Duhau/671d0588dc7ae7bfd774ce63e21bac10d66a268a.

45 Brady, Howard Thomas, *Mirrors and Mazes: A guide through the climate change debate*, 2nd Edition, Creative Commons, Canberra, Australia, 2017, p. 105, ISBN:10: 1-5466-2911-4.

46 Brady, Howard Thomas, *Mirrors and Mazes: A guide through the climate change debate*, 2nd Edition, Creative Commons, Canberra, Australia, 2017, p. 105, ISBN:10: 1-5466-2911-4.

47 Spencer, Roy, "New Satellite Upper Troposphere Product: Still No Tropical HotSpot," *Roy Spencer. Ph. D.*, 21 May 2015. https://www.drroyspencer.com/2015/05/new-satellite-upper-troposphere-product-still-no-tropical-hotspot/.

48 Alexander, Ralph B., "Latest Computer Models Run Almost as Hot as Before," *Science Under Attack,* 22 February 2021. https://www.scienceunderattack.com/blog/2021/2/22/latest-computer-climate-models-run-almost-as-hot-as-before-71?format=amp.

49 McKitrick, Ross and Christy, John, "Pervasive Warming Bias in CMIP6 Tropospheric Layers," *Earth and Space Science*, September 2020, Volume 7, Issue 9, 15 July 2020. https://doi.org/10.1029/2020EA001281.

50 Plummer, Dianne, "Evaluating The Evidence For Climate Change – And Its Impacts," *Forbes*, 31 January 2025. https://www.forbes.com/sites/dianneplummer/2025/01/31/climate-change-is-it-the-greatest-crisis-or-a-global-deception/.

51 Hausfather, Zeke et al., "Evaluating the Performance of Past Climate Model Projections," *Geophysical Research Letters,* Volume 47, Issue 1, 4 December 2019. https://doi.org/10.1029/2019GL085378.

52 Voosen, Paul, "New climate models predict a warming surge," *Science*, News, 16 April 2019. https://www.science.org/content/article/new-climate-models-predict-warming-surge.

53 Zhu, Jiang et al., "High climate sensitivity in CMIP6 model not supported by paleoclimate," *Nature Climate Change,* Volume 10, pp. 378-379, 30 April 2020. https://doi.org/10.1038/s41558-020-0764-6.

54 McKitrick, Ross and Christy, John, "Pervasive Warming Bias in CMIP6 Tropospheric Layers," *Earth and Space Science*, September 2020, Volume 7, Issue 9, 15 July 2020. https://doi.org/10.1029/2020EA001281.

55 Christy, John R., "Climate Models for Policy? ... a Bridge Too Far," 14th International EIKE Climate and Energy Conference, European Institute for Climate and Energy (EIKE), 17 February 2022, minute 27:00. https://youtu.be/XltFOh7Cg2U?si=GBuFwI3SV9vpjYyk.

56 Climate Depot, "CANCELLED, Nobel Prize-Winning Dr. John Clauser—who recently declared climate science a 'pseudoscience'—has his IMF talk abruptly cancelled," *Watts Up With That,* 23 July 2023. https://wattsupwiththat.com/2023/07/23/report-nobel-prize-winning-scientist-dr-john-clauser-who-recently-declared-climate-science-a-pseudoscience-has-his-imf-talk-abruptly-canceled/.

Chapter 11 - The Impact of Clouds

1 Quaas, J. et al., "Aerosol indirect effects—general circulation model inter-

comparison and evaluation with satellite data," *Atmospheric Chemistry and Physics*, Volume 9, Issue 22, 16 November 2009. https://doi.org/10.5194/acp-9-8697-2009.

2 Pruppacher, H. R. and Klett, J. D., *Microphysics of Clouds and Precipitation,* Second Edition, Springer, 31 December 1976, p. 976, ISBN-13 978-0792344094.

3 Brady, Howard Thomas, *Mirrors and Mazes: A guide through the climate change debate,* Second Edition, Creative Commons, Canberra, Australia, 2017, p. 106, ISBN: 10: 1-5466-2911-4.

4 Quinn, P. K. et al., "Sea spray aerosol makes up a small fraction of marine cloud condensation nuclei," NOAA Library. https://repository.library.noaa.gov/view/noaa/61835/noaa_61835_DS1.pdf?utm_source=chatgpt.com.

5 Charlson, Robert J. et al., "Oceanic phytoplankton, atmospheric sulfur, cloud albedo, and climate," *Nature*, 326, pp. 655-661, 22 April 1987. https://doi.org/10.1038/326655a0 . See also *Wikipedia*, "Cloud Condensation Nuclei (CCNs)," https://en.m.wikipedia.org/wiki/Cloud_condensation_nuclei.

6 CERN, "CLOUD challenges current understanding of aerosol particle formation in polar and marine regions," 15 December 2023. https://home.cern/news/news/physics/cloud-challenges-current-understanding-aerosol-particle-formation-polar-and.

7 Twomey, S., "The Influence of Pollution on the Shortwave Albedo of Clouds," *Journal of the Atmospheric Sciences,* Volume 34, Issue 7, pp. 1149-1152, 1 July 1977. https://doi.org/10.1175/1520-0469(1977)034%3C1149:TIOPOT%3E2.0.CO;2.

8 Kirkby, Jasper et al., "Role of sulfuric acid, ammonia, and galactic cosmic rays in atmospheric aerosol nucleation," *Nature*, volume 476, pp. 429-433, 24 August 2011. https://doi.org/10.1038/nature10343.

9 Wang, Gehui et al., "Atmospheric sulfate aerosol formation enhanced by interfacial anions," *PNAS Nexus*, Volume 4, Issue 3, March 2025, 24 February 2025. https://doi.org/10.1093/pnasnexus/pgaf058.

10 Ehn, Mikael et al., "A large source of low-volatility secondary organic aerosol," *Nature*, Volume 506, pp. 476-479, 27 February 2014. https://doi.org/10.1038/nature13032.

11 Enghoff, Martin B., "Aerosol nucleation induced by a high energy particle beam," *Geophysical Research Letters,* Volume 38, Issue 9, 12 May 2011. https://doi.org/10.1029/2011GL047036.

12 Brady, Howard Thomas, *Mirrors and Mazes: A guide through the climate change debate,* Second Edition, Creative Commons, Canberra, Australia, 2017, p. 106, ISBN: 10: 1-5466-2911-4.

13 Svensmark, Henrik,et al., "Atmospheric ionization and cloud radiative forcing," *Nature Scientific Reports,* Volume 11, Article Number 19668, 11 October 2021. https://doi.org/10.1038/s41598-021-99033-1.

14 Rahmanifard, Fatemeh, "Cosmic rays increase remarkably as solar activity shows persistent decline, resembles Dalton minimum of 1790-1830," *THE WATCHERS*, "Watch the World Evolve and Transition." https://watchers.news/2020/08/12/cosmic-rays-increase-remarkably-as-solar-activity-shows-persistent-decline-resembles-dalton-minimum-of-1790-1830/#:~:-text="During%20the%20next%20solar%20cycle%2C%20we%20could%20see,as%20GCRs%20present%20a%20hazard%20for%20space%20missions.

15 IPCC, "1. The Climate System: An Overview," lead coordinating author, Baede, A.P.M., p. 90. https://www.ipcc.ch/site/assets/uploads/2018/03/TAR-01.pdf#:~:text=On%20the%20other%20hand%2C%20most%20clouds%20are,-compensates%20for%20the%20greenhouse%20effect%20of%20clouds.

16 Ramanathan, V. et al., "Cloud-Radiative Forcing and Climate: Results from the Earth Radiation Budget Experiment," *Science*, Volume 243, Issue 4887, pp. 57-63, 6 January 1989. https://doi.org/10.1126/science243.4887.57.

17 Stephens, Graeme L. et al., "An update on Earth's energy balance in light of the latest global observations," *Nature Geoscience,* Volume 5, pp. 691-696, 23 September 2012, p. 693. https://doi.org/10.1038/ngeo1580.

18 L'Ecuyer, Tristan S. et al., "Reassessing the Effect of Cloud Type on Earth's Energy Balance in the Age of Active Spaceborne Observations. Part I: Top of Atmosphere and Surface," *Journal of Climate,* Volume 32, Issue 19, pp. 6197-6217, 1 October 2019. https://doi.org/10.1175/JCLI-D-18-0753.1.

19 IPCC, *The Physical Science Basis, Contribution of Working Group 1 to the Fifth Assessment Report of the Intergovernmental Panel on Climate Change,* "Chapter 7 - Clouds and Aerosols," Cambridge University Press, Cambridge, UK, Section 7.2.1.2, pp. 580-582, 30 September 2013. https://www.ipcc.ch/site/assets/uploads/2018/02/WG1AR5_Chapter07_FINAL-1.pdf.

20 IPCC, *The Physical Science Basis, Contribution of Working Group 1 to the Fifth Assessment Report of the Intergovernmental Panel on Climate Change,* "Chapter 7 - Clouds and Aerosols," Cambridge University Press, Cambridge, UK, Section 7.2.6, pp. 587, 30 September 2013. https://www.ipcc.ch/site/assets/uploads/2018/02/WG1AR5_Chapter07_FINAL-1.pdf.

21 IPCC, *The Physical Science Basis, Contribution of Working Group 1 to the Fifth Assessment Report of the Intergovernmental Panel on Climate Change,* "Executive Summary - Clouds and Aerosols," Cambridge University Press, Cambridge, UK, p. 574, 30 September 2013. https://www.ipcc.ch/site/assets/uploads/2018/02/WG1AR5_Chapter07_FINAL-1.pdf.

22 IPCC, *The Physical Science Basis, Contribution of Working Group 1 to the*

Fifth Assessment Report of the Intergovernmental Panel on Climate Change, "Chapter 7 - Clouds and Aerosols," Cambridge University Press, Cambridge, UK, Section 7.2.5.1, p. 588, 30 September 2013. https://www.ipcc.ch/site/assets/uploads/2018/02/WG1AR5_Chapter07_FINAL-1.pdf.

23 McKitrick, Ross and Christy, John, "Pervasive Warming Bias in CMIP6 Tropospheric Layers," *Earth and Space Science*, September 2020, Volume 7, Issue 9, 15 July 2020. https://doi.org/10.1029/2020EA001281.

24 Vinós, Javier, *Solving the Climate Puzzle: The Sun's Surprising Role*, Critical Science Press, Madrid, 2023, p. 72, ISBN: 978-84-12867-7-0.

25 Nikolov, Ned and Zeller, Karl F., "Role of Earth's Albedo Variations and Top-of-the-Atmosphere Energy Imbalance in Recent Warming: New Insights from Satellite and Surface Observations," *MDPI Geomatics,* Volume 4, Issue 3, pp. 311-341, 20 August 2024. https://doi.org/10.3390/geomatics4030017.

26 Nikolov, Ned and Zeller, Karl F., "Role of Earth's Albedo Variations and Top-of-the-Atmosphere Energy Imbalance in Recent Warming: New Insights from Satellite and Surface Observations," *MDPI Geomatics,* Volume 4, Issue 3, pp. 311-341, 20 August 2024. https://doi.org/10.3390/geomatics4030017.

27 Nikolov, Ned and Zeller, Karl F., "Role of Earth's Albedo Variations and Top-of-the-Atmosphere Energy Imbalance in Recent Warming: New Insights from Satellite and Surface Observations," *MDPI Geomatics,* Volume 4, Issue 3, pp. 311-341, 20 August 2024. https://doi.org/10.3390/geomatics4030017.

28 Vinós, Javier, *Solving the Climate Puzzle: The Sun's Surprising Role*, Critical Science Press, Madrid, 2023, p. 34, ISBN: 978-84-12867-7-0. See also Tselioudis, George et al., "Contraction of the World's Storm-Cloud Zones the Primary Contributor to the 21st Century Increase in the Earth's Sunlight Absorption," *Geophysical Research Letters*, Volume 52, Issue 11, 8 June 2025. https://doi.org/10.1029/2025GL114882.

29 Nikolov, Ned and Zeller, Karl F., "Role of Earth's Albedo Variations and Top-of-the-Atmosphere Energy Imbalance in Recent Warming: New Insights from Satellite and Surface Observations," *MDPI Geomatics,* Volume 4, Issue 3, pp. 311-341, 20 August 2024. https://doi.org/10.3390/geomatics4030017.

30 Qualglia, Ilaria and Visioni, Daniele, "Modeling 2020 regulatory changes in international shipping changes in international shipping emissions helps explain anomalous 2023 warming," *EDS Letters*, Volume 15, Issue 6, pp. 1527-1441, 28 November 2024. https://doi.org/10.5194/esd-15-1527-2024.

31 L'Ecuyer, Tristan S. et al., "Reassessing the Effect of Cloud Type on Earth's Energy Balance in the Age of Active Spaceborne Observations. Part I: Top of Atmosphere and Surface," *Journal of Climate,* Volume 32, Issue 19, pp. 6197-6217, 1 October 2019. https://doi.org/10.1175/JCLI-D-18-0753.1.

32 IPCC, *Climate Change 2021: The Physical Science Basis: Contribution of*

Working Goup I to the Sixth Assessment Report of the Intergovernmental Panel on Climate Change, "Summary for Policymakers," Cambridge University Press, Cambridge, UK and New York, NY, USA, Section A.2, p. 5, 9 August 2021. https://www.ipcc.ch/report/ar6/wg1/downloads/report/IPCC_AR6_WGI_SPM.pdf.

33 Goessling, Helge F. et al., "Recent global temperature surge intensified by record-low planetary albedo," *Science*, volume 387, Number 6729, pp. 68-73, 5 December 2024. https://doi.org/10.1126/science.adq7280.

34 Pruppacher, H. R. and Klett, J. D., Microphysics of Clouds and Precipitation, Second Edition, Springer, 31 December 1976, p. 976, ISBN-13 978-0792344094: See also, Quaas, J. et al., "Aerosol indirect effects—general circulation model intercomparison and evaluation with satellite data," *Atmospheric Chemistry and Physics,* Volume 9, Issue 22, 16 November 2009. https://doi.org/10.5194/acp-9-8697-2009.

35 Wentz, Frank, "How Much More Rain Will Global Warming Bring?", *Science*, Volume 317, Issue 5835, pp. 233-235, 13 July 2007. https://doi.org/10.1126/science.1140746.

36 McCoy, Daniel T. et al., "Untangling causality in midlatitude aerosol-cloud adjustments," *European Geosciences Union*, Volume 20, Issue 7, pp. 4085-4103, 6 April 2020. https://doi.org/10.5194/acp-20-4085-2020.

37 Twomey, S., "The Influence of Pollution on the Shortwave Albedo of Clouds," *Journal of the Atmospheric Sciences,* Volume 34, Issue 7, pp. 1149-1152, 1 July 1977. https://doi.org/10.1175/1520-0469(1977)034%3C1149:TIOPOT%3E2.0.CO;2.

38 Yuan, Tianle et al., "Abrupt reduction in shipping emissions as inadvertent geoengineering termination shock produces substantial radiative warming," *Nature, Communications Earth & Environment,* Volume 5, Article 281, 30 May 2024. https://doi.org/10.1038/s43247-024-01442-3.

39 Qualglia, Ilaria and Visioni, Daniele, "Modeling 2020 regulatory changes in international shipping changes in international shipping emissions helps explain anomalous 2023 warming," *EDS Letters*, Volume 15, Issue 6, pp. 1527-1441, 28 November 2024. https://doi.org/10.5194/esd-15-1527-2024.

40 Myhre, Gunnar et al., "New estimates of radiative forcing due to well mixed greenhouse gases," *Geophysical Research Letters,* Volume 25, Issue 14, pp. 2715-2718, 15 July1998. https://doi.org/10.1029/98GL01908 ; see also the "Logarithmic Decline of CO_2 Marginal Warming" section of Chapter 9 for the proper use of the Myhre equations.

41 Wentz, Frank, "How Much More Rain Will Global Warming Bring?", *Science*, Volume 317, Issue 5835, pp. 233-235, 13 July 2007. https://doi.org/10.1126/science.1140746.

42 L'Ecuyer, Tristan S. et al., "Reassessing the Effect of Cloud Type on Earth's Energy Balance in the Age of Active Spaceborne Observations. Part I: Top of Atmosphere and Surface," *Journal of Climate,* Volume 32, Issue 19, pp. 6197-6217, 1 October 2019. https://doi.org/10.1175/JCLI-D-18-0753.1.

43 *Climate4you.com,* "Total cloud cover vs. global surface temperature," "Climate and Clouds, Cloud cover effect on climate," https://climate4you.com/ClimateAndClouds.htm#TotalCloudCoverVersusGlobalTemperature.

44 For ISCCP cloud measurements see Nikolov, Ned and Zeller, Karl F., "Role of Earth's Albedo Variations and Top-of-the Atmosphere Energy Imbalance in Recent Warming: New Insights from Satellite and Surface Observations," *MDPI Geomatics*, Volume 4, Issue 3, pp. 311-341, 20 August 2024. https://doi.org/10.3390/geomatics4030017; For MODIS satellite cloud measurements see Tselioudis, George et al., "Contraction of the World's Storm-Cloud Zones the Primary Contributor to the 21st Century Increase in the Earth's Sunlight Absorption," *Geophysical Research Letters*, Volume 52, Issue 11, 8 June 2025. https://doi.org/10.1029/2025GL114882.

45 Nikolov, Ned and Zeller, Karl F., "Role of Earth's Albedo Variations and Top-of-the-Atmosphere Energy Imbalance in Recent Warming: New Insights from Satellite and Surface Observations," *MDPI Geomatics*, Volume 4, Issue 3, pp. 311-341, 20 August 2024. https://doi.org/10.3390/geomatics4030017 See also, Qualglia, Ilaria and Visioni, Daniele, "Modeling 2020 regulatory changes in international shipping changes in international shipping emissions helps explain anomalous 2023 warming," *EDS Letters,* Volume 15, Issue 6, pp. 1527-1441, 28 November 2024. https://doi.org/10.5194/esd-15-1527-2024.

46 Lindzen, Richard et al., "Does the Earth Have an Adaptive Infrared Iris?", *Bulletin of the American Meteorological Society,* Volume 82, Issue 2, pp. 417-431, 1 March 2001. https://doi.org/10.1175/1520-0477(2001)082%3C0417:DTEHAA%3E2.3.CO;2.

47 Climate Discussion NEXUS, "Time for an Eye Exam," *Science Notes*, 25 January 2023. https://climatediscussionnexus.com/2023/01/25/time-for-an-eye-exam/.

48 Vahrenholt, Fritz and Lüning, Sebastion, *The Neglected Sun: Why the Sun Precludes Climate Catastrophe*, The Heartland Institute, Arlington Heights, IL, 2015, p. 260, ISBN 13-978-1-934791-54-7.

49 Climate Discussion NEXUS, "Time for an Eye Exam," *Science Notes,* 25 January 2023. https://climatediscussionnexus.com/2023/01/25/time-for-an-eye-exam/.

50 IPCC, *The Physical Science Basis, Contribution of Working Group 1 to the Fifth Assessment Report of the Intergovernmental Panel on Climate Change,* "Chapter 7 - Clouds and Aerosols," Cambridge University Press, Cambridge,

UK, Section 7.2.5.2, p. 589, 30 September 2013. https://www.ipcc.ch/site/assets/uploads/2018/02/WG1AR5_Chapter07_FINAL-1.pdf.

51 Yu, Lu et al., "Assessment of long-wave radiative effect of nighttime cirrus based on CloudSat and CALIPSO measurements and single-column radiative transfer simulations," *Journal of Quantitative Spectroscopy and Radiative Transfer,* Volume 221, pp. 87-97, December 2018. https://doi.org/10.1016/j.jqsrt.2018.09.019.

52 *Wikipedia*, "Atmospheric Optics," cloud coloration. https://en.wikipedia.org/wiki/Atmospheric_optics.

53 CO2 Coalition, "Nobel Laureate John Clauser Elected to CO2 Coalition Board of Directors," CO2 Coalition, 5 May 2023. https://co2coalition.org/publications/nobel-laureate-john-clauser-elected-to-co2-coalition-board-of-directors/.

54 IPCC, *The Physical Science Basis, Contribution of Working Group 1 to the Fifth Assessment Report of the Intergovernmental Panel on Climate Change,* "Chapter 7 - Clouds and Aerosols," Cambridge University Press, Cambridge, UK, Section 7.1.2, p. 576, 30 September 2013. https://www.ipcc.ch/site/assets/uploads/2018/02/WG1AR5_Chapter07_FINAL-1.pdf

55 McKitrick, Ross and Christy, John, "Pervasive Warming Bias in CMIP6 Tropospheric Layers," *Earth and Space Science,* September 2020, Volume 7, Issue 9, 15 July 2020. https://doi.org/10.1029/2020EA001281.

56 Brady, Howard Thomas, *Mirrors and Mazes: A guide through the Climate Change Debate,* Second Edition, Creative Commons, Canberra, Australia, 2017, p. 108, ISBN: 10: 1-5466-2911-4.

57 Heller, Tony, "Dr. William Gray funding cut," *@HellerClimate,* X post, March 13, 2024, https://x.com/TonyClimate/status/1768106051650699491.

58 CO2 Coalition, "Nobel Laureate John Clauser Elected to CO2 Coalition Board of Directors," CO2 Coalition, 5 May 2023. https://co2coalition.org/publications/nobel-laureate-john-clauser-elected-to-co2-coalition-board-of-directors/.

59 Shea, M.A. and Smart, D. F., "Geomagnetic cutoff rigidities and geomagnetic coordinates appropriate for the Carrington flare Epoch," *Advances in Space Research,* Volume 38, Issue 2, pp. 209-214, 15 September 2006. https://doi.org/10.1016/j.asr.2005.03.156.

60 Curry, Judith A. et al., "Overview of Arctic Cloud and Radiation Characteristics," *Journal of Climate,* Volume 9, Issue 8, 1 August 1996. https://doi.org/10.1175/1520-0442(1996)009%3C1731:OOACAR%3E2.0.CO;2.

61 Town, Michael S. et al., "Cloud Cover Over the South Pole from Visual Observations, Satellite Retrievals, and Surface-Based Infrared Radiation Measure-

ments," *Journal of Climate,* Volume 20, Issue 3, pp. 544-559, 1 February 2007. https://doi.org/10.1175/JCLI4005.1.

Chapter 12 - Other Climate Forcings

1 Chung, Chul Eddy et al., "Global anthropogenic aerosol direct forcing derived from satellite and ground-based observations," *Journal of Geophysical Research: Atmosphers,* Volume 110, Issue D24, 23 December 2005. https://doi.org/10.1029/2005JD006356.

2 L'Ecuyer, Tristan S. et al., "Reassessing the Effect of Cloud Type on Earth's Energy Balance in the Age of Active Spaceborne Observations. Part I: Top of Atmosphere and Surface," *Journal of Climate,* Volume 32, Issue 19, pp. 6197-6217, 1 October 2019. https://doi.org/10.1175/JCLI-D-18-0753.1.

3 Charlson, R. J. et al., "Oceanic phytoplankton, atmospheric sulfur, cloud albedo, and climate," *Nature*, Volume 326, Issue 6114, pp. 655-661, 22 April 1987. https://doi.org/10.1038/326655a0.

4 Green, Tamara K. and Hatton, Angela D., "The CLAW Hypothesis: A New Perspective on the Role of Biogenic Sulphur in the Regulation of Global Climate," *Oceanography and Marine Biology: An Annual Review*, Volume 52, pp. 315-336, September 2014, p. 327. https://www.researchgate.net/publication/265384642_The_CLAW_hypothesis_a_new_perspective_on_the_role_of_biogenic_sulphur_in_the_regulation_of_global_climate.

5 Ayers, G. P. and Gras, J. L., "Seasonal relationship between cloud condensation nuclei and aerosol methanesulphonate in marine air," *Nature*, Volume 353, pp. 834-835, 31 October 1991. https://doi.org/10.1038/353834A0.

6 Andreae, Meinrat O. et al., "Biogenic sulfur emissions and aerosols over the tropical South Atlantic: 3. Atmospheric dimethylsulfide, aerosols, and cloud condensation nuclei," *Journal of Geophysical Research: Atmospheres,* Volume 100, Issue D6, pp. 11335-11356, 20 June 1995. https://doi.org/10.1029/94JD02828.

7 Ardyna, Mathieu and Arrigo, Kevin Robert, "Phytoplankton dynamics in a changing Arctic Ocean," *Nature Climate Change,* Volume 10, pp. 892-903, 25 September 2020. https://doi.org/10.1038/s41558-020-0905-y.

8 Sciare, J. et al., "Interannual variability of atmospheric dimethylsulfide in the southern Indian Ocean," *Journal of Geophysical Research,* Volume 105, Issue D21, pp. 26369 to 26377, 16 November 2000. https://doi.org/10.1029/2000jD900236.

9 Gordon, Hamish et al., "Causes and importance of new particle formation in the present-day and preindustrial atmospheres," *Journal of Geophysical Research: Atmospheres,* Volume 122, Issue 16, pp. 8739-8760, 10 July 2017. https://doi.org/10.1002/2017JD026844.

10 Thomas, M. A. et al., "Quantification of DMS aerosol-cloud-climate interactions using the ECHAM5-HAMMOZ model in a current climate scenario," *Atmospheric Chemistry and Physics,* Volume 10, pp. 7425-7438, 10 August 2010, p. 7425. https://doi.org/10.5194/acp-10-7425-2010.

11 Twomey, S., "The Influence of Pollution on the Shortwave Albedo of Clouds," *Journal of the Atmospheric Sciences,* Volume 34, Issue 7, pp. 1149-1152, 1 July 1977. https://doi.org/10.1175/1520-0469(1977)034%3C1149:TIOPOT%3E2.0.CO;2.

12 Albrecht, Bruce A., "Aerosols, Cloud Microphysics, and Fractional Cloudiness," *Science,* Volume 245, Issue 4923, pp. 1227-1230, 15 September 1989. https://doi.org/10.1126/science.245.4923.1227.

13 McCoy, Daniel et al., "Natural aerosols explain seasonal and spatial patterns of Southern Ocean cloud albedo," *Science Advances,* Volume 1, No. 6, 17 July 2015. https://doi.org/10.1126/sciadv.1500157.

14 Thomas, M. A. et al., "Quantification of DMS aerosol-cloud-climate interactions using the ECHAM5-HAMMOZ model in a current climate scenario," *Atmospheric Chemistry and Physics,* Volume 10, pp. 7425-7438, 10 August 2010, p. 7425. https://doi.org/10.5194/acp-10-7425-2010.

15 Wang, Hai et al., "Atmosphere teleconnections from abatement of China aerosol emissions and exacerbate Northeast Pacific warm blob events," *PNAS,* Volume 121, Issue 21, e2313797121, 6 May 2024. https://doi.org/10.1073/pnas.2313797121.

16 *Our World in Data,* "Global sulfur dioxide (SO_2) emissions by world region, Sulfur dioxide emissions from all sectors." https://ourworldindata.org/grapher/so-emissions-by-world-region-in-million-tonnes.

17 International Maritime Organization, "IMO 2020—cleaner shipping for cleaner air," 20 December 2019. https://www.imo.org/en/MediaCentre/PressBriefings/pages/34-IMO-2020-sulphur-limit-.aspx.

18 International Maritime Organization, "IMO 2020—cleaner shipping for cleaner air," 20 December 2019. https://www.imo.org/en/MediaCentre/PressBriefings/pages/34-IMO-2020-sulphur-limit-.aspx.

19 Yuan, Tianle et al., "Abrupt reduction in shipping emissions as inadvertent geoengineering termination shock produces substantial radiative warming," *Nature, Communications Earth & Environment,* Volume 5, Article 281, 30 May 2024. https://doi.org/10.1038/s43247-024-01442-3.

20 Yuan, Tianle et al., "Abrupt reduction in shipping emissions as inadvertent geoengineering termination shock produces substantial radiative warming," *Nature, Communications Earth & Environment,* Volume 5, Article 281, 30 May 2024. https://doi.org/10.1038/s43247-024-01442-3.

21 Wang, Hai et al., "Atmosphere teleconnections from abatement of China aerosol emissions and exacerbate Northeast Pacific warm blob events," *PNAS*, Volume 121, Issue 21, e2313797121, 6 May 2024. https://doi.org/10.1073/pnas.2313797121.

22 Wu, You-Ting et al., "Stronger Arctic amplification from anthropogenic aerosols than from greenhouse gases," *Climate and Atmosphere,* Volume 7, Article number 142, 30 June 2024. https://doi.org/10.1038/s41612-024-00696-0.

23Wu, You-Ting et al., "Stronger Arctic amplification from anthropogenic aerosols than from greenhouse gases," *Climate and Atmosphere*, Volume 7, Article number 142, 30 June 2024, p. 1. https://doi.org/10.1038/s41612-024-00696-0.

24 Quaglia, Ilaria and Visioni, Daniele, "Modeling 2020 regulatory changes in international shipping emissions help explain anomalous 2023 warming," *ESD Letters,* Volume 15, Issue 6, ESD, 15, pp. 1527-1541, 28 November 2024. https://doi.org/10.5194/esd-15-1527-2024.

25 Gettelman, A. et al., "Has Reduced Ship Emissions Brought Forward Global Warming?" *Geophysical Research Letters,* Volume 51, Issue 15, e2024GL109077, 12 August 2024. https://doi.org/10.1029/2024GL109077.

26 Gordon, Hamish et al., "Causes and importance of new particle formation in the present-day and preindustrial atmospheres," *Journal of Geophysical Research: Atmospheres,* Volume 122, Issue 16, pp. 8739-8760, 10 July 2017. https://doi.org/10.1002/2017JD026844.

27 Schneider, David P. et al., "Climate response to large, high-latitude and low-latitude volcanic eruptions in the Community Climate Model," *Journal of Geophysical Research: Atmospheres*, Volume 114, Issue D15, 1 August 2009. https://doi.org/10.1029/2008JD011222.

28 *Wikipedia*, "Year Without a Summer." https://en.wikipedia.org/wiki/Year_Without_a_Summer.

29 *Wikipedia*, "Year Without a Summer." https://en.wikipedia.org/wiki/Year_Without_a_Summer.

30 Jenkins, Stuart et al., "Tonga eruption increases chance of temporary surface temperature anomaly above 1.5°C," *Nature Climate Change,* 12 January 2023, pp. 127-129. https://doi.org/10.1038/s41558-022-01568-2.

31 Juncker, Martin et al., "Long-Term Climate Impacts of Large Stratospheric Water Vapor Perturbations," *Journal of Climate,* Volume 37: Issue 17, pp. 4507-4521, 14 August 2024. https://doi.org/10.1175/JCLI-D-23-0437.1.

32 Robock, Alan, "Volcanic eruptions and climate change," *Reviews of Geophysics,* Volume 38, Issue 2, pp. 191-219, 1 May 2000. https://doi.org/10.1029/1998RG000054.

33 Hegerl, Gabriele C. et al., "Detection of volcanic, solar, and greenhouse gas signals in paleo-reconstructions of Northern Hemispheric temperature," *Geophysical Research Letters*, Volume 30, Issue 5, 12 March 2003, Figure 3. https://doi.org/10.1029/2002GL016635; See also, Vinós, Javier, *Solving the Climate Puzzle: The Sun's Surprising Role*, Critical Science Press, Madrid, 2023, p. 164, ISBN: 978-84-125867-7-0.

34 Schneider, David P. et al., "Climate response to large, high-latitude and low-latitude volcanic eruptions in the Community Climate Model," *Journal of Geophysical Research: Atmospheres,* Volume 114, Issue D15, 1 August 2009. https://doi.org/10.1029/2008JD011222.

35 Vinós, Javier, *Solving the Climate Puzzle: The Sun's Surprising Role*, Critical Science Press, Madrid, 2023, p. 165, ISBN: 978-84-125867-7-0.

36 Zhong, Y. et al., "Centennial-scale climate change from decadally-paced explosive volcanism: a coupled sea ice-ocean mechanism," *Climate Dynamics,* Volume 37, pp. 2373-2387, 31 December 2010. https://doi.org/10.1007/s00382-010-0967-z.

37 Miller, Gifford H. et al., "Abrupt onset of the Little Ice Age triggered by volcanism and sustained by sea-ice/ocean feedbacks," *Geophysical Research Letters,* Volume 39, Issue 2, 31 January 2012. https://doi.org/10.1029/2011GL050168.

38 Schneider, David P. et al., "Climate response to large, high-latitude and low-latitude volcanic eruptions in the Community Climate Model," *Journal of Geophysical Research: Atmospheres,* Volume 114, Issue D15, 1 August 2009. https://doi.org/10.1029/2008JD011222.

39 Crowley, T. J. et al., "Technical details concerning development of a 1200 yr proxy index for global volcanism," *Earth System Science Data,* Volume 5, pp. 187-197, 23 May 2013. Figure 8., pg. 194. https://doi.org/10.5194/essd-5-187-2013.

40 Crowley, T. J. et al., "Technical details concerning development of a 1200 yr proxy index for global volcanism," *Earth System Science Data,* Volume 5, pp. 187-197, 23 May 2013. Figure 8., pg. 194. https://doi.org/10.5194/essd-5-187-2013.

41 Robock, Alan, "Volcanic eruptions and climate change," *Reviews of Geophysics,* Volume 38, Issue 2, pp. 191-219, 1 May 2000. https://doi.org/10.1029/1998RG000054.

42 Hegerl, Gabriele C. et al., "Detection of volcanic, solar and greenhouse gas signals in paleo-reconstructions of Northern Hemispheric temperature, *Geophysical Research Letters,* Volume 30, Issue 5, 12 March 2003, Figure 3. https://doi.org/10.1029/2002GL016635 See also Vinós, Javier, *Solving the Climate Puzzle, The Sun's Surprising Role,* Critical Science Press, Madrid, 2023, ISBN:

978-84-125867-7-0, pg. 164.

43 IPCC, *The Physical Science Basis, Contribution of Working Group I to the Sixth Assessment Report on the Intergovernmental Panel on Climate Change,* “Chapter 7 - The Earth’s Energy Budget, Climate Feedbacks and Climate Sensitivity,” Cambridge University Press, Cambridge UK and New York, NY, USA, Section 7.4.2.4, p. 961, 9 August 2021, p. 961. www.ipcc.ch/report/ar6/wg1/chapter/chapter-7/.

44 Wunsch, Carl, “What Is the Thermohaline Circulation?” *Science*, Volume 298, pp. 1179-1181, 8 November 2002. https://www.science.org/doi/10.1126/science.1079329.

45 Zhong, Y. et al., “Centennial-scale climate change from decadally-paced explosive volcanism: a coupled sea ice-ocean mechanism,” *Climate Dynamics,* Volume 37, pp. 2373-2387, 31 December 2010. https://doi.org/10.1007/s00382-010-0967-z.

46 Crowley, T. J. et al., “Technical details concerning development of a 1200 yr proxy index for global volcanism,” *Earth System Science Data,* Volume 5, pp. 187-197, 23 May 2013, Figure B, p. 194. https://doi.org/10.5194/essd-5-187-2013.

47 Schneider, David P. et al., “Climate response to large, high-latitude and low-latitude volcanic eruptions in the Community Climate Model,” *Journal of Geophysical Research: Atmospheres,* Volume 114, Issue D15, 1 August 2009. https://doi.org/10.1029/2008JD011222.

48 Crowley, T. J. et al., “Technical details concerning development of a 1200 yr proxy index for global volcanism,” *Earth System Science Data,* Volume 5, pp. 187-197, 23 May 2013. Figure 8., pg. 194. https://doi.org/10.5194/essd-5-187-2013.

49 Rensberger, Moyce, “Aztec Sacrifices Laid to Hunger, not Just Religion,” *The New York Times*, 19 February 1977. https://www.nytimes.com/1977/02/19/archives/aztec-sacrifices-laid-to-hunger-not-just-religion.html See also *Mexicolore*, “The human cost of famine,’ AZTECS. https://www.mexicolore.co.uk/aztecs/aztec-life/human-cost-of-famine.

50 Parker, Geoffrey, *Global Crisis: War, Climate Change and Catastrophe in the Seventeenth Century*, Yale University Press, April 30, 2013, pp. 904, pg. xxv, ISBN: 978-0300-153231.

51 Hessayon, Ariel, “The Little Ice Age and the River Thames frost fairs of the seventeenth century,” *Historical Essays,* 18 February 2022. https://arielhessayon.substack.com/p/the-little-ice-age-and-the-river.

52 Vinós, Javier, *Solving the Climate Puzzle, The Sun’s Surprising Role*, Critical Science Press, Madrid, 2023, p. 173, ISBN: 978-84-125867-7-0.

53 Fagan, Brian, *The Little Ice Age, How Climate Made History, 1300-1850,* Basic Books, New York, 2000, p. 83, ISBN: -10-465-02272-3.

54 White, Sam, *A Cold Welcome,* Harvard Press, Cambridge, Massachusetts, 2017, pg. 161. ISBN 9780674244900. See also, White, Sam, "A Cold Welcome: The Little Ice Age and America's Colonial Beginnings," The National Socio-Environmental Synthesis Center, 25 October 2016. https://www.youtube.com/watch?v=OXrZxYs9gdM&t=1601s Minute 31:45 to 32:11.

55 Crowley, T. J. et al., "Technical details concerning development of a 1200 yr proxy index for global volcanism," *Earth System Science Data,* Volume 5, pp. 187-197, 23 May 2013. Figure B., pg. 194. https://doi.org/10.5194/essd-5-187-2013.

56 Royal Museums Greenwich, "Submarine volcanoes." https://www.rmg.co.uk/stories/ocean/submarine-volcanoes.

57 Plimer, Ian, "Professor Ian Plimer on Green Murder," The Christopher Dawson Centre for Cultural Studies, 26 April, 2022. https://youtu.be/tx-QcX0fm5bs?si=GcfGnk95KjGZNrel; See also, Plimer, Ian, "Ian Plimer, a Geologist's View of Climate Change at Heartland Institute Climate Conference," The Heartland Institute, 7 March 2023. https://youtu.be/tK4LNIvlc-CY?si=aZ3fbBSVX2qwKpya.

58 Viterito, Arthur, "Viterito/Kamis/Yim/Catt: Impacts of Geothermal Energy on Climate," *Tom Nelson Podcast,* #181, 23 December, 2023. https://www.youtube.com/watch?v=lbDbA32fNek.

59 Loose, Brice et al., "Evidence of an active volcanic heat source beneath the Pine Island Glacier," *Nature Communications*, Volume 9, Article number 2431, 22 June 2018. https://doi.org/10.1038/s41467-018-04421-3.

60 Otosaka, Inès, N. et al. "Mass balance of the Greenland and Antarctic ice sheets from 1992 to 2020," *Earth System Science Data,* Volume 15, Issue 4, pp. 1597-1616, 20 April 2023. https://doi.org/10.5194/essd-15-1597-2023; This study shows East Antarctica gained mass from 1992 to 2020, while West Antarctic lost significant mass. From 2021-2023, Antarctica, as a whole, gained 108 gigatons per year of ice concentrated mostly in East Antarctica as West Antarctica continued to lose mass. See also Wang, Wei, "Spatiotemporal mass change rate analysis from 2002 to 2023 over the Antarctic Ice Sheet and four glacier basins in Wilkes-Queen Mary Land," *Earth Sciences,* Volume 68, pp. 1086-1099, 19 March 2025. https://doi.org/10.1007/s11430-024-1517-1.

61 May, Andy, "The Sun-Cloud Effect: The Winter Gatekeeper Hypothesis (III). Meridional Transport, the Most Fundamental Climate Variable," *Andy May Petrophysicist,* 16 August 2022. https://andymaypetrophysicist.com/2022/08/16/the-sun-climate-effect-the-winter-gatekeeper-hypothesis-iii-meridional-transport-the-most-fundamental-climate-variable/.

62 Koonin, Steven, "Hot or Not: Steven Koonin Questions Conventional Climate Science and Methodology," *Uncommon Knowledge,* Hoover Institution, 15 August 2023. https://www.youtube.com/watch?v=l90FpjPGLBE.

Chapter 13 - The Oceans' Impact on Climate

1 Svensmark, Henrik et al., "Atmospheric Ionization and Cloud Radiative Forcing," *Nature Scientific Reports,* Volume 11, Article Number 19668, 11 October 2021. https://doi.org/10.1038/s41598-021-99033-1.

2 Derakshani, Max, "ENSO Warming vs. CO_2 Warming," *Tom Nelson Podcast,* #89, 31 March 2023. https://www.youtube.com/watch?v=oo8lyL6lYQU.

3 Vahrenholt, Fritz and Lüning, Sebastion, *The Neglected Sun, Why the Sun Precludes Climate Catastrophe,* The Heartland Institute, Arlington Heights, IL, 2015, p, 137, ISBN 13-978-1-934791-54-7. Citations: Scafetta, N., "Empirical evidence for a celestial origin of the climate oscillations and its implications," *Journal of Atmospheric and Solar-Terrestrial Physics,* Volume 72, pp. 951-970, 2010. https://doi.org/10.1016/j.gsf.2023.101650; Scafetta, Nicola and West, Bruce J., "Phenomenological reconstructions of the solar signature in the Northern Hemisphere surface temperature records since 1600," *Journal of Geophysical Research,* Volume 112, Issue D24, 03 November 2007. https://doi.org/10.1029/2007JD008437; Scafetta, Nicola. and West, Bruce J., "Is climate sensitive to solar variability?" *Physics Today,* Volume 61, Issue 3, pp. 50-51, March 2008. https://doi.org/10.1063/1.2897951; Beer, J., Mende, W. and Stellmacher R., "The role of the sun in climate forcing," *Quaternary Science Reviews,* Volume 19, Issues 1-5, pp. 403-415, 1 January 2000. https://doi.org/10.1016/S0277-3791(99)00072-4; Rao, U. R., "Contribution of changing galactic cosmic ray flux to global warming." *Current Science,* Volume 100, Issue 2, pp. 223-225, 25 January 2011. https://www.jstor.org/stable/24073049; Shaviv, Nir J., "On climate response to changes in the cosmic ray flux and radiative budget," *Journal of Geophysical Research,* Volume 110, Issue A8, pp. 1-15, 23 August 2005. https://doi.org/10.1029/2004JA010866; Shaviv, Nir J., "Using the oceans as a calorimeter to quantify the solar radiative forcing," *Journal of Geophysical Research,* Volume 113, Issue A11, pp. 1-13, 4 November 2008. https://doi.org/10.1029/2007JA012989; El-Borie, Ali et al., "Spectral analysis of solar variability and their possible role on global warming," *Journal of Environmental Protection,* Volume 1, Number 2, June 2010. http://dx.doi.org/10.4236/jep.2010.12014; De Jager, C. and Duhau, Silvia, "The variable solar dynamo and the forecast of solar activity; influence on terrestrial surface temperature." Global Warming in the 21st Century, In J. M. Cossia (ed), Nova Science, Hauppauge, NY, pp. 77-106, January 2011. https://www.semanticscholar.org/paper/THE-VARIABLE-SOLAR-DYNAMO-AND-THE-FORECAST-OF-SOLAR-Jager-Duhau/671d0588dc7ae7bfd774ce63e21bac10d66a268a

4 Soon, Willie et al., "The Detection and Attribution of Northern Hemisphere

Land Surface Warming (1850-2018) in Terms of Human and Natural Factors: Challenges of Inadequate Data," *Climate*, Volume 11, Issue 9, 28 August 2023, p. 179. https://doi.org/10.3390/cli11090179; See also, Alexander, Ralph B., "The Sun Can Explain 70% or More of Global Warming, Says New Study," *Science Under Attack,* 18 September 2023. https://www.scienceunderattack.com/blog/2023/9/18/the-sun-can-explain-70-or-more-of-global-warming-says-new-study-138?format=amp; and Soon, Willie et al., "Controversy surrounding the Sun's role in climate change," *Climate Etc.,* 2023. https://judithcurry.com/2023/09/10/controversy-surrounding-the-suns-role-in-climate-change/?amp=1.

5 Vinós, Javier, May, Andy, "The Sun-Climate Effect: The Winter Gatekeeper Hypothesis (III). Meridional transport, the most fundamental climate variable," 16 August, 2022, *Andy May Petrophysicist.* https://andymaypetrophysicist.com/2022/08/16/the-sun-climate-effect-the-winter-gatekeeper-hypothesisiii-meridional-transport-the-most-fundamental-climate-variable/.

6 Girihagama, Lakshika and Nof, Doron, "Why is the ocean surface slightly warmer than the atmosphere?" *Journal of Marine Research,* Yale Peabody Museum, Yale University, Volume 76, pp. 23-46, 2018. https://elischolar.library.yale.edu/journal_of_marine_research/455/.

7 Wong, Elizabeth W. and Minnett, Peter J., "The Response of the Ocean Thermal Skin Layer to Variations in Incident Infrared Radiation," *Journal of Geophysical Research: Oceans,* Volume 123, Issue 4, pp. 2475-2493, 23 March 2018. https://doi.org/10.1002/2017JC013351.

8 Shaw, Joseph A. et al., "Scanning infrared radiometer for measuring the air-sea temperature difference," *Applied Optics,* Volume 40, No. 27, pp. 4807-4815, 20 September 2001. https://doi.org/10.1364/AO.40.004807.

9 Steele, Jim, "Why the Sun, Not CO_2, Heats the Oceans Revisiting the Debate: Does Greenhouse Back-Radiation Warm the Oceans?" *Watts Up With That?,* 31 July 2022. https://wattsupwiththat.com/2022/07/31/why-the-sun-not-co2-heats-the-oceans-revisiting-the-debate-does-greenhouse-back-radiation-warm-the-oceans/.

10 Vinós, Javier, *Climate of the Past, Present and Future: A Scientific Debate,* 2nd Edition, Critical Science Press, September 2022, p. 35, ISBN: 9788412586701.

11 Watson, Andrew J. et al., "Revised estimates of ocean-atmosphere CO_2 flux are consistent with ocean carbon inventory," *Nature Communications,* Volume 11, p. 4422, 4 September 2020. https://doi.org/10.1038/s41467-020-18203-3.

12 Minnett, P. J. et al., "Half a century of satellite remote sensing of sea-surface temperature," *Remote Sensing of Environment,* Volume 233, November 2019. https://doi.org/10.1016/j.rse.2019.111366.

13 Wong, Elizabeth W. and Minnett, Peter J., "The Response of the Ocean Thermal Skin Layer to Variations in Incident Infrared Radiation," *Journal of Geophysical Research: Oceans*, Volume 123, Issue 4, pp. 2475-2493, 23 March 2018. https://doi.org/10.1002/2017JC013351.

14 Steele, Jim, "Why the Sun, Not CO_2, Heats the Oceans Revisiting the Debate: Does Greenhouse Back-Radiation Warm the Oceans?" *Watts Up With That?*, 31 July 2022. https://wattsupwiththat.com/2022/07/31/why-the-sun-not-co2-heats-the-oceans-revisiting-the-debate-does-greenhouse-back-radiation-warm-the-oceans/.

15 Fairall, C. W. et al., "Cool-skin and warm-layer effects on sea surface temperature," *Journal of Geophysical Research,* Volume 101, Number C1, pp. 1295-1308, 15 January 1996, p. 1296. https://doi.org/10.1029/95JC03190.

16 Yan, Yunwei et al., "Tropical Cool-Skin and Warm-Layer Effects and Their Impact on Surface Heat Fluxes," *Journal of Physical Oceanography*, pp. 45-62, January 2024, Figure 5. https://doi.org/10.1175/JPO-D-23-0103.1.

17 Dai, Aiguo and Deng, Jiechun, "Arctic Amplification Weakens the Variability of Daily Temperatures Over Northern Middle-High Latitude," *Journal of Climate*, Volume 34, 1 April 2021. https://doi.org/10.1175/JCLI-D-20-0514.1; See also, Martinez, A. and Iglesias, G., "Global wind energy resources decline under climate change," *Energy*, Volume 288, 1 February 2024. https://doi.org/10.1016/j.energy.2023.129765.

18 Minnett, P. J. et al., "Half a century of satellite remote sensing of sea-surface temperature," *Remote Sensing of Environment,* Volume 233, November 2019. https://doi.org/10.1016/j.rse.2019.111366.

19 Yan, Yunwei et al., "Tropical Cool-Skin and Warm-Layer Effects and Their Impact on Surface Heat Fluxes," *Journal of Physical Oceanography*, pp. 45-62, January 2024, Figure 6. https://doi.org/10.1175/JPO-D-23-0103.1.

20 Wong, Elizabeth and Minnett, Peter, "The Response of the Ocean Thermal Skin Layer to Variations of Incident Infrared Radiation," *Journal of Geophysical Research: Oceans,* Volume 123, Issue 4, pp. 2475-2493, 23 March 2018. https://doi.org/10.1002/2017JC013351.

21 McVicar, Tim et al., "Global review and synthesis of trends in observed terrestrial near-surface wind speeds: implications for evaporation," *Journal of Hydrology,* Volume 416-417, pp. 182-205, 24 January 2012. https://doi.org/10.1016/j.jhydrol.2011.10.024.

22 Lindsey, Rebecca, "Climate and Earth's Energy Budget," *NASA Earth Observatory,* 14 January 2009. https://earthobservatory.nasa.gov/features/EnergyBalance.

23 *Wikipedia*, "Earth's Energy Budget." https://en.wikipedia.org/wiki/Earth%27s_energy_budget.

24 Wong, Elizabeth and Minnett, Peter, "The Response of the Ocean Thermal Skin Layer to Variations of Incident Infrared Radiation," *Journal of Geophysical Research: Oceans,* Volume 123, Issue 4, pp. 2475-2493, 23 March 2018. https://doi.org/10.1002/2017JC013351.

25 Lindsey, Rebecca, "Climate and Earth's Energy Budget," *NASA Earth Observatory,* 14 January 2009. https://earthobservatory.nasa.gov/features/EnergyBalance.

26 Plimer, Ian, "Professor Ian Plimer on Green Murder," *The Christopher Dawson Centre for Cultural Studies,* April 26, 2022, 58 minutes, 40 seconds. https://youtu.be/txQcX0fm5bs?si=GcfGnk95KjGZNrel.

27 Ott, Markus, "What Warms the Oceans?" *Tom Nelson Podcast,* 19 March 2023. https://youtu.be/m9PCgCGo17w?si=aQi-y6xbfyAb6Mi4.

28 Ott, Markus, "What Warms the Oceans?" *Tom Nelson Podcast,* 19 March 2023. https://youtu.be/m9PCgCGo17w?si=aQi-y6xbfyAb6Mi4.

29 *Wikipedia,* "Atmospheric Optics," Cloud Coloration. https://en.wikipedia.org/wiki/Atmospheric_optics.

30 *Wikipedia,* "Earth's Energy Budget." https://en.wikipedia.org/wiki/Earth%27s_energy_budget.

31 CO2 Coalition, "Nobel Laureate John Clauser Elected to CO2 Coalition Board of Directors," CO2 Coalition, 5 May 2023. https://co2coalition.org/publications/nobel-laureate-john-clauser-elected-to-co2-coalition-board-of-directors/.

32 Viterito, Arthur, "Viterito/Kamis/Yim/Catt: Impacts of Geothermal Energy on Climate," *Tom Nelson Podcast,* #181, 23 December, 2023. https://www.youtube.com/watch?v=lbDbA32fNek.

33 *Climate4you,* "Temperature in Polar regions; Arctic and Antarctica; Polar regions as key regions for global climate change." https://climate4you.com.

34 Parkinson, Claire L., "A 40-y record reveals gradual Antarctica sea ice increases followed by decreases at rates far exceeding the rates seen in the Arctic," *PNAS,* Volume 116, Issue 29, 1 July 2019, pp. 14414 to 14423. https://doi.org/10.1073/pnas.1906556116.

35 Soden, Brian J., Held, Isaac M., "An Assessment of Climate Feedbacks in Coupled Ocean-Atmosphere Models," *Journal of Climate,* 3354-3360, 15 July 2006. https://doi.org/10.1175/JCLI3799.1.

36 IPCC, *The Physical Science Basis, Contribution of Working Group I to the Sixth Assessment Report on the Intergovernmental Panel on Climate Change,* "Chapter 7 – The Earth's Energy Budget, Climate Feedbacks and Climate Sensitivity," Cambridge University Press, Cambridge, UK and New York, NY, USA, Section 7.4.2.4, Section 7.4.2.4, 9, p. 961, 9 August 2021. www.ipcc.ch/report/

ar6/wg1/chapter/chapter-7/.

37 Domeisen, Daniela I.V., "Estimating the Frequency of Sudden Stratospheric Warming Events from Surface Observations of the North Atlantic Oscillation," *Journal of Geophysical Research: Atmospheres,* Volume 124, Issue 6, pp. 3180 to 3194, 4 March 2019. https://doi.org/10.1029/2018JD030077.

38 NOAA Global Monitoring Laboratory reports ~1900 to ~2000 ppb of methane in Barrow Alaska for 2023 to 2024. https://gml.noaa.gov.

39 NOAA Global Monitoring Laboratory reports ~1800 to ~1850 at the South Pole for 2023 to 2024. https://gml.noaa.gov.

40 Myhre, Gunnar et al., "New estimates of radiative forcing due to well-mixed greenhouse gases," *Geophysical Research Letters*, Volume 25, Issue 14, pp. 2715-2718, 15 July 1998. https://doi.org/10.1029/98GL01908.

41 Viterito, Arthur, "Viterito/Kamis/Yim/Catt: Impacts of Geothermal Energy on Climate," *Tom Nelson Podcast*, #181, 23 December 2023. https://www.youtube.com/watch?v=lbDbA32fNek.

42 Audet, Alexander C., "Correspondence Among Mid-Latitude Glacier Equilibrium Line Altitudes, Atmospheric Temperatures, and the Westerly Wind Fields," *Geophysical Research Letters,* Volume 49, Issue 23, 14 November 2022. https://doi.org/10.1029/2022GL099897.

43 Vinós, Javier, *Climate of the Past, Present and Future: A Scientific Debate,* 2nd Edition, Critical Science Press, September 2022, ISBN: 9788412586701.

44 Hartmann, Dennis, *Global Physical Climatology,* Second Edition, 2016, Amsterdam, Elsevier, pp. 175-176, ISBN 978-0-12-328531-7.

45 Hartmann, Dennis, *Global Physical Climatology,* Second Edition, 2016, Amsterdam, Elsevier, p. 167, ISBN 978-0-12-328531-7.

46 *Wikipedia*, "Hadley Cell; Energetics and Transport." https://en.wikipedia.org/wiki/Hadley_cell.

47 Vinós, Javier, *Solving the Climate Puzzle: The Sun's Surprising Role*, Critical Science Press, Madrid, 2023, p. 105, ISBN: 978-84-125867-7-0.

48 Arnfield, John A. and Gentilli, Joseph, "Poleward transfer of heat," *Britannica*, 28 March 2025. https://britannica.com/science/climate-meteorology/Poleward-transfer-of-heat.

49 Wunsch, Carl, "What Is the Thermohaline Circulation?" *Science*, Volume 298, 8 November 2002, pp. 1179-1181. https://www.science.org/doi/10.1126/science.1079329.

50 May, Andy, "Meridional Transport, the most fundamental climate variable," *Andy May Petrophysicist,* October 2022. https://andymaypetrophysicist.com/2022/10/24/meridional-transport-the-most-fundamental-climate-vari-

able/; See also, May, Andy, "CO_2-driven climate models of the IPCC are inadequate," *Tom Nelson Podcast,* #34, 24 October 2022. https://www.youtube.com/watch?v=6aNkmXArlZk.

51 Vinós, Javier, *Solving the Climate Puzzle: The Sun's Surprising Role*, Critical Science Press, Madrid, 2023, ISBN: 978-84-125867-7-0.

52 Peixoto, José, Oort, Abraham, *Physics of Climate,* American Institute of Physics, 1992, 520 pp., ISBN: 978088318711.

53 Lyle, Mitchell et al., "Pacific Ocean and Cenozoic evolution of climate," *Reviews of Geophysics,* Volume 46, Issue 2, 19 April 2008. https://doi.org/10.1029/2005RG000190.

54 Vinós, Javier, *Solving the Climate Puzzle: The Sun's Surprising Role,* Critical Science Press, Madrid, 2023, p. 133, ISBN: 978-84-125867-7-0.

55 Yang, Haijun et al., "Decomposing the meridional heat transport in the climate system," *Climate Dynamics*, Volume 44, pp. 2751-2768, 21 October 2014. https://doi.org/10.1007/s00382-014-2380-5.

56 Derakhshani, Max, "ENSO Warming vs. CO_2 Warming, *Tom Nelson Podcast,* #89, 31 March 2023. https://www.youtube.com/watch?v=oo8lyL6lYQU.

57 Zhang, Rong et al., "A Review of the Role of the Atlantic Meridional Overturning Circulation in Atlantic Multidecadal Variability and Associated Climate Impacts," *Reviews of Geophysics*, Volume 57, Issue 2, pp. 316-375, 29 April 2019. https://doi.org/10.1029/2019RG000644.

58 Dilley, David, "Global Warming Will Be Dead By 2030," *Tom Nelson Podcast,* #216, 14 May 2024. https://www.youtube.com/watch?v=NeFePI1nW1Y.

59 Chylék, Petr et al., "Isolating the anthropogenic component of Arctic warming," *Geophysical Research Letters,* Volume 41, Issue 10, pp. 3569 to 3576, 28 May 2014. https://doi.org/10.1002/2014GL060184.

60 Chylék, Petr et al., "Isolating the anthropogenic component of Arctic warming," *Geophysical Research Letters,* Volume 41, Issue 10, pp. 3569 to 3576, 28 May 2014. https://doi.org/10.1002/2014GL060184.

61 Chylék, Petr et al., "Isolating the anthropogenic component of Arctic warming," *Geophysical Research Letters,* Volume 41, Issue 10, pp. 3569 to 3576, 28 May 2014. https://doi.org/10.1002/2014GL060184.

62 Schalamon, Forina Roana et al., "The role of large-scale atmospheric patterns for recent warming periods in Greenland from 1900-1915," *Weather and Climate Dynamics*, Volume 6, Issue 3, pp. 1075-1088, 7 October 2025. https://doi.org/10.5194/wcd-6-1075-2025.

63 Smithsonian Astrophysical Observatory under NASA, April 2021, https://ui.adsabs.harvard.edu/abs/2021EGUGA..23.2582K/abstract.

64 Lupo, Anthony R., “The Role of the Oceans,” *Climate and Energy,* edited by Beisner, Calvin E. and Legates, David R., Regnery Publishing, Washington DC, 2024, p. 143, ISBN:978-1-68451-267-6.

65 Qian, Weihong, and Zhu, Yafen, “Climate Change in China from 1880 to 1998 and Its Impact on the Environmental Condition,” *Nature Climate Change,* Volume 50, pp. 419-444, September 2001. https://doi.org/10.1023/A:1010673212131.

66 Lupo, Anthony R., “The Role of the Oceans,” *Climate and Energy,* edited by Beisner, Calvin E. and Legates, David R., Regnery Publishing, Washington DC, 2024, p. 141, ISBN:978-1-68451-267-6.

67 Dilley, David, “Global Warming Will Be Dead By 2030,” *Tom Nelson Podcast*, #216, 14 May 2024. https://www.youtube.com/watch?v=NeFePI1nW1Y.

68 Lupo, Anthony R., “The Role of the Oceans,” *Climate and Energy*, edited by Beisner, Calvin E. and Legates, David R., Regnery Publishing, Washington DC, 2024, p. 141, ISBN:978-1-68451-267-6.

69 Donat, Markus et al., “Warm oceans caused hottest Dust Bowl years in 1934/1936,” *AERC Centre of Excellence for Climate System Science*, 4 May 2015. https://www.sciencedaily.com/releases/2015/05/150504101248.htm.

70 Brown, Geoff, “120 Years of Climate Doom: The 1970s Ice Age Scare,” *CLIMATE DISPATCH,* 24 May 2017. https://climatechangedispatch.com/120-years-of-climate-scares-1970s-ice-age-scare/.

71 HSDL, “Study of Climatological Research as It Pertains to Intelligence Problems.” https://hsdl.org/c/abstract/?docid=725433.

72 Kukla, George and Matthews, Robley, “When Will the Present Interglacial End?” *Science*, Volume 178, Number 4057, 13 October1972. https://doi.org/10.1126/science.178.4057.190.

73 Maximus, Fabius, “An Important Letter Sent to the President About the Danger of Climate Change,” *TheStreet,* 21 October 2009. https://www.thestreet.com/economonitor/us/an-important-letter-sent-to-the-president-about-the-danger-of-climate-change.

74 Lupo, Anthony R., “The Role of the Oceans,” *Climate and Energy*, edited by Beisner, Calvin E. and Legates, David R., Regnery Publishing, Washington DC, 2024, p. 136, ISBN:978-1-68451-267-6.

Chapter 14 - Summary and Implications

1 Adams, John, “Facts are stubborn things …” *BrainyQuotes.* https://www.brainyquote.com/quotes/john_adams_134175.

2 Twain, Mark, “It is easier to fool people…” *Goodreads.* htpps://ww.goodreads.com/quotes/584507-it-is-easier-to-fool-people-than-to-convince-them-that.

3 Feynman, Richard, "...If it disagrees with experiment, it's wrong," Cornell University, 1964. https://youtu.be/b240PGCMwV0.

4 Chang, Edmund K. et al., "Observed and projected decrease in Northern Hemiphere extratropical cyclone activity in summer and its impact on maximum temperature," *Geophysical Research Letters,* Volume 43, pp. 2200-2208, 12 March 2016. https://doi.org/10.1002/2016GL068172.

5 Tuleya, Robert E., "Impact of Upper-Tropospheric Temperature Anomalies and Vertical Wind Shear on Tropical Cyclone Evolution Using an Idealized Version of the Operational GFDL Hurricane Model," *Journal of the Atmospheric Sciences,* Volume 73, Issue 10, pp. 3803-3820, 1 October 2016, p. 3803. https://doi.org/10.1175/jas-d-16-0045.1.

6 Gertler, Charles G. and O'Gorman, Paul A., "Changing available energy for extratropical cyclones and associated convection in Northern Hemisphere summer," *PNAS,* Volume 116, Issue 10, pp. 4105-4110, 19 February 2019. https://doi.org/10.1073/pnas.1812312116.

7 Bony, S., et al., "How well do we understand and evaluate climate change feedback processes?" *Journal of Climate Change,* Volume 19, Issue 15, pp. 3445-3482, 2006. https://doi.org/10.1175/JCLI3819.1.

8 Tu, Shifei et al., "Decreasing trend in destructive potential of tropical cyclones in the South India Ocean since the mid-1990s" *Nature Communications Earth & Environment,* Volume 5, p. 543, 30 December 2024. https://doi.org/10.1038/s43247-024-01683-2.

9 Trouet, V. et al., "North Atlantic storminess and Atlantic Meridional Overturning Circulation during the last Millennium: Reconciling contradictory proxy records of NAO variability," *Global and Planetary Change,* Volumes 84-85, pp. 48-55, p. 51-53, March 2012. https://doi.org/10.1016/j.gloplacha.2011.10.003.

10 Behringer, Wolfgang, *A Cultural History of Climate,* Polity Press, Cambridge, UK, 2010, p. 88, ISBN 13: 978-0-7456-4528-5.

11 Tan, L. et al., "Climate patterns in North Central China during the last 1800 years and their possible driving force," *Climate of the Past,* Volume 7, Issue 3, pp. 685-692, 4 July 2011, p. 688. https://doi.org/10.5194/cp-7-685-2011.

12 IPCC, *Climate Change 2021: The Physical Science Basis. Contributions of Working Group I to the Sixth Assessment Report of the Intergovernmental Panel on Climate Change,* "Chapter 12 - Climate Change Information for Regional Impact and for Risk Assessment," Cambridge University Press, Cambridge, UK and New York, NY, USA, Table 12.12, p. 1856, 9 August 2021. https://doi.org/10.1017/9781009157896.014.

13 Shihab, Hussein, "Threat to Islands," *Canberra Times,* September 26, 1988.

14 Heller, Tony, "Children Just Won't Know What an Honest Scientist Is," *Real Climate Science*, 28 February 2018. https://realclimatescience.com/2018/02/children-just-wont-know-what-an-honest-scientist-is/#gsc.tab=0.

15 *Albuquerque Journal*, "Climate Outlook: No Maple Syrup," December 2001. https://www.newspapers.com/search/results/?keyword=Albuquerque+Journal%2C+"Climate+Outlook%3A+No+Maple+Syrup%2C"+December+2001.

16 Markley, Edward J., "The Arctic Is Screaming: Ominous Arctic Melt Worries Experts," The Select Committee on Energy Independence and Global Warming, Representative Edward J. Markey, Chairman. https://www.markey.senate.gov/imo/media/globalwarming/resources/articles_id=0018.html.

17 Gore, Al, "Al Gore Nobel Prize Lecture," The Nobel Prize. https://www.nobelprize.org/prizes/peace/2007/gore/lecture/.

18 Vidal, John, "Arctic expert predicts final collapse of sea ice within four years," *The Guardian*, 17 September 2012. https://www.theguardian.com/environment/2012/sep/17/arctic-collapse-sea-ice.

19 Amos, Jonathan, "Arctic summers ice-free by 2013," *BBC News*, 12 December 2007. http://news.bbc.co.uk/2/hi/7139797.stm.

20 Heller, Tony, "NASA Ice-Free Prophesy Update," *Real Climate Science*, 11 June 2018. https://realclimatescience.com/2018/06/nasa-ice-free-prophesy-update/#gsc.tab=0.

21 McMahon, Jeff, "We Have Five Years to Save Ourselves from Climate Change, Harvard Scientist Says," *Forbes*, January 15, 2018. https://www.forbes.com/sites/jeffmcmahon/2018/01/15/carbon-pollution-has-shoved-the-climate-backward-at-least-12-million-years-harvard-scientist-says/.

22 Balmakov, Roman, "Inconvenient Truth: 32 Climate Predictions Proven False—Facts Matter," *The Epoch Times*, 2024. https://www.youtube.com/watch?v=E1e5HAZo4iw&t=763s.

23 Morrison, Chris, "Islands That Climate Alarmists Said Would Soon 'Disappear' Due to Rising Sea Found to Have Grown in Size," *Climate-Science.press*, 8 April 2024. https://climate-science.press/2024/04/08/islands-that-climate-alarmists-said-would-soon-disappear-due-to-rising-sea-found-to-have-grown-in-size/.

24 Myhre, Gunnar et al., "New estimates of radiative forcing due to well mixed greenhouse gases," *Geophysical Research Letters*, Volume 25, Issue 14, pp. 2715-2718, 15 July 1998, p. 2718. https://doi.org/10.1029/98GL01908.

25 Stefan-Boltzmann Law Calculator, *BYJU'S*. https://byjus.com/stefan-boltzmann-law-calculator/.

26 *...and Then There's Physics*, "Effective emission height," 5 March 2014. https://andthentheresphysics.wordpress.com/2014/03/05/effective-emis-

sion-height/; See also, Wilson, Derrek J. and Gea-Banacloche, Julio, "Simple model to estimate the contribution of atmospheric CO_2 to the Earth's greenhouse effect," *American Journal of Physics,* Volume 80, Issue 4, pp. 306-315, 1 April 2012, p. 314. https://doi.org/10.1119/1.3681188.

27 Jones, Nicola, "How the World Passed a Carbon Threshold and Why It Matters," *YaleEnvironment360,* 26 January 2017. https://e360.yale.edu/features/how-the-world-passed-a-carbon-threshold-400ppm-and-why-it-matters?_bhlid=38186d09213e68ec77ed9db9b24aa48cba37d1c1&utm_campaign=when-the-glitter-of-an-economic-boom-clashes-with-the-environment&utm_medium=newsletter&utm_source=thedeepview.

28 Rosane, Olivia, "A File of Shame: World on Track for 3.2 Degrees Celsius of Warming, Latest IPCC Report Warns," *EcoWatch,* 4 April 2022. https://www.ecowatch.com/ipcc-report-climate-change.html.

29 IPCC, *The Physical Science Basis, Contribution of Working Group I to the Sixth Assessment Report on the Intergovernmental Panel on Climate Change,* "Chapter 7 - The Earth's Energy Budget, Climate Feedbacks and Climate Sensitivity," Cambridge University Press, Cambridge UK and New York, NY, USA, 7, Section 7.4.2.2, p. 957, 9 August 2021. http://www.ipcc.ch/report/ar6/wg1/chapter/chapter-7/.

30 IPCC, *The Physical Science Basis, Contribution of Working Group I to the Sixth Assessment Report on the Intergovernmental Panel on Climate Change,* "Chapter 7 - The Earth's Energy Budget, Climate Feedbacks and Climate Sensitivity," Cambridge University Press, Cambridge UK and New York, NY, USA, Section 7.4.2.4, p. 961, 9 August 2021. www.ipcc.ch/report/ar6/wg1/chapter/chapter-7/.

31 *Goodreads,* "W.C. Fields Quotes, Quotable Quote." https://www.goodreads.com/quotes/340982-if-you-can-t-dazzle-them-with-brilliance-baffle-them-with.

32 Each 1°C increase produces 7% increase in water vapor see: IPCC, The Physical Science Basis, Contribution of Working Group I to the *Sixth Assessment Report on the Intergovernmental Panel on Climate Change,* "Chapter 8 - Water Cycle Changes," Cambridge University Press, Cambridge UK and New York, NY, USA Section 8.2.1, p. 1065, 9 August 2021. https://www.ipcc.ch/report/ar6/wg1/downloads/report/IPCC_AR6_WGI_Chapter08.pdf; Clouds and precipitation scale with water vapor, see: Wenz, Frank J. et al., "How Much More Rain Will Global Warming Bring?" *Science,* Volume 317, 13 July 2007. https://doi.org/10.1126/science.1140746.

33 IPCC, *The Physical Science Bais, Contribution of Working Group 1 to the Fifth Assessment Report of the Intergovernmental Panel on Climate Change,* "Chapter 7 - Clouds and Aerosols," Cambridge University Press, Cambridge, UK, Section 7.7.2.1.2, pp. 580-582, 30 September 2013. https://www.ipcc.ch/

site/assets/uploads/2018/02/WG1AR5_Chapter07_FINAL-1.pdf.

34 L'Ecuyer, Tristan S. et al., "Reassessing the Effect of Cloud Type on Earth's Energy Balance in the Age of Active Spaceborne Observations. Part I: Top of Atmosphere and Surface," *Journal of Climate*, Volume 32, Issue 19, pp. 6197-6217, 1 October 2019. https://doi.org/10.1175/JCLI-D-18-0753.1.

35 *Climate4you.com,* "Total cloud cover vs. global surface temperature," Climate and clouds, cloud cover effect on climate. https://climate4you.com/ClimateAndClouds.htm#TotalCloudCoverVersusGlobalTemperature.

36 IPCC, *The Physical Science Basis, Contribution of Working Group 1 to the Fifth Assessment Report of the Intergovernmental Panel on Climate Change,* "Executive Summary - Clouds and Aerosols," Cambridge University Press, Cambridge, UK, Executive Summary, p. 574, 30 September 2013. https://www.ipcc.ch/site/assets/uploads/2018/02/WG1AR5_Chapter07_FINAL-1.pdf.

37 Goodreads, "W.C. Fields Quotes, Quotable Quote." https://www.goodreads.com/quotes/340982-if-you-can-t-dazzle-them-with-brilliance-baffle-them-with.

38 Clauser, John, "Nobel Laureate John Clauser: Climate Models Miss Key Variable," American Thought Leaders, *The Epoch Times,* 6 September 2023. https://youtu.be/CvqIqy8dUvA?si=2lkV0c6ehw3FqcRs.

39 CO_2 Coalition, "Nobel Laureate John Clauser Elected to CO_2 Coalition Board of Directors," CO_2 Coalition, 5 May 2023. https://co2coalition.org/publications/nobel-laureate-john-clauser-elected-to-co2-coalition-board-of-directors/.

40 McKitrick, Ross and Christy, John, "Pervasive Warming Bias in CMIP6 Tropospheric Layers," *Earth and Space Science*, September 2020, Volume 7, Issue 9, 15 July 2020. https://doi.org/10.1029/2020EA001281.

41 IPCC, "Climate change widespread, rapid, and intensifying," 9 August 2021. https://www.ipcc.ch/2021/08/09/ar6-wg1-20210809-pr/.

42 Historical Greenhouse Gas Levels in the Atmosphere, https://www.eea.europa.eu/data-and-maps/daviz/atmospheric-concentration-of-carbon-dioxide-5/download.table.

43 Myhre, Gunnar et al., "New estimates of radiative forcing due to well mixed greenhouse gases," *Geophysical Research Letters,* Volume 25, Issue 14, pp. 2715-2718, 15 July 1998. https://doi.org/10.1029/98GL01908.

44 Nelson, Michael and Nelson, David B., "Decoupling CO_2 from Climate Change," *International Journal of Geosciences*, Volume 15, pp. 246-269, 26 March 2024. https://doi.org/10.4236/ijg.2024.153015.

45 Nelson, Michael and Nelson, David B., "Decoupling CO_2 from Climate Change," *International Journal of Geosciences,* Volume 15, pp. 246-269, 26

March 2024. https://doi.org/10.4236/ijg.2024.153015.

46 Wrightstone, Gregory, "Fact #27, CO_2 rose after the Second World War, but temperature fell," CO2 Coalition. https://co2coalition.org/facts/ ; Temperatures from National Center for Atmospheric Research Staff,, "Global Surface temperature data: HadCRUT4 and CRUTEM4. https://climatedataguide.ucar.edu/cliate-data/global-surface-temperature-data-hadcrut4-and-crutem4; CO_2 emissions from Boden T. A. et al.. "Global and National Annual CO_2 Emissions from Fossil Fuel Burning 1751-2014," Wolfram Data Repository, 1 April 2002. https://datarepository.wolframcloud.com/resources/dzviovich_Global-and-National-Annual-CO_2-Emissions-from-Fossil-Fuel-Burning-1751-2014/.

47 Qualglia, Ilaria and Visioni, Daniele, "Modeling 2020 regulatory changes in international shipping changes in international shipping emissions helps explain anomalous 2023 warming," *EDS Letters*, Volume 15, Issue 6, pp. 1527-1441, 28 November 2024. https://doi.org/10.5194/esd-15-1527-2024.

48 Vinós, Javier, *Solving the Climate Puzzle: The Sun's Surprising Role*, Critical Science Press, Madrid, 2023, p. 34, ISBN: 978-84-12867-7-0.

49 Vinós, Javier, "Periodicities in solar variability and climate change: A simple model," *Energy Matters, Energy, Environment and Policy,* posted by Evan Mearns, 11 May 2016. https://euanmearns.com/periodicities-in-solar-variability-and-climate-change-a-simple-model.

50 Steinhilber, F. et al., "9,400 years of cosmic radiation and solar activity from ice cores and tree rings," *Proceedings of the National Academy of Sciences,* Volume 109, Issue 16, pp. 5967-5971, 2 April 2012. https://doi.org/10.1073/pnas.1118965109.

51 Schmutz, Werner K., "Changes in the Total Solar Irradiance and Climatic Effect," *Journal of Space Weather and Space Climate*, Volume 11, Article number 40, 22 July 2021. https://doi.org/10.1051/swsc/2021016.

52 Georgieva, Katya and Vereteneko, Svetlana, "Solar influences on the Earth's atmosphere: solved and unsolved questions," *Frontiers in Astronomy and Space Sciences,* Volume 10, 21 December 2023, pp. 11-12. https://doi.org/10.3389/fspas.2023.1244402.

53 Vinós, Javier, *Solving the Climate Puzzle: The Sun's Surprising Role,* Critical Science Press, Madrid, 2023, p. 269, ISBN: 978-84-125867-7-0.

54 Svensmark, Henrik and Friis-Christensen, Eigil, "Variation of cosmic ray flux and global cloud coverage—a missing link in solar-climate relationships," *Journal of Atmospheric and Solar-Terrestrial Physics,* Volume 59, Issue 11, pp. 1225-1232, July 1997. https://doi.org/10.1016/S1364-6826(97)00001-1.

55 Ljungqvist, Fredrik, "A new reconstruction of temperature variability in the extra-tropical Northern Hemisphere during the last two millennia," *Geograf-*

iska Annaler: Series A, Physical Geography, Volume 92, Issue 3, pp. 339-351, 6 September 2010. https://doi.org/10.1111/j.1468-0459.2010.00399.x.

56 Yogi Berra Museum and Learning Center, "Yogi-isms." https://yogiberramuseum.org/about-yogi/yogisms/.

57 Crok, Marcel and May, Andy, *The Frozen Climate Views of the IPCC, an analysis of AR6,* CLINTEL Foundation, Amsterdam, The Netherlands, 2023, ISBN: 979-8-89074-861-4.

58 *Wikipedia,* "Atlantic Multidecadal Oscillation." https://en.wikipedia.org/wiki/Atlantic_multidecadal_oscillation.

59 Qian, Weihong, and Zhu, Yafen, "Climate Change in China from 1880 to 1998 and Its Impact on the Environmental Condition," *Nature Climate Change,* Volume 50, pp. 419-444, September 2001. https://doi.org/10.1023/A:1010673212131.

60 Parker, Geoffrey, *Global Crisis: War, Climate Change, and Catastrophe in the Seventeenth Century,* Yale University Press, 30 April 2013, pp. 904, p. 124, ISBN: 978-0300-153231.

61 White, Sam, *A Cold Welcome,* Harvard Press, Cambridge, Massachusetts, 2017, pp. 246-247, ISBN 9780674244900.

62 Rensberger, Moyce, "Aztec Sacrifices Laid to Hunger, Not Just Religion," *The New York Times,* 19 February 1977. https://www.nytimes.com/1977/02/19/archives/aztec-sacrifices-laid-to-hunger-not-just-religion.html; See also, *Mexicolore,* "The human cost of famine," AZTECS. https://www.mexicolore.co.uk/aztecs/aztec-life/human-cost-of-famine.

63 Deming, Edwards W., "In God we trust …" *A–Z Quotes,* W. Edwards Deming Quotes. https://www.azquotes.com/author/3858-W_Edwards_Deming.

64 Zhu, Zaichun et al., "Greening of the Earth and its drivers," *Nature Climate Change,* Volume 6, pp. 791-755, 25 April 2016. https://doi.org/10.1038/nclimate3004; See also, Piao, Shilong et al., "Characteristics, drivers and feedbacks of global greening" *Nature Reviews,* Volume 1, pp. 14-27, 9 December 2019. https://doi.org/10.1038/s43017-019-0001-x.

65 Taylor, Charles A. and Schlenker, Wolfram, "Environmental Drivers of Agricultural Productivity Growth: CO_2 Fertilization of US Field Crops," National Bureau of Economic Research, Cambridge, MA, January 2023. https://www.nber.org/system/files/working_papers/w29320/w29320.pdf.

66 Klotzbach, Philip J. et al., "Trends in Tropical Cyclone Activity 1990–2021," *Geophysical Research Letters,* Volume 49, Issue 6, e2021GL095774, 28 March 2022. https://doi.org/10.1029/2021GL095774.67 Zuohao, Cao and Cai, Huaqing, "Trend Analysis of U.S. Tornado Activity Frequency," *Atmosphere,* Volume 13, p. 498, 2022. https://doi.org/10.3390/atmos13030498.

68 IPCC, *The Physical Science Basis, Contribution of Working Group I to the Sixth Assessment Report on the Intergovernmental Panel on Climate Change,* "Chapter 11 - Weather and Climate Extreme Events in a Changing Climate," Cambridge University Press, Cambridge UK and New York, NY, USA, Section 11.5.2, p 1568, 9 August 2021. https://www.ipcc.ch/report/ar6/wg1/downloads/report/IPCC_AR6_WGI_Chapter11.pdf.

69 The United States Environmental Protection Agency (EPA) Climate Change Indicators Heat Waves. EPA: "Climate Change Indicators: High and Low Temperatures," 19 January 2021. https://19january2021snapshot.epa.gov/climate-indicators/climate-change-indicators-high-and-low-temperatures_.html.

70 Vicente, Servio M. et al., "Global drought trends and future projections," *Philosophical Transactions,* Royal Society A, 380:20201085, 18 August 2022. https://doi.org/10.1098/rsta.2021.0285.

71 Zhao, Qi et al., "Global regional and national burden of mortality associated with non-optimal ambient temperatures from 2000 to 2019: a three-stage modeling study," *The Lancet*, Volume 5, Issue 7, E415-E425, July 2021. https://doi.org/10.1016/S2542-5196(21)00081-4.

72 Lomborg, Bjorn, "Climate Change Is Not an Apocalyptic Threat—Let's Address It Smartly," Hoover Institution, Stanford University, September 2024. https://www.hoover.org/sites/default/files/research/docs/Lomborg_ClimateChange_web_240903.pdf.

73 Hille, Karl B., "Carbon Dioxide Fertilization Greening the Earth, Study Finds," NASA, 26 April. https://www.nasa.gov/feature/goddard/2016/carbon-dioxide-fertilization-greening-Earth.

74 Chen, Xin et al., "The global greening continues despite increased drought stress since 2020," *Global Ecology and Conservation*, Volume 49, January 2024. https://doi.org/10.1016/j.gecco.2023.e02791.

75 Zhu, Zaichun et al., "Greening of the Earth and its drivers," *Nature Climate Change,* Volume 6, pp. 791-755, 25 April 2016. https://doi.org/10.1038/nclimate3004; See also, Piao, Shilong et al., "Characteristics, drivers, and feedbacks of global greening," *Nature Reviews,* Volume 1, pp. 14-27, 9 December 2019. https://doi.org/10.1038/s43017-019-0001-x.

76 US Global Change Research Program (USGCRP), "Climate Science Special Report: Fourth National Climate Assessment, Volume 1", Chapter 6 ("Temperature Changes in the United States"), USGCRP, Washington, D.C. https://science2017.globalchange.gov/chapter/6/.

77 The Fourth National Climate Assessment (NCA4), Chapter 6, "Temperature Change in the United States," https://science2017.globalchange.gov/chapter/6/.

78 Lindzen, Richard, "What historians will definitely wonder about ..." *A–Z Quotes,* Authors, Richard Lindzen. https://www.azquotes.com/quote/1386476.

79 Tol, Richard S., “The Economic Impacts of Climate Change,” *Review of Environmental Economics and Policy,* Volume 12, Number 1, Winter 2018, pp. 4-25. https://doi.org/10.1093/reep/rex027.

80 *Aljazeera*, “Japan's elderly population rises to record 36.25 million,” 16 September 2024. https://www.aljazeera.com/economy/2024/9/16/japans-elderly-population-rises-to-record-36-25-million.

81 Wall Street Journal Editorial Board, “The Real Cost of the Inflation Act Subsidies, $1.2 trillion,” *The Wall Street Journal,* Opinion, 24 March 2023. https://www.wsj.com/articles/inflation-reduction-act-subsidies-cost-goldman-sachs-report-5623cd29.

82 Lawder, David, “Yellen says $3 trillion needed annually for climate financing, far more than current level,” *Reuters*, 27 July 2024. https://www.reuters.com/sustainability/sustainable-finance-reporting/yellen-says-3-trillion-needed-annually-climate-financing-far-more-than-current-2024-07-27/.

83 Environmental Protection Agency, “ICYMI: Administrator Lee Zelden Finds Gold Bars from EPA at Stacey Abrams' Connected Group, Biden-Harris Ethical Red Flags,” *EPA News Release,* 21 February 2025. https://www.epa.gov/newsreleases/icymi-administrator-lee-zeldin-finds-gold-bars-epa-stacey-abrams-connected-group-biden.

84 *Goodreads*, “W. C. Fields Quotes, Quotable Quote.” https://www.goodreads.com/quotes/485284-you-can-fool-some-of-the-people-some-of-the.

85 Zeldon, Lee, “Lee Zeldon Bombshell! Biden's EPA Slush Fund Exposed,” *Pod Force One with Miranda Devine*, 20 August 2025. https://www.youtube.com/watch?v=dEGndA2UcJY, Minutes 7:00 to 8:30.

86 Environmental Defense Fund, home page, https://www.edf.org/.

87 Bravendar, Robin, “Meet the top-paid green group bosses,” *E&E News*, by Politico, 29 April 2024. https://www.eenews.net/articles/meet-the-top-paid-green-group-bosses/.

88 Economic Research Institute, “Rebecca Kujawa, President and Chief Executive Officer of NEER,” 2023 fiscal year. https://www.erieri.com/executive/salary/rebecca-kujawa-b6ku.

89 United States Securities and Exchange Commission, “Schedule 14A, Enphase Energy, Inc.”, 19 March 2024, p. 48. https://www.sec.gov/Archives/edgar/data/1463101/000146310124000049/enph-20240404.htm?utm_source=chatgpt.com.

90 Tol, Richard S., “The Economic Impacts of Climate Change,” *Review of Environmental Economics and Policy,* Volume 12, Number 1, Winter 2018, pp. 4-25. https://doi.org/10.1093/reep/rex027.

91 Zhu, Zaichun et al., “Greening of the Earth and its drivers,” *Nature Climate*

Change, Volume 6, pp. 791-795, 25 April 2016. https://doi.org/10.1038/nclimate3004; See also, Piao, Shilong et al., "Characteristics, drivers and feedbacks of global greening," *Nature Reviews*, Volume 1, pp. 14-27, 9 December 2019. https://doi.org/10.1038/s43017-019-0001-x.

92 Fitzner, Zach, "The environmental impacts of solar and wind energy," *earth.com,* 5 December 2018. https://www.earth.com/news/environmental-impacts-solar-wind-energy/.

93 University of Nottingham, "Power Grid May Be Destabilized By Domestic Renewable Energy," *Technology Networks,* Applied Sciences, Science News, 5 March 2022. https://www.technologynetworks.com/applied-sciences/news/power-grid-may-be-destabilized-by-domestic-renewable-energy-sources-359204.

94 Lomborg, Bjorn, "The enormous cost of green policies," X post, 4 January 2025. https://x.com/BjornLomborg/status/1875529420049424746.

95 Bettarelli, Luca et al., "Energy Inflation and Inequality," *Energy Economics,* Volume 124, August 2023. https://doi.org/10.1016/j.eneco.2023.106823.

96 Heymann, Eric, "German industrial production: The decline is not over yet," Deutsch Bank Research, 13 February 2024. https://www.dbresearch.de/PROD/RPS_DE-PROD/PROD0000000000531908/German_industrial_production%3A_The__decline_is_not_.PDF?undefined&realload=oaSmAyQUxIjYyK8o7xesJBiNfEysYVJTpW4dexd2fdI168vJfisv9nhqb04CijZA.

97 Graham, Shaun, "The Financial Impact of Increased Consumption and Rising Electricity Rates in Datacenter Facilities Spending," *IDC*, September 2024. https://my.idc.com/getdoc.jsp?containerId=US52548324.

98 Goovaerts, Diane, "Where are renewables in the data center power conversation?" *Fierce-Network*, 25 August 2025. https://www.fierce-network.com/cloud/where-are-renewables-data-center-power-conversation.

99 California Independent System Operator, "Root Cause Analysis, Mid-August 2020 Extreme Heat Wave," California Public Utilities Commission, California Energy Commission, 13, January 2021. https://www.caiso.com/Documents/Final-Root-Cause-Analysis-Mid-August-2020-Extreme-Heat-Wave.pdf?utm_source=chatgpt.comme.

100 Churchhill, Winston, International Churchill Society, Quotes, 20 December 2017. https://winstonchurchill.org/resources/quotes/truth-is-incontrovertible/.

Index

A

B

C

D

E

F

G

H

L

M

N

O

T

U

V

W

Y

Z

MORE INFORMATION & RESOURCES

CONNECT WITH THE AUTHOR

Have questions, comments, or want to continue the conversation? We'd love to hear from you.

Website: www.WhyNoClimateCrisis.com

Email: trkurz1955@outlook.com

WHAT YOU'LL FIND ON THE WEBSITE

- Data, charts, and graphs referenced throughout this book
- Summary of the surprising evidence you should know
- Recommended reading list for further research
- Responses to common objections and frequently asked questions

SHARE THIS BOOK

If this book has given you new perspective, consider sharing it with others:

- Give a copy to a friend, family member, or colleague open to exploring the evidence
- Recommend it to your book club, school, or community group

- Leave an honest review on Amazon, Goodreads, or your favorite bookseller
- Share your thoughts on social media — tag us so we can see your response!

A NOTE FROM THE AUTHOR

The goal of this book has never been to tell you what to think — only to encourage you to look carefully at the evidence yourself. I welcome disagreement, debate, and dialogue. Whether you finished this book as a skeptic or as a convert, the most important thing is that you are asking questions. Please reach out. Let's keep the conversation going.

www.WhyNoClimateCrisis.com | trkurz1955@outlook.com